ENVIRONMENTAL SCIENCE, ENGINEERING AND TECHNOLOGY

# ENERGY-ENVIRONMENT-ECONOMICS

# ENVIRONMENTAL SCIENCE, ENGINEERING AND TECHNOLOGY

Additional books in this series can be found on Nova's website under the Series tab.

Additional e-books in this series can be found on Nova's website under the e-book tab.

ENVIRONMENTAL SCIENCE, ENGINEERING AND TECHNOLOGY

# ENERGY-ENVIRONMENT-ECONOMICS

RIAD BENELMIR
EDITOR

nova publishers
*New York*

For permission to use material from this book please contact us:
Telephone 631-231-7269; Fax 631-231-8175
Web Site: http://www.novapublishers.com



**Library of Congress Cataloging-in-Publication Data**

ISBN: 978-1-62948-862-2

*Published by Nova Science Publishers, Inc. † New York*

# CONTENTS

# PREFACE

This book sheds light on the interaction between the utilization of energy and the environment, as well as the economic aspects involved with this utilization. It is a vehicle for an international exchange and dissemination of ideas in the multidisciplinary field of energy-environment-economics between research scientists, engineers, economists, policy makers, and others concerned about these issues.

In: Energy-Environment-Economics
Editor: Riad Benelmir

ISBN: 978-1-62948-862-2


*Chapter 1*

# ADAPTABILITY OF REFRIGERATION HYBRID SYSTEM TO LOW TEMPERATURE SOURCES

***Nahla Bouaziz*[*], *Ridha Ben Iffa*[†] *and Lakhdar Kairouani*[‡]**
Unité de Recherche Energétique & Environnement,
Ecole Nationale d'Ingénieur de Tunis. Université de Tunis EL Manar,
le Belvedere, Tunisia

## ABSTRACT

The present study deals with a compression-absorption machine. The proposed hybrid cooling system uses water-ammonia as a working fluid and operates at three pressure levels. The absorber is at an intermediate pressure (Pint) taken between the evaporator pressure (PEV) and the condenser pressure (PCD), unlike the single stage machine which works between two pressure levels. The proposed new system is studied and compared to the conventional machine. In order to evaluate the performance of the invoked machine, a procedure based on the MAPLE software is set up to compute accurately the thermodynamic properties of the working fluid. The analyses of the numerical results highlight that the performance of the novel proposed configuration is better than that relative to the conventional system. The study reveals the great impact of the intermediate pressure on the performance improvement and on reducing the generator temperature allowing the system to work at low enthalpy. In fact, for an evaporator temperature and a condenser temperature fixed respectively at -10°C and 40°C, the proposed hybrid refrigeration cycle operates at a generator temperature TGE=75°C and the installation's COP is about 0.56. While for the same conditions, the single stage machine COP cannot exceed 0.51 with a generator temperature of about 135°C. Consequently, our enhanced novel configuration presents the opportunity to operate at low enthalpy sources.

**Keywords:** Absorption, Ammonia, COP, Hybrid, Refrigeration

---

[*] E-mail address: nahlabouaziz@yahoo.fr.
[†] E-mail address: r.iffa@yahoo.fr.
[‡] E-mail address: Lakdar_kairouani@yahoo.fr.

# 1. Introduction

Refrigeration is one of the leading consumers of electric power in the word. It is encountered in everyday life and it is greatly used in human activity. Its applications are numerous in many fields and sectors, such as agriculture, industry and commercial sector [1], [2]. In addition, refrigeration industry complies with the new rules of ozone protection and thus it contributes significantly in reducing chlorofluocarbon's emissions [3], [4]. In this context, our investigation is carried out with a particular attention on the optimization of the system performance and the minimization of energy consumption of such refrigeration cycles [5], [6]. This study presents a new approach to enhance the refrigerating cycle COP and to operate at low enthalpy sources [7], [8]. We propose a novel configuration of a hybrid machine using NH3-H2O and operating at three pressure levels. The proposed machine is compared to the conventional refrigerating system.

# 2. Modelisation

Several authors have studied different models of absorption machines using H2O/NH3 and LiBr/H2O as working fluids and geothermal energy and solar energy as energy sources [9], [10]. The absorption machine is constituted by six main elements evaporator, absorber, pump, generator, condenser and expansion valve. Some systems are composed with simple stage machine [11] and others are formed by a succession of stages with various component associations and sometimes inserting other new components [12], [13]. To model an absorption cooling cycle, it is sufficient to model each component of the refrigeration cycle [14], [15]. In the following section, we present the simple effect absorption machine and we develop a novel absorption hybrid configuration. We will explain and quantify it's adaptability to low-enthalpy sources. The configurations object of this work are followed by the representation of the corresponding cycle on the Oldham diagram.

## 2.1. Conventional Model

The components of the conventional system are the evaporator, the absorber, the pump, the generator, the condenser and two valves: one is used for the poor solution return to the absorber and the other for the refrigerant relaxation in the evaporator. It is well known that a single stage cooling system works between two pressure levels: evaporator pressure (*PEV*) and condenser pressure (*PCD*) [16]. We have shown in a previous work that the conventional absorption machine, using NH3-H2O as a working fluid, cannot operate with a generator temperature below 100°C with evaporator and condenser temperature fixed respectively to -10 °C and 40°C respectively [17]. Indeed, its coefficient of performance (*COP*) cannot exceed 0.51. We keep in mind that the theoretical *COP* is calculated by considering the following assumptions:

- The temperature at each main component is uniform.
- The refrigerant vapour is in thermodynamic equilibrium with the rich solution at the absorber entrance and the poor solution leaving the generator.
- The fluid at the generator outlet can be considered in a saturated state with a steam quality less than unity [18].

In this study, the conventional absorption cycle parts are shown in Figure 1 and the corresponding cycle on the Oldham diagram is represented on Figure 2.

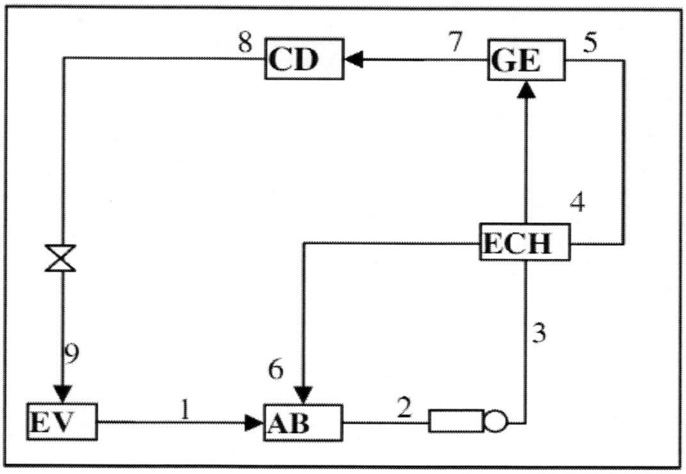


Figure 1. Conventional cycle.

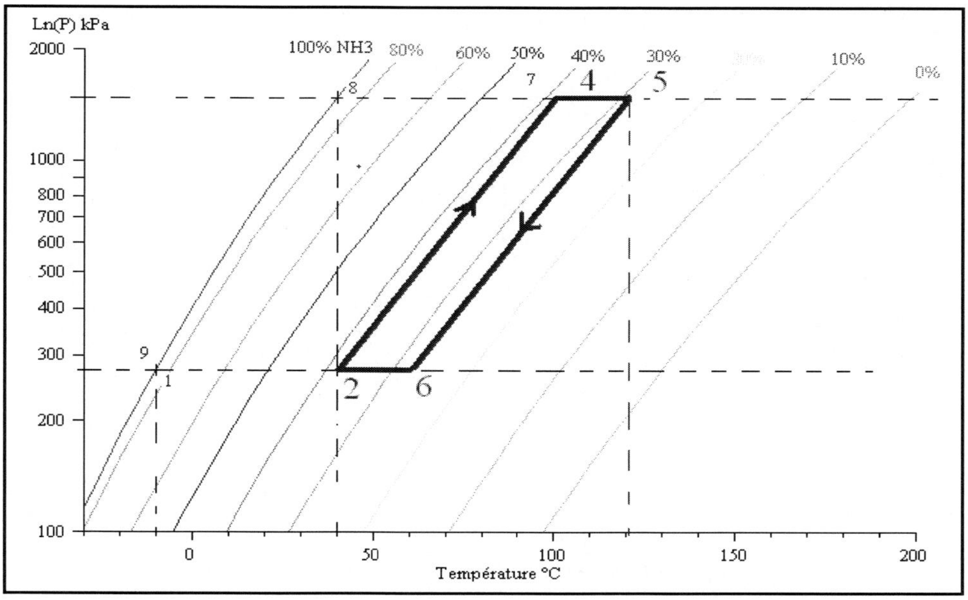


Figure 2. Diagram of Oldham for conventional machine.

Bellow, we propose the novel enhanced refrigeration system operating at three pressure levels. This machine is composed by an evaporator, an absorber, a generator, a condenser and a compressor inserted between the evaporator and the absorber. The cycle parts are represented in Figure 3 and the corresponding cycle on the Oldham diagram is represented on Figure 4.

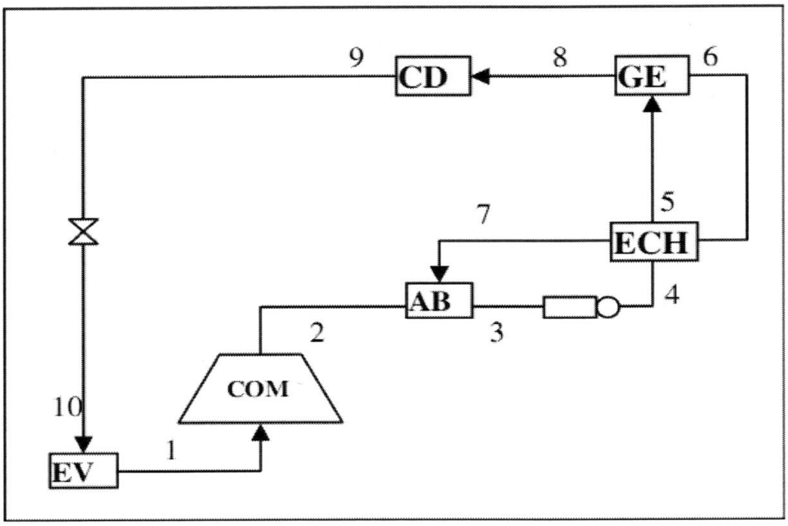


Figure 3. Proposed hybrid configuration.

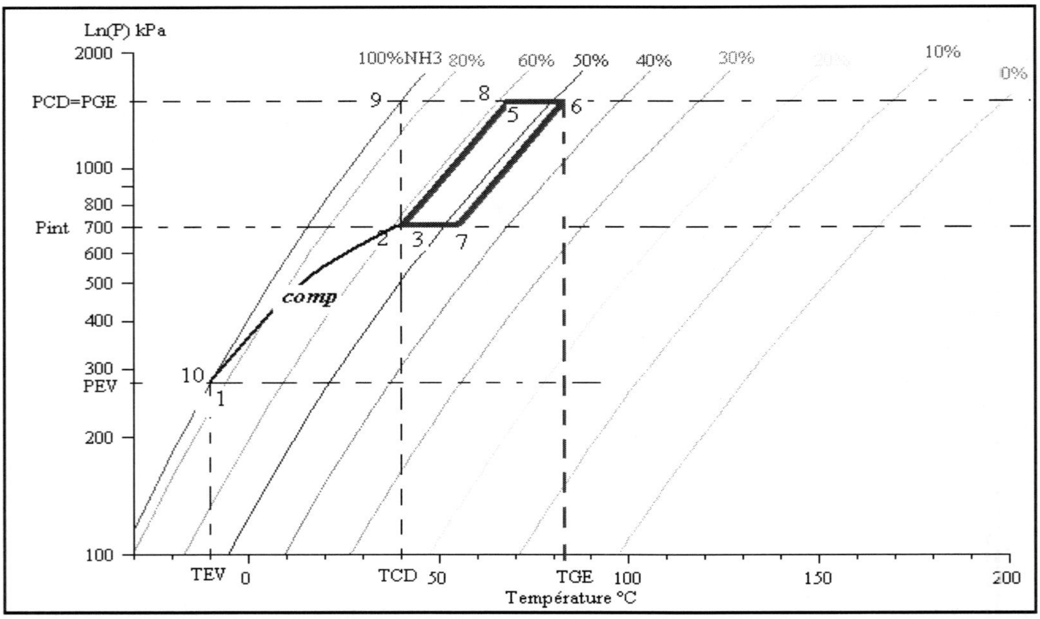


Figure 4. Diagram of Oldham for proposed configuration.

## 2.2. Mathematical Model

In the following, we develop the energy and mass balance. The specific solution circulation factor represents the mass of solution per Kg of refrigerant vapour evaporated in the generator.

$$f = \frac{1 - x_5}{x_4 - x_5}. \tag{1}$$

It carries the energy balances in each component of the installation, exchanging heat or work with the external environment. Neglecting the rectifier, we obtain:

| Condenser (8-9) | Evaporator (10-1) | Generator (5-6;8) | Absorber |
|---|---|---|---|
| $\dot{m}_9 = \dot{m}_8 = \dot{m}_{NH3}$ | $\dot{m}_{10} = \dot{m}_1 = \dot{m}_{NH3}$ | $\dot{m}_5 = \dot{m}_6 + \dot{m}_8$ | $\dot{m}_3 = \dot{m}_2 + \dot{m}_7$ |
| $T_9 = T_{CD}$ | $T_1 = T_{EV}$ | $T_8 = T(x_5, P_{CD})$, $T_6 = T_{GE}$ | $T_3 = T_{CD}$ |
| $h_9 = h(T_{CD})$ | $h_{10} = h_9$, $h_1 = h(T_{EV})$ | $h_6 = h(x_6, P_{CD})$, $h_5 = h(x_5, P_{CD})$ | |
| $P_9 = P(T_{CD}) = P_{CD}$ | $P_{10} = P_1 = P(T_{EV}) = P_{EV}$ | $P_6 = P_5 = P_8 = P_{CD}$ | |
| | $\dot{Q}_{EV} = \dot{m}_{NH3}(h_1 - h_{10})$ | $x_6 = x(T_6, P_{CD})$, $x_5 = x_6 + \Delta x$ | $x_3 = x_7 + \Delta x$ |
| $\dot{Q}_{CD} = \dot{m}_{NH3} \cdot (h_8 - h_9)$ | | $\dot{Q}_{GE} = \dot{m}_{NH3}(h_8 + (f-1) \cdot h_6 - f \cdot h_5)$ | $\dot{Q}_{AB} = \dot{m}_{NH3}(h_2 + (f-1) \cdot h_7 - f \cdot h_3)$ |

## 2.3. Compressor

In order to determine the compressor power, we consider the following hypothesis:

- At the evaporator outlet (EV), vapour ammonia is assimilated as an ideal gas, so we have: $T_1 \times P_1^{(k-1)/k} = T_2 \times P_2^{(k-1)/k}$.
- In an ideal case, adiabatic process, it consumes power given by:

$$\dot{Q}_{is} = cp_{NH3} \times (T_2 - T_1).$$

Considering the isentropic efficiency, the real power is defined as follows:

$$\dot{Q}_{réel} = \frac{\dot{Q}_{is}}{\eta_{is}}.$$

where: $\dot{Q}_{réel} = h_2 - h_1$.

So we can deduce from the two previous expressions the value of the steam enthalpy at the compressor outlet: $h_2 = h_1 + \frac{\dot{Q}_{is}}{\eta_{is}}$.

The term performance is determined as follows [6]:

$$\eta_{is} = 0.874 - 0.0135 \cdot \tau \text{ and } \tau = \frac{P_2}{P_1}.$$

We deduce the *COP* from the generator conservation:

$$COP = \frac{\dot{Q}_{Ev}}{\dot{Q}_{GE}} = \frac{h_1 - h_{10}}{h_8 + (f-1).h_6 - f.h_5 + h_2 - h_1}.$$

## 3. Results and Discussion

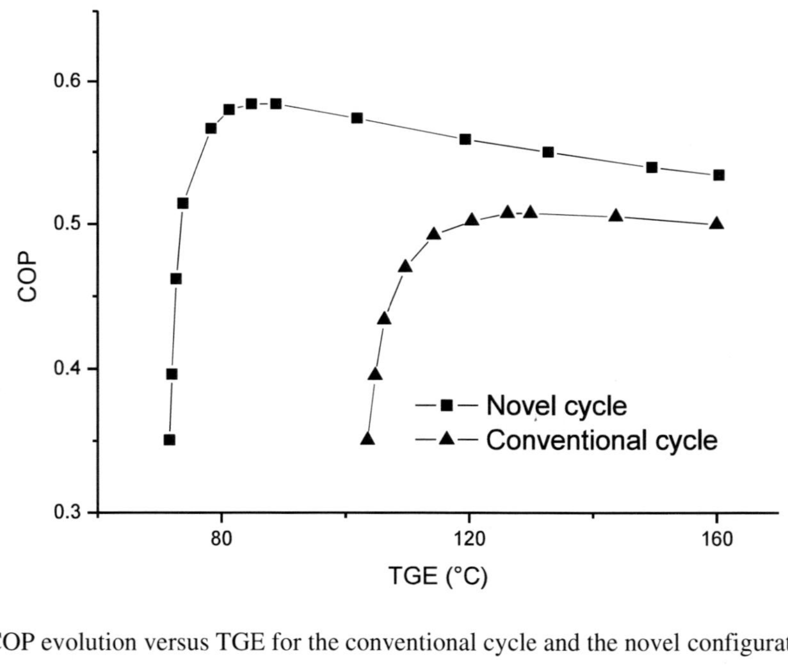


Figure 5. COP evolution versus TGE for the conventional cycle and the novel configuration.

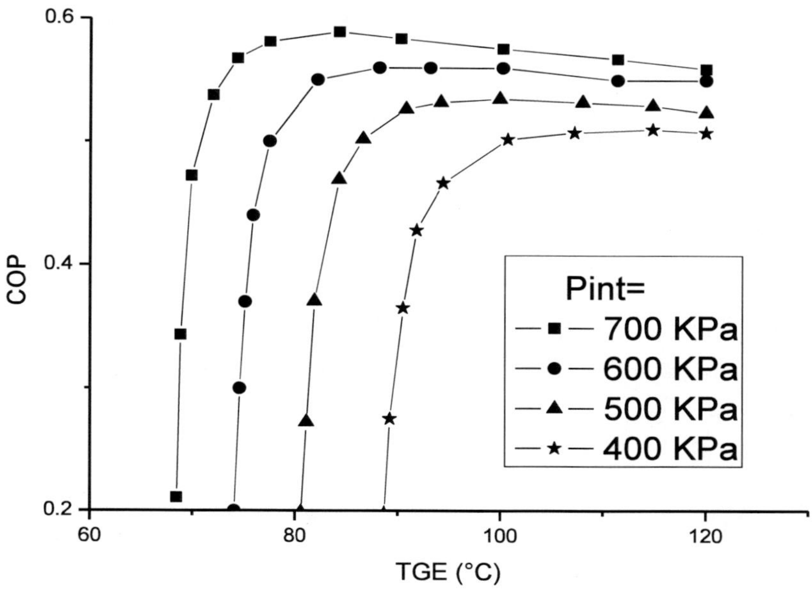


Figure 6. COP Evolution versus TGE with TEV = -10°C and TCD = 40°C for different intermediate pressure.

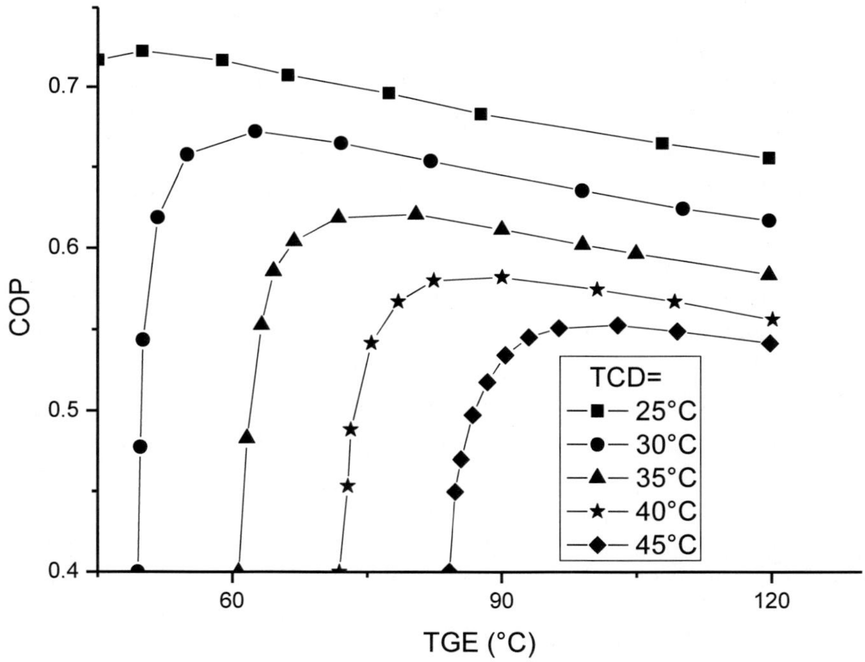


Figure 7. COP evolution versus TGE with TEV = -10°C and Pint = 650kPa for different TCD.

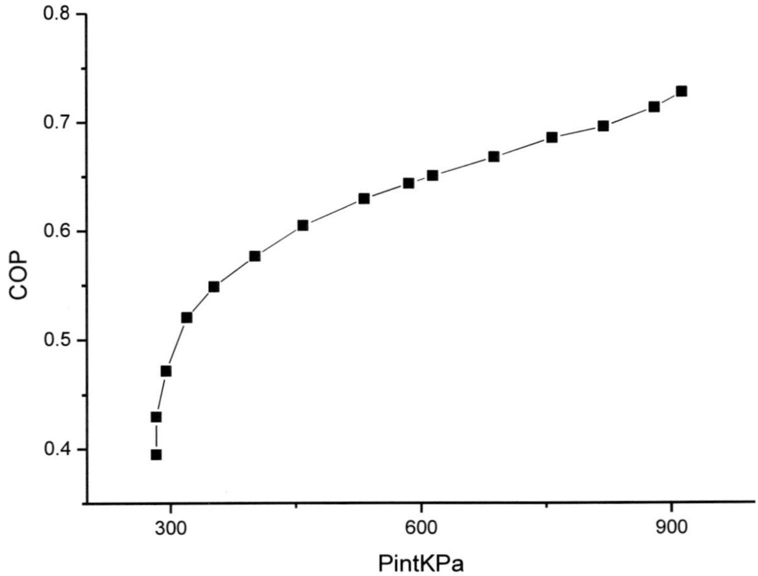


Figure 8. COP evolution versus Pint for TGE = 80°C, TCD = 30°C and TEV = -10°C.

For a single stage machine, the *COP***'s** evolution depends mainly on three parameters: the evaporation temperature necessary to produce the desired cold, the condensation temperature and the generator temperature. For the new installation, a new parameter reveals interesting to study the performance of the cycle proposed in this study, it's the pressure of the absorber that is an intermediate pressure (Pint). The numerical simulation, developed in the present investigation carries out a comparative study of the two systems performances.

Figure 5 shows the *COP*'s evolution versus the generator temperature, for an evaporator temperature and a condenser temperature fixed respectively to -10°C and 40°C. For the novel configuration, the absorption pressure is fixed to 650 kPa.

The analysis of Figure 5 shows that for the new proposed absorption machine, the coefficient of performance is higher than that relative to the classic system. It's about 0.58. While for the conventional configuration, the *COP* cannot exceed 0.51 [6]. On the other hand, the operating temperature is relatively low, around 70°C compared to that relative to the simple effect cycle which is about 105°C. In addition, even for high temperature generator (*TGE*>105°C) the *COP* continues to be higher than the conventional cycle. For a condensation temperature of 40°C and an evaporation temperature of -10°C, Figure 6 presents a family of curves showing the *COP* evolution versus the generator temperature for different condenser temperature. We note that the new machine works on the whole domain of temperature and its coefficient of performance increases if the intermediate pressure increases. Besides, the operating temperature (TGE) decreases if the pressure of the absorber increases. The enhanced configuration object of this investigation allows consequently a gain of energy and gives the possibility to operate at low enthalpy sources.

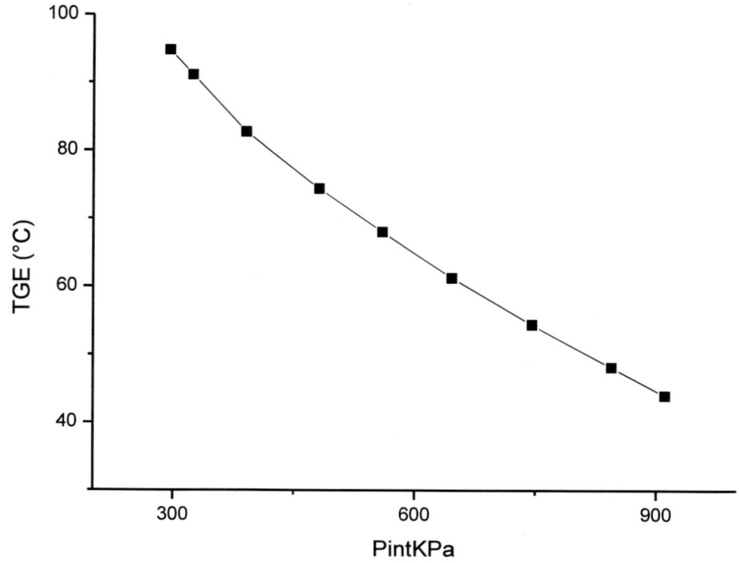


Figure 9. TGE versus Pint.

Figure 8 shows the *COP* evolution versus the intermediate pressure (pressure of the absorber) for TGE = 80°C, an evaporator temperature and a condenser temperature fixed relatively to -10°C and 30°C. We recall that under these conditions, the *COP* of the conventional system does not exceed 0.4.

Setting mass fraction variation $\Delta x$ at 10% and for the same conditions, we note that the intermediate pressure has a great effect on reducing operating temperature of the machine since the conventional cycle can start only from a temperature of 95°C (see Figure 9).

## CONCLUSION

In this investigation, we proposed an absorption cooling machine using water-ammonia as a working fluid and operating at three pressure levels. The proposed hybrid cycle, where the absorber is at an intermediate pressure taken between the evaporator and the condenser pressure is analysed and compared to the conventional cycle. The numerical simulation highlights the improved performance of the novel system. It showed the double advantage of the hybrid absorption configuration. It presents the opportunity to operate at low enthalpy source with a better *COP*.

In fact, for an evaporator temperature and a condenser temperature fixed respectively to -10°C and 40°C, the compression-absorption machine operates at a generator temperature, TGE=75°C with an installation's *COP* about 0.56 while for the same conditions, the one stage machine, functioning at two pressure levels has a COP which cannot exceed 0.51 with a generator temperature of about 135°C. The numerical simulation results, carried out in this investigation show also that the intermediate pressure has not only influence on the *COP* enhancement of the refrigeration cycle but also it has a great impact in reducing the generator temperature. Consequently, it contributes to the adaptability of refrigeration systems to low temperature allowing the use of waste heat in many industrial and mobile applications.

## NOMENCLATURE

| Symbol | Quantity | SI Unit |
|---|---|---|
| *COP* | Coefficient of Performance | |
| f | Circulation ratio | |
| h | Specific enthalpy | (J/kg) |
| $\dot{m}$ | Mass flow rate | (kg/s) |
| P | Pressure | (bar, Pa) |
| $\dot{Q}$ | Heat-transfer rate | (W) |
| T | Temperature | (K, °C) |
| x | mass fraction | |
| Subscripts | | |
| AB | Absorber | |
| CD | Condenser | |
| COM | Compressor | |
| EV | Evaporator | |
| GE | Generator | |
| η | Efficiency % | |

## REFERENCES


[1] M. Thioye, Amélioration de la performance des machines frigorifiques à absorption par l'utilisation de cycles à absorption et désorption étagés. *Int J. Refrig*. 20, No. 2, (1997).136-145.

[2] S. B. Riffat, G. Qiu, Comparative investigation of thermoelectric air-conditioners versus vapour compression and absorption air-conditioners. *App. Th. Eng*. 24 (2004) 1979–1993.

[3] A. Keçeciler, H.I. Acar, A. Dogan, Thermodynamic analysis of the absorption refrigeration system with geothermal energy: an experimental study. *Energy Conversion and Management* 41 (2000) 37-48.

[4] A.T. Bulgan, Use of low temperature energy sources in aqua-ammonia absorption refrigeration systems. *Energy Conversion and Management* 38 (1997) 1431-1438.

[5] S. Göktun, Performance analysis of a heat engine driven combined vapor compression-absorption-ejector refrigerator. *Energy Convers. Mgmt* 41 (2000) 1885-1895.

[6] A. Laouir, P. legoff, J. M. Hornt, Cycle de frigopompes à absorption en cascades matérielles, détermination du nombre d'étages optimal pour le mélange ammoniac-eau. *Int J. Refrig*. 25 (2002) 136-148.

[7] M. T. Syed, M. A. Siddiqui, Performance and economic study of the combined absorption/compression heat pump. *Energy Convers. Mgmt* 40; (1999) 575-591.

[8] R. Fathi, C. Guemimi, S. Ouaskit, An irreversible thermodynamic model for solar absorption refrigerator. *Renewable Energy* 29 (2004) 1349-1365.

[9] S. N. Mumah, S. S. Adefila, E. A. Arinze, *Energy convers. Mgmt*. 35, N 8 (1994) 737-750.

[10] S. G. Alvares, Ch. Trepp, Simulation of solar driven aqua-ammonia absorption refrigeration system. *Int J. Refrig*. 10 (1987) 40-49.

[11] M. Robert, Tozer, Fundamental thermodynamics of ideal absorption cycles. *Int J. Refrig*. 20, (1997) 120-135.

[12] S. Arh, B. Gaspersic, Development and comparison of different advanced absorption cycles. *Rev. Int. Froid*. 13, (1990) 41-50.

[13] B. Sahina, A. Kodal, Thermoeconomic optimization of double-stage combined refrigeration system: a finite-time approach. *Int J. Refrig*. 25, (2002) 872-877.

[14] M. Fukuta, T. Yanagisawaa, H. Iwatab, K. Tadab, Performance of compression/absorption hybrid refrigeration cycle with propane/mineral oil combination. *Int J. Refrig*. 25 (2002) 907-915.

[15] D. Sun, I. W. Eames, S. Aphornratana, Evaluation of a novel combined ejector-absorption refrigeration cycle-I: computer simulation. *Int. J. Refrig*. Vol. 19, N 3 (1996) 172-180.

[16] Y. T. Kang, Y. Kunugi, T. Kashiwagi, Review of advanced absorption cycles: performance improvement and temperature lift enhancement. *Int J. Refrig*. 23 (2000) 388-401.

[17] L. Kairouani, E. Nehdi, R. Ben Iffa, Thermodynamic Investigation of Two-Stage Absorption Refrigeration System Connected by a Compressor. *Am. J. App. Sc* 2, 6; (2005) 1036-1041.

[18] R. J. Romero, L. Guillen, I. Pilatowsky, Monomethylamine-water vapour absorption refrigeration system. *App. Th. Eng*. 25 (2005) 867-876.

In: Energy-Environment-Economics
Editor: Riad Benelmir

ISBN: 978-1-62948-862-2


*Chapter 2*

# WIND-POWERED DESALINATION BY MECHANICAL VAPOR COMPRESSION

***Imen Chermiti*[*], *Nejib Ben Hidouri and Ammar Brahim***
Gabes University,
Engineers National School of Gabes
Chemical and Process Engineering Department
Applied Thermodynamics Unit
Gabes, Tunisia

## ABSTRACT

The origin and continuation of mankind is based on water. Water is one of the most abundant resources on earth, covering three-fourths of the planet's surface and about 97% of the earth's water is of high salinity. However, the increase of the world population associated to the agricultural and industrial developments involved an increase in the demand of fresh water:

Desalination seems to be the solution of this acute problem and the key to meet the future challenges for producing fresh water. Desalination requires large amounts of energy, today mostly from fossil fuels. It is therefore favorable to rely on renewable energies like wind energy, an environmentally-friendly energy source, to desalinate seawater.

The main objective of this paper is the study of the possibility to install a single-effect mechanical vapor compression unit (MVC unit) using wind power in the region of Gabes. The paper includes an estimation of the wind energy potential in this region and a study of the influence of the chosen operating temperatures on the compressor specific power consumption of the MVC unit.

**Keywords**: Fresh water, Desalination, Wind power, Mechanical vapor compression, MVC

---

[*] E-mail address: chermitii@yahoo.fr

# NOMENCLATURE

| | |
|---|---|
| A | Area, $m^2$ |
| BPE | Boiling point elevation, °C |
| c | scale parameter, m/s |
| Cp | Specific heat at constant pressure, kJ/kg °C |
| Cr | Compression ratio |
| E | Energie, kJ/kg |
| F | Function of repartition |
| f | Density of probability |
| g | Power probability density |
| H | Enthalpy, kJ/kg |
| k | form parameter |
| L | Latent heat for evaporation, kJ/kg |
| n | Polytropic coefficient |
| NEA | Non equilibrium allowance, °C |
| P | Power, W |
| p | Pressor, kPa |
| Q | Mass flow rate, kg/s o, |
| S | Entropy, kJ/kg °C |
| T | Temperature, °C |
| ΔT | Temperature drop, °C |
| t.o.g | Top of the ground |
| U | heat transfer coefficient, $kW/m^2 °C$ |
| V | Wind velocity, m/s |
| v | specific volume, $m^3/kg$ |
| W | Specific power consumption, $kWh/m^3$ |
| x | Salinity, ppm |
| Z | Ground roughness, m |
| Z | Height, m |

***Greek symbols:***

| | |
|---|---|
| γ | Isentropic coefficient |
| η | Output or efficiency, % |
| ρ | Density, $kg/m^3$ |
| Γ | Gamma function |
| τ | Friction coefficient |

***Subscripts:***

| | |
|---|---|
| a | Aerogenerator |
| av | Available |
| b | Brine |
| c | Cut |
| d | Starting, distillate |
| e | Used or evaporated |
| ex | Extractable |
| f | feed |
| i | height of reference |
| is | Isentropic |
| j | Desired height |
| n | Nominal |
| o | Outlet brine |
| p | Demister or polytropic |
| s | Start or superheated compressed vapor |
| th | Theorical |
| u | Useful |
| v | Saturated vapor above demister |
| vp | Saturated vapor below demister |
| sw | Intake see water |

# 1. INTRODUCTION

Tunisia is a small country located on the southern border of the Mediterranean basin characterized by an arid to a semi-arid climate and has very limited water resources, an important part of which is being brackish and unsuitable for drinking or irrigation.

The need for adopting desalination to solve the fresh water shortage problem in Tunisia, in general, and in the south of Tunisia, in particular, is becoming increasingly evident and water desalination, which is currently pursued as a major component of an overall strategy is considered to meet future water demand and enhance the quality of drinking water in various parts of the country. Moreover, Tunisia is notable for its extensive research aimed at coupling renewable energy sources with desalination technology to produce more economical and environmentally friendly systems.

Solar and wind energies have been explored, as each kind provides a potentially vast renewable energy source for Tunisia [1]. Among the renewable energy types, wind energy seems to be one of the most promising with a very high international growth rate. Recently, many researches are focused on the development and the optimization of the wind turbines as one of the most successful technique of recuperation of wind energy. During the last years, Tunisia has embarked upon wind energy programs and many studies are made to evaluate the wind characteristics and the wind power potential in Tunisia. In this case, the evaluation of the wind energy potential of "Sidi Daoud" site (north east region of Tunisia) based on measured data collected between 2000 and 2004 was made by Ben Amar et al. [2]. The study concerns an estimation of the energy potential for the first wind farm installed in the site and consists of 32 aerogenerators (MADE AE-32) of 10.56 MW capacity.

Elamouri and Ben Amar [3] elaborated a database for the users of the wind power and evaluated wind energy potential in the whole Tunisian territory. An experimental study of wind velocity was reported by Dahmani and al. [4] concerning "Borj-Cedria" site (region near the capital Tunis). Further, Dahmani and al. [4] performed a simulated study of two types of wind turbine in the site: XL. 1 of 'Bergey wind power' company and Whisper 200 of 'Southwest wind power' company. Dahmouni and al. [5] studied the wind resource in the gulf of Tunis during the year 2008 and estimated the annual production of the wind turbine Enercon E82. Many works are still required in order to optimize the use and coupling of wind energy with desalination systems.

In this paper, the possibility of using the wind energy for a mechanical vapor compression desalination unit (MVC) is performed. An estimation of the wind potential available in Gabes region (south east region of Tunisia) is determined, and then it is interested to study the specific consumption of the MVC unit and the influence of the chosen operating temperatures on it.

## 2. Estimation of the Wind Potential Energy in the Site of Gabes

Dimensioning of the wind turbine used to operate the compressor of the unit requires a statistical analysis of the considered site (Gabes region). For this reason, the available and the extractable wind power are given.

### 2.1. Weibull Method

The Weibull method is the most frequently used method for modeling the wind speed distribution. The probability density function indicates the percentage of time for which the wind flows with specific wind speed, it is given by:

$$\mathbf{f(V,k,c)} = \left(\frac{k}{c}\right)\left(\frac{V}{c}\right)^{\mathbf{k-1}} \mathbf{e}^{\left[-\left(\frac{V}{c}\right)^{k}\right]}. \tag{1}$$

where V is the wind velocity, *c* and *k* mean Weibull scale and shape parameters. Basically, the scale parameter, c, indicates how windy a wind location under consideration is, whereas the shape parameter, *k*, indicates how peaked the wind distribution is (if the wind speeds tend to be very close to a certain value, the distribution will have a high *k* value and be very peaked) **[6]**. The cumulative distribution of the Weibull function is given by [7]:

$$\mathbf{F(V,k,c) = 1 - e}^{\left[-\left(\frac{V}{c}\right)^{k}\right]}. \tag{2}$$

## 2.2. Wind Energy Modeling

✓ ***Wind Power Estimation***

- *Wind available power:*

The available wind power on the site undergoes a succession of losses in cascades and only one part called useful power can be used [8].

For a period of measurement, the mean wind power density in the site is given by the following expression:

$$\langle P_{av}\rangle = \frac{1}{2}\,\rho\langle V^3\rangle. \tag{3}$$

The cubical mean velocity $\langle V^3\rangle$ in the site is calculated referring to the frequency of each velocity and it is given as a function of the probability density function [8]:

$$\langle V^3\rangle = \int_0^{\infty} f(V)V^3 dV = c^3\,\Gamma\left(1+\frac{3}{k}\right), \tag{4}$$

The mean velocity is given by the following expressions [9]:

$$\langle V\rangle = \int_0^{\infty} f(V)dV = c\,\Gamma\left(1+\frac{1}{k}\right). \tag{5}$$

where Γ is the defined Gamma function given for any reality x positive not null by [6]:

$$\Gamma(x) = \int_0^{\infty} \varphi^{x-1}e^{-\varphi}dt\ . \tag{6}$$

- *Wind extractable power:*

The maximum power $P_{ex}$ that can be extracted from a wind turbine is limited and cannot exceed more than 16/27 of the available wind power, this is known as the Betz' limit.

The mean extractable power $P_{ex}$ is given by:

$$\langle \mathbf{P_{ex}}\rangle = \frac{\mathbf{16}}{\mathbf{27}}\,\mathbf{P_{ex}} = \mathbf{0.37}\langle \mathbf{V^3}\rangle\,. \tag{7}$$

- *Wind useful power:*

Each wind turbine is characterized by it starting velocity $V_s$, it nominal velocity $V_n$ and it cut velocity $V_c$. The rotor of a wind turbine is designed to operate in a speed range of the going wind from $V_s$ to $V_n$ and turns at a quasi-constant rotational speed. From speed equal to $V_c$, the machine is automatically stopped for safety reasons, for that wind velocities lower than $V_s$ and higher than $V_c$ are not considered. The mean useful power in the site is given by the following expression:

$$\langle \mathbf{P_u} \rangle = \frac{1}{2}\rho \, \langle \mathbf{V_u^3} \rangle. \tag{8}$$

where $\langle \mathbf{V_u^3} \rangle$ is the cubical mean useful velocity in the site over the period of measurement.

The mean useful wind can be expressed as **[4]**:

$$\langle \mathbf{P_u} \rangle = \int_0^{\infty} \mathbf{g(V)dV}\,. \tag{9}$$

where g (V) is the power probability density for each wind velocity V, it is given by the following expression:

$$\mathbf{g(V)} = \frac{1}{2}\rho \, \mathbf{V^3 \, f(V)}. \tag{10}$$

The cubical mean useful velocity can be expressed as a function of the wind velocity and the turbine limit velocities ($V_s$, $V_n$ and $V_c$) and it is given by [8]:

$$\langle V_u^3 \rangle = \int_{V_s}^{V_n} \mathbf{V^3} \left(\frac{k}{c}\right)\left(\frac{V}{c}\right)^{k-1} \mathbf{e}^{\left[-\left(\frac{V}{c}\right)^{k}\right]} \mathbf{dV} + \mathbf{V_n^3} \int_{V_n}^{V_c} \left(\frac{k}{c}\right)\left(\frac{V}{c}\right)^{k-1} \mathbf{e}^{\left[-\left(\frac{V}{c}\right)^{k}\right]} \mathbf{d\,V}. \tag{11}$$

- *Wind used power:*

The mean power really used is given by:

$$\langle \mathbf{P_e} \rangle = \eta \langle \mathbf{P_u} \rangle\,. \tag{12}$$

The efficiency of the wind turbine is given by [8]:

$$\eta = \frac{\langle \mathbf{P_n} \rangle}{\langle \mathbf{P_{th}} \rangle} = \frac{\langle \mathbf{P_n} \rangle}{\frac{1}{2}\rho \, \mathbf{A_a} \langle \mathbf{V_n^3} \rangle}\,. \tag{13}$$

where $\langle \mathbf{P_n} \rangle$ is the mean nominal power given as a characteristic of the wind turbine and $\langle \mathbf{P_{th}} \rangle$ is the mean electric power calculated for a wind velocity equal to the nominal velocity $V_n$.

✓ ***Wind Data Extrapolation***

Wind velocities are measured at 10 m height t.o.g and wind turbines are generally installed at altitudes higher or equal to 25 m [8].

As known, the wind velocity increases with altitude. Therefore, it is necessary to estimate the wind velocities at the various turbine heights for studying the feasibility of wind energy use [5].

Figure 1 shows variations of mean wind velocity versus altitude.

As can be seen from this figure, wind velocity considerably increases for altitudes less than or equal to 20 meters, then a small evolution starting from 30 m and more is obtained (it is useless to go more than 30m except if necessary).

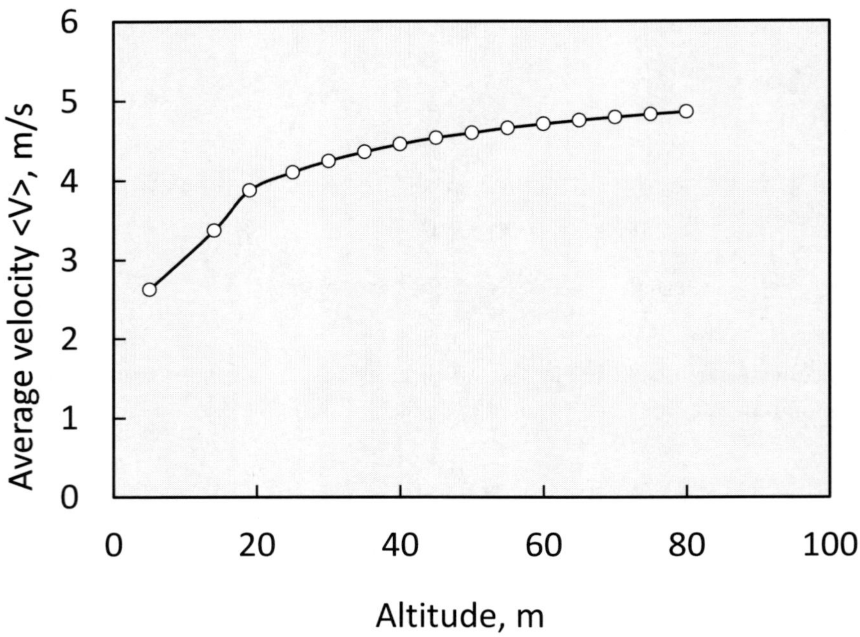


Figure 1. Wind velocity versus altitude.

Thus, the wind velocity and the characteristics of the site must be corrected and extrapolated.

There are many methods that can be used to adjust the wind velocity from one level to another level.

In this case, Tchinda et al. [10] established a relationship between wind velocity at 10 m height t.o.g and wind velocity at heights other than 10 m, it is given by:

$$\mathbf{V_j} = \mathbf{V_i}\left(\frac{\mathbf{Z_j}}{\mathbf{Z_i}}\right)^{\tau}. \tag{14}$$

where $V_i$ is the wind speed at height $Z_i$=10 m above ground level, $V_j$ is the wind speed at height $Z_j$ above ground level and $\tau$ the friction coefficient. The friction coefficient is a constant ranging between 0.1 to 0.4 [11]. The most used formula to extrapolate the characteristics of the site is the modified power method [8], [5], knowing the Weibull parameters to the $Z_i$ height, the parameters $c_j$ and $k_j$ are related to $c_i$ and $k_i$ by the following relations:

$$c_j = c_i\left(\frac{Z_j}{Z_i}\right)^{\alpha}, \tag{15}$$

$$k_j = \left[\frac{k_i}{1-0.0881\,\mathrm{Ln}\left(\frac{Z_j}{Z_i}\right)}\right], \tag{16}$$

$$\alpha = \frac{1}{\ln\left(\frac{\sqrt{Z_i Z_j}}{Z_o}\right)} - 0.0881\,\ln\left(\frac{c_i}{6}\right). \tag{17}$$

where α is a constant that depends on Z and c and $Z_o$ is the roughness of the site.

## 2.3. Wind Measurement Representations

In this section, data recorded on the considered site is treated. The objective is to evaluate the most important characteristics of wind energy. The classical and statistical treatments provide the estimation of the mean wind velocity, the wind velocity distribution function and its characteristics and the mean wind power density available and extracted in the site of Gabes.

✓ ***Classical Representation***

The data used for the wind assessment was recorded by National Institute of Meteorology (INM) concerning the site of Gabes during the year 2007. The histogram of Figure 2 illustrates monthly mean wind velocity at 10 m height t.o.g. It can be seen from this figure that wind mean speed in the site of Gabes ranges between 2.3 and 4.2 m/s. The highest monthly mean wind speeds occur during the months of January- April, with a maximum occurring in April. Generally, lower wind speeds are observed in the remainder of the year, with a minimum in November. It has risen from 2.3 m/s in November to 4.2 m/s in April. The average annual wind speed is found to be 2.93 m/s. Conforming to this value, a recent study was made to evaluate the wind resource in Tunisia [3].

This study reveals that the Tunisian territory can be divided into three groups of sites. The region of Gabes belongs to the second group of average wind potential, of which the annual mean speed lies between 2.7 and 3.6 m/s.

✓ ***Statistical Representation***

To evaluate the wind intensity in a site, measured wind velocity data are commonly available in time series format, in which each data point represents either an instantaneous sample wind speed or an average wind speed over a period of time [5].

Table 1 is an example of such data in which wind velocity data are available in frequency distribution format for different ranges.

**Table 1. Wind velocity classes**

| *Range* | *Velocity (m/s)* | *Duration (day)* | *Frequency (%)* |
|---|---|---|---|
| *1* | 0-1 | 11 | 3,0054 |
| *2* | 1-2 | 77 | 21,0382 |
| *3* | 2-3 | 108 | 29,5081 |
| *4* | 3-4 | 96 | 26,2295 |
| *5* | 4-5 | 51 | 13,9344 |
| *6* | 5-6 | 17 | 4,6448 |
| *7* | 6-7 | 5 | 1,3661 |
| *8* | 7-8 | 1 | 0,2732 |

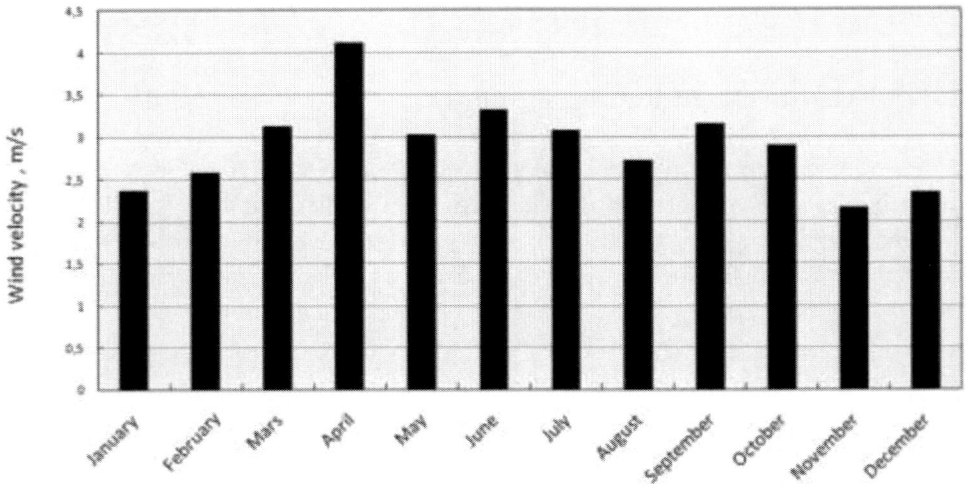


Figure 2. Monthly distribution histogram of wind velocity in the site of Gabes.

## 2.4. Characteristics of the Site of Gabes

The energy intensity which could be extracted from the wind resource in the site of Gabes is determined by the knowledge of the wind characteristics in this area. The two parameters of Weibull *c* and *k* are determined using the graphical method [12]. The application of such method requires that the wind speed data be in cumulative frequency distribution format (Table 1). The determination of the Weibull distribution parameters requires the logarithmic transformation of equation (2) that gives:

$$\mathrm{Ln}\left(-\mathrm{Ln}\left(1-\mathrm{F(V)}\right)\right)=\mathrm{k\,Ln(V)}-\mathrm{k\,Ln(c)}. \tag{18}$$

A plot of $\mathrm{Ln}\left(-\mathrm{Ln}\left(1-\mathrm{F(V)}\right)\right)$ versus $\mathrm{Ln(V)}$ presents a straight line. The gradient of the line is (k), and the intercept with the y axis is $\left(-\mathrm{k}\,\mathrm{Ln(c)}\right)$. The values of k and c are reported in Table 2. These parameters enable the establishment of the classified frequencies histogram and the Weibull distribution function f (V) curve as functions of the wind velocity for Z = 10 m (see Figure 3). This figure illustrates the good recovery of meteorological data by the Weibull distribution.

**Table 2. Characteristics of the site at 10m and 25m t.o.g.**

| Site of Gabes | Roughness $Z_o$ | Friction coefficient τ | Z = 10 m | Z = 25 m |
|---|---|---|---|---|
| k [−] | 0,4 | 0,25 | 2,929 | 3,18 |
| c [m/s] | | | 3,718 | 4,6 |
| <V> [m/s] | | | 3,3 | 4,12 |
| $<V^3>$ [$m^3/s^3$] | | | 45,85 | 87,61 |

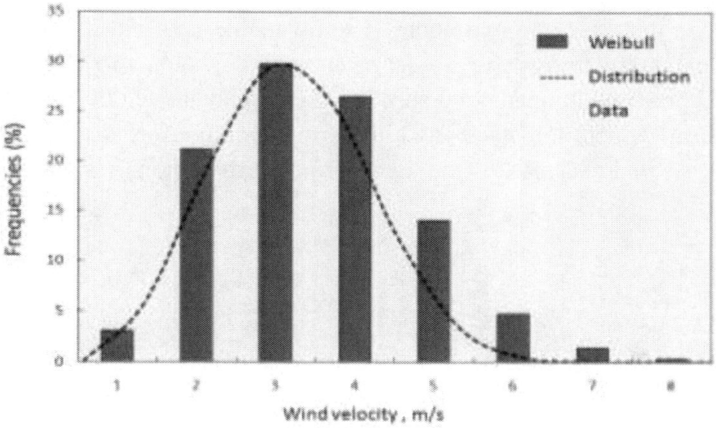


Figure 3. Frequency-velocity histogram of the site of Gabes.

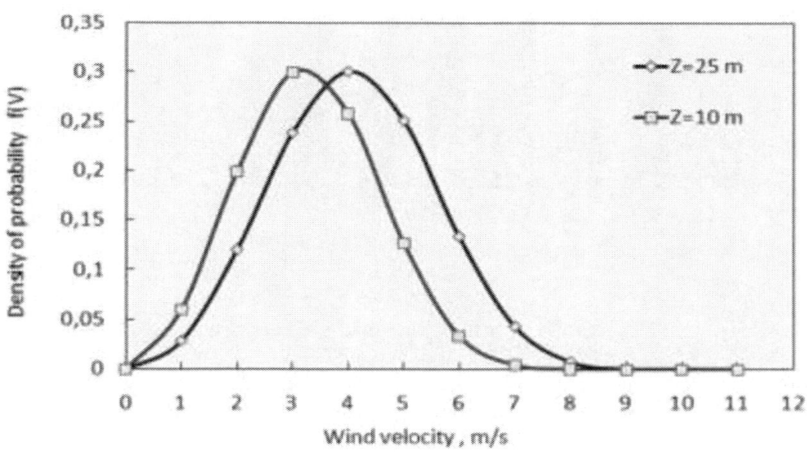


Figure 4. Weibull distribution at 10 and 25m.

Figure 4 illustrates the Weibull distribution of Gabes region given by the density of distribution function versus wind velocity for 10 and 25 m t.o.g. As shown in Figure 4, the Weibull function for Z = 10 m reaches its maximum value at about 3 m/s, then it decreases as velocity increases until being practically equal to zero for velocities higher than 7m/s. The same behavior is observed for the Weibull function at Z = 25 m. In this case, the maximum density of distribution function value is obtained at 4 m/s and will be cancelled at 8 m/s. These results show that the installation of the wind turbine at altitudes higher than 25 m, considerably improves the wind characteristics.

Values of the roughness $Z_0$ and the friction coefficient $\tau$ are relative to the site's nature [11].

## 2.5. Evaluation of the Wind Potential in the Site of Gabes

Figure 5 illustrates an histogram of the available and the extractable monthly wind energy at the altitude of 25 m which is calculated on the basis of the meteorological data. The monthly variation of the available and the extractable power shows maximum values in the month of April. In this case, the maximum mean available power density value is about 104 $W/m^2$ and the maximum mean extracted power value is equal to 60 $W/m^2$. These values correspond to the maximum mean wind speed as mentioned above (classical representation).

The minimum values are observed during the month of November where the mean available power is equal to 20 $W/m^2$ and the mean extractable power is about 12 $W/m^2$.

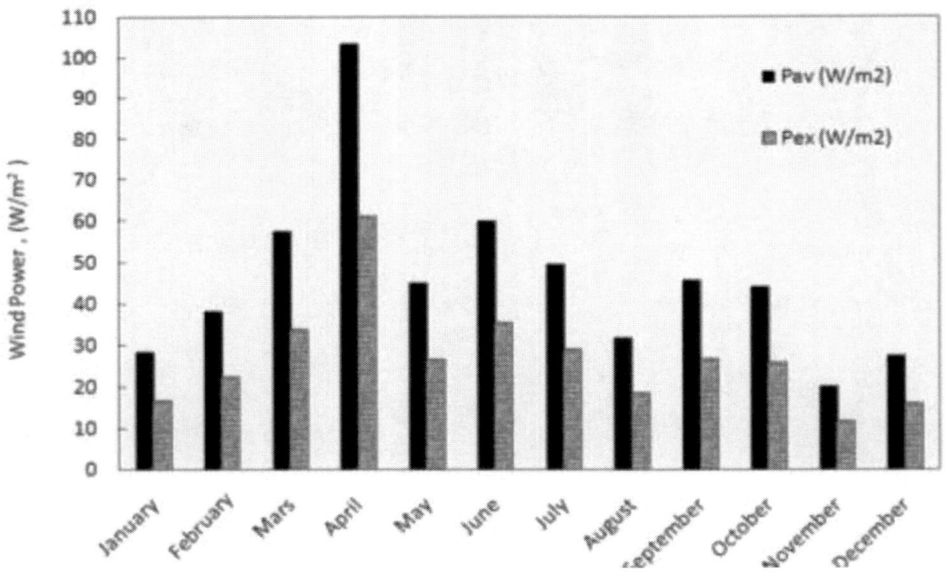


Figure 5. Histogram of monthly available and extractable wind energy in the site of Gabes at 25m.

The annual wind power energy in the site can be estimated with two ways: the first is based on the meteorological data (Meteorological) and the second on the values of the mean velocity <V> and the cubical mean velocity $\langle V^3 \rangle$ (Weibull). The values of the annual wind energy (available and extractable) at 25m evaluated by the two methods are presented in

Figure 6. The annual extractable mean energy is the average of the extractable monthly wind energies. As shown in Figure 6, values calculated by the Weibull method are higher than those given by meteorological calculation. Referring to the Weibull calculations, the mean annual available wind potential is equal to 55 W/m$^2$ whereas the extractable wind potential is about 32W/m$^2$.

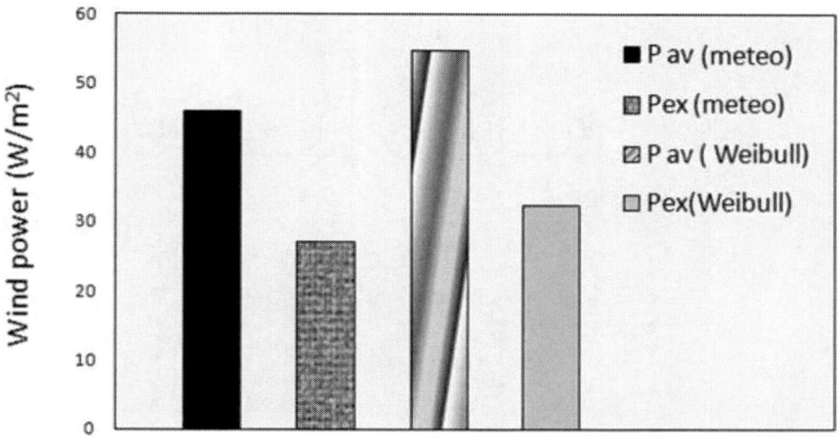


Figure 6. Histogram of annual available and extractable wind energy in the site of Gabes at 25m.

# 3. Wind-Powered Desalination by Mechanical Vapor Compression System

## 3.1. Process Description

A schematic diagram of the single-effect mechanical vapor compression (MVC) system is shown in Figure 7.

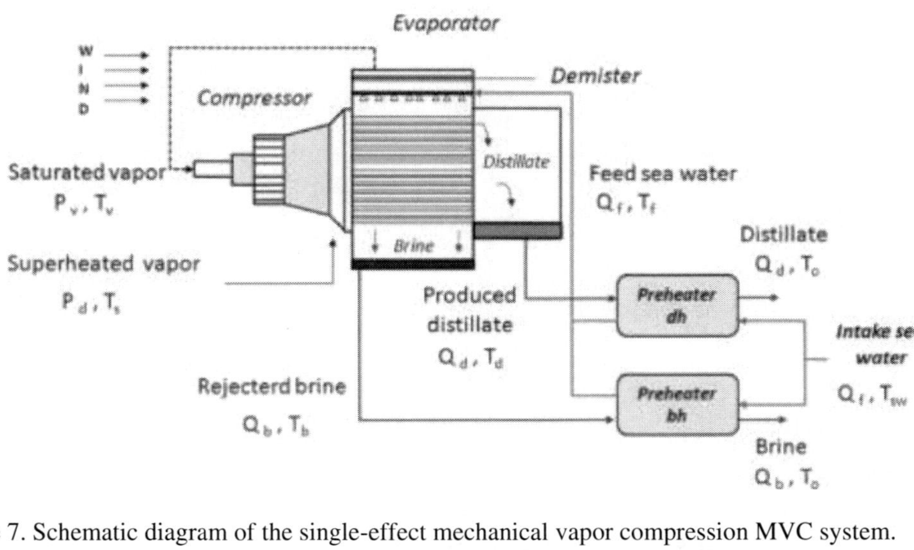


Figure 7. Schematic diagram of the single-effect mechanical vapor compression MVC system.

In this paper, the studied MVC system includes an evaporator, two feed preheaters and a vapor compressor. Feed water is pumped and taken through two heat exchangers in parallel, each one of them heated by the product water and brine, respectively, which are discharged from the evaporator. The two preheaters are designed to achieve a small temperature difference between the brine and distillate stream and the feed of sea water (Figure 8).

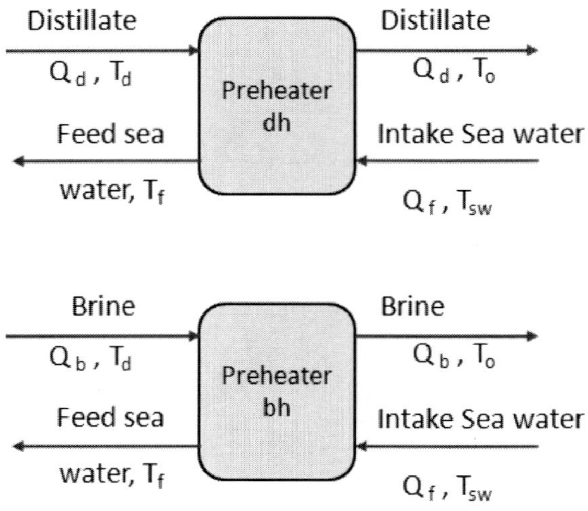


Figure 8. Distillate and brine preheaters.

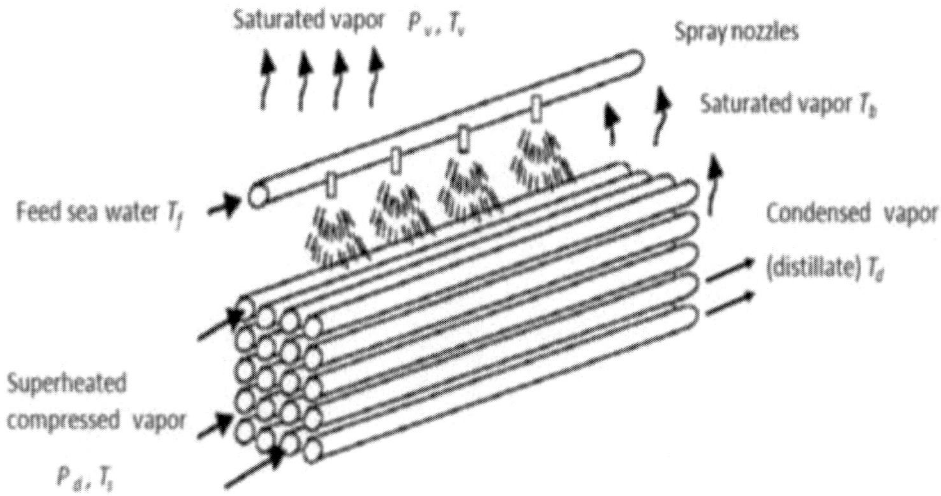


Figure 9. Schematic of the evaporator tubes.

Seawater is then pumped and will be pulverized above the evaporator tubes and then the evaporation will start (Figure 9). The formed vapor crosses the demister and will be directly forwarded to the compressor. The compression process increases the saturation temperature of the vapor and adds a quantity of superheating. The compressed vapor runs out inside the evaporator tubes where it condenses and releases its superheating and its latent heat with the brine through the external surface of the evaporator tubes.

## 3.2. Mathematical Modeling of the MVC System

✓ *Adopted Assumptions*

To model the MVC desalination system, two models are evaluated according to the compression nature: the simple model which supposes that compression is isentropic and the detailed model if compression is supposed polytropic.

To simplify the calculation, certain assumptions are adopted [13]:

- Same outlet temperatures in the two preheaters.
- A null salt concentration in the distillate.
- A vapor specific heat $Cp_v = 1.8$ kJ/ kg °C.
- A polytropic efficiency $\eta_p = 0.8$.

✓ *Detailed Model Equations*

The mass balance of salt is given by:

$$\mathbf{Q_f = Q_d + Q_b = Q_d + \frac{X_f}{X_b} Q_f} \,. \tag{19}$$

The preheater and evaporator energy equations are also given by:

$$\mathbf{Q_f C_{p_f}(T_f - T_{sw}) = Q_d C_{p_d}(T_d - T_o) + Q_b C_{p_b}(T_b - T_o)} \,, \tag{20}$$

$$\mathbf{Q_f C_{p_f}(T_b - T_f) + Q_d L_{vp} = Q_d C_{p_v}(T_s - T_d) + Q_d L_d} \,. \tag{21}$$

where $T_f$, $T_o$ and $T_{sw}$ are the feed seawater temperature, the outlet temperature of the brine and distillate and the temperature of intake see water, respectively.

It is necessary to take in consideration that the vapor temperature below the demister $T_{vp}$ is lower than the brine boiling temperature $T_b$ due to the boiling point elevation (BPE) and the non equilibrium allowance (NEA) [14]:

$$\mathbf{T_{vp} = T_b - BPE - NEA} \,. \tag{22}$$

The vapor temperature above the demister $T_v$ is reduced due to temperature losses $\Delta T_p$ [13]:

$$\mathbf{T_v = T_{vp} - \Delta T_p} \,. \tag{23}$$

Generally, evaporators used in this type of unit are falling film evaporators. This type of evaporator consists of a horizontal tube and falling film in a vessel where sea water feed is sprayed through nozzles. The evaporator heat transfer area can be expressed in terms of the sensible and latent heat of the condensing vapor. It can be also expressed as the sensible heat of the feed stream and the latent heat for evaporation:

$$\mathbf{A_e\, U_e\, (T_d - T_b) = Q_d L_d + Q_d C_{p_v}(T_s - T_d) = Q_d L_{vp} + Q_f C_{p_f}(T_b - T_f)}. \tag{24}$$

The selected compressor is a centrifugal compressor generally designed for the MVC units.

In the majority of compressors, the variations of kinetic and potential energies are considered negligible. At first approximation, compression is supposed adiabatic:

$$\mathbf{W = C_p(T_s - T_v)}\,. \tag{25}$$

Isentropic work of the compressor is given as a function of inlet saturation pressure $P_v$ and outlet saturation pressure $P_d$:

$$\mathbf{W_{is} = W(\Delta S = 0) = \frac{P_v v_v \gamma}{(\gamma - 1)}\left[(p_d/p_v)^{\frac{\gamma-1}{\gamma}} - 1\right]}. \tag{26}$$

where $\gamma$ is the isentropique coefficient.

The assumption of an isentropic work is not the true thermodynamic way followed by gases in the majority of the compressors and compression is neither isothermal nor adiabatic, but polytropic.

The compression ratio Cr is defined in terms of the inlet saturation pressure $p_V$ and the outlet saturation pressure $p_d$:

$$\mathbf{Cr = p_d/p_v}\,. \tag{27}$$

Consequently, for a real compression and in order to take in consideration the friction losses, a polytropic way of gas evolution following the relation $pv^n$ =cte is defined and the polytropic work is then given by:

$$\mathbf{W_p = \frac{P_v v_v n}{(n-1)}\left[Cr^{\frac{n-1}{n}} - 1\right]}. \tag{28}$$

The real compressor work is related to the polytropic work by the following relation:

$$\mathbf{\eta_p = \frac{W_p}{W} = \frac{W_p}{H_s - H_v}}. \tag{29}$$

where $\eta_p$ is the polytropic effectiveness which is often fixed by the manufacturers.
This effectiveness is used to convert calculation in the polytropic case into adiabatic one:

$$\mathbf{n = \frac{\eta_p \gamma}{1 + \eta_p \gamma - \gamma}}. \tag{30}$$

The isentropic work of the compressor is given as the enthalpy difference of the compressed superheated vapor and the inlet vapor:

$$\mathbf{W_{is} = H_{is} - H_v}\,. \tag{31}$$

The isentropic superheated temperature of compressed vapor $T_{is}$ is expressed in terms of the vapor temperature $T_v$ and the compression ratio Cr:

$$\mathbf{T_{is} = (T_v + 273.15)(Cr)^{\frac{\gamma-1}{\gamma}} - 273.15}\,. \tag{32}$$

The polytropic work of the compressor is then given by the following expression:

$$\frac{\mathbf{W_p}}{\mathbf{W_{is}}} = \frac{\frac{n}{(n-1)}\left[\mathbf{Cr}^{\frac{n-1}{n}} - 1\right]}{\frac{\gamma}{(\gamma-1)}\left[\mathbf{Cr}^{\frac{\gamma-1}{\gamma}} - 1\right]} = \eta_p \frac{\left[\mathbf{Cr}^{\frac{n-1}{n}} - 1\right]}{\left[\mathbf{Cr}^{\frac{\gamma-1}{\gamma}} - 1\right]}\,. \tag{33}$$

The enthalpy of the compressed superheated vapor $H_s$ is expressed in terms of the saturation enthalpy of the compressed vapor $H_d$, the isentropic sensible heat, and the vapor enthalpy $H_v$:

$$\mathbf{H_s - H_v = \left[(H_d - H_v) + C_{pv}(T_{is} - T_d)\right]} \frac{\left[\mathbf{Cr}^{\frac{n-1}{n}} - 1\right]}{\left[\mathbf{Cr}^{\frac{\gamma-1}{\gamma}} - 1\right]}\,. \tag{34}$$

The superheated temperature is obtained as a function of the compressed superheated vapor $H_s$ and the saturation enthalpy of the compressed vapor $H_d$:

$$\mathbf{T_s = \frac{H_s - H_d}{C_{pv}} + T_d}\,. \tag{35}$$

### 3.3. Detailed Model Solution

Calculation starts with the specification of the inlet parameters of the model. The first solution step involves calculations of the feed and brine flow rates $Q_f$ and $Q_b$. This is followed by iterative solution of equations (20)-(35) based on Newton's method which requires an initial guess for the feed seawater temperature $T_f$, and the outlet temperature of the brine and distillate $T_o$ is then carried on. The choice of these parameters must hold in account that temperature difference between ($T_b$ - $T_f$), ($T_d$ - $T_b$) and ($T_o$ - $T_{sw}$) should be between 1 and 5°C [13]. The inlet parameters of the model are presented in Table 3.

**Table 3. Operational parameters of the MVC unit**

| | |
|---|---|
| Feed sea water salinity $x_f$, ppm | 35000 |
| Brine salinity $x_b$, ppm | 70000 |
| Intake sea water temperature $T_{sw}$, °C | 25 |

## 3.4. Results and Discussions

- ***Influence of $T_b$ and $\Delta T$ on the Superheated Temperature $T_s$***

Variations in the superheated temperature are shown in Figure 10. As illustrated, the superheated temperature $T_s$ varies over a range of 58.7°C - 120°C.

At fixed temperature difference $\Delta T$, superheated temperature increases with the brine boiling temperature $T_b$ and for fixed value of $T_b$, the superheated temperature $T_s$ increases with $\Delta T$.

A low superheating is obtained for high brine boiling temperature $T_b$ because the low values of the latent heat are obtained for the high temperatures.

- ***Influence of $T_s$, $T_b$ and $\Delta T$ on the Specific Power Consumption $W_p$***

The variations of the specific power consumption of the MVC unit as function of $T_b$, $T_s$ and $\Delta T$ are shown in Figure 11 and Figure 12.

Calculations are made as a function of the brine boiling temperature $T_b$ (ranging between 50°C and 100°C) and the temperature difference $\Delta T$ between brine temperature $T_b$ and the saturation temperature of the compressed vapor $T_d$ (ranging between 1 °C and 4 °C).

- ***As Shown, the Specific Power Consumption Varies over a Range of 3.8-13 kWh/m$^3$***

It decreases at higher boiling temperatures $T_b$ due to the decrease in specific volume of inlet vapor at higher temperatures and it increases at larger temperature differences.

It is found that at $T_b$=60°C, the specific power consumption will decrease from 11.84 to 4.6 kWh/m $^3$ if the temperature difference $\Delta T$ is decreased for 3°C. This means a percent decrease of about 61% in the specific power consumption of the compresssor for 3°C increase in the temperature difference $\Delta T$.

It is clear that $\Delta T$ is a parameter that should be selected carefully when designing the MVC unit.

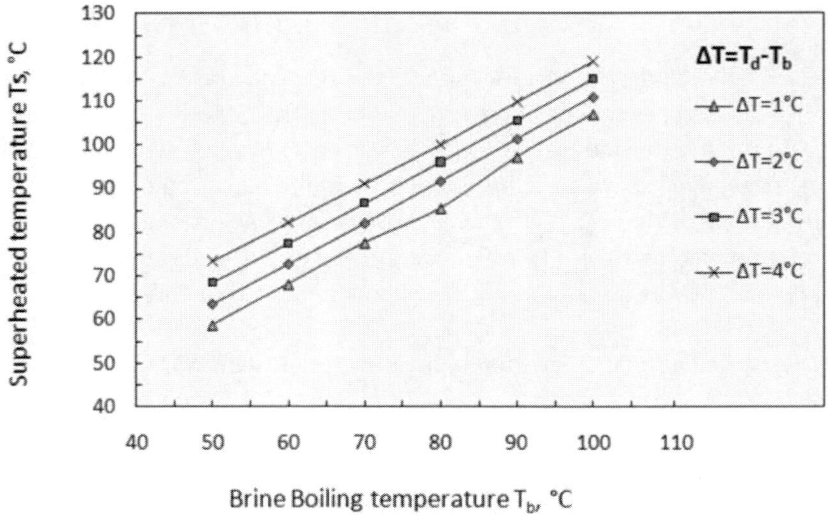


Figure 10. Superheated compressed vapor temperature $T_s$ versus $T_b$ and $\Delta T$.

We can note also that the specific power consumption $W_p$ increases at higher difference between brine temperature $T_b$ and the superheated temperature $T_s$. This is because of the higher energy needed to increase the temperature of the compressed steam over a larger range.

The maximum specific power consumption is about 13 kWh/m $^3$ and it is obtained for a brine boiling temperature equal to 50°C and for $\Delta T$ = 4°C, this value is consistent with values reported in the literature [13], [14].

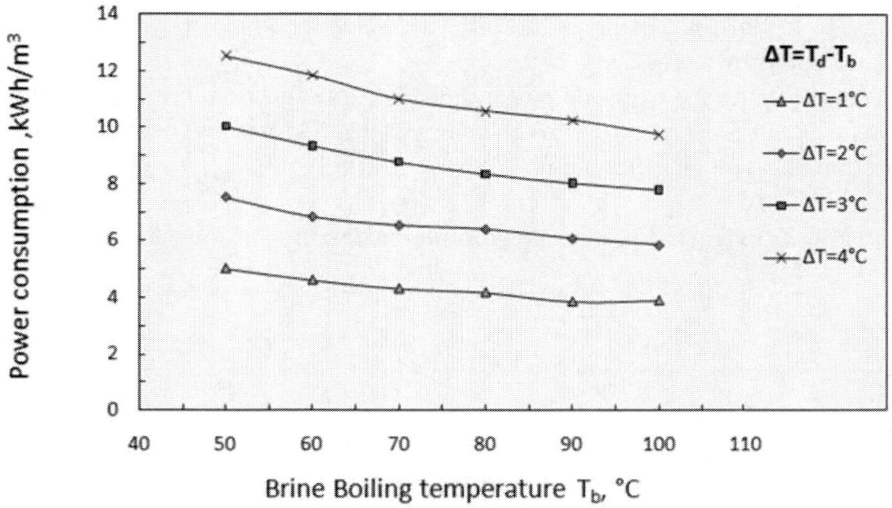


Figure 11. Power consumption versus $T_b$ and $\Delta T$.

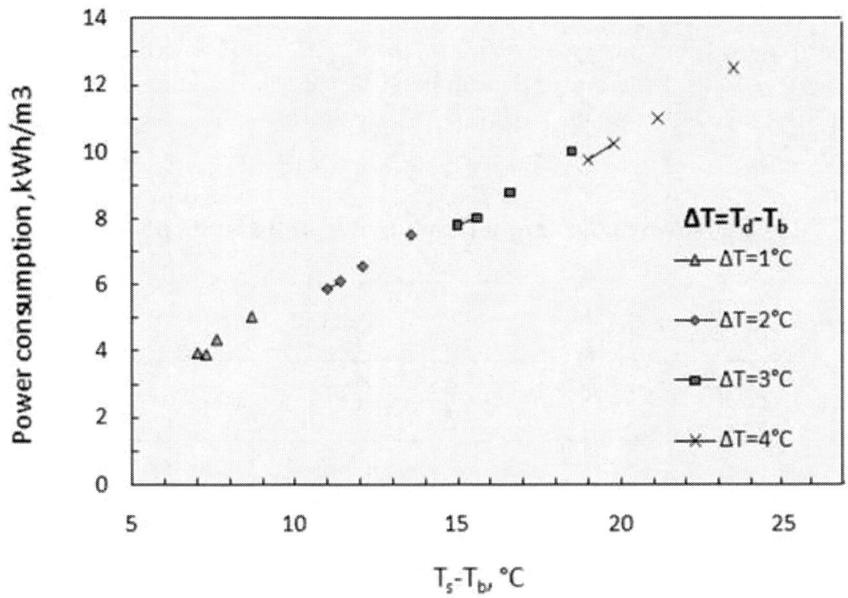


Figure 12. Power consumption versus $T_s$- $T_b$ and $\Delta T$.

## 4. DISCUSSIONS

The north region of Tunisia appears as a zone particularly favorable for exploitation of wind energy conversion [3], whereas this study shows that only wind turbine of weak powers can be used in the site of Gabes and wind energy conversion systems with simple technology having a starting speed of 2 to 3 m/s can be considered.

As a part of our estimation of the wind power extracted in the site of Gabes, we determine this power as a function of the rotor diameter of small wind turbine varying in a range from 2 to 7m for small wind turbines (Exel of the company 'Bergey Windpower', Inclin of the company Bornay….) [15].

Table 4 illustrates the extracted power in the site as a function of the rotor diameter. It is clear that the compressor specific work increases as the diameter increases and can reach 1.24 kW corresponding to a diameter of 7m.

**Table 4. Extracted power consumption as function of the rotor diameter**

| *Rotor diameter (m)* | 2 | 3 | 5 | 7 |
|---|---|---|---|---|
| *Extracted power consumption (kW)* | 0,10 | 0,23 | 0,63 | 1,24 |

In the other hand and in order to minimize the maintenance of the unit and compressor and to prevent scaling and corrosion potentials, the recommended temperature at low operating evaporator is 50°C. Meanwhile, in accordance with this evaporator operating temperature, a realistic evaporator temperature difference 3°C is a compromised temperature difference across the heat transfer areas and the required compressor work [16].

We report in Table 5 the power consumption of the MVC unit as a function of the unit productivity. As shown, to produce 10 $m^3$ per day, for example, the MVC system should require 4.18 kW.

**Table 5. Power consumption as function of the unity production**

| *Power consumption ($m^3/d$)* | 2,09 | 4,18 | 6,27 | 8,36 |
|---|---|---|---|---|
| *Production ($m^3/d$)* | 5 | 10 | 15 | 20 |

It is clear that the power possible to be extracted from the studied site is not enough to drive the MVC desalination unit.

The values of the mean wind energy extracted in the site of Gabes are between 12 $W/m^2$ in November and 60 $W/m^2$ in April, compared to the monthly mean solar radiation available and received by a horizontal plane (566.23 $W/m^2$ [17]), the maximum wind energy available is really inferior to the maximum of solar energy.

For that, coupling these two renewable energies can be useful to emphasize the operation of the MVC desalination units in the region of Gabes.

## CONCLUSION

In the first part of this paper, an estimation of the wind energy available has been prepared and the wind speed frequencies are considered to discuss possible applications of this renewable energy in the site of Gabes. This study is imperative for the dimensioning of any aerogenerator.

The calculation of the mean values of the wind speeds for the year 2007, monthly and annual mean of the extractable wind energy by turbines give an idea of the wind potential available in the site.

The characteristics of the site at 10 m and the density of recoverable annual average power to a 25 m height by using the extrapolation of the Weibull parameters are determined.

As shown for a wind annual average velocity of 3 m/s, the average recoverable wind potential can reach $32 W/m^2$ at 25 m.

In the second part of the paper, we are concentrated to determine the power consumption of the MVC unit.The compressor specific consumption varies from 4 kWh $/m^3$ to 13 kWh $/m^3$.

We note clearly a strong dependence of this power with the temperature difference $\Delta T$ between the brine temperature $T_b$ and the saturation temperature of the compressed vapor $T_d$, the most important parameter that should be optimized to obtain a minimal energy consumption.

The study relized in this paper shows that the power available in the site is not enough to drive the MVC desalination unit and the availability of the solar resource make the use of hybrid wind–solar systems to drive the desalination unit a promising alternative.

We hope that with the political evolution of the country, renewable energies will experience some support in order to improve the results presented in this paper.

## REFERENCES


[1] BenJemaa F., Houcine I., & Chahbani M. H. (1998). Desalination in Tunisia: Past experience and future prospects. *Desalination, 116*, 123-134.

[2] Ben Amar F., Elamouri M., & Dhifaoui R. (2008). Energy assessment of the first wind farm section of Sidi Daoud, Tunisia. *Renewable Energy, 33*, 2311-2321.

[3] Elamouri M., & Ben Amar F. (2008).Wind energy potential in Tunisia. *Renewable Energy, 33*, 758–768.

[4] Dahmouni A. W., Ben Salah M., Askri F., Aloui F., & Ben Nasrallah S. (2008). Wind energy potential in the site of Borj -Cedria in Tunisia. *Renewable Energies Review*, 101-108.

[5] Dahmouni A. W., Ben Salah M., Askri F., Kerkeni C., & Ben Nasrallah S. (2010). Wind energy in the Gulf of Tunis, Tunisia. *Renewable and Sustainable Energy Reviews, 14*, 1303-1311.

[6] Weisser D. (2003). A wind energy analysis of Grenada: an estimation using the 'Weibull' density function. *Renewable Energy, 28,* 1803-1812.

[7] Kavak A. E., & Akpinar S. (2004). Determination of the wind energy potential for Maden-Elazig, Turkey. *Energy Conversion and Management, 45*, 2901-2914.

[8] Merzouk N. K., & Benyoucef B. (2006). Perspectives du pompage éolien dans le sud Algérien. *Agronomic Researche Review: Renewable Energies Development Center, 17.*

[9] Merad L., Benyoucef B., Boussoukaia T., & Omari B. (2003). Etude d'un procédé thermique de dessalement d'eau de mer par aérogénérateur à Rendement élevé. *Renewable Energy: $11^{th}$ International days of Thermic),* 13-19.

[10] Tchinda R., Kendjio J., Kaptouom E., & Njomo D. (2000). Estimation of mean wind energy available in far north Cameroon. *Energy Conversion & Management, 4,* 1917-1929.

[11] Geoffroy A., & Bouzidi Y., Collecte, organisation, traitement, analyse de mesures éoliques et modélisation énergétique, Training course Report: Laboratory of solar and wind evaluation, Santa Maria University, Chile.

[12] Seguro J. V., & Lambert T. W. (2000). Modern estimation of the parameters of the Weibull wind speed distribution for wind energy analysis. *Journal of Wind Engineering and Industrial Aerodynamics, 85,* 75-84.

[13] Ettouney H., & El Dessouky H. T. (2002). *Fundamentals of salt water desalination.* Elsevier.

[14] Ettouny H. (2006). Design of single-effect mechanical vapor compression. *Desalination, 190,* 1-15.

[15] Catalogue des fabricants d'aérogénérateurs de petite et moyenne puissance (2002). *Envirement and Energy controle Agency: Renewable Energy Departement* .ADEME.

[16] Karameldin A., Lotfy A., & Melchemar S. (2002). The Red Sea area wind driven mechanical vapor compression desalination system. *Desalination, 153,* 47-53.

[17] Mechlouch R. F, & Ben Brahim A. (2001). Modélisation de la Densité du Flux Solaire Global de la Ville de Gabès. *Renewable energy review, 4,* 1-9.

In: Energy-Environment-Economics
Editor: Riad Benelmir

ISBN: 978-1-62948-862-2


*Chapter 3*

# EFFECT ANALYSIS OF CONSTRUCTIVE QUALITY AND ORIENTATION ON THE ENERGY CONSUMPTION OF THE BUILDINGS IN THE NORTH OF MOROCCO

***S. Babbah*[1*], *I. R. Maestre*[2†], *A. Draoui*[1], *P. Cubillas*[2] and *F. J. G. Gallero*[3]**

[1]Laboratoire d'Energétique – Equipe des Transferts Thermiques et Energétique (ETTE)
FST de Tanger, Tanger, Marocco
[2]Departamento de Máquinas y Motores Térmicos
Escuela Politécnica Superior de Algeciras (Universidad de Cádiz)
Algeciras, Spain
[3]Departamento de Física Aplicada
Escuela Politécnica Superior de Algeciras (Universidad de Cádiz)
Algeciras, Spain

## ABSTRACT

The majority of the buildings existing in the North of Morocco have a badly construction adapted to the climate of the area. This has, as a result, a strong energy consumption both in winter and in summer which is not desirable economically, and it also has a negative impact on the environment. The aim, within the framework of this work, is to obtain a thermal comfort level with an energy consumption reduced by analyzing the influences of constructive quality and the orientation of the building for its energy needs. For that, we carried out a description and analyzed various types of buildings of the area in the North of Morocco. This enabled us to release certain types of residences representative from the point of view of the construction materials used and typology.

* E-mail address: babbahsomia@yahoo.fr.
† E-mail address: ismael.maestre@uca.es.

# 1. INTRODUCTION

Morocco is confronted with a strong energy constraint due to its total dependence (more than 97%) from the outside for its provision. The sector of the buildings (residential and tertiary) represents more than 25% of Moroccan national consumption in energy and emits 30% of gases to produce the Greenhouse Effect responsible for the global warming of the planetary climate. These numbers tend to increase due to the urban and industrial developments which Morocco has known these last years. It is, consequently, essential to reduce the consumption of energy through an improvement of energy efficiency particularly in the sector of the building.

The study presented in this work is the results obtained within the framework of the common research project (PCI): “Characterization of the energy demand of the buildings in the North of Morocco” (A/3047/05) financed by the Spanish Agency for International Cooperation (AECI) in 2006 and 2007. One of the main objectives of this project was to make specific proposals for a possible future Moroccan legislation for the thermics of buildings. Thus, our objective consists in looking for numerical simulations to use as a simple means to ensure an acceptable thermal comfort level with an energy consumption reduced, by studying the influence of the envelope and the orientation of the building on its energy consumption. In this study, simulations were carried out on two different types of constructions of buildings mostly used in the North of Morocco. Our numericals simulations were carried out using the software eQUEST based on the engine of calculation DOE2.2. [1], [3]-[5]. Before carrying out these simulations, we initially created a database of the climatic data of our area of study.

# 2. CLIMATIC CHARACTERIZATION

## 2.1. Constitutions of the Weather Data

In our study, as a preliminary stage, we limited ourselves to characterize the climate of the main cities in the North of Morocco: Tangier, Tétouan, Nador and Chaouen. This choice is due primarily to the importance of the population of these cities. We have only the experimental weather data of Tangier (for years: 2003, 2004, 2005 and 2006) which we obtained through our own recorders installed at the weather station at the Ibn Batouta airport of Tanger. For the other cities, we decided to generate synthetic years starting from the propotional data of NASA Surfaces Meteorology and Solar Energy, available on their Web site address: http://eosweb.larc.gov/sse/.

## 2.2. Classifications of the Climatic Zones

The differentiation of the climatic zones was carried out by calculating the climatic severity of winter and summer for each geographical area studied. The climatic severity of an area is the quotient between the energy request for a building in this area, and the corresponding one for the same building in an area of reference. For our study, we chose the town of Madrid, Spain as a reference. We then defined a climatic severity for summer and another for winter, and we combined the Degree-Day and the solar radiation of the area, in order to show that when two areas have the same climatic severity during winter (SCw), the energy request for heating of the same building located in these two areas is appreciably equal. This is also applicable for the climatic severity in summer (SCs) in the event for cooling of the building.

## 3. Descriptions of the Studied Residences

We chose a typical construction considered as most widespread in of the North of Morocco [7].

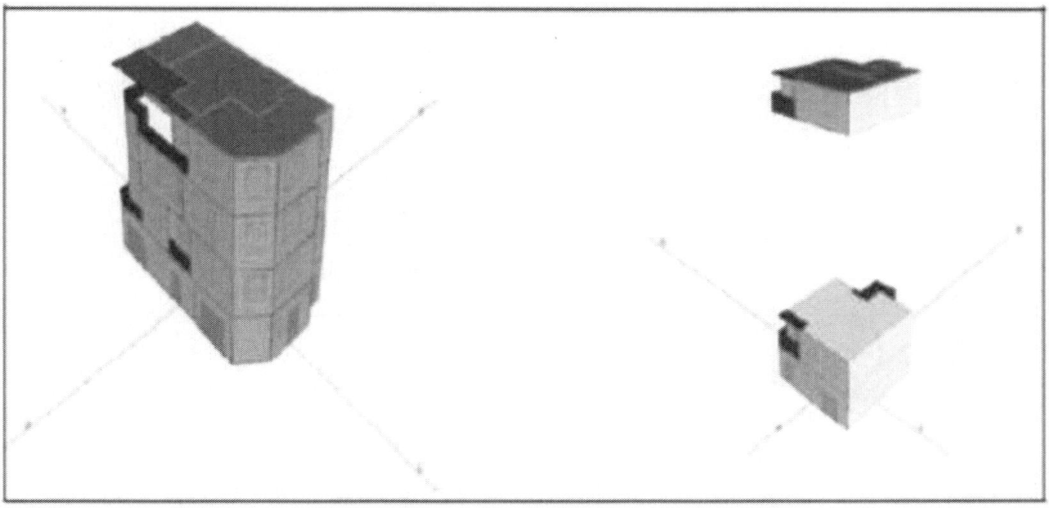

a b

Figure 1a. View in 3D of the bldg. with 4 floors. Figure 1b. View in 3D of the bldg. with 10 floors.

### 3.1. Building Type 1 (Bldg. 1)

It is a house made up of four levels (one ground floor plus 3 floors) (see figure 1-a) with the following distribution:

- The ground floor is of a surface of 146 $m^2$ mainly for regular commercial business.
- The first, the second and the third floor have the same surface of 140 $m^2$ and are intended to live in. They consist of several rooms, living rooms, kitchen, bathroom and toilets.

### 3.2. Building Type 2 (Bldg. 2)

- It is a building of ten floors (one ground floor + 9 floors) (see Figure1-b) mainly for living and which floors have the same distribution as those of the first type with a surface of 140m² for each floor.
- The height of the floors is of 3 meters for the two types of building which has an average occupation of 6 people in each house.

## 4. Compositions of the Walls of the Buildings

The envelope of the building plays a thermal part of separation between the internal and external environments and it also intervenes for storage of heat like a distributor of the latter to the interior and external air. We made up a database of the various thermophysical characteristics of various construction materials mostly used in buildings in the North of Morocco [2], [6]. The composition of the various walls used for the two types of buildings in reference is presented in the following table (Table 1):

**Table 1. Composition of the Walls for two types of the bldg. in reference**

| *Types of walls* | *Materials* | *Thickness (m)* |
|---|---|---|
| *External walls* | Coating Cement | 0,015 |
| | Red brick | 0,075 |
| | Blade of air | 0,1 |
| | Red brick | 0,075 |
| | Coating Cement | 0,015 |
| *Interior walls* | Coating Cement | 0,01 |
| | Red brick | 0,075 |
| | Coating Cement | 0,01 |
| *Floors* | Stone | 0,1 |
| | Concrete | 0,2 |
| | Coating Cement | 0,05 |
| | Tiling | 0,004 |
| *Roofs* | Tiling | 0,004 |
| | Coating Cement | 0,015 |
| | Concrete | 0,2 |
| | Red brick | 0,075 |
| | Coating Cement | 0,015 |

We used for the buildings in reference a simple glazing. For the opaque walls, we considered the alternatives most used in the sector of construction of the buildings in Morocco with the incorporation of an isolating element (5 cm then 10 cm) in all the opaque walls, namely for both the exterior and interior walls, the floor and the roof to be able to carry out the various possible alternatives of their compositions.

# 5. RESULTS

In our study, we chose to show a series of graphs which illustrate the influence of certain parameters on the behavior of the unit building/system. We limited ourselves, in this presentation, to the analysis of the influences of construction materials and the orientation for each climatic zone in the North of Morocco on the consumption of final energy i.e on the maximum values of heating (in winter) and cooling (in summer).

## 5.1. Influence of the Constructive Quality of Materials

Table 2 presents the various cases of constructive quality (compositions of the various walls) studied. In Figures 2 and 3, we presented the variation of the consumption of final energy in (KWh/m²/year) and the maximum values consumed by the heating and the cooling (W/m²) according to the change of the constructive quality of the walls of the buildings (External Walls, Floors and Roofs).

**Table 2. Various case studies of constructive quality (see Figures 2 and 3)**

| *"Case" of the bldg. Type 1* | *"Case" of the bldg. Type 2* | *Constructive quality of the walls of the buildings (external Walls, Floors and Roofs)* |
|---|---|---|
| 11 | 21 | Composition of the walls in reference of each building (***without insulation***) |
| 12 | 22 | Incorporation of ***5 cm insulation*** in the composition of the walls in reference of each building |
| 13 | 23 | Incorporation of ***5 cm insulation*** in the composition of the walls in reference each building + ***double glazing*** |
| 14 | 24 | Incorporation of ***10 cm insulation*** in the composition of the walls in reference of each building |
| 15 | 25 | Incorporation of *10 cm insulation* in the composition of the walls in reference of each buildings + ***double glazing*** |

During the modification of constructive quality (incorporation of insulator in the opaque walls of the buildings: Walls, Floors and Roofs), we observed that the variation of final energy was accentuated than the maximum values consumed by the heating and the cooling. This shows the fact that the improvement of constructive quality (Walls, ceilings, floors, windows) which gives a positive contribution for the mode of heating, has a negative contribution for the mode of cooling and conversely. Indeed, when the insulation is increased, one reduces the requirements in heating and increases in consumption for cooling.

As the total intake, per year of the buildings, is not so dependent on this constructive quality (incorporation of insulator of 5 cm and 10 cm), the composition of the walls of the two types of buildings in reference remains most satisfactory (10 cm of the blade of air) with a minimum of the economic costs in the construction.

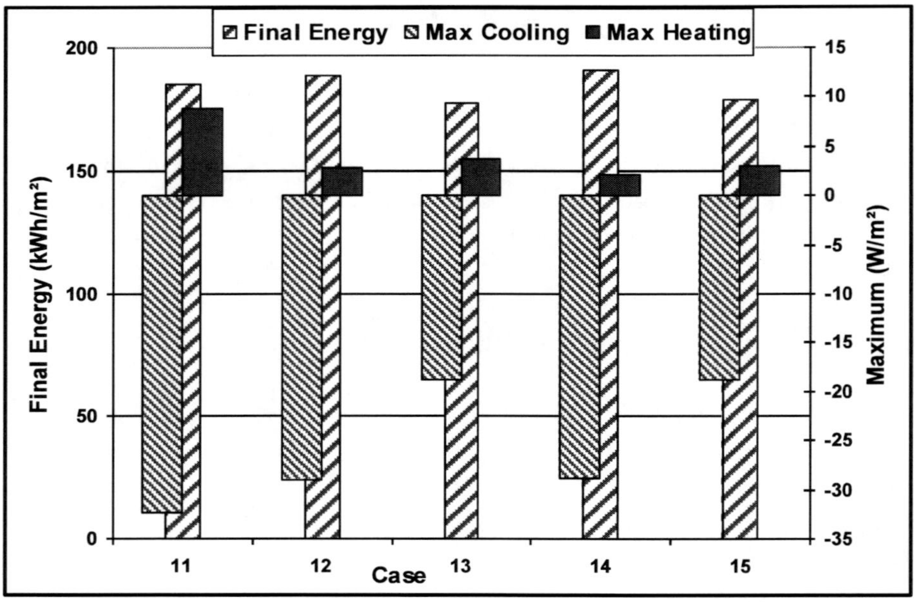


Figure 2. Variation of final energy and the maximum values of the needs for heating and cooling according to the constructive quality of the walls for the Bldg. type 1.

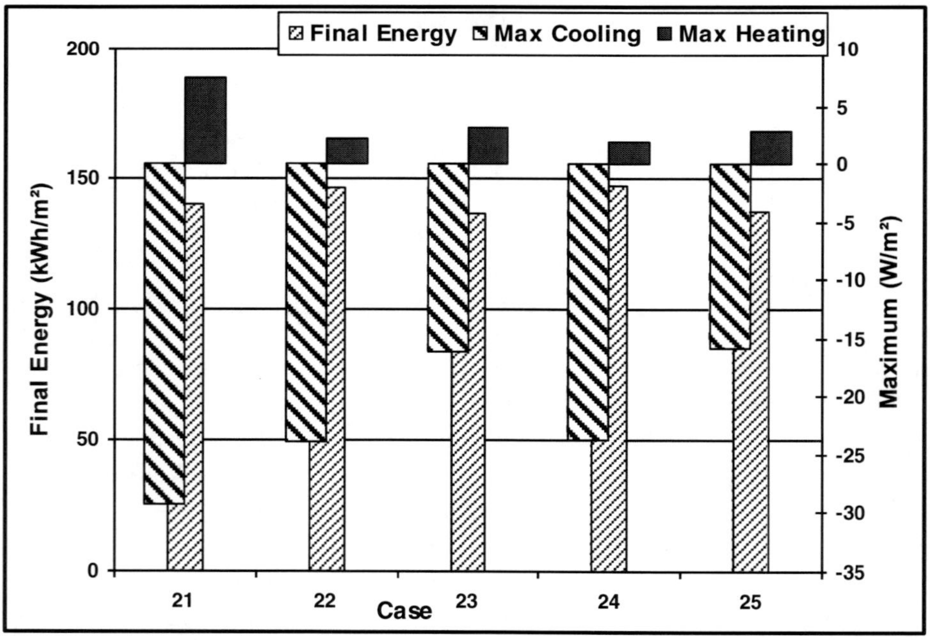


Figure 3. Variation of final energy and the maximum values of heating and cooling according to the constructive quality of the walls for the Bldg. type 2.

## 5.2. Influence of the Orientation of the Buildings

The climatic zones that we presented are zones 1, 2, 3 and 4 which represent the main cities in the North of Morocco: Tangier, Tétouan, Chaouen and Nador.

In the case of building type 1, variations of final consumption energy according to the orientation of the building to the North, South, East and the West, show us clearly that for all the orientations, we will be able to predict to save energy compared to the orientation in reference (Northern case). However, this saving in energy does not prove significant (Figure 4).

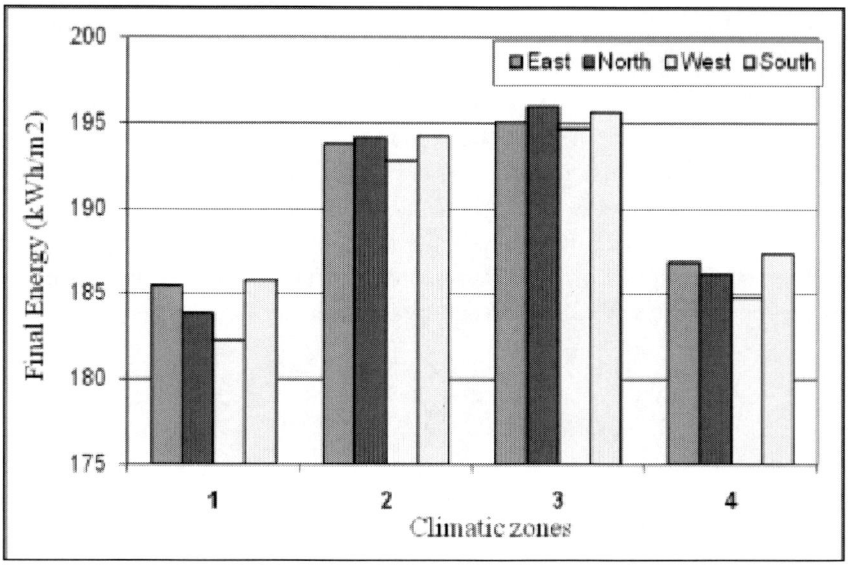


Figure 4. Variation of final energy according to the orientation for each climatic zone (bldg. type1).

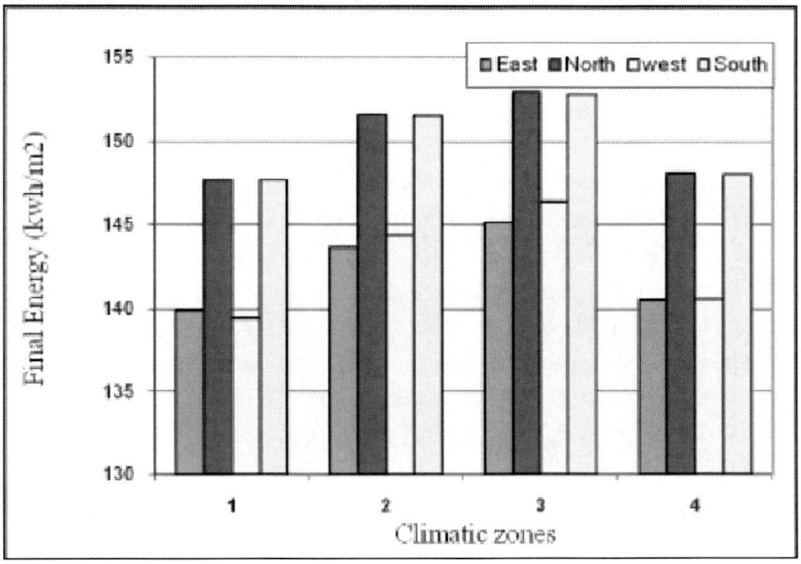


Figure 5. Variation of final energy according to the orientation for each climatic zone (bldg. type 2).

However, in the case of building type 2, the orientations "East" and "West" present (for all the climatic zones) a significant reduction of the final energy consumption with savings varying in 6,56% (Western orientation in climatic zone 2) to 9,44% (Western orientation in climatic zone 1). The energy saving obtained is due mainly to the reduction of the consumption of cooling due to the climate moderated by the Mediterranean Sea for the summer period.

## CONCLUSION

Concerning the thermal quality of construction materials of the studied buildings, it can be assured that energy consumption is influenced little by the improvement of this quality. Basically due to the fact that one based oneself on a situation of thermal quality acceptable.

It is well-known that the orientation of constructions plays an important role in their energy consumption. Thus, this consumption depends mainly on the constructive typology of the facade, its glazing, etc. As a conclusion, we will say that the result of our study highlighted the aspects which have a significant impact on the energy consumption of buildings. Whereas, for the favorable orientations, we can obtain a saving of the final energy near to 10%, whenever we have an important total glazed surface.

## ACKNOWLEDGMENT

The authors wish to thank the AECI (Spanish Agency of International Cooperation) for their financial support for this scientific project and research for the years 2006 and 2007.

## REFERENCES

ASHRAE Standard 90.1-2004: "*Energy Standard for Buildings except Low-Rise Residential Buildings, SI Edition*". American Society of Heating, Refrigerating and Air-Conditioning Engineers. Atlanta, GA.

A. El Bakkouri, "Caractéristique thermophysique et mécanique de quelques matériaux locaux utilisés dans l'isolation et la construction au Maroc: Le plâtre, le liège et la brique creuse". *Thèse de 3ème Cycle*, Faculté des Sciences, Tétouan, Maroc, 1999.

J. J. Hirsch & Associates. *eQuest (Quick energy analysis simulation tool).* Versión 3.52. California 2004. Disponible en http://www.doe2.com.

J. Harry, Jr. Sauer, et al., "*Principles of heating ventilating and air conditioning*", 2001.

J. J. Hirsch & Associates, *DOE-2.2. Building Energy Use and Cost Analysis Program* – Volume1: Basic, Volume2: Dictionary and Volume 3: Topics Lawrence Berkeley National Laboratory.

«*Réglementation Thermique 2000* », Centre Scientifique et Technique du bâtiment, 2000.

S. Babbah, A. Draoui, Ch. Menezo, R. Yezou, J. Benabdelouahab, « Evaluation Energétique des Bâtiments au Nord du Maroc ». *11ème JITH, 16-18 novembre 2005*, Tanger, Maroc.

Web Site: http://eosweb.larc.gov/sse/.

In: Energy-Environment-Economics
Editor: Riad Benelmir

ISBN: 978-1-62948-862-2


*Chapter 4*

# A COMPARATIVE MODELING OF AN EXPERIMENTAL SINGLE-ZONE BUILDING WITH BOND GRAPHS AND TRNSYS

***A. Merabtine, S. Mokraoui, R. Benelmir and D. Descieux***
Faculty of Sciences and Technology – Lab. LERMAB
Henri Poincaré University; France

## ABSTRACT

In this work, after a brief presentation of state of the art on building thermal modelling, we show the interest of having an open and flexible environment to develop and test new models. The Bond Graphs technique, used to represent the building dynamic thermal behavior over a single zone building structure, is compared with the building analysis tool TRNSYS and with experimental data. An evaluation of indoor temperature and heating loads was carried out in order to check our Bond Graphs and TRNSYS models. Validation procedure has shown very good agreement between measured data and both Bond Graphs and TRNSYS simulations.

**Keywords:** Bond Graphs, TRNSYS, building simulation

## NOMENCLATURE

| | |
|---|---|
| A | surface of the constructive element, [m²] |
| C | thermal capacity, [J/K] |
| c | specific heat, [J/kgK] |
| e | thickness, [m] |
| h | convective coefficient, [W/m²K] |
| k | thermal conductivity, [W/mK] |
| MSe | Modulated temperature source, [°C] |
| N | number of materials |

| | |
|---|---|
| $R_{cd\_cv\_ext}$ | thermal resistance _ conduction and outside convection heat transfer, [K/J] |
| $R_{cv\_int}$ | thermal resistance _ inside convection heat transfer, [K/J] |
| V | volume, [$m^3$] |

***Subscripts***

| | |
|---|---|
| air | Ambient air |
| element | constructive |
| ext | external |
| int | internal |

***Symbols***

| | |
|---|---|
| $\rho$ | density, [kg/ $m^3$] |
| $\rho_{air}$ | air density, [kg/ $m^3$] |

## 1. Introduction

The rapidly growing energy use gave rise to exhaustion of energy resources and heavy environmental impact. The building sector alone stands for the most energy consuming sector in France with 43% of final energy consumption and 25% of greenhouse gases emission [1]. Thus, efforts should be taken in new buildings construction and existent buildings thermal renovation in order to satisfy low energy building criterions [2]. To meet this objective, it is necessary to use a representation for building analysis in order to model both qualitative and quantitative dynamic aspects. Nowadays, several software packages such as ESP, DOE2, PLEIADES, TAS [3] have been developed to carry out parametrical studies during the design of low energy efficient buildings. Some commercial software packages are based on a simplified steady state models. These ones benefit in reliability and time, but are poor in accuracy, so they are often applied at the preliminary design stage. Detailed dynamic models benefit in accuracy, they are often applied in the analysis of annual energy consumption and performance of building HVAC systems. Regarding the building system as a whole where dynamic exchanges of heat and mass transfer occur, the models can be mainly categorized into physical models, data driven models and gray models.

Physical modeling, also called forward modeling, begins with a description of the building system or component of interest and defines the building according to its physical description. Most simulation models are based on these principals, such as EnergyPlus [4], DOE-2 [5], TAS [6], ESP-r [7] and TRNSYS [8]. These models require a large number of parameters as inputs for the simulation and the process of collecting a physical description is time consuming. Data driven dynamic models are often described by regression techniques [9], [10] where a standard parametric model is adapted to measured data obtained from an experiment on the building process. In this kind of models, it is generally necessary to acquire data over a long period of time with widely varying conditions in order to train the models for accurate predictions under all conditions. Also, these models do not reflect any specific physical structure. Gray models [11], [12] are a combination of model based on physical laws and with a parametric identification procedure using a limited number of parameters having a definite physical meaning. The thermal network model using electrical analogy parameters is an example of gray models. Gray models can represent the physical properties of building system and predict its thermal behavior and consumption and are suitable when we are facing a nonlinear process such as solar radiation processes. They are considered as simplified physical models which can represent properly the physical properties of the building system. Therefore, much of such simplified models are commonly developed to overcome the heaviness of detailed physical models. However, there are few unified models whatsoever detailed physical or gray models which can be used for representing and analyzing different building sub-systems simultaneously [13].

In this work, a simplified graphical modeling tool, which in some extent can be considered in halfway between detailed physical and gray models, has been developed. This model is based on Bond Graphs approach. This approach has the potential to display explicitly the nature of power in a building system, such as a phenomenon of storage, processing and dissipating energy. This approach responds well to needs of dynamic modeling of the building with respect to the following characteristics [14], [15]:

- an energetic approach that allows a decomposition of the building into subsystems that exchange power;
- a graphical representation in order to visualize the power transfer and the causality;
- An inherent flexibility easing model extension allowing consideration of other details not taken into account at the first stage of modelling;
- Writing systematic mathematical differential equations from the model Bond Graphs.

Modeling and simulation of the thermodynamic behavior of a building structure is studied primarily by Cellier [16]. His Bond Graphs model was compared with two state-of-the-art building analysis programs: CALPAS 3 and DOE-2. The results for temperature profiles and heat flow rates have been found to be very similar. This analyze has been extended by Nebot et al. [17], [18] by application of Bond Graphs in order to model the heat and humidity within greenhouses. Yu et al. [19] compare a Bond Graphs model of an HVAC system in a building with block-wise Simulink approach. It is shown that combination with those two approaches, to realize complicated models of building HVAC system for the application of model-based fault detection and diagnostic, is a good solution.

In order to demonstrate the credibility of Bond graph modeling, this approach is compared with the simulation tool TRNSYS. A fundamental distinction between TRNSYS and bond graph methods lies in the way in which each method describes the system. In TRNSYS, every system is modeled using specified laws governing all involved phenomenon. Whereas, Bond graph models take advantage of the general physical law of energy conservation such as: dissipation (represented with a resistance R), accumulation (Capacity C) and inertia (I). Therefore, in that point of view, Bond graph can be used as a unified language for all areas of physics. In the other hand, it can be stated that TRNSYS method is visibly more detailed than Bond graph one in the sense that it uses a complete phenomenological approach.

The present study describes an investigation of the thermal behaviour of a single-zone building. Our model is developed by use of Bond Graphs approach. For validation, a comparison between Bond Graphs model, TRNSYS program and measured data is made.

After a description of the topology and characteristics of the real building system, this paper begins with an overview of a simulation tools including a Bond Graphs's history, characteristics and development which can be applied in the real single-zone. The description of TRNSYS environment and modeling are also included so as to explain and cite the various "Types" used to achieve a comprehensive model. Next, some simulation results obtained with the two models are discussed. A comparison between the results and both measured data, in order to validate our Bond Graphs model, is made. Finally, the last section offers a short outlook on the results and future research.

## 2. Modeling Versus Experimentation

The first model is based on a graphical modular approach using Bond Graphs which is developed in this work with the 20-sim software [20]. The second model is based on the well known TRNSYS environment [8]. Both models numerical results will be compared to experimental from a single-zone building.

### 2.1. The experimental Single-zone Building

The experimental building is a workshop space of 2050 m$^2$ assimilated to a single-zone building located in Nancy (France) (Figure 1). It is a heated space with the following dimensions: 9 × 25 × 82 m (height × width × length). Table 1 presents the building envelope material properties.

Figure 1. The experimental single-zone building.

In Table 1, the U-value coefficients are calculated including the internal and external heat transfer coefficients, which have respectively the values of 3 and 17 W/m$^2$K.

The heating system consists of a natural gas boiler allowing up to 200 kW of heating rate. The heat supply is insured by 8 air heaters. The temperature is controlled by means of a thermostatic valve system and a temperature sensor located at each blowing orifice of the air heaters. In order to validate our model, accurate temperature measurements have been performed during three successive days through a platinum resistance sensor which provides results with an uncertainty not higher than ± 0.01 K. During the tests, the set-point temperature is fixed to 17 °C for day time (from 8 am to 8 pm) and 15 °C for night time. The exterior temperature variation file has been generated by means of a local weather station providing maximum and minimum temperatures of the day.

**Table 1. Material specifications of the building**

| Material | Thermal conductivity (W/m.K) | Thickness (m) | U value (W/m$^2$.K) |
|---|---|---|---|
| Exterior wall (inside to outside) | | | 0.433 |
| Concrete block | 1.05 | 0.2 | |
| Rock wool | 0.04 | 0.08 | |
| Steel | 46 | 0.001 | |
| | | | |
| Roof (inside to outside) | | | 0.477 |
| Asbestos-cement | 0.95 | 0.075 | |
| Glass wool | 0.04 | 0.075 | |
| Bituminous roofing | 0.5 | 0.01 | |
| | | | |
| Floor | | | 3.522 |
| Concrete | 1.76 | 0.2 | |

## 2.2. The Bond Graphs Approach

Bond Graphs were introduced in 1961 by Paynter [21] who thought that “energy and power alone are the fundamental dynamical variables, the ultimate currency of all physical interaction and transaction”. Multi-port elements, which at the time were common in circuit diagram, were combined with bi-directional connectors to model the flow of energy between abstract systems [22]. His work was extended by Karnopp and Rosenberg [23] in order to develop a standard reference on Bond Graphs notation for the professional engineer. Their motivation to use Paynter’s energy notation is due to the ability to analyze multiple domains concurrently using a single formalism.

Modelling of physical systems with Bond Graphs experienced enormous developments. However, the modelling of the thermal and thermodynamic systems remains still an open field. In the other hand, these types of process, present in industries at the risk which require, for their knowledge and their control, of the increasingly precise and exploitable models. Its advantages are the capacity to reveal the relationships among the system, subsystems and elements that can easily be invoked and linked by use of the connect statement.

The application of the Bond Graphs for modelling of thermal systems is interesting mainly due to the possibility of adding or removing elements in an object oriented way. For instance, we could add dissipative elements R for the heat transfer phenomena, capacitive elements C for the heat capacities and inertia elements for inertia phenomena.

In the case of heat transfer modelling within the building, one can use either the temperature- entropy pair (T,S) to constitute a true Bond Graphs or temperature-heat flow pair (T,Q) for a pseudo-thermal Bond Graphs. The use of the latter is however more convenient for the modelling of this type of system.

In this work we show how the Bond Graphs model can be used for modelling, simulation and optimization of dynamic systems in order to improve the building's energy efficiency. Their modelling can be carried out using unified approach able to highlight the physical nature of the thermal system. A comparison between the results and both measured data and those generated by TRNSYS software, is made to validate our model.

The model of the present building invokes smaller models (sub-models) that are related to the constructive elements of the building: external walls, roof and slab.

First, the sub-models have been created to describe the physics of heat transfer considering conduction and convection. Heat flow through these elements is affected by convection and conduction phenomena as well as by outdoor/indoor temperature difference. The sub-models describing these phenomena are shown in Figure 2.

Where:

$$R_{cv\_\mathrm{int}} = \frac{1}{h_{\mathrm{int}} A}, \tag{1}$$

$$C_{element} = c_{element} \times \sum_{i=1}^{N} \left( \rho_{material\_i} \times V_{material\_i} \right), \tag{2}$$

$$c_{element} = \frac{\sum_{i=1}^{N} \rho_{material\_i} \times V_{material\_i} \times c_{material\_i}}{\sum_{i=1}^{N} \rho_{material\_i} \times V_{material\_i}}, \tag{3}$$

$$R_{cd\_cv\_ext} = \frac{e_{material}}{kA} + \frac{1}{h_{ext} A}. \tag{4}$$

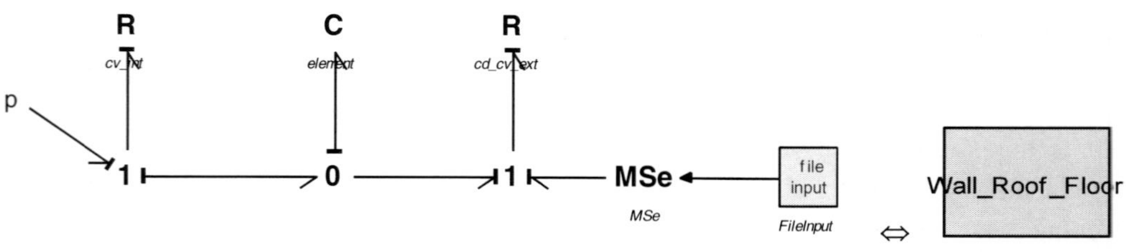


Figure 2. Pseudo-Bond Graphs for walls, roof and floor submodels.

In the next step, all elements at the boundaries of building are connected to the thermal capacity $C_{zone}$ which takes the form:

$$C_{zone} = \rho_{air} \times V_{building} \times c_{air}. \tag{5}$$

where:

$\rho_{air} = 1.29$ [kg/m$^3$] at 20 °C; $V_{building} = 18450$ [m$^3$]; $c_{air} = 1004$ [J/kg.K].

The PID controller drives the heating process in order to minimize the difference between the indoor set and measured temperatures [20].

The pseudo-Bond Graphs of the building can be represented in Figure 3.

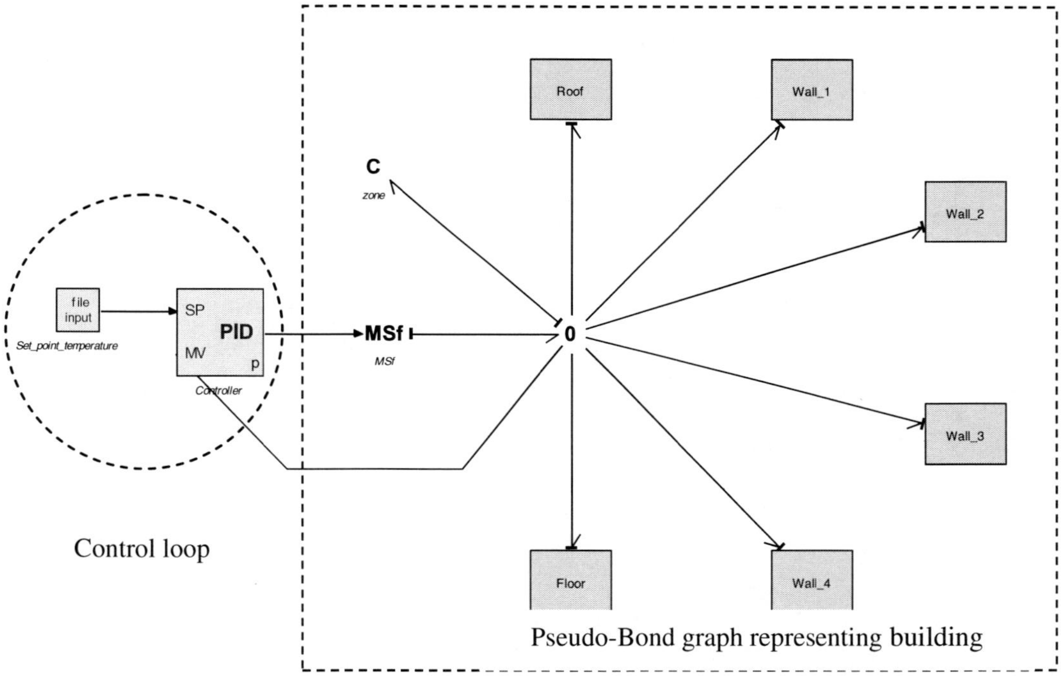


Figure 3. The Pseudo-Bond Graphs model of the building.

## 2.3. TRNSYS Modeling

TRNSYS (version 16) is designed to simulate the transient performance of thermal energy systems. This program was developed by the Solar Energy Laboratory of the University of Wisconsin [8].TRNSYS relies on modular approach to solve large systems of equations described by FORTRAN subroutines. In this paper the multi-zone building model (Type 56) is used.

The present building heating system is modeled with the TRNSYS program. The temperature regulation loop is modeled using a PID controller in order to fit the actual behavior of the building. For comparison purpose with Bond Graphs simulation, it was assumed no occupancies in the building neither any other internal gains. The schematic diagram of the TRNSYS model is shown in Figure 4.

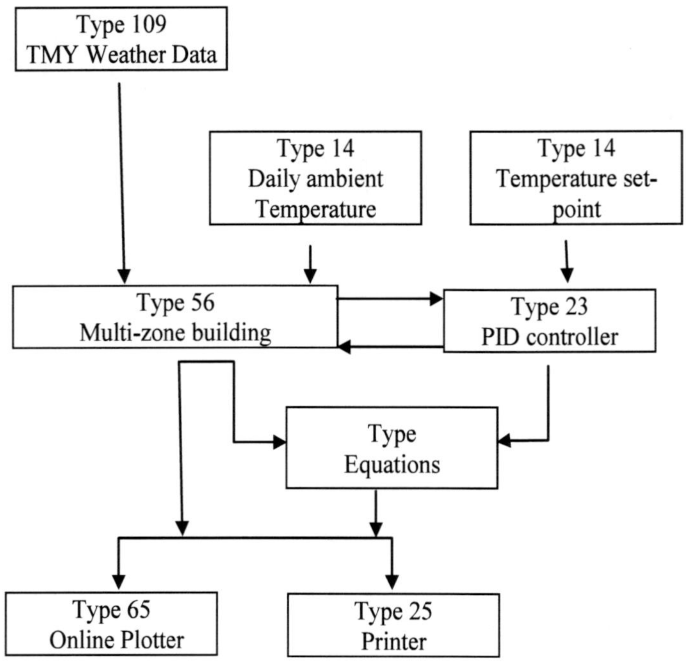


Figure 4. TRNSYS information flow diagram for building modeling.

## 3. Simulation Results

TRNSYS and Bond Graphs models were run using measured weather data for three successive days of the winter period (February 5th to 7th, 2009). Heating load are calculated and compared between the two models and a comparison against measured indoor temperature is carried out. Figure 5 represents the distribution of the outdoor temperature, the set point temperature and the measured indoor temperature.

The set point temperature is fixed to 15 °C for night time and 17 °C for day time, whereas, the outdoor temperature fluctuates between the minimum and the maximum of the considered day. We can observe the important fluctuations of the measured internal temperature around the set point which can be estimated to about ± 0.5 °C, this is due to the control system quality.

Figure 6 shows a comparison between the calculated indoor temperatures with the Bond Graphs approach and the TRNSYS modeling against the experimental data.

Obviously, a good agreement is observed between the simulated data and the measured ones. The models reproduce properly the thermal behavior highlighted by the experimental data. We can, especially, denote a good harmony between the two simulation models, which demonstrates the potential of the new Bond Graphs approach and its ability to model the building thermal behavior systems.

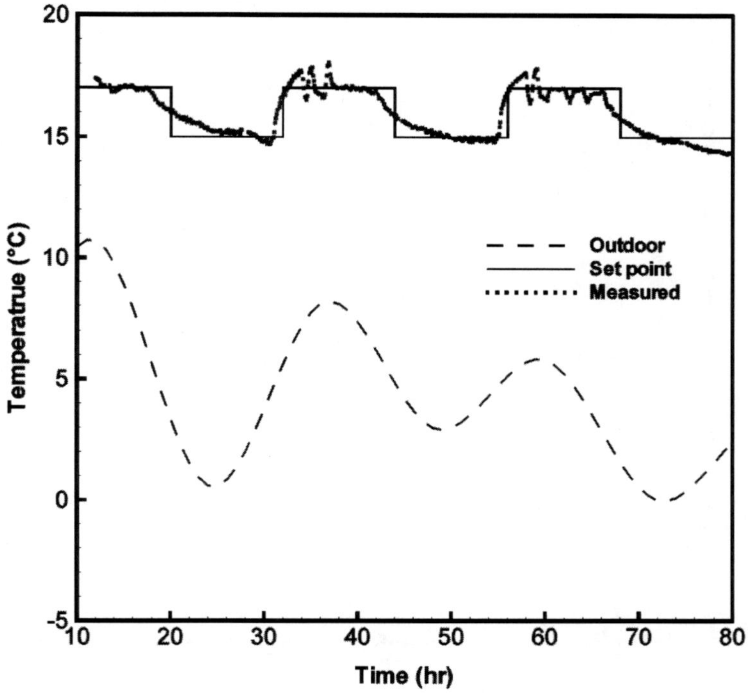


Figure 5. Outdoor, set point and measured indoor temperatures.

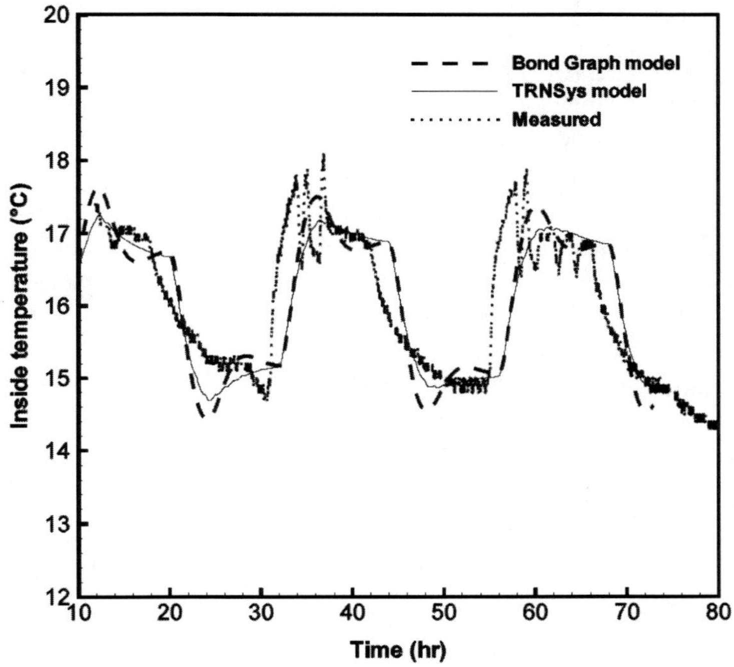


Figure 6. Comparison between measured and simulated indoor temperature data.

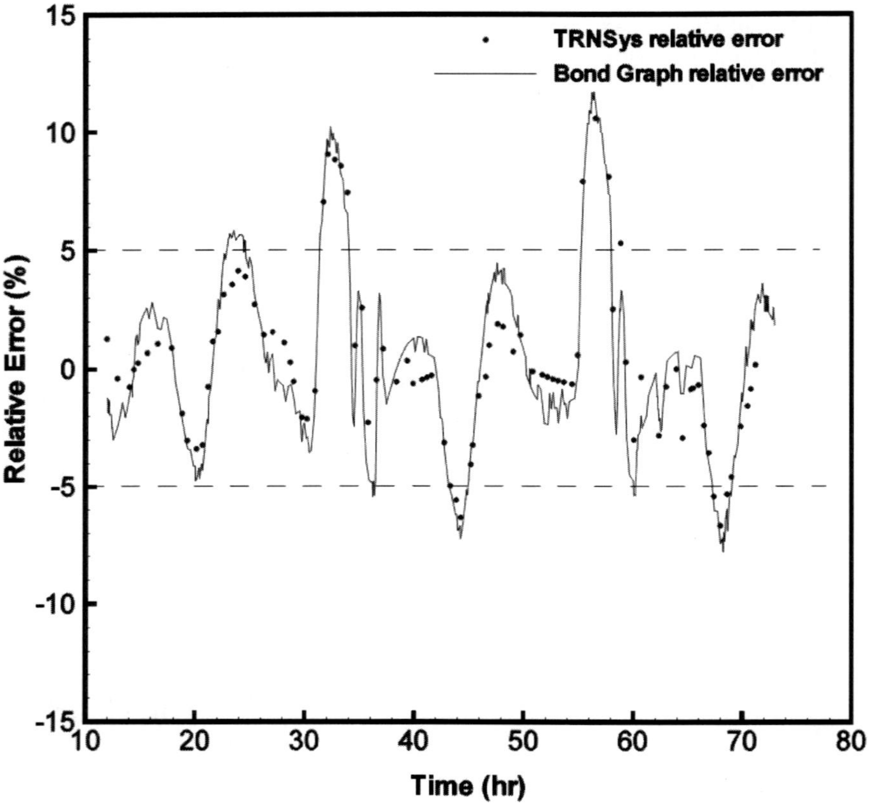


Figure 7. Curves of the relative deviation from measurements.

In order to assess the reliability of the Bond Graphs and the TRNSYS models, we carried out a comparison of the relative and absolute deviations on the calculated indoor temperatures from the measured ones between the two simulation models. Figure 7 and Figure 8 report these variations as a function of simulation time.

As it can be seen from Figure 7 and Figure 8, the relative and absolute deviationsare almost between ± 5 % and ± 1°C, which is considered as a satisfactory result regarding the number of measured data (step time of 10 seconds). Furthermore, we observe almost the same pattern between the two models, which let us assume that the two modeling approach have the same behavior. The average relative deviation between the calculated and the measured indoor temperatures is about 3.63 % and 2.46 %, respectively, for the Bond Graphs and the TRNSYS models.

Otherwise, the required heat flow has been calculated with the two models and a comparison has been made and exhibited on the plot of Figure 9.

It is shown from Figure 9, that the calculated heat flow curves have the same behavior. The mean absolute deviation between the two models is estimated to about 3.82 kW which matches with a relative deviation of 4.2 % regarding the average value of the heating flow. Integration of the above curves yields the total heating load of the building over the simulation time which are estimated to 5792 kWh and 5491 kWh, respectively by the Bond Graphs and TRNSYS models.

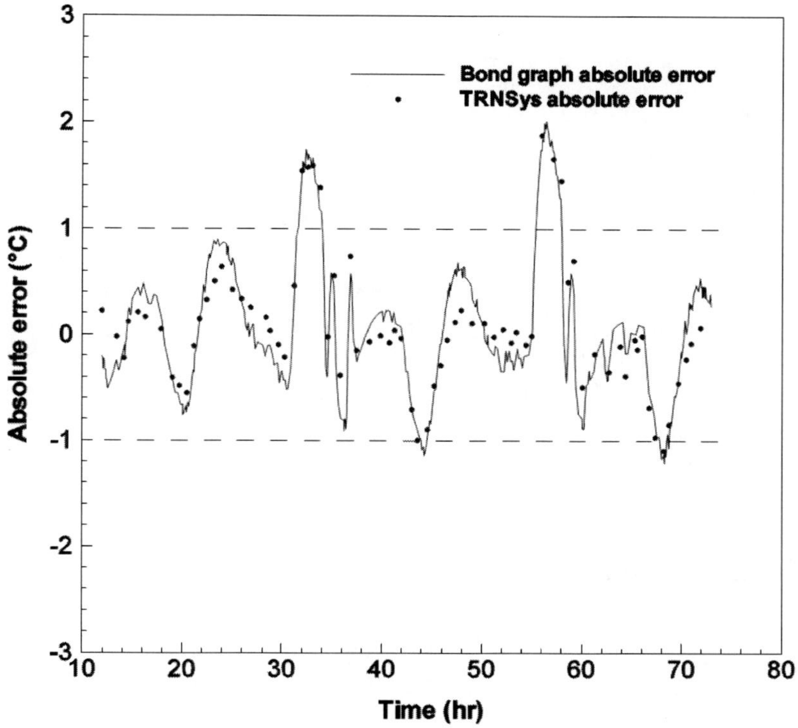


Figure 8. Curves of the absolute deviation from measurements.

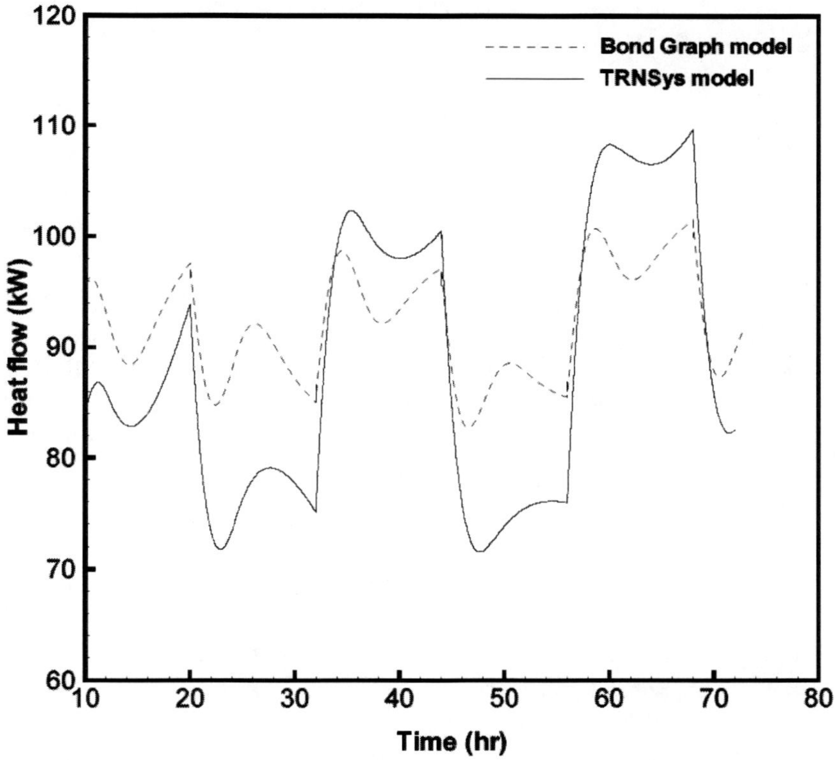


Figure 9. Required heating rate change over three days.

## Conclusion

A building analysis model was developed using Bond Graphs and TRNSYS program to describe the thermal behavior of the building. Bond Graphs approach brings some highlights on the applicability of graphical techniques to simulate and analyze building energetic performance. This Bond Graphs model was compared both to TRNSYS model and experimental data. The results are very satisfactory. Both Bond Graphs and TRNSYS models reach the thermal behavior of the tested building throughout the provided measured indoor temperature and lead to a very closed results for heating loads. Although the developed Bond Graphs model is still in its first stage of development, it provides real potential for dealing with such energy systems. Subsequently, further improvements will be attached to this model especially by implementing solar radiation processes.

Bond Graph is confirmed as a structured approach for modeling and simulation of multidisciplinary system [24]. With its flexible and extensible characteristics, it is potentially suitable to use it for the optimization of energy consumption in buildings to evaluate different alternative options and control the energy efficiency.

## References


[1] International Energy Agency, *Key World Energy Statistics*, 2009.

[2] Chlela F. Developpement d'une méthodologie de conception de bâtiments à basse consommation d'énergie. *Master thesis*. France : La Rochelle University 2008.

[3] Crawley DB, Hand JW, Kummert M, Griffith BT. Contrasting the capabilities of building energy performance simulation programs. in: Proc. of the Ninth Building Simulation Conference (IBPSA'05), Montreal, Canada 2005; vol. 1:231-238.

[4] Crawley DB. EnergyPlus: energy simulation program. *ASHRAE Journal* 2000; 42(4):49-56.

[5] Lawrence Berkeley Laboratory. DOE-2 engineering manual version 2.1C. Berkeley, CA 1982.

[6] Gough M. Component based building energy system simulation. *International Journal of Ambient Energy* 1986;7:137-43.

[7] Clark JA. The ESP system towards a new generation of building energy analysis program. In: *Building Energy Simulation Conference*, Seattle, WA 1985.

[8] Klein SA. TRNSYS – A transient system simulation program. Engineering Experiment Station Report 38-13. Solar Energy Laboratory, University of Wisconsin-Madison 2000.

[9] Kalogirou S, Neocleous C, Schizas CN. Heating load estimation using artificial neural networks. In: *Proc. CLIMA 2000 Conference*, Brussels, Belgium 1997.

[10] Minoru K, Charles D, John M. Hourly thermal load prediction for the next 24 h by ARIMA, EWMA, LR and an artificial neural network. *ASHRAE Trans* 1995; 101(1):186-200.

[11] Deque F, Ollivier F, Poblador A. Grey boxes used to represent buildings with a minimum number of geometric and thermal parameters. *Energy and Buildings* 2000; 31:29-35.

[12] Sohlberg B. Grey box modelling for model predictive control of heating process. *Journal of Process Control* 2003; 13:225-38.

[13] Tsai JJ-H, Gero JS. A qualitative energy-based unified representation for buildings. *Automation in construction* 2010; 19(1): 20-42.

[14] Ould Bouamama B, Duaphin-Tanguy G. Modelisation par Bond Graphs: Element de base pour l'énergétique. *Technique de l'ingénieur* 2006; BE8280.

[15] Ould Bouamama B, Duaphin-Tanguy G. Modelisation par Bond Graphs: Application aux systèmes énergétiques. *Technique de l'ingénieur* 2006; BE8281.

[16] Weiner M, Cellier FE. Modelling and simulation of a solar energy system by use of Bond Graphs. Proc. 1st SCS inH. *Conference on Bond Graphs modelling and simulation,* San Diego, CA 1993. pp. 301-306.

[17] Nebot A, Cellier FE. Simulation of heat and humidity budgets of Biosphere 2 without air conditioning. *Ecological engineering* 1999; 13(1-4):333-356.

[18] Cellier FE, Nebot A. Bond Graphs modelling of heat and humidity budgets of Biosphere 2. *Environmental modelling and software* 2006; 21(11):1598-1606.

[19] Yu B, Van paassen AHC. Simulink and Bond Graphs modelling of an air-conditioning room. *Simulation modelling Practice and theory* 2004;12(1):61-76.

[20] Controllab Products BV. 20-sim for Microsoft Windows version 4.1. http://20sim.com

[21] Paynter H. *Analysis and design of engineering systems.* The MIT press 1961.

[22] Denmann W. Towards the automated modelling and formal verification of analog designs. *Master thesis.* Canada : Concordia University 2009.

[23] Karnopp D, Rosenberg RC. Analysis and simulation of multiport systems: The Bond Graphs approach to physical system dynamics. The MIT press 1968.

[24] J. Marachlian, R. Benelmir, A. El Bakkali, G. Olivier. Exergy based simulation model for vehicle HVAC operation. *Applied Thermal Engineering*, vol. 31, Iss. 5, pp. 696-700, 2011.

In: Energy-Environment-Economics
Editor: Riad Benelmir
ISBN: 978-1-62948-862-2


*Chapter 5*

# BOND GRAPH APPROACH FOR MODELING AND ANALYSIS OF TRANSIENT HEAT CONDUCTION IN THE PROSPECT OF ENERGY BUILDING APPLICATION

***S. Mokraoui[1], A. Merabtine[1], R. Benelmir[1] and N. Laraqi[2]***
[1]Faculty of Sciences and Technology – Lab. LERMAB
Henri Poincaré University, Vandoeuvre-Les-Nancy, France
[2]GTE Departement, Lab. TIE, Paris, France

## ABSTRACT

Modeling of building thermal behavior needs effective tools. This is particularly true when conduction of heat through walls and/or slabs has to be computed. This paper represents a framework of a new methodology in Building energy simulation. Hence, it will give emphasis to a new modeling approach for solving the transient heat conduction equation in finite medium with different boundary conditions in order to apply it for modeling heat transfer in the different elements of buildings structure. The developed modeling approach is performed using the Bond graph technique, a graphical modeling language which is perfectly accustomed with dynamic systems involving energy transfer. With this original model, two transient heat conduction problems corresponding to the most practical cases in building envelope, such a heat transfer through vertical walls, roofs and slabs, have been treated. The validation procedure consists of comparing the results obtained with this model with analytical solution. It has shown very good agreement between measured data and Bond Graphs model simulation. The Bond Graphs technique is then used to model the building dynamic thermal behavior over a single zone building structure and compared with a set of experimental data. An evaluation of indoor temperature was carried out in order to check our Bond Graphs model.

**Keywords:** Bond graph, transient heat conduction, plane wall, simulation, building

## NOMENCLATURE

| | |
|---|---|
| $A$ | surface of the constructive element, [m²] |
| $C$ | thermal capacity, [J/K] |
| $c$ | specific heat, [J/Kg.K] |
| $e$ | effort variable |
| $f$ | flow variable |
| $L$ | thickness, [m] |
| $h_i$, $h_e$ | inside and outside convective coefficients, [W/m²K] |
| $k$ | thermal conductivity, [W/m.K] |
| $Q$ | heat rate [J] |
| $\dot{Q}$ | heat flow [W] |
| $R$ | thermal resistance inside the wall, [K/W] |
| $R_{cvi}$, $R_{cve}$ | inside and outside thermal resistances, [K/W] |
| $Se$ | temperature source, [°C] |
| $T$ | temperature [°C] |
| $V$ | volume, [$m^3$] |

*Subscripts*

| | |
|---|---|
| *0* | initial |
| *cve* | outside convection |
| *cvi* | inside convection |
| *e* | external |
| *i* | internal |
| *in* | entering |
| *l* | layer |
| W | *wall* |

*Symbols*

| | |
|---|---|
| $\alpha$ | diffusivity [m²/s] |
| $\beta$ | eigenvalue |
| $\rho$ | density, [kg/ $m^3$] |

## 1. INTRODUCTION

Building envelope is one of the most important systems affecting energy efficiency of the building. Proper whole building thermal modeling requires accurate descriptions of the building envelope components. The heating loads of the buildings depend on the thermal transmittance of envelope components, mainly the transmittance by heat conduction through walls, roofs and floors. Hence, in order to evaluate heat storage and losses, transient heat conduction through walls has become an area of interest in building process. This trend has been motivated by the fast development of computer technology which allows solving such problems quickly and accurately. In fact, with increasing capacity of computers, dynamic thermal modeling of building has been intensive. The first attempts of dynamic simulation were analogical models constructed of real resistors and capacitors. The advantage was simplicity and higher computational speed comparing with the models running in computers of early stage. Afterward, several approaches such finite difference and direct analytical methods emerged making modeling more and more complicated and still CPU time consuming. That is why, an interesting category of models, called Gray models, have been subject for several investigations in order to develop simple and fast running design tools.

Gray models are based on physical laws and an identification procedure using a limited number of parameters having a definite physical meaning. The thermal network model using electrical analogy parameters is an example of gray models. Gray models can represent the physical properties of building system and predict its thermal behavior and consumption and are suitable for nonlinear processes such as solar radiation. They are considered as simplified physical models which can represent properly the physical properties of the building system (for a review of principles see [1]).

Among the several gray models, Déqué et al. [2] developed a gray box model that can predict the thermal behavior and energy consumption of buildings starting from a minimum of geometrical and physical parameters. Their model give emphasize, especially, to the technological approach by ensuring the visibility of the physical system. This feature provides a high flexibility. Lorenz and Massy [3], Tindale [4] and Gouda et al. [5] use the lumped parameter method involving the break-up of construction elements into a number of uniform temperature elements about which an energy balance can be expressed. A nonlinear constrained optimization method has been used for reducing the model order of building elements by Gouda et al. [5].

Another methodology belonging to the Gray box modeling technique consists in the use of a graphical language by means of simplified graphical modeling tool. This model is based on Bond Graphs approach. This approach has the potential to display explicitly the nature of power in a building system, such as a phenomenon of storage, processing and dissipating energy. This approach responds well to needs of dynamic modeling of the building with respect to the following characteristics:

- an energetic approach that allows a decomposition of the building into subsystems that exchange power;
- a graphical representation in order to visualize the power transfer and the causality;
- an inherent flexibility easing model extension allowing consideration of other details not taken into account at the first stage of modelling;
- writing systematic mathematical differential equations from the modelBond Graphs.

This method has been carried out in vehicle thermal design [24] as well as in thermal building design [8, 9] where the results seem to be similar to those obtained with the analysis tools DOE-2 and CALPAS 3 [9]. The same findings have been given by Yu et al. [8] when they compare the bond graph results with those of Matlab-Simulink tools.

It is worth noting, that in the bond graph methodology, as in several gray box models, the modeling approach involves four distinctive levels. A technological level where is constructed the architecture of the system by the assembly of different components, which can be identified in the "real" system (heat exchanger, walls, radiator …). In the physical level, the modeling uses an energy description of the physical phenomena based on fundamental concepts of physics in which the internal physical mechanisms are specified in order to describe the behavioral aspects of the system. For instance, dissipation, transformation, storage of energy are physical aspects of a thermal system. In the mathematical level, we specify the exact nature of the relations between variables, parameters, constants by use of mathematical equations which describe the system behavior. Finally, the algorithmic level allows solving the mathematical equation in order to simulate the system behavior.

In this article, we describe how bond graph approach can be used to model the heat transfer by conduction in a plane walls with the objective of providing the best modeling scheme. In section 2, a representation of the bond graph methodology is carried out. Next, the bond graph model for two cases study regarding the boundary conditions type is developed in section 3. Section 4 provides the analytical solution for the above mentioned cases. Finally, a comparison between the results of bond graph and analytical models is carried out; followed by a building case study where validation of assembled Bond Graph model is achieved by its results with experimental data of indoor temperature.

## 2. BOND GRAPH METHODOLOGY

Modeling of thermal behavior of building is a difficult task. An alternative simplified method to the dedicated analytical approaches is the Bond Graph modeling method.

The bond graph technique is based on a graphical formalism. It is well suited for modeling physical processes and multidisciplinary dynamic engineering systems including features and components involved in different energy domains [10]. Its philosophy is founded on a systematic and common way representation of power flow between the model's components. Paynter [11] started the bond graph formalism and used it for modeling dynamic multiport systems. He suggested that energy and power are the fundamental dynamic variables which characterize all physical interactions.

In Bond graph modeling, the interaction between two components is modeled by a bond with a semi-arrow in the end. The power is represented as product of two physical quantities, one extensive, the other intensive. These two power conjugated variables are called *effort* and *flow* and are denoted by the letters $e$ and $f$ (Figure 1).

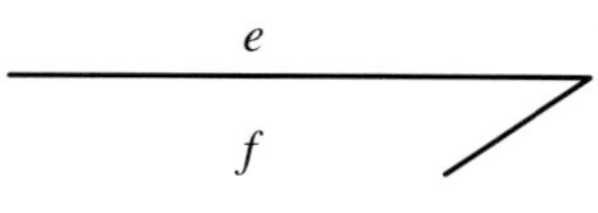


Figure 1. Bond graph link representation.

The selection of the two physical quantities is specific for each physical domain. For instance, in electrical domain, we use a voltage $u$ as effort variable and a current $i$ as flow variable. In thermal domain, the effort variable is represented by a temperature $T$ and the flow variable by an entropy flow $\dot{S}$.

A classification of Bond Graph elements can be made up by the number of ports; ports are places where interactions with other processes take place. There are one port elements symbolizing inertial elements (I), capacitive elements (C), resistive elements (R), effort sources (Se) and flow sources (Sf), and two ports elements representing transformer elements (TF) and gyrator elements (GY). The elements I, C, and R are passive elements because they convert the supplied energy into stored or dissipated energy. The sources Se and Sf are active elements because they supply power to the system. The bonds are inter-linked by two type junction elements (0 and 1-junctions) which serve to connect I, C, R, and source elements. At the 0-junction the flow adds up to zero while all efforts are equal, and at the 1-junction all effort variables add up to zero while all flows are equal. The concept of causality is an important concept embedded in Bond Graph theory. This refers to cause and effect relationship. Thus, as part of the Bond Graph modeling process, a causality assignment is implicitly introduced (Karnopp et al. [12].

Causality assignment is independent of the power flow direction. This leads to the description of Bond Graphs in the form of state – space equation. The sources (Se and Sf) have fixed causality, the dissipative element (R) has free causality depending on the causality of the other elements of Bond Graph, and the storage elements (I, C) have preferential causality, that is integral causality or derivative causality, but it is always desirable that C and I elements to be in integral causality. Transformers, gyrators and junction elements have constrainedly causality.

As explained previously, the power variables of the thermal system are the temperature $T$ and the entropy flow $\dot{S}$, but it is known that entropy flow cannot be measured directly [13]. It is easier to use heat flow $\dot{Q}$ as flow variable which is a measurable variable. However, a product of temperature and heat flow is not the power transferred between ports. This has enabled researchers to introduce pseudo-bond graphs method. Their advantage is that modeling of thermal systems becomes easier.

## 3. The Bond Graph Method for Solving Transient Heat Conduction Problems

In this section, we present the bond graph method used for solving one-dimensional unsteady heat conduction problem in a finite medium subject to asymmetric, time-independent, boundary conditions. We consider two cases: the case of a plane wall with two different convective conditions at each surface boundary of the wall; the case of the wall with one convective condition at one surface boundary, the second surface is kept at constant temperature. In the first case (Figure 2), comparable to that of heat transfer through the wall of building, the absolute temperature $T$ may be chosen as an effort variable and the heat flow as a flow variable.

Lumped parameter assumption is usually adopted for this kind of cases. This is realized by splitting the wall into a number of layers, where, temperature and the thermo physical properties are assumed homogeneous. Each layer stores and conducts heat simultaneously. The external ones are subject to convection heat exchange with inside and outside surrounding.

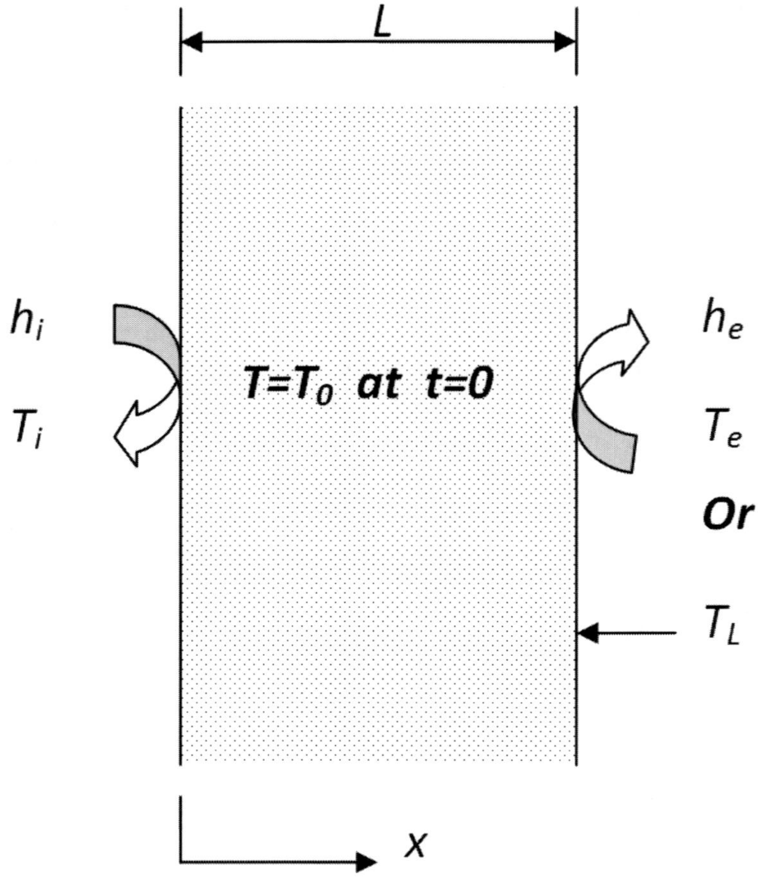


Figure 2. Representation of the heat transfer problem in a wall.

Thus, the word pseudo bond graph representation of the heat conduction into the wall constituted of four layers is shown in Figure 3. This figure indicates the technological level of modeling. The corresponding electrical representation is shown in Figure 4.

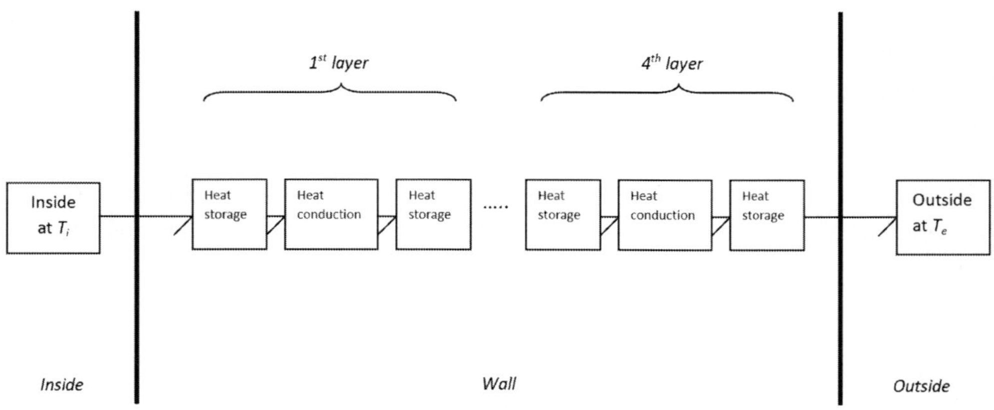


Figure 3. Word pseudo-bond graph representation of the wall.

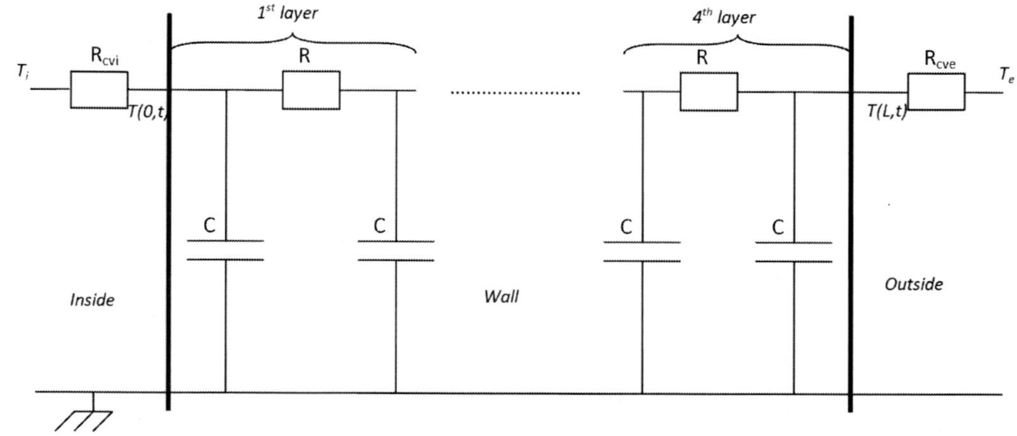


Figure 4. Electrical representation of the wall.

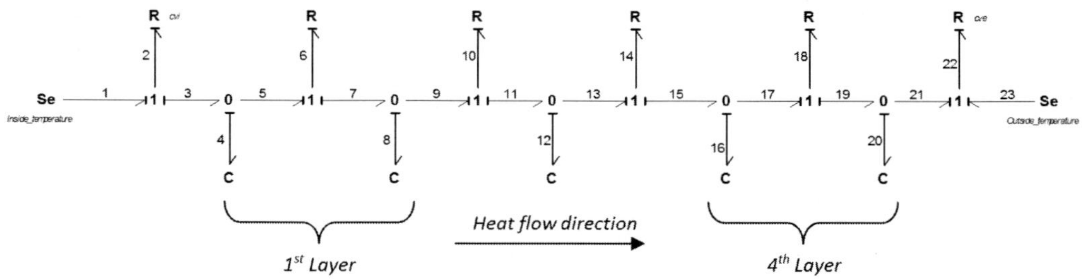


Figure 5. Pseudo-bond graph model of the wall.

In Figure 5, *the physical level* of modeling by a bond graph is pointed out. The inside and outside temperatures, $T_i = 21°C$ and $T_e = 0°C$, are modeled by the effort sources $S_{e1}$ and $S_{e1}$. The convective boundary conditions are represented by two 1-junctions: $1_{1,2,3}$ and $1_{22,23,24}$ related respectively to the resistance elements: $R_{cvi}$ and $R_{cve}$ representing the inside and outside resistances to heat transfer by convection. These two 1-junctions are characterized by equality to zero of the sum of effort variables (temperatures). Therefore the following relations are obtained:

$$T_1 = T_2 + T_3 \text{ or } T_i = T_2 + T(0,t), \tag{1}$$

$$T_{22} = T_{21} + T_{23} \text{ or } T_{22} = T(L,t) - T_e. \tag{2}$$

In Bond Graph modeling, The constitutive equation related to R-elements is: $f = \dfrac{e}{R}$. Thus, the dynamic equations of the two resistance elements: $R_{cvi}$ and $R_{cve}$ are given by

$$f_2 = \frac{e_2}{R_2} \Rightarrow \dot{Q}_2 = \frac{T_2}{R_2} = \frac{T_i - T(0,t)}{R_{cvi}} = \dot{Q}_i, \tag{3}$$

$$f_{22} = \frac{e_{22}}{R_{22}} \Rightarrow \dot{Q}_{22} = \frac{T_{22}}{R_{22}} = \frac{T(L,t) - T_e}{R_{cve}} = \dot{Q}_e. \tag{4}$$

Where $R_{cvi} = \dfrac{1}{h_i A}$ and $R_{cve} = \dfrac{1}{h_e A}$.

Through boundary surfaces of the wall, the heat flow entering and leaving the wall are represented by two 0-junctions: $0_{1,2,3}; 0_{20,21,22}$ characterized by the equality to zero of the sum of flow variable (heat flow). The effort variables are equals for each of those junctions. Therefore,

$$T_3 = T_4 = T_5 = T(0,t), \tag{5}$$

$$\dot{Q}_3 = \dot{Q}_i = \dot{Q}_4 + \dot{Q}_5, \tag{6}$$

$$T_{19} = T_{20} = T_{21} = T(L,t), \tag{7}$$

$$\dot{Q}_{21} = \dot{Q}_e = \dot{Q}_{19} + \dot{Q}_{20}, \tag{8}$$

For each layer inside the wall, the heat quantity is decomposed into two parts: the first part is dissipated by conduction; modeled by a 1-junction related to R-element representing the conductive resistance, whereas the second part is stored by the layer; modeled by a 0-junction related to C-element representing the thermal capacity. Then the dynamic equations for those junctions are given by

$$\text{For } 1^{\text{st}} \text{ layer: } \begin{cases} 0_{3,4,5} : \begin{cases} \dot{Q}_4 = \dot{Q}_3 - \dot{Q}_5 \\ T_3 = T_4 = T_5 \end{cases} \\ 1_{5,6,7} : \begin{cases} \dot{Q}_5 = \dot{Q}_6 = \dot{Q}_7 \\ T_6 = T_5 - T_7 \end{cases} \\ 0_{7,8,9} : \begin{cases} \dot{Q}_8 = \dot{Q}_7 - \dot{Q}_9 \\ T_8 = T_7 = T_9 \end{cases} \end{cases}. \tag{9}$$

$$\text{For } 2^{\text{nd}} \text{ layer: } \begin{cases} 0_{7,8,9}: \begin{cases} \dot{Q}_8 = \dot{Q}_7 - \dot{Q}_9 \\ T_8 = T_7 = T_9 \end{cases} \\ 1_{9,10,11}: \begin{cases} \dot{Q}_9 = \dot{Q}_{11} = \dot{Q}_{13} \\ T_{10} = T_9 - T_{11} \end{cases} \\ 0_{11,12,13}: \begin{cases} \dot{Q}_{12} = \dot{Q}_{11} - \dot{Q}_{13} \\ T_{11} = T_{12} = T_{13} \end{cases} \end{cases} . \tag{10}$$

$$\text{For } 3^{\text{rd}} \text{ layer: } \begin{cases} 0_{11,12,13}: \begin{cases} \dot{Q}_{12} = \dot{Q}_{11} - \dot{Q}_{13} \\ T_{11} = T_{12} = T_{13} \end{cases} \\ 1_{13,14,15}: \begin{cases} \dot{Q}_{13} = \dot{Q}_{14} = \dot{Q}_{15} \\ T_{14} = T_{13} - T_{15} \end{cases} \\ 0_{15,16,17}: \begin{cases} \dot{Q}_{16} = \dot{Q}_{15} - \dot{Q}_{17} \\ T_{15} = T_{16} = T_{17} \end{cases} \end{cases} . \tag{11}$$

$$\text{For } 4^{\text{th}} \text{ layer: } \begin{cases} 0_{15,16,17}: \begin{cases} \dot{Q}_{16} = \dot{Q}_{15} - \dot{Q}_{17} \\ T_{15} = T_{16} = T_{17} \end{cases} \\ 1_{17,18,19}: \begin{cases} \dot{Q}_{17} = \dot{Q}_{18} = \dot{Q}_{19} \\ T_{18} = T_{17} - T_{19} \end{cases} \\ 0_{19,20,21}: \begin{cases} \dot{Q}_{20} = \dot{Q}_{19} - \dot{Q}_{21} \\ T_{19} = T_{20} = T_{21} \end{cases} \end{cases} . \tag{12}$$

The constitutive equations corresponding to R and C-elements inside the wall are expressed as

$$T_4 = \frac{1}{C_4}\int_0^t \dot{Q}_4 dt + T_4(0), \tag{13}$$

$$T_{12} = \frac{1}{C_{12}}\int_0^t \dot{Q}_{12} dt + T_{12}(0), \tag{14}$$

$$T_{16} = \frac{1}{C_{16}}\int_0^t \dot{Q}_{16} dt + Q_{16}(0), \tag{15}$$

$$T_{20} = \frac{1}{C_{20}} \int_0^t \dot{Q}_{20} dt + Q_{20}(0), \quad (16)$$

$$\dot{Q}_6 = \frac{T_6}{R_6}, \quad (17)$$

$$\dot{Q}_{10} = \frac{T_{10}}{R_{10}}, \quad (18)$$

$$\dot{Q}_{14} = \frac{T_{14}}{R_{14}}, \quad (19)$$

$$\dot{Q}_{18} = \frac{T_{18}}{R_{18}}. \quad (20)$$

Where:

$C_4$, $C_8$, $C_{12}$, $C_{16}$ and $C_{20}$ are thermal capacities for each layer; $C_4 = C_8 = C_{12} = C_{16} = C_{20} = \frac{C_W}{5}$.

And $C_W = \rho V c$ is the heat capacity of the wall, with, $\rho$ denotes the density, $V$ the volume and $c$ the specific heat capacity.

$R_6$, $R_{11}$, $R_{14}$ and $R_{19}$ are the conductive resistances for each layer; $R_6 = R_{11} = R_{14} = R_{19} = R_W$.

$R_W = \frac{L}{kA}$ is the global conductive resistance for wall.

Thereafter, thanks to mathematical step of modeling it can be easier to calculate $T(0,t)$, $T(L,t)$, $\dot{Q}_i$ and $\dot{Q}_e$ witch are evaluated by using the above equations.

In the second case, associated to the case of heat transfer through a roof a building, the same approach is used. At $x = L$, we fix the temperature $T_L$. In Bond graph model, this condition is approximated by tending the convection coefficient $h_e$ toward infinity, i.e., $R_{cve}$ becomes negligible, hence, the effort source $S_e$ corresponding to the outside temperature becomes equal to $T_L$.

# 4. Analytical Solution

We consider a heat conduction problem in a one-dimensional finite region, as illustrated in Figure 2. We use the separation of variables method [14].

The mathematical formulation of this problem is as follows.

*Case 1.* Heat conduction in a wall with convective conditions on both sides

Energy balance

$$\frac{\partial^2 T}{\partial x^2} = \frac{1}{\alpha}\frac{\partial T}{\partial t} \text{ for } 0 < x < L,\ t > 0. \tag{21}$$

Initial condition

$$T(x,t) = T_0 \text{ for } t = 0,\ 0 \le x \le L. \tag{22}$$

Boundary conditions

$$-k\frac{\partial T}{\partial x} = h_i\left[T_i - T\right] \text{ for } x = 0,\ t > 0, \tag{23}$$

$$-k\frac{\partial T}{\partial x} = h_e\left[T - T_e\right] \text{ for } x = L,\ t > 0. \tag{24}$$

*Case 2.* Heat conduction in a wall with convective condition on one side and fixed temperature on the other side, the mathematical formulation is as follows:

Energy balance

$$\frac{\partial^2 T}{\partial x^2} = \frac{1}{\alpha}\frac{\partial T}{\partial t} \text{ for } 0 < x < L,\ t > 0. \tag{25}$$

Initial condition

$$T(x,t) = T_0 \text{ for } t = 0,\ 0 \le x \le L. \tag{26}$$

Boundary conditions

$$-k\frac{\partial T}{\partial x} = h_i\left[T_i - T\right] \text{ for } x = 0,\ t > 0, \tag{27}$$

$$T(x,t) = T_L \text{ for } x = L,\ t > 0. \tag{28}$$

Among various available methods to solve this kind of problems there is the method of separation of variables which is most effective and straightforward to apply when both the differential equation and boundary conditions are homogenous. In our case, we have non-homogenous boundary conditions, then, a direct application of separation of variables method is not possible. In many cases, a non-homogenous problem can be splited up into several simpler problems. In our case, we shall split this problem into two parts:

- a steady-state problem defined by the temperature variable $U(x)$;
- a homogenous time-dependent problem defined by the temperature variable $V(x,t)$.

Consequently, the temperature $T(x,t)$ is expressed as

$$T(x,t) = U(x) + V(x,t). \tag{29}$$

The solution of first sub-problem is straightforward

$$U(x) = Ax + B. \tag{30}$$

where parameter *A* and *B*, are determined using the boundary conditions of each case.

$$A = \frac{T_e - T_i}{L + \frac{1}{H_e} + \frac{1}{Hi}};\ B = T_i + \frac{T_e - T_i}{1 + H_i L + \frac{H_i}{H_e}} \text{ for case 1;}$$

$$A = \frac{1}{L}\left[T_L - \frac{H_i T_i L + T_L}{H_i L + 1}\right];\ B = \frac{H_i T_i L + T_L}{H_i L + 1} \text{ for case 2.}$$

With, $H_i = \frac{h_i}{k}$ and $H_e = \frac{h_e}{k}$

The second sub-problem can be treated using separation of variables method. Assume temperature $V(x,t)$ have the following form

$$V(x,t) = \Psi(x)\Gamma(t). \tag{31}$$

Where, $\Psi(x)$ is a space function and $\Gamma(t)$ is a time function. The solution for the variable $V(x,t)$ is given in the form:

$$V(x,t)=\sum_{m=1}^{\infty}\frac{C_m}{N_m}\Psi(\beta_m,x)e^{-\alpha\beta_m t}\ . \tag{32}$$

where,

$\Psi(\beta_m,x)$ are the eigenfunctions; $\beta_m$ are the eigenvalues wich can be determined by solving the transcendental equation; $N_m$ is Norm of the eigenfunction and $C_m$ is the constant of integration for each case. The expressions for those parameters are given by:

*Case 1.*

$$\Psi(\beta_m,x)=\beta_m\cos(\beta_m x)+H_i\sin(\beta_m x);$$

$$N_m=\frac{1}{2}\left[\left(\beta_m^2+H_i^2\right)\left(L+\frac{H_e}{\beta_m^2+H_e^2}\right)+H_i\right];$$

$$C_m=\sin(\beta_m L)\left[T_0-B-A\left(L+\frac{H_i}{\beta_m^2}\right)\right]-$$

$$-\cos(\beta_m L)\left[\frac{A}{\beta_m}+\frac{H_i}{\beta_m}(T_0-AL-B)\right]+\frac{A}{\beta_m}+\frac{H_i}{\beta_m}(T_0-B).$$

*Case 2.*

$$\Psi(\beta_m,x)=\sin(\beta_m(L-x));$$

$$N_m=\frac{1}{2}\left[\frac{L(\beta_m^2+H_i^2)+H_i}{\beta_m^2+H_i^2}\right];$$

$$C_m=\frac{1}{\beta_m}\left[T_0-AL-B-(T_0-B)\cos(\beta_m L)\right]+\frac{A}{\beta_m^2}\sin(\beta_m L).$$

Finally, the complete solution for the temperature $T(x,t)$ is expressed, for each case, in the form:

$$T(x,t)=(Ax+B)+\sum_{m=1}^{\infty}\frac{C_m}{N_m}\Psi(\beta_m,x)e^{-\alpha\beta_m^2 t}\ . \tag{33}$$

## 5. Simulation Results

### 5.1. Validation

In this section, the performed simulations of transient heat conduction through walls will be illustrated by calculating the surface temperatures, heat fluxes and energy stored quantities.

The thermo – physical and geometric characteristics of the chosen wall are presented in Table 1. This table includes also the boundary conditions values.

In order to validate our Pseudo-Bond graph model, we have compared all simulation results with those of analytical method. Figure 6(a) exhibits the results of temperature variations with time for three positions in the wall (x = 0 m; 0.1 m and 0.2 m). It shows that there is a good agreement between bond graph and both analytical calculations. Figure 6(b) indicates the magnitude of absolute deviations on the calculated temperatures between bond graph model and analytical method. It can be observed from Figure 6(b) that the maximum of deviations occurs at the first hours then tend towards zero, for each position in the wall, after about 20 hours of simulation.

**Table 1. Characteristics data of the studied heat conduction problems**

| | | |
|---|---|---|
| Geometric parameters | Wall Thickness; $L$ *(m)* | 0.2 |
| | Wall heat transfer section; $A$ *($m^2$)* | 1 |
| Thermo-physical parameters | Wall material | Concrete block |
| | Thermal Conductivity; $k(W.m^{-1}.K^{-1})$ | 0.963 |
| | specific Heat *capacity; c* $(J.Kg^{-1}.K^{-1})$ | 650 |
| | Density; $\rho\ (Kg.m^{-3})$ | 1300 |
| Inside and outside conditions | Inside convection coefficient; $h_i\ (W.m^{-2}.K^{-1})$ | 3 |
| | Outside convection coefficient; $h_e\ (W.m^{-2}.K^{-1})$ | 17.8 |
| | Inside temperature; $T_i\ (°C)$ | 21 |
| | Wall : Outside temperature; $T_e\ (°C)$<br>Fixed end temperature; $T_L\ (°C)$ | 0<br>12 |
| | Initial temperature; $T_0\ (°C)$ | 12 |

The Pseudo-Bond graph determined profiles of heat stored and heat flow leaving and entering the plane wall are compared in Figures 7(a) and 7(b) with their analytical counterparts. Agreement between the two simulation results is shown to be satisfactory. The steady state is reached after 20 hours of simulation. Furthermore, these graphs clearly illustrate the fact that the heat flow leaving is more important than the heat flow entering the wall. This can be explained by the important temperature gradient between the outside and the wall $\left(\left[T_e - T(L,t)\right] > \left[T_i - T(0,t)\right]\right)$. Moreover, the heat stored by the wall (or, in this case, evacuated) decreases to reach a value of -1.08 *MJ* /m² at the steady state.

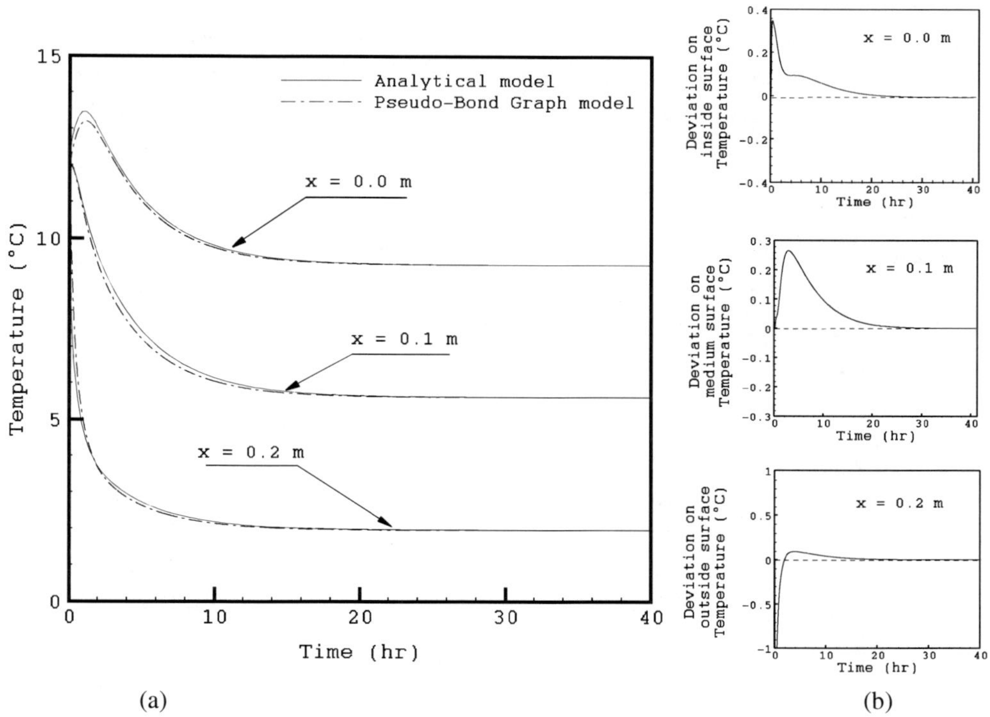


Figure 6. Temperature profiles in the wall (a) and the corresponding absolute deviations between Bond graph model and analytical method (b). (case 1: two convective boundary conditions).

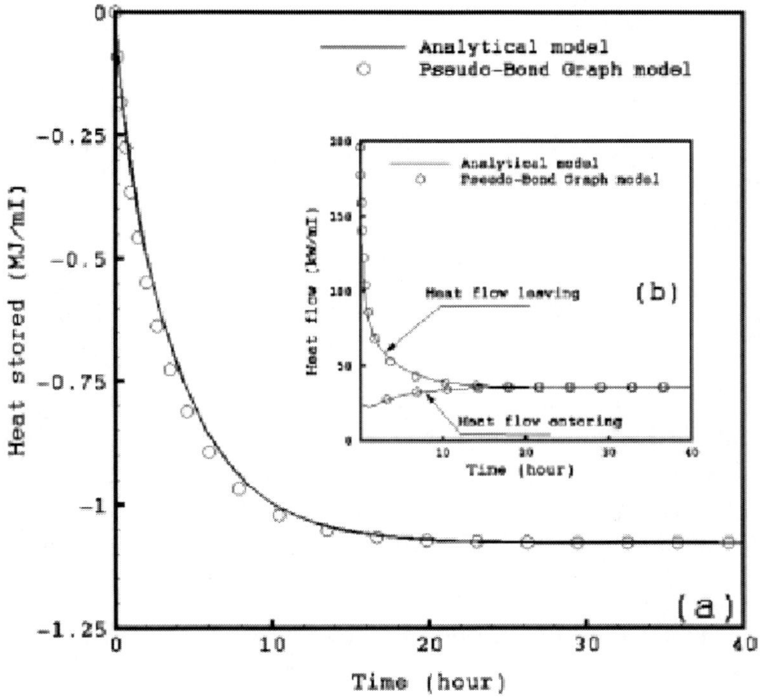


Figure7. Heat stored by the wall (a) and heat flows leaving and entering the wall (b). (case 1: two convective boundary conditions).

Results related to the heat conduction in the wall subject to a fixed temperature condition in one side and one convective condition on the other are depicted hereafter. Temperature profiles plots, for three different positions in the wall (x = 0 m; x = 0.1 m; x = 0.2 m), are shown in Figure 8(a). The corresponding deviations plots, between analytical and bond graph methods, on the calculated temperatures are presented in Figure 8(b). Obviously, good agreement can be observed between the analytical predictions and those obtained by the pseudo Bond Graph model. From Figure 8(a), we can analysis the thermal behaviour of the wall. Indeed, the achieved steady state temperature decreases more and more when going away from inside surface boundary to the outside one. Upon observing Figure 8(b), it can be noticed that the deviations on the calculated temperature does not exceed *± 0.4 °C* for the boundary surface *x = 0 m and ±0.025 °C* for *x = 0.1 m.*

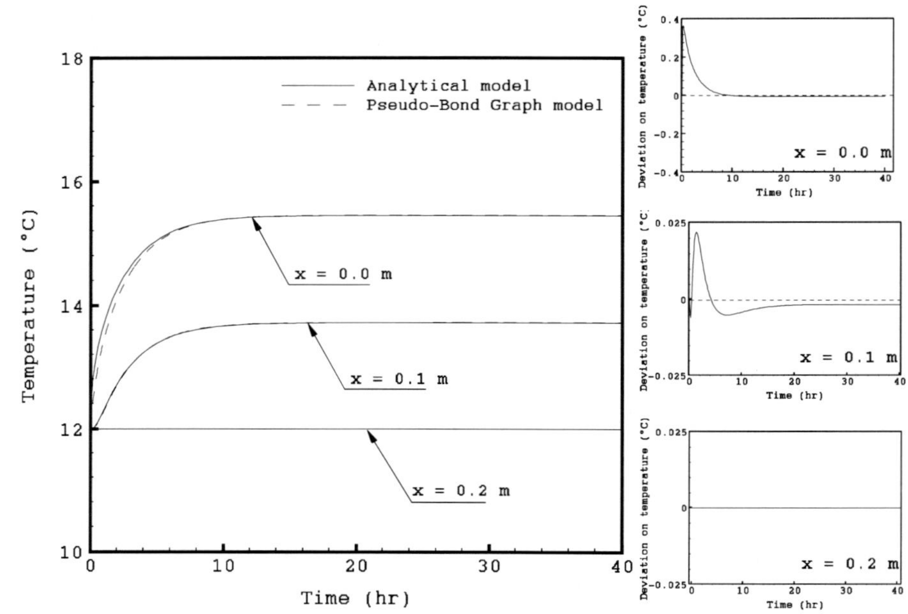


(a) (b)

Figure 8. Temperature profiles in the wall (a) and the corresponding absolute deviations between Bond graph model and analytical method (b). (case 2: one convective and one fixed temperature conditions at each boundary).

Thereafter, the heat stored into the wall, as well as the heat flows entering and leaving the wall is illustrated in Figure 9. From Figure 9(b), the leaving flows (heat losses) decrease rapidly at the first times then increase gradually with time before reaching the steady-state. Alternatively, the entering flows (heat gains) decrease continually until reaching the same steady-state. It is also indicated, from Figure 9(a), that the heat stored by the wall decreases firstly then increases with time because the inside gradient of temperature $[T_i - T(0,t)]$ becomes more important than the outside one. However, for the first moments, the calculated bond graph heats and flux shows some disparities with the analytical ones. Upon comparing these results it can be stated that the pseudo-bond graph model agrees accurately with analytical method.

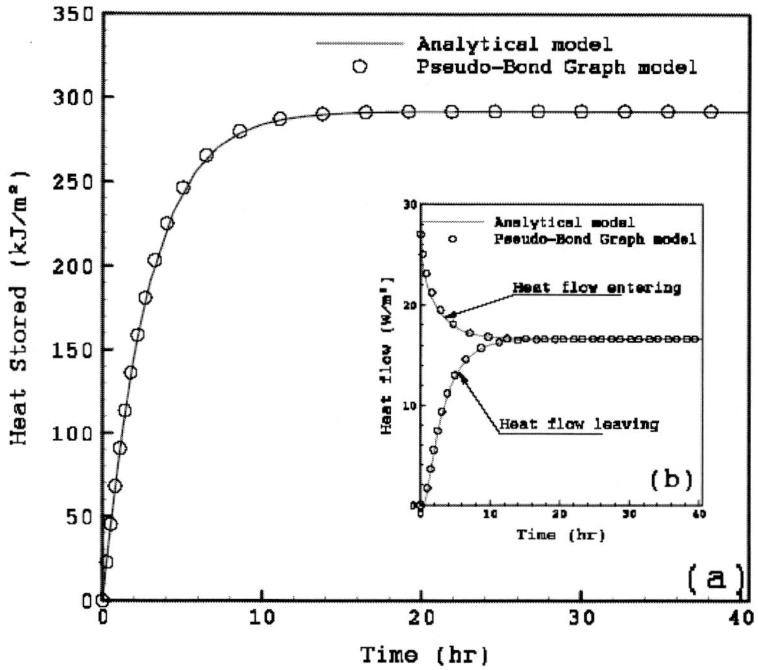


Figure 9. Heat stored by a wall (a) and heat flows leaving and entering the wall (b). (case 2: one convective and one fixed temperature conditions at each boundary).

## 5.2. Test Case

The experimental building is a workshop space of 2050 $m^2$ assimilated to a single-zone building located in Nancy (France). It is a heated space with the following dimensions: 9×25×82m (height × width × length). Table 2 presents the building envelope material properties.

**Table 2. Material specifications of the building**

| Material | Thermal conductivity (W/m.K) | Thickness (m) | U value ($W/m^2.K$) |
|---|---|---|---|
| Exterior wall (inside to outside) | | | 0.433 |
| Concrete block | 1.05 | 0.2 | |
| Rock wool | 0.04 | 0.08 | |
| Steel | 46 | 0.001 | |
| Roof (inside to outside) | | | 0.477 |
| Asbestos-cement | 0.95 | 0.075 | |
| Glass wool | 0.04 | 0.075 | |
| Bituminous roofing | 0.5 | 0.01 | |
| Floor | | | 3.522 |
| Concrete | 1.76 | 0.2 | |

In Table 2, the U-value coefficients are calculated including the internal and external heat transfer coefficients, which have respectively the values of 3 and 17 $W/m^2K$.

The heating system consists of a natural gas boiler allowing up to 200 kW of heating rate. The heat supply is insured by 8 air heaters. The temperature is controlled by means of a thermostatic valve system and a temperature sensor located at each blowing orifice of the air heaters. In order to validate our model, accurate temperature measurements have been performed during three successive days through a platinum resistance sensor which provides results with an uncertainty not higher than ± 0.01 K. During the tests, the set-point temperature is fixed to 17 °C for day time (from 8 am to 8 pm) and 15 °C for night time. The local is unoccupied and we consider that solar gain is negligible because of closing windows during this period. The exterior temperature variation file has been generated by means of a local weather station providing maximum and minimum temperatures of the day.

The model of the present building invokes smaller models (sub-models) that are related to the constructive elements of the building: external walls, roof and slab.

First, the sub-models have been created to describe the physics of heat transfer considering conduction and convection. Heat flow through these elements is affected by convection and conduction phenomena as well as by outdoor/indoor temperature difference.

In the next step, all elements at the boundaries of building are connected to the thermal capacity $C_{zone}$ which takes the form:

$$C_{zone} = \rho_{air} \times V_{building} \times c_{air}\,. \tag{34}$$

The PID controller drives the heating process in order to minimize the difference between the indoor set and measured temperatures. The pseudo-Bond Graphs of the building can be represented in Figure 10.

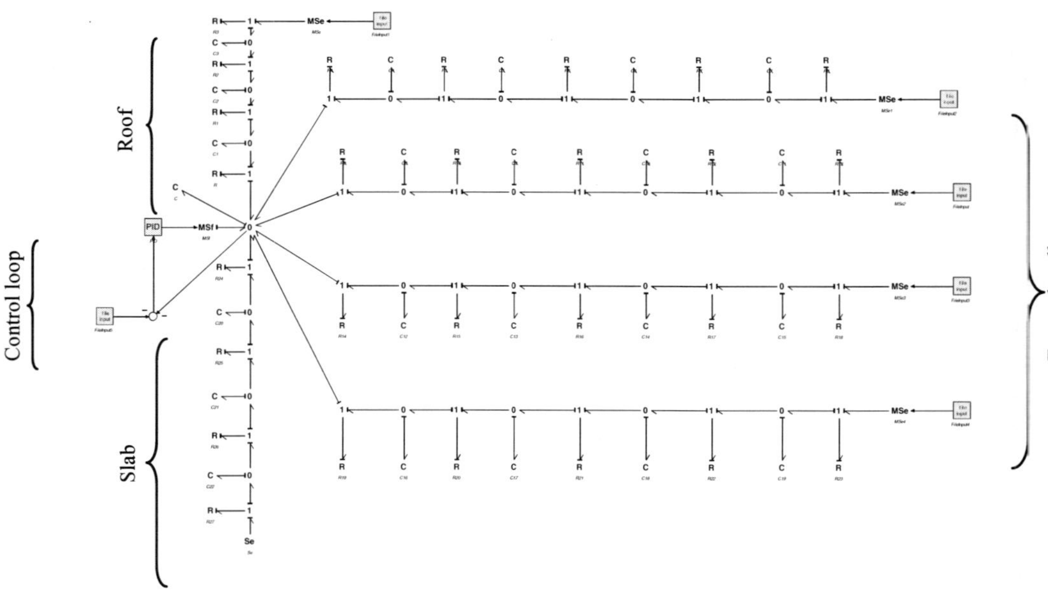


Figure 10. The pseudo-Bond Graph model of the building.

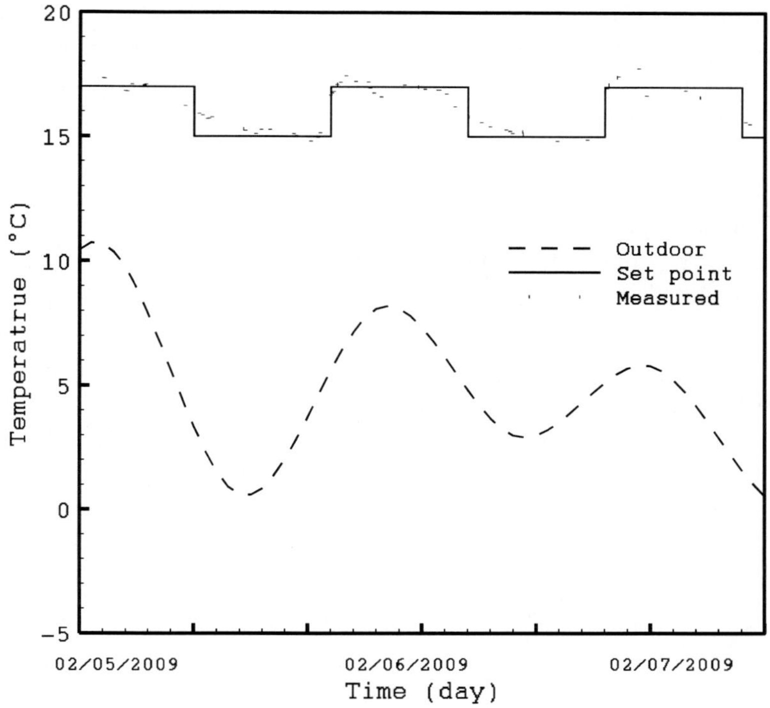


Figure 11. Outdoor, set point and measured indoor temperatures.

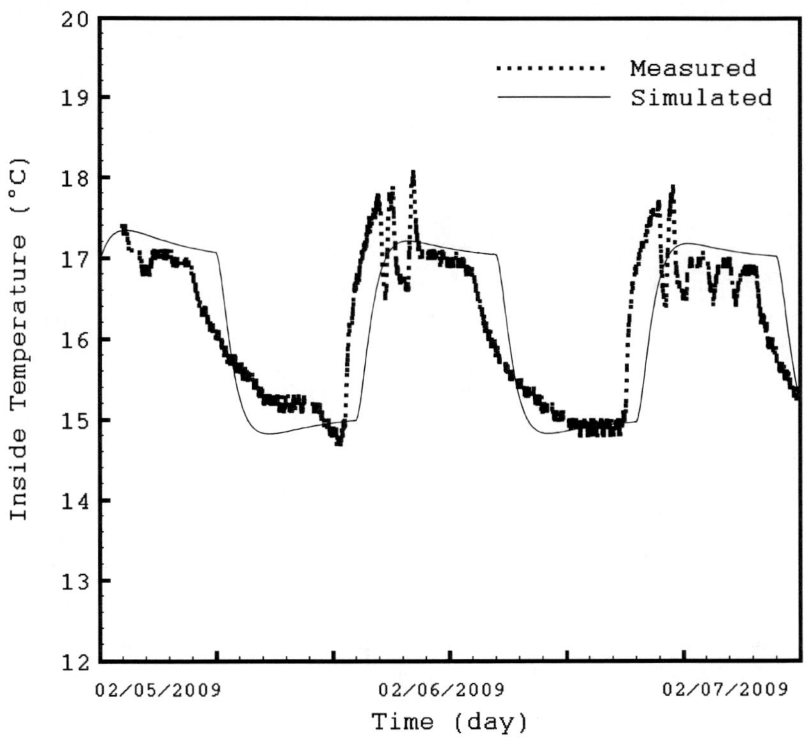


Figure 12. Comparison between measured and simulated indoor temperature data.

Pseudo-Bond Graph model was run using measured weather data for three successive days of the winter period (February $5^{th}$ to $7^{th}$, 2009). Heating load are calculated and compared between the two models and a comparison against measured indoor temperature is carried out. Figure 11 represents the distribution of the outdoor temperature, the set point temperature and the measured indoor temperature.

The set point temperature is fixed to 15 °C for night time and 17 °C for day time, whereas, the outdoor temperature fluctuates between the minimum and the maximum of the considered day. We can observe the important fluctuations of the measured internal temperature around the set point which can be estimated to about ± 0.5 °C, this is due to the control system quality.

Figure 12 shows a comparison between the calculated indoor temperatures with the Bond Graphs approach against the experimental data.

Obviously, a good agreement is observed between the simulated data and the measured ones. The models reproduce properly the thermal behavior highlighted by the experimental data. We can, especially, denote a good harmony between the two simulation models, which demonstrates the potential of the new Bond Graphs approach and its ability to model the building thermal behavior systems.

## CONCLUSION

In this paper the pseudo-Bond graph model of one-dimensional transient heat conduction through plane walls has been developed. Two kinds of problems, regarding boundary conditions, were selected: The first heat conduction problem involves two different convective conditions at each boundary surface of the wall which symbolizes a roof or a vertical wall in a building. The second problem deals with heat conduction in a wall which is subject to one convective condition in one boundary surface and a fixed temperature condition in the other side; this symbolizes heat transfer in a slab.

In the graphical approach modeling, lumped capacity assumption has been adopted with decomposition of the wall into several layers. The developed pseudo-bond graph model has a fairly simple form. In order to validate our model, a comparison between simulation results obtained with Pseudo-Bond graph model and those obtained with analytical method has been carried out. Contrasting the results demonstrates the relevance and the usability of the bond graph method for heat transfer applications in building while keeping a satisfactory level of accuracy. Furthermore, it is possible to increase the accuracy of the results by adding supplementary layers.

Subsequently, the developed wall pseudo-bond graph models are assembled in a single building model in order to simulate and analyze the building thermal behavior. Application of this model demonstrates satisfactory results. Indeed, the calculated indoor temperatures with bond graph model are very closed to the provided measured ones. Hence, Bond graph approach provides real potential for dealing with energy building systems.

Finally, it is noteworthy that this investigation focused on cases which involve time-independent boundary conditions on both sides of the wall with the absence of heat exchange by radiation. Thus nothing can be inferred about the reliability of the Bond graph methodology regardless of solar radiation and apply it on a real building. What we suggest as prospects of this study is to account for radiations exchange including solar gains in a way close to reality as discussed for instance by Weiner *et al.* [9] and Yu *et al.* [15].

## REFERENCES


[1] Sohlberg B. Grey box modelling for model predictive control of heating process. *Journal of Process Control* 2003; 13:225-38.

[2] Deque F, Olivier F, Poblador A. Grey boxes used to represent buildings with a minimum number of geometric and thermal parameters. *Energy and Buildings* 2000; 31:29-35.

[3] Lorenz F, Massy G. Méthode d'évaluation de l'économie d'énergie apportée par l'intermittence du chauffage dans les bâtiments. Traitement par différences finies d'un model à deux constantes de temps. Rapport No. GM820130-01. Faculté des sciences appliquées, université de Liège, Liège, Belgium (in French).

[4] Tindale A. Third-order lumped parameter simulation method. *Building Services Engineering Research & Technology* 1993 14(3):87-97.

[5] Gouda M.M., Danaher S., Underwood C.P. Building thermal model reduction using nonlinear constrained optimization. *Building and environment* 2002 37 (12) 1255-1265.

[6] Cellier FE, Nebot A. Bond Graphs modelling of heat and humidity budgets of Biosphere 2. *Environmental modelling and software* 2006; 21(11):1598-1606.

[7] Weiner M, Cellier FE. Modelling and simulation of a solar energy system by use of Bond Graphs. *Proc. 1st SCS inH. Conference on Bond Graphs modelling and simulation,* San Diego, CA 1993. pp. 301-306.

[8] Borutzky W. Bond Graph methodology: *Development and Analysis of Multidisciplinary Dynamic System Models.* Springer-Verlag, London 2010. ISBN: 978-1-84882-881-0.

[9] Paynter N. Analysis and design of engineering systems. MIT Press 1961).

[10] Karnopp D, Rosenberg R. *System dynamics: A unified approach.* Wiley Intersciences. New York 1990.

[11] Nebot A, Cellier FE. Simulation of heat and humidity budgets of Biosphere 2 without air conditioning. *Ecological engineering* 1999; 13(1-4):333-356.

[12] Özisik M.N. *Heat Conduction.* 2nd edition, Wiley, New York 1993. ISBN: 0-471-53256-8.

[13] Yu B, Van paassen AHC. Simulink and Bond Graphs modelling of an air-conditioning room. *Simulation modelling Practice and theory* 2004;12(1):61-76.

[14] J. Marachlian, R. Benelmir, A. El Bakkali, G. Olivier, « Exergy based simulation model for vehicle HVAC operation », *Applied Thermal Engineering*, vol. 31, Iss. 5, pp. 696-700, 2011.

In: Energy-Environment-Economics
Editor: Riad Benelmir

ISBN: 978-1-62948-862-2


*Chapter 6*

# HIGH TEMPERATURE GAS-COOLED REACTORS

***Piyush Sabharwall*, EungSoo Kim and Hans Gougar***

Idaho National Laboratory, US

## ABSTRACT

The next generation nuclear reactor will likely be a helium cooled gas reactor with an outlet temperature of about 750–800°C for the first of a kind, further increased to 900–950°C for the nth of a kind. These reactors will not only produce electricity, they will also provide process heat for applications such as hydrogen production, coal gasification, etc. Helium was selected as the coolant for this reactor because it is inert and relatively easy to handle, has a low macroscopic neutron cross section, and can be operated at high temperatures without high pressurization. In order to overcome the inherent disadvantage of lower heat transfer and heat transport characteristics of gas coolants and obtain higher thermodynamic efficiencies (Brayton Cycle), it is necessary to operate the fuel elements at highest temperatures as possible (fuel centerline temperature being the limit) and permit high gas temperature rise in the reactor by reducing the mass flow rate and pressurizing the gas.Gas reactors because of the capability of transmutation of fertile transuranics, could also be utilized for deep burn, which provides profound benefits in terms of better use of uranium resources, reduction in long term nuclear materials proliferation risks and high level storage requirements. In this study the general HTGR characteristics are mentioned along with its operating conditions and accident scenario's such as water and air ingress, along with current HTGR challenges.

## 1. INTRODUCTION

A variety of sources, including nuclear, hydro, and fossil power plants, are being considered to meet the growing world-wide need for energy. Nuclear energy is one of the front runners in meeting this demand. The next generation of nuclear reactors are receiving increased attention because of their inherent passive safety features, which are necessary to ensure that the key parameters for reactor safety remain within safe boundary values in all

* E-mail address: Sabharp20@gmail.com.

circumstances.Self control of primary processes refers to a reactors inherent safety, whereas passive safety features are vital in operation when strong deviation of normal process behavior occurs.

For the last decade or so the nuclear industry has focused its efforts primarily on:

- High System efficiency.
- Use of nuclear process heat for industrial applications.
- Inherent safety.
- Achieving higher burn up rates for fuels.
- Extended life-time.
- Non-proliferation resistance.

The high temperature gas-cooled rector (HTGR) is attractive to the corporate sector because of its ability to produce power and provide process heat for various industrial applications. The proposed HTGR is a helium-cooled, graphite moderated, thermal neutron spectrum reactor that will operate at about 7MPa and a reactor outlet temperature (ROT)of about 750°Cfor the first of a kind, which will increase to about 900-950°C for the nth of a kind.

The HTGR technology is established by eight formerly built reactors that vary in size, outlet temperature, primary fluid, and purpose: Dragon (Prismatic Reactor; United Kingdom), Arbeitsgemeinschaft Versuchsreaktor (AVR; Pebble Bed Reactor; Germany),Thorium Hochtemperatur Reaktor (THTR; Pebble Bed Reactor; Germany), High Temperature Test Reactor (HTTR; Prismatic Reactor; Japan), High Temperature Reactor-10 (HTR-10; Pebble Bed Reactor; People's Republic of China), Peach Bottom 1 (Prismatic Reactor; United States), and Fort St. Vrain (FSV; Prismatic Reactor; United States). The lessons learned from these reactors that most impact the next generation nuclear reactor are (Beck et al. 2010):

- Ingress events such as moisture ingress and air ingress
- Primary coolant bypass flow and impurities
- Fission products release resulting from events beyond design basis
- Fission products release resulting from normal operations.

These lessons learned will greatly benefit future generations of HTGRs and assist in ensuring safe, reliable, and high-performance operation.

The two reactor concepts presently being studied for next generation of HTGRs are the prismatic reactor and the pebble bed reactor. The prismatic reactor has cylindrical fuel compacts stacked inside the channels drilled into the hexagonal graphite blocks. The fuel blocks are stacked firmly against each other in columns that form an annulus between an inner reflector and an outer reflector, both of which consist of rings of unfueled graphite blocks. This annular core configuration ensures inherent safety under all accident and transient conditions. During transients, the graphite reflector mass acts as an important temporal heat sink and storage device to maintain fuel temperatures below the peak fuel limiting values (~1250°C) that may damage the fuel. Despite its larger physical size, a graphite-moderated, gas-cooled reactor is smaller neutronically than a light water reactor (LWR) of the same power level, and hence is more stable against xenon-induced power oscillations. The power stability of the gas turbine modular helium reactor (GT-MHR) has been studied extensively, and calculations have demonstrated that the 10-block high, 600 MWt GT-MHR is stable in the axial, radial, and azimuthal directions against xenon oscillations (MacDonald et al. 2003). Prismatic reactors are batch-loaded – additional fuel must be loaded into the blocks to maintain criticality throughout the cycle. This results in excess reactivity at the beginning of the cycle that has to be held down by burnable poisons or control rod insertion.

The pebble bed reactor consists of an annular vat filled with fuel spheres or pebbles that are dropped in at the top of the vat and removed at the bottom. This continuous online refueling reduces the required shut-down frequency and allows operation with very little excess reactivity, further enhancing safety. Helium coolant is blown through the interstitial void that makes up about 39%of the pebble bed volume (Gougar 2010). The pebble design feature tailored to the specific reactor is the moderator-to-fuel ratio, which is adjusted by properly selecting the radius of the fueled zone within the pebble (MacDonald et al. 2003).

The overall layout, functionality, and behavior of these two HTGRs are quite similar, the primary difference being fuel geometry and mobility. Thermal fluid analyses and fuel temperature data determine if the core is operating within design specifications. These analyses are dominated by the thermal coupling between the solid structures and the coolant. Neutronic feedback is captured with simple reactivity coefficients or quasi-static kinetic modules to generate time-dependent power data (Gougar2010). Both prismatic and pebble bed versions of the HTGR possess the same general functionality and bulk design. tristructural isotropic (TRISO) particles are embedded in a graphite matrix to form fuel elements that occupy a tall annular or cylindrical region inside the vessel and are surrounded by graphite reflector blocks. Helium is circulated through cooling pathways in the core to carry off the fission energy and be converted into the desired energy product. Reactivity is controlled by varying helium inventories or inserting control rods into the side reflector. The behavior of materials, neutronics, and thermal fluids is largely the same and can thus be addressed with the same general modeling approach. However, key differences exist in the fuel geometry, requiring different heat transfer correlations, fuel management techniques, and other modeling assumptions that prohibit a prismatic core simulator from being used to model a pebble bed core and vice versa. The coolant in both reactor concepts flows downwards in the core, mainly for the following reasons:

- The temperature difference across the core is about 400 to 500°C.It is therefore better to have the control rods in the lower temperature end.
- The HTGR vessel is very tall. If forced convection is lost because of a pressurized loss-of-coolant event (no breaks), getting natural circulation in a tall chimney needs to be helped/enhanced. The cooler gas with higher density will help push and circulate the hotter gas down and out the duct and in turn draw the cooler helium inside the reactor vessel/outer reflector gap region in a siphoning mode. This sets up the driving conditions to start natural circulation and get a chimney-effect flow, thereby avoiding stagnation.
- Minimizing heat loss as coolant leaves the reactor vessel after going through the core, if the coolant were to flow upwards, more heat would be lost to the vessel wall thereby raising its temperature and leading to some material issues.

- Avoiding serious consequences in the water-ingress accident. The preferred PCU (steam Rankine cycle) location is therefore lower than the reactor vessel.

# 2. HTGR System Characteristics

The HTGR is one of six reactor concepts recommended by the Generation IV Technology Roadmap for further development (Department of Energy [DOE] 2002). Using high temperature gas coolant, the HTGR can achieve (1) high thermodynamic efficiency, (2) inherited safety features, and (3) industrial applications. The following sections present some general features of the current HTGR systems.

## 2.1. Coolant

Nuclear reactor coolant is used to remove heat from the nuclear reactor core and transfer it to electrical generators. The general benefits of using gases as nuclear coolants are as follows:

- Easy to handle
- Low macroscopic neutron cross sections
- Plentiful and cheap
- Operable at high temperatures without higher pressurization.

The disadvantages of using gases as nuclear coolants are:

- Low heat transfer performance (an order of magnitude lower than water)
- High pumping requirements (uses between 8 to 20% of gross plant power).

To partially overcome the inherent disadvantages of gas coolants and at the same time obtain attractive thermodynamic efficiencies, it is necessary to operate the fuel elements at the highest temperatures possible (commensurate with metallurgy), permit the gas to flow at a mass-flow rate, and pressurize the gas. Several gas coolants, including air, carbon dioxide, nitrogen, hydrogen, and helium, have been used or considered in the gas cooled reactor systems (Melese and Katz 1985):

- *Air* was an obvious choice for the early low-power reactors and for the first production reactors in France and the United Kingdom. However, it is no longer used because of its poor thermal performance and nitriding issues.
- *Carbon dioxide* ($CO_2$) was widely used as a reactor coolant at the early stage of gas cooled reactor technology because of its good thermal performance and availability. However, $CO_2$ can be easily decomposed, leaving a free radical in the system that increases corrosion. Chemical reactions between graphite and $CO_2$also limited the ROTbelow650°C.

- *Nitrogen* has not been used much because of its relatively high neutron cross section, nitriding problems at high temperature, and poorer thermal performance.
- *Hydrogen* can provide an excellent thermal performance but has only been used for special applications such as rocket propulsion because of its serious chemical reaction with oxygen and diffusion through structural metals.
- *Helium* is currently considered the best coolant option for the HTGR technology because of its good thermal performance, excellent radiation stability, and inertness. Helium-neutron reaction cross sections are also quite low. Only the $He_3$ isotope has the relatively large cross section for thermal neutrons of about 1 barn. Its abundance in naturally occurring helium, however, is so low (0.00013 percent) that the induced radioactivity is negligible. Serious radioactivity in a helium coolant comes from other gaseous impurities. Those that may exist in high-grade helium are hydrogen (about 5 ppm by mass), water vapor (about 50 ppm), and air (about 75ppm) with its various constituents, including argon.

Table 1 summarizes thermal-hydraulic properties of some major gas coolants (helium, air, $CO_2$, $H_2O$, argon) at typical HTGR operating conditions (700°C and 7.0 MPa). According to this comparison, helium shows better thermal characteristics and higher thermal conductivity and heat capacity than the others with similar viscosity. Its only disadvantage is considered to be its low density.

**Table 1. Comparisons of Water and Gas Coolant Properties (NIST Chemistry Web-book)**

| Coolant | k (W/m K) | P (kg/m$^3$) | Cp (J/kg K) | M (Pa s) |
|---|---|---|---|---|
| He (700°C, 7 MPa) | 0.36 | 3.43 | 5189 | 4.50E-05 |
| Air (700°C, 70 MPa) | 0.07 | 21.48 | 1187 | 4.39E-05 |
| $CO_2$ (700°C, 70 MPa) | 0.07 | 37.54 | 1242 | 4.07E-05 |
| $H_2O$ (700°C, 70 MPa) | 0.1 | 15.9 | 2385 | 3.68E-05 |
| Ar (700°C, 70 MPa) | 0.03 | 49 | 532 | 4.30E-05 |

## 2.2. General Design and Operating Conditions

The key operating parameters and design features for the recent HTGR concepts (GT-MHR by General Atomics, ANTARES by Areva, and PBMR by Westinghouse) are shown in Table 2.The primary coolant is helium and the designed thermal power ranges from 500 to 600 MWt.

The GT-MHR and ANTARES are based on the prismatic core design; the PBMR is based on the pebble bed core design. Power conversion systems are the steam Rankine cycle for the GT-MHR and ANTARES and steam/gas turbine cycles for the PBMR. Average power density ranges from 4.8 W/cm$^3$ for the PBMR to 6.5 W/cm$^3$ for the GT-MHR. Cores are designed to be annular shape surrounded by inner and outer reflectors made of graphite. The reactor core outlet temperature is 750°C for all concepts and the operating pressures are 7.0, 5.0, and 9.0 MPa for the GT-MHR, ANTRES, and PBMR, respectively. The designed lifetimes are 60 years. All the conceptual designs incorporate passive safety features.

**Table 2. Operating Parameters Comparison for the NGNP Designs and FSV HTGR (Crozier (2008), AREVA(2008), Koekemoer (2009))**

| Condition or Feature | GT-MHR (General Atomics) | ANTARES (AREVA) | PBMR (Westinghouse) |
|---|---|---|---|
| Power Output [MWt] | 550 – 600 | 565 | 500 |
| Average Power Density (W/cm$^3$) | 6.5 | – | 4.8 |
| Moderator | Graphite | Graphite | Graphite |
| Core Geometry | Annular | Annular | Annular |
| Reactor Type | Prismatic | Prismatic | Pebble Bed |
| Safety Design Philosophy | Passive | Passive | Passive |
| Plant Design Life (Years) | 60 | 60 | 60 |
| Core Outlet Temperature (°C) | 750 | 750 | 750 |
| Core Inlet Temperature (°C) | 322 | 325 | 280 |
| Coolant Pressure (Mpa) | 7 | 5 | 9 |
| Coolant Flow Rate (kg/s) | – | 282 | 204 |
| Secondary Outlet Temperature (°C) | 540 | 550 | 700/541 |
| Secondary Inlet Temperature (°C) | 200 | – | 267/217 |
| Secondary Fluid | Steam | Steam | He, Steam |
| Secondary Coolant Flow Rate (kg/s) | – | 141 | 204 |

## 2.3. Core Design and Fuel

This section briefly summarizes general core designs for the HTGR and unique fuel designs.

### *2.3.1. Prismatic and Pebble Bed Core*

The HTGR has two different types of core designs: prismatic core design and pebble bed core design as follows (MacDonald 2003):

- The *prismatic core design* consists of hexagonal graphite blocks as shown in Figure 1. Approximately one-third of these blocks are fuel blocks arranged in an annular core. The remaining two-thirds of the blocks form an inner and outer reflector about the annulus. During transients, large mass and heat capacity of the graphite reflectors play an important role as temporal heat sink and storage maintaining fuel temperatures below the limitation (~1600°C). The blocks are stationary during reactor operation, but at the end of each power cycle, every block can be replaced if needed. The Peach Bottom reactor, FSV, HTTR (in Japan), and GT-MHR are designed to incorporate the prismatic core designs.

- The *pebble bed core design* consists of graphite pebbles that are dropped in at the top and removed at the bottom, they flow slowly through the core region shown in Figure 2.This design configuration introduces several unique advantages with continuous online refueling, reducing the frequency of required shutdowns. Also, the pebble bed reactor operates with very little excess reactivity, substantially enhancing safety. The fuel enrichment is typically only 8%. Pebble bed reactors of 300 MWt or less have been shown analytically to be passively safe, but still need to be demonstrated at higher power levels (600 MWt). The PMBR (in South Africa) and the HTR-10 (in China) are pursuing the pebble bed core design.

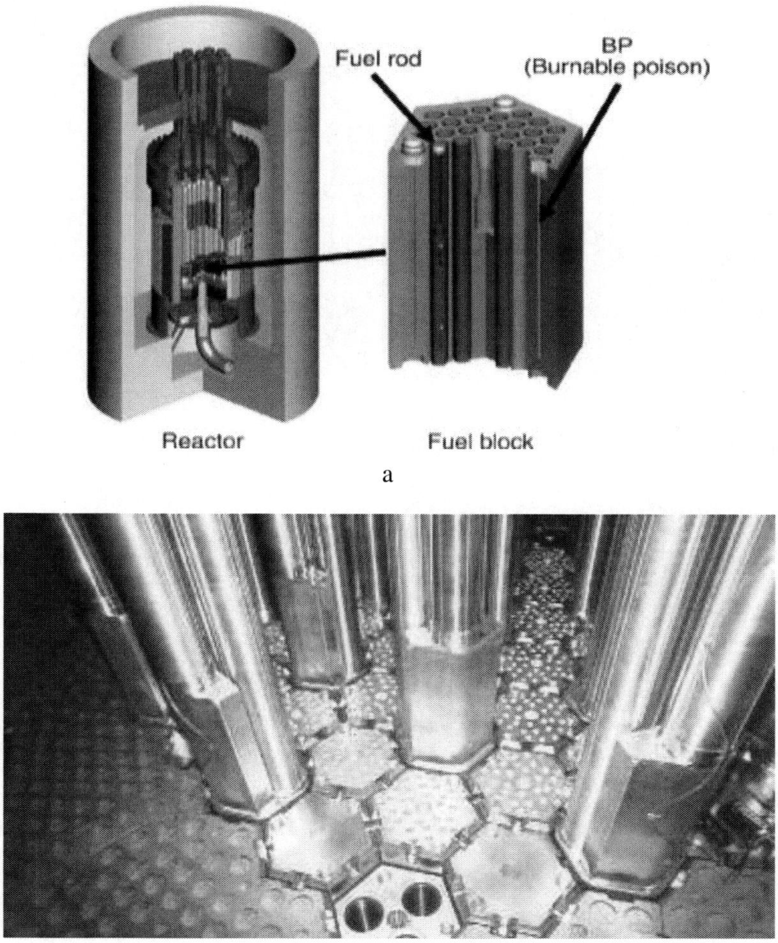


a

b

Figure 1. Prismatic core of HTTR: (a) HTTR core schematics; (b) photograph of HTTR core (courtesy of JAEA).

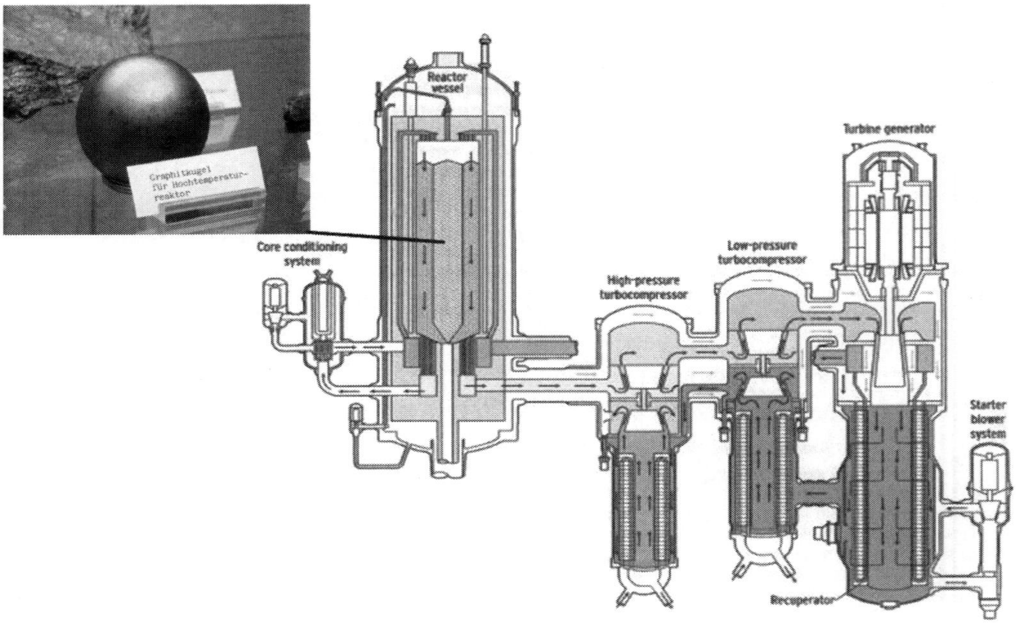


Figure 2. Pebble bed reactor design (PBMR) (courtesy of Westinghouse).

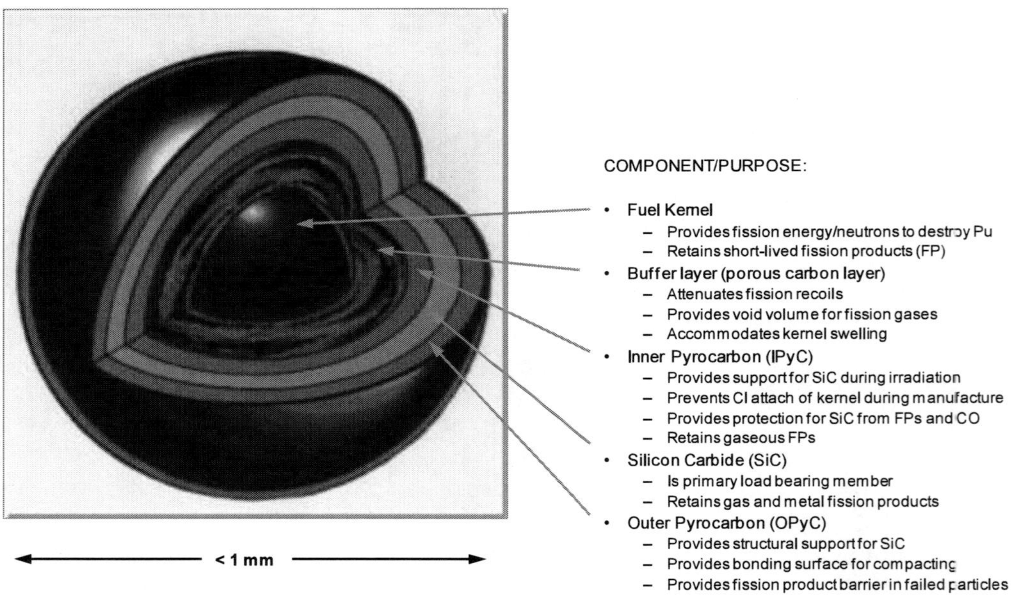


Figure 3. TRISOfuel particle (courtesy of General Atomics).

### *2.3.2. Tristructural Isotropic Fuel*

HTGR fuel is made of TRISO coated particles – a type of microfuel particle demonstrated in Germany and elsewhere. A TRISO particle is a spherical layered composite about 1 mm in diameter as shown in Figure 3. It consists of a fuel kernel composed of $UO_X$ (sometimes UC or UCO) in the center, coated with four layers of three isotropic materials. Fuel kernels are encapsulated in silicon carbide (ceramic) coatings, forming the Fuel Particles shown in Figure 3. The four layers are a porous buffer layer made of carbon, a dense inner layer of pyrolytic carbon (PyC), a middle ceramic layer of SiC to retain fission products at elevated temperatures and give the TRISO particle more structural integrity, and a dense outer layer of PyC. TRISO fuel particles are designed not to crack from process stresses (such as differential thermal expansion or fission gas pressure) at temperatures beyond 1600°Cas shown in Figure 4, and therefore can contain the fuel in the worst of accident scenarios in a properly designed reactor.

The AVR was the first nuclear reactor to use TRISO fuels. The THTR-300is the first nuclear powerplant to use it. Currently, TRISO fuel compacts are being used in the HTGR experimental reactors such as the HTR-10 in China, and the HTTR in Japan.

The dimensions of the fuel and approximate densities of each layer are shown in Table 3. The kernel dimensions are based on the GT-MHR design. The buffer layer and the TRISO coating layer dimensions are the same as that used by the Germans in their successful gas reactor program in the 1970s and 80s.

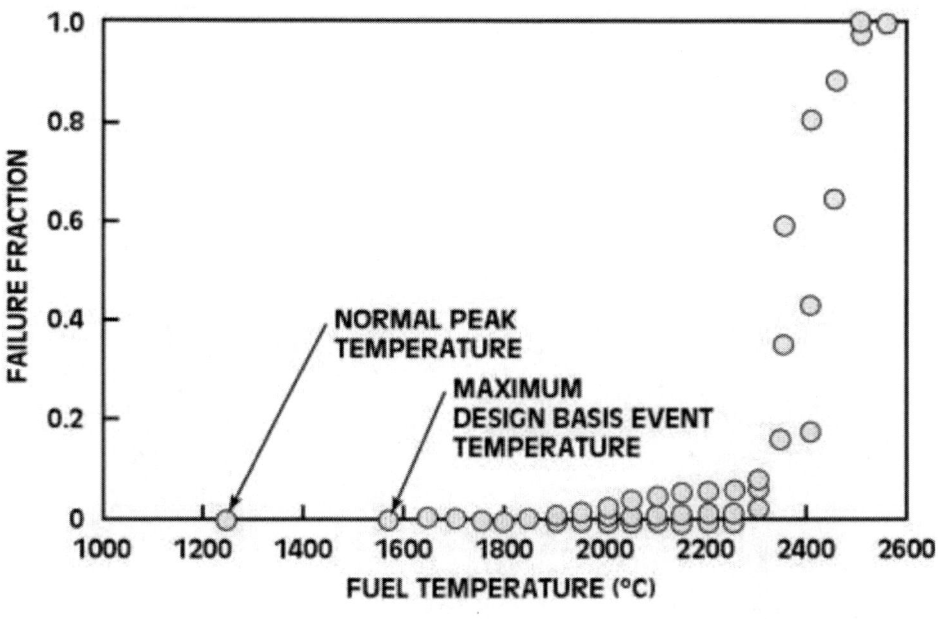


Figure 4. Fuel temperature vs. failure fraction of TRISO fuel (courtesy of General Atomics).

### *2.3.3. Graphite Moderator and Reflector*

Graphite is used in HTGRs as moderator and reflector. Japan and China have both constructed experimental (small-scale) graphite moderated high temperature reactors (HTTR for Japan and HTR-10 for China) but a commercial graphite-moderated reactor has not been constructed in the world since the 1980s. The last graphite reactor constructed in the United States was the helium-cooled HTGR at FSV, Colorado in the late 1970s. The design and construction of a commercial HTGR therefore requires the reestablishment of the nuclear graphite supply chain, including reliable coke sources, experienced graphite manufacturers, and the generation of sufficient graphite properties and environmental effects data to facilitate graphite core design and licensing.

**Table 3. TRISO particle properties and dimensions (MacDonald 2003)**

| Parameters | | Value |
|---|---|---|
| Kernal | Composition | UCO |
| | Diameter | 350 microns |
| | Density | >10.5 g/cc |
| Buffer | Thickness | 100 microns |
| | Density | ~1 g/cc |
| IPyC | Thickness | 40 microns |
| | Density | ~1.9 g/cc |
| SiC | Thickness | 35 microns |
| | Density | ~3.2 g/cc |
| OPyC | Thickness | 40 microns |
| | Density | ~1.9 g/cc |

Figure 5. Graphite block for prismatic reactor (Pappano 2009).

## 2.4. Power Conversion System

The HTGR can provide various power conversion system options because of its high temperature operations above 700°C. They generally include steam turbine system, gas turbine system, and supercritical $CO_2$ system. Figure 6 shows some main thermodynamic cycle efficiencies with different temperature and configurations. As shown in this figure, at low temperature less than 700°C, the steam Rankine cycle shows good efficiency compared to the other cycles. However, at high temperature above 700°C, the efficiency of gas turbine cycle and supercritical $CO_2$ cycle is much higher than the steam cycle. In the current pressurized water reactor (PWR) systems, overall system efficiency ranges around 33–35% and can reach up to 55% for the theoretical gas turbine and supercritical $CO_2$ cycle as shown in Figure 6.

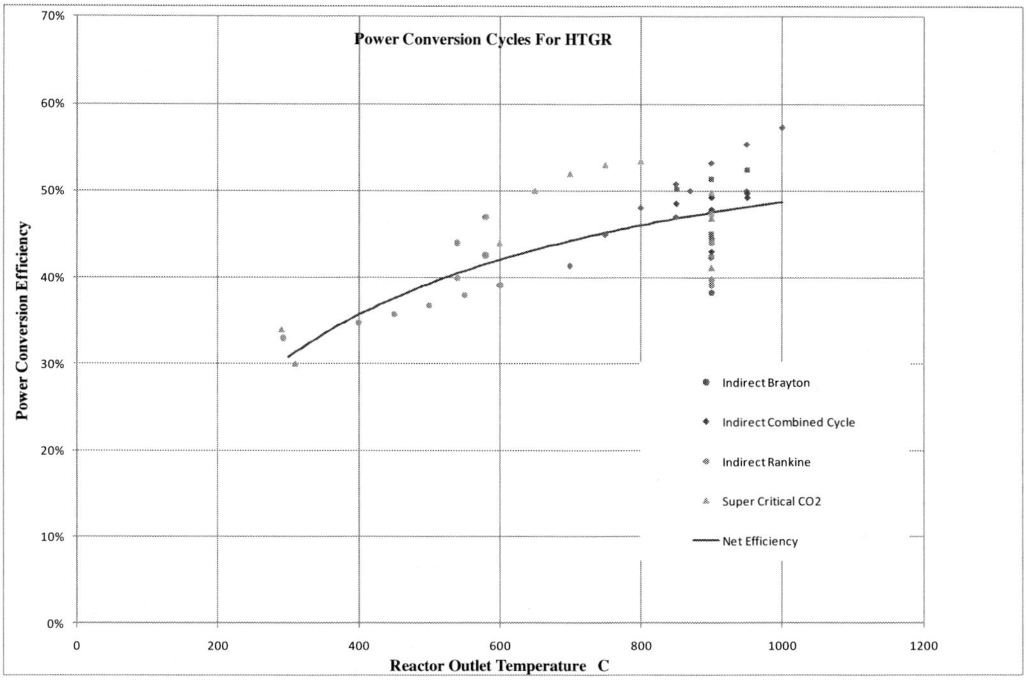


Figure 6. Cycle efficiency vs. turbine inlet temperature (McKellar et al. 2009).

### *2.4.1. Steam Turbine System (Rankine Cycle)*

The steam turbine system is the most common power generation system today, generating about 80% of all electric power used throughout the world, including virtually all solar thermal, biomass, coal, and nuclear power plants. The steam turbine system is based on a Rankine cycle that consists of the following four process steps (DOE 1992):

1. Heat is supplied to the steam generator (boiler) where liquid is converted to steam or vapor.
2. The vapor is expanded adiabatically in the turbine to produce a work output.
3. Vapor leaving the turbine enters the condenser where heat is removed and the vapor is condensed into the liquid state.
4. Saturated liquid is delivered to the condensate pump, then to the feed pump, and the high pressure liquid is finally delivered to the to the steam generator.

The efficiency of a Rankine cycle is usually limited by the working fluid. Without the pressure reaching super critical levels for the working fluid, the temperature range is quite small because of high operating pressure and material issues. The turbine inlet temperature is generally less than 400°C for nuclear reactor and 500°C for coal-fired power stations. This low turbine entry temperature makes the Rankine cycle often available for a bottoming cycle in the combined-cycle gas turbine system. General, Rankine cycle efficiency is 33-35% for nuclear power plant and about 42% for a modern coal-fired-power plants, while the ideal Carnot efficiency is about 60%. One of the principal advantages the Rankine cycle is that it requires relatively little work for compression work to drive the pump, since the working fluid at this point is liquid. By condensing the fluid, the work required by the pump consumes only 1-3% of the turbine power and contributes to a much higher efficiency for a real cycle (DOE 1992).

To improve the cycle performance, reheat or regeneration is applied. In the reheat cycle, two turbines work in series. The first accepts vapor from the boiler at high pressure. After the vapor has passed through the first turbine, it reenters the boiler and is reheated before passing through a second, lower pressure turbine. Among other advantages, this prevents the vapor from condensing during its expansion, which can seriously damage the turbine blades, and improves the cycle efficiency as more of the heat flow into the cycle occurs at higher temperature. In the regenerative cycle, the working fluid exiting from the condenser (possibly as a subcooled liquid) is heated by steam tapped from the hot portion of the cycle to improve efficiency. The regenerative features effectively raise the nominal cycle heat input temperature, by reducing the addition of heat from the boiler/fuel source at the relatively low feed water temperatures.

### *2.4.2. Gas Turbine System (Brayton Cycle)*

A gas turbine system is based on the Brayton cycle. Because of its high efficiency at high temperature and compactness, the Brayton cycle has been extensively investigated as a potential replacement of the conventional steam Rankine cycle for the next generation reactors. Gas turbines are usually operated on an open cycle, but it can also be operated on a closed cycle. The gas Brayton cycle consists of the following steps:

1. Isentropic compression in a compressor.
2. Constant pressure heat addition in a heat exchanger.
3. Isentropic expansion in a turbine.
4. Constant pressure heat rejection in a heat exchanger.

The most proposed gas turbine cycle in the nuclear plant is the helium Brayton cycle, which can achieve 45-48% overall efficiency at around 900°C. Lots of methods are used to improve the basic Brayton cycle efficiency, but the following are generally used:

- Reheat.
- Increasing Pressure Ratio.
- Intercooling.
- Regeneration.
- Steam Rankine Bottoming.

Though the closed helium Brayton cycle is currently in the development stage, it is expected to become a promising future technology because of its high efficiency and simplicity.

### *2.4.3. Gas Turbine System (Supercritical $CO_2$Cycle )*

Helium Brayton cycle generally require core outlet temperature around 900°C. The high temperature environment required for this cycle is one of the main challenges for mechanical structure integrity. To overcome this issue, a supercritical $CO_2$ cycle was proposed that can provide high efficiency even at lower temperatures ranging from 500-700°C. The supercritical $CO_2$ recompression cycle was first recommended by Angellino (1967) and proposed by Dostal et al. (2004) for nuclear power plants. A main advantage of this system originates from its reduced compression work, compared to an ideal gas such as helium, because of its liquid-like behavior above critical point. All gas coolants have specific temperatures called critical points (7 MPa, 31°C for $CO_2$ gas) over which the gas does not change to liquid, even though very high pressures are applied. Above this point (supercritical region), the fluid behaves like a gas but has a very small density variation like a liquid, requiring only small compressing power. Lots of research is currently being conducted in this field.

## 2.5. Other System Components

### *2.5.1. Reactor Pressure Vessel*

The reactor pressure vessel (RPV) provides containment and structure for the reactor components, including the reactor core and the control rods, capable of withstanding the temperatures generated by reaction heat. Figure7 represents a typical reactor vessel and support system.

The following functions are performed by the RPV Systems (Collins 2009):

- House and support the components of the reactor core, reactor internals, and reactor support structure
- Maintain positioning relative to the control rods
- Contain the primary coolant inventory within a leak tight pressure boundary
- Maintain the integrity of the coolant pressure boundary.

### *2.5.2. Cross Vessel*

A cross vessel in the HTGR transfers fluids from the RPV to the intermediate heat exchanger (IHX) or the steam generator. This cross vessel is generally made of the same materials as the IHX and steam generator. The cross vessels include a concentric duct (primary hot gas duct) that separates the hot (core exit) and the cold (core inlet) gas flow streams. Hot (750-800°C) helium coolant is transferred via the hot duct from the core, while cold helium coolant (280-340°C) is returned via the cold duct to the core. The insulation is intended to reduce heat losses to the core inlet cold gas stream, and can be remotely removed and replaced (if needed) during the plant lifetime. The cross vessel piping consists of four main components: support structure, hot and cold ducts, insulating materials, and insulation liner.

Figure 7. Typical HTGR Vessel and Support System (Collins2009).

### *2.5.3. Intermediate Heat Exchanger*

An IHX is one of the essential components in the indirect HTGR system. It transfers heat from the reactor primary loop to the secondary loop. The IHX is likely the component most critically impacted by increased temperatures. The outlet gas temperature does not affect material selection for the IHX in the range of 750-900°C, but it may affect the design allowable stresses. Various types of heat exchangers are considered as the IHX:

- *Tubular Heat Exchanger (Shell-and-Tube or Helical Coil)* The shell-and-tube heat exchanger is the most common type found in industry. This exchanger is generally built of a bundle of round tubes mounted in a cylindrical shell with the tube axis parallel to that of the shell as shown in Figure 8(a). One fluid flows inside the tubes and the other fluid flows across and along the tubes. The major components of this exchanger are tubes (or tube bundles), shell, front-end head, rear-end head, baffles, and tube sheets (Shah and Sekulic 2003). Fluid is distributed to the tubes through a manifold and tube sheet. To increase heat transfer efficiency, further modifications to the flow paths of the outer and inner fluids can be accomplished by adding baffles to the shell to increase fluid contact with the tubes, and by creating multiple flow paths or passes for the fluid flowing through the tubes (Sherman and Chen 2008). These heat exchangers are used for gas-liquid heat transfer applications, primarily when the operating temperature and/or pressure is very high (Shah and Sekulic 2003).Helical coil heat exchangers are shell and tube type heat exchangers that consist of tubes spirally wound into bundles and fitted into a shell as shown in Figure 8(b). The spiral geometry of the tubes results in a higher heat transfer rate than that for a straight tube (Shah and Sekulic 2003). Because of the tube bundle geometry, a considerable amount of surface can be accommodated inside the shell. These heat exchangers are used for gas to liquid heat transfer applications, primarily when the operating temperature and/or pressure is very high.

- *Printed Circuit Heat Exchanger (PCHE)* The PCHE is a relatively new concept that has only been commercially manufactured by Heatric™ since 1985. As the name PCHE implies, they are manufactured using the same technique used to produce standard printed circuit boards for electronic equipment as shown in Figure 9. In the first step of the manufacturing process, the fluid passages are photochemically etched into the metal plate. Normally, only one side of each plate is etched-out. The etched-out plates are thereafter joined by diffusion bonding, which is the second step and results in extremely strong all-metal (strong bond having 90-95% parent metal strength) heat exchanger cores. Plates for primary and secondary fluids are stacked alternately and formed into a module. Modules may be used individually or joined with others to achieve the needed energy transfer capacity between fluids. The diffusion bonding process allows grain growth, thereby essentially eliminating the interface at the joints, which in turn gives the parental metal strength. Because of the use of diffusion bonding, the expected lifetime of the heat exchanger exceeds that of heat exchangers based on a brazed structure (Dewson and Thonon 2003).

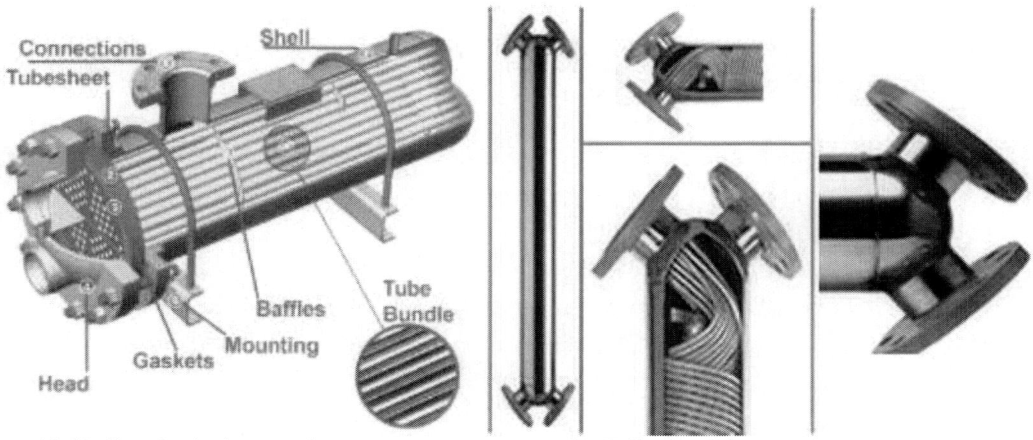


(a) Shell-and-tube heat exchanger. (b) Helical-coil heat exchanger

Figure 8. Tubular heat exchangers.

Table 4 shows operating conditions for the tubular heat exchanger and the PCHE.

**Table 4. Usual operating ranges (can be wider with special materials)**

| Exchanger Type | Temperature [°C] | Pressure [bar] |
|---|---|---|
| Tubular | -25 ~ 650 | 300/1400 |
| PCHE | -200 ~ 900 | 1000 |

Figure 9. Printed circuit heat exchanger (Courtesy of HEATRIC).

# 3. HTGR SAFETY

The HTGR is a nuclear reactor, so safety is one of the most important issues for consideration of the designs. The following design features are generally considered important to achieving high safety:

- Highly reliable and less complex shutdown and decay heat removal system.
- Longer time constant.
- Simplified safety system.
- Minimization of potential for severe accidents and their consequences.
- Reduced radioactive exposure to plant personnel in the normal/abnormal conditions.
- Multiple barriers against radiation release and reduced potential for severe accident consequences.
- Proven technology.

## 3.1. Passive Safety Features

The current HTGR incorporates the following passive safety features sufficient to meet safety requirements:

- Large negative temperature coefficient
- Passive decay heat removal system independent of coolant
- No AC powered safety-related system
- No operator action required
- Insensitivity to incorrect operator action.

Passive safety in the HTGR is achieved by incorporating the following design characteristics:

- Helium coolant is neutronically transparent and chemically inert. It is also single phase, which does not change its phase significantly like water in the PWR.
- Graphite has high temperature stability and very large heat capacity, becoming a large source of heat energy storage, resulting in long response time.
- The fission product is contained inside the coated fuel particle.

The final safety goal for a nuclear reactor is to protect or minimize radioactive product. To meet this goal, the HTGR adapted the concept of multiple barriers. The HTGR has three barriers to prevent radionuclide release:(1) fuel elements, (2) helium coolant, and (3) the reactor building itself.

In both normal and off-normal conditions, most of the radionuclide is retained in the fuel element. In order for radionuclides to be released from the fuel, it must pass through multiple barriers: fuel kernel, PyCanSiC fuel coating layers, and fuel compact/matrix graphite. These fuel coating layers are the most important of the barriers. Once the radionuclide is released to the helium coolant, it is retained in the reactor pressure boundary consisting of metallic structures. The final barrier is the reactor building surrounding the helium pressure boundary as shown in Figure 10. This structure protects the helium pressure vessel and reactor cavity cooling system (RCCS) from external hazards. Confinement, which allows venting, is better for HTGR economics and safety containment, which retains pressure.

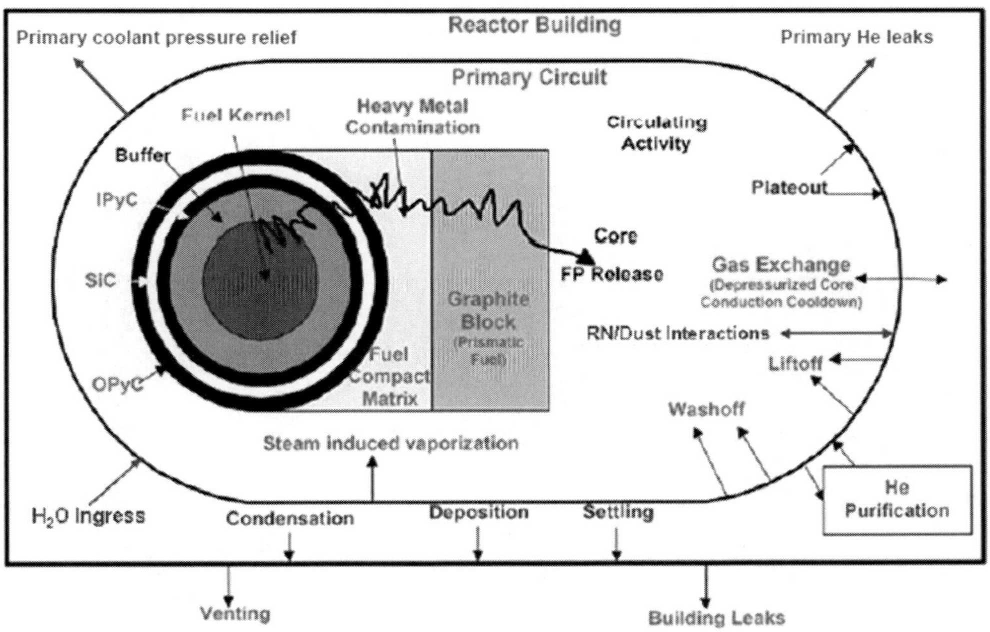


Figure 10. Multiple barrier for radionuclide in the HTGR (General Atomics 1988).

A main issue in HTGR off-normal conditions, such as loss of coolant/flow accidents, is effectively removing core decay heat. Figure 11 shows the heat transfer paths from the HTGR core to the RCCS. Decay heat in an HTGR can be passively removed because of:

- *Small power density:* Power density is low, limiting decay heat which sufficiently suppresses the temperature increase.
- *High heat capacity graphite:* Graphite can store large amounts of energy from the reactor core with less temperature increase than metals. It also slows reactor response in accident situations.
- *High heat conduction and radiation:* The graphite core conducts heat and maintains very high temperatures so the heat can be effectively transferred by conduction and removed by radiation.
- *Uninsulated reactor vessel:* The reactor vessel is not insulated, so in an accident condition, the decay heat is removed by radiation heat transfer at the reactor vessel.
- *RCCS:* The HTGR RCCS finally removes heat from the vessel by radiation and natural circulation. Generally, the RCCS is operated in natural circulation mode.

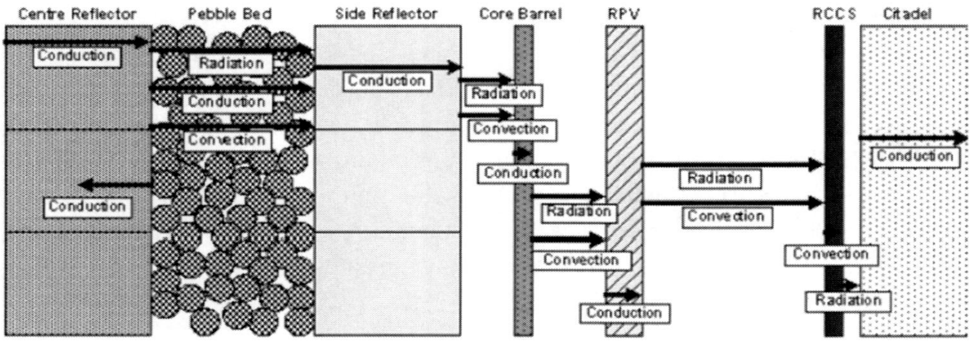


Figure 11. Pebble bed reactor passive heat transfer path for annular core design (General Atomics 1998).

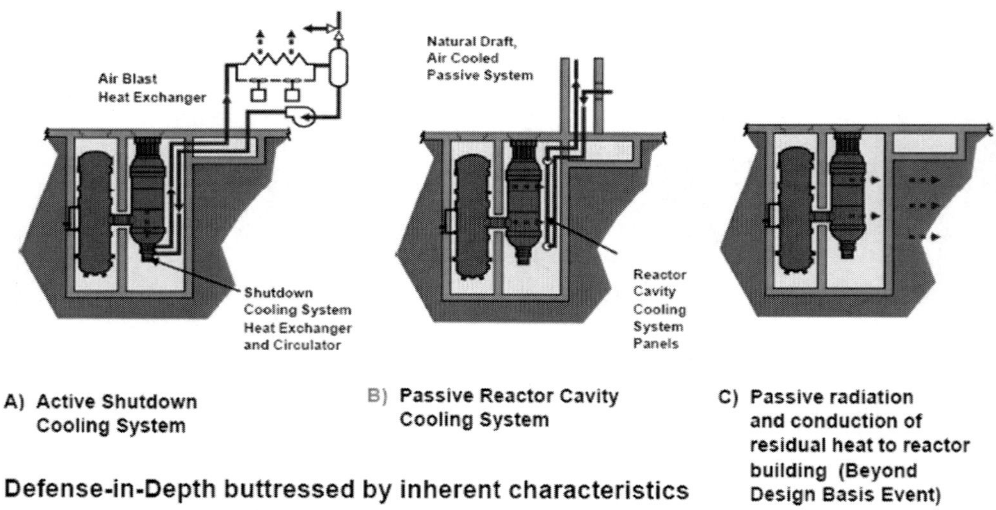


Figure 12. Basic concept of the RCCS (General Atomics 1992).

## 3.2. HTGR Safety System

The RCCS is one of the HTGRs most important safety systems because it removes decay heat from the reactor core. The RCCS consists of cooling structures surrounding the reactor vessel that remove heat from the vessel by radiation and natural convection. The working RCCS fluid can be air or water flow. The RCCS is designed to operate in the passive mode. Figure 12 shows some general RCCS concepts.

Two independent HTGR systems control reactivity for reactor shutdown: control rod and reverse shutdown system. Each system can maintain HTGR subcriticality. A large negative temperature coefficient also provides inherent safety feature.

## 3.3. Accidents

This section briefly summarizes two potentially major accidents should they occur in the HTGR: air-ingress and water-ingress.

### *3.3.1. Air Ingress Accident*

An air ingress accident is currently considered to be the most serious potential accident in the HTGR. This accident is followed by a pipe break in the primary system. After the pipe breaks, the helium coolant is rapidly blown out of the primary vessel through depressurization. If pressures between the inside and outside vessels equilibrate, air in the outside vessel will ingress into the reactor. Once air gets into the inside reactor vessel, the following will occur:

- The graphite will start to oxidate; this exothermic reaction can provide additional heat to the reactor core by increasing core temperature.
- Graphite oxidation will significantly degrade the structural integrity of the reactor core.
- Graphite oxidation can produce CO and $CO_2$ gases that can be released to the atmosphere.

An air ingress accident scenario was previously considered to be followed by:

- Pipe break.
- Depressurization.
- Diffusion (slow air-ingress).
- Natural Circulation (rapid air-ingress).

The main air-ingress mechanism in this scenario was considered to be molecular diffusion, which is very slow (Takeda 1996 and Oh et al. 2006). According to this scenario, onset-of-natural circulation (ONC) took more than 100 hours after the pipe broke, and rapid air-ingress started. During that time, the reactor core was sufficiently cooled down, and the core maximum temperature is maintained well below the critical limit of ~1600°C.

However, recent studies showed that the previous scenario was incorrect because of the effects of density gradient driven flow. The potential for density-gradient governed stratified air to ingress into the HTGR following a large break loss of coolant accident was first described in the NGNP Methods Technical Program (Schultz et al. 2006) and extensively studied by Oh et al. (2010). According to their studies, molecular diffusion has a very minor effect on the air-ingress accident, and density gradient driven flow dominates the air-ingress process. The air-ingress accident therefore follows this procedure:

- Pipe break.
- Depressurization.
- Stratified flow – Stage 1 (driving force = molecular weight gradient).
- Stratified flow – Stage 2 (driving force = temperature gradient) or molecular diffusion.
- Onset-natural circulation.

Subsequent to the break in the hot duct hypothesized in depressurized conduction cool-down, air present in the reactor cavity will enter the reactor vessel. Because of the significantly higher molecular weight and lower initial temperature of the reactor cavity air, the air-helium mixture in the cavity is always heavier than the helium discharging from the reactor vessel via the break into the reactor cavity. Once the air-helium mixture enters the reactor vessel, it will pool at the bottom of the lower plenum then move from the lower plenum into the core via diffusion and the density-gradient induced by heating. When density-gradient-driven stratified flow is considered a contributing phenomena for air ingress into the reactor vessel, much earlier natural circulation occurs in the reactor vessel based on two factors. Density-gradient-driven stratified flow is a much more rapid mechanism (at least one order of magnitude) for moving air into the reactor vessel lower plenum than diffusion. Consequently, the diffusion dominated phase begins with a much larger flow area and a much shorter distance for air to move into the core than earlier scenarios which attribute all air ingress from the reactor cavity into the core to diffusion only. Figure 13 shows the mechanism of the density gradient flow in the air-ingress accident.

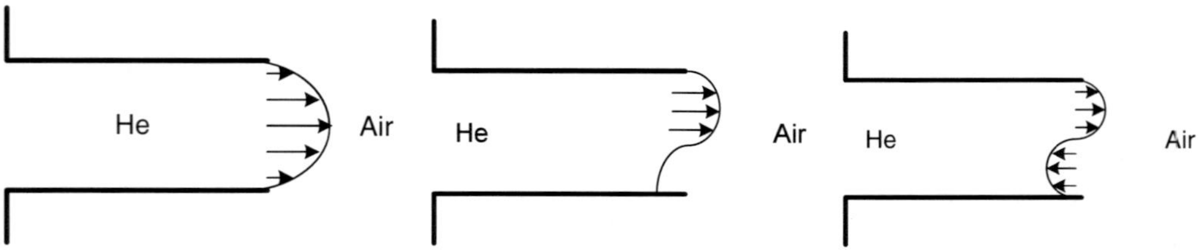


(a) Depressurization.

(b) Onset of density driven flow (no flow at the bottom of the break).

(c) Density driven flow (Reverse flow at the bottom of the pipe).

Figure 13. Density-driven induced stratified flow (Oh et al. 2010).

### *3.3.2. Water Ingress Accident*

A water ingress accident is initiated by a steam generator tube rupture or leakage in the HTGR. The water ingress is a design basis accident generally based on the following scenario (Lohnert 1992):

- A double-ended guillotine break of a heating tube in steam generator occurs accidentally.
- Protective action is triggered in the HTGR within 10 seconds. During this time, all the control rods drop and blowers are shut down, the steam generator isolation values close, and the steam generator water rapidly drains and ingresses into the primary vessel. The maximum penetrated water volume depends on the system design and successful operation of shut-down valve.
- Pressure in the primary vessel increases from steam evaporation. The pressure should be maintained below vessel maximum pressure limit. If the pressure exceeds the limit, a relief valve will open. To minimize this effect, an emergency cooler is operated to condense water and the gas purification plant is operated to remove $H_2$ and CO.

Water ingress results in the following issues (Hosegood 1988):

- *Positive Radioactivity:* Most high temperature reactor (HTR) systems are significantly under-moderated for economic reasons such that accidental water or steam ingress can increase the reactivity of the system, resulting in a:
  - negative contribution from the absorption by hydrogen
  - positive contribution from the softening of the neutron energy spectrum, which increases the effective fission cross section and reduces resonance capture in the fertile materials.
  - positive contribution because of reduced neutron leakage out of the core region caused by the presence of water in the coolant channel.
- *Chemical Reaction and Corrosion:* Chemical reaction between water (or steam) and graphite is endothermic reaction and not relevant from the corrosion point of view. But, it may pose a considerable hazard with regard to pressure increase and production of explosive gas mixtures.
- *Pressure Increase:* The water ingress can highly increase primary vessel pressure by evaporation.

The AVR was the first pebble bed high temperature reactor in the world and was operated from 1967 to 1988 in Juelich, Germany. The focus of this experimental reactor was to test the pebble bed core and many different types of pebble shaped fuel elements (Moorman 2008). In May 1978, a water leak accident occurred in the AVR following a small leak in full power operation; however, the leak was not recognized so the operation continued. The reactor was eventually shut down for other reasons, at which time it was learned that 27.5 tons of water had leaked into the primary reactor. The water was removed, the steam generator tube was plugged, and the plant was restarted. The following lessons were learned from this experience (Moorman 2008):

- Water leakage did not damage the core elements
- Water ingress is a possible scenario, but it can be controlled
- 3% of water volume leads to a positive reactivity, but it did not happen in the AVR
- The steam generator should not be positioned at a higher level than the reactor

## 4. High Temperature Reactor Core

The excellent characteristics of the HTGR result from the combination of helium, graphite, and coated particle fuel. It is possible to not only run the reactor in conjunction with the most advanced steam power plant, but also to contemplate the direct use of the helium in gas turbines and to consider the application of the very hot helium to carry out important industrial processes, giving the HTR great development potential (Massimo 1976).The HTGR can change from a low enriched uranium cycle with a very high burnup to a thorium cycle, possibly sacrificing burnup for a high conversion ratio to match changing circumstances with respect to the cost and availability of uranium ore. Reactors with prismatic fuel can be refueled either on-load or off-load; this choice determines the frequency of these operations because off-load refueling cannot take place too frequently. In pebble-bed reactors the refueling is continuous. The fuel pebbles can flow many times through the reactor before being discarded, depending on burnup and the physical condition of the pebble (Massimo 1976). The HTR system must be able to shut down the reactor and keep it in under critical condition any moment of its life, compensating also the reactivity introduced by the worst possible accident. Besides that, enough excess reactivity must be provided to regulate the system and allow for Xenon override and burnup compensation. The control rod requirement is dependent on the reactor composition, and therefore has to be calculated as a function of burnup. In reactors with continuous reloading, the control requirement has to be calculated during the running-in period and in the equilibrium condition (Massimo 1976).

Control rods are often classified according to their function as explained by Massimmo:

- Regulating rods are used for automatic control
- Shim rods are used for compensating slow reactivity changes (e.g., burnup)
- Safety rods are used for scram.

## 5. Burnup, Deep-Burn and Transmutation

HTGR is a thermal reactor using helium as a coolant (essentially transparent to neutrons and does not degrade neutron energies) and graphite as the moderator. Since helium does not interact with neutrons, the temperature feedback is the only significant contributor to the power coefficient. This provides for the stable operation of the reactor (Rodriguez and Baxter 2001). Other key HTGR features are its TRISO fuel particles with ceramic (SiC) coatings that can withstand much higher temperatures and provide large thermal margins to ensure fuel integrity during loss of coolant events. The spherical shape of the TRISO particle is able to accommodate the production of fission gas products within the coated particles with lower resultant internal pressure. The composite effect of various buffer layers is that the particles are able to tolerate high levels of irradiation and allow high burnup.

## 5.1. Burnup

The burnup calculations deal with the time evolution of reactor parameters over long periods involving the complete lifetime of the reactor. For these calculations the time derivative of the neutron flux can be neglected and the static form of the Boltzmann equation (or of the diffusion equation) can be used. For each isotope it is possible to write a balance equation relating the loss and production contributions to its concentration (depletion equations) as explained by Massimo (1976).

$$\frac{dN_k}{dt} = \Phi \sum_{i=l}^{m} N_i \sigma_{fi} y_{ik} + \Phi \sum_{s=r}^{q} N_s \sigma_{as} \gamma_{sk} + \\ + \sum_{j=n}^{p} N_j \lambda_j \alpha_{jk} - \lambda_k N_k - \Phi N_k \sigma_{ak}$$

where

$N_k = atomic\ concentration\ of\ isotope\ k$,

$\Phi = flux$,

$\sigma_{fi} = isotope\ i\ fission\ cross-\mathrm{sec}tion$,

$\sigma_{ai} = isotope\ i\ absorption\ cross-\mathrm{sec}tion$,

$\lambda_i = isotope\ decay\ cons\ \mathrm{tan}\ t$,

$y_{ik} = yield\ of\ isotope\ k\ due\ to\ a\ fission\ in\ isotope\ i$,

$\gamma_{sk} = probability\ that\ a\ neutron\ absorption\ in\ isotope\ s\ produces\ isotope\ k$,

$\alpha_{jk} = probability\ that\ the\ decay\ of\ isotope\ j\ produces\ isotope\ k$.

The first summation (index *i*) extends to all fissionable isotopes. The second summation (index *s*) extends to those isotopes that can produce isotope *k* after a neutron absorption. The third summation (index *j*) extends to those isotopes whose decay product can be isotope *k*.

All cross-sections here are one-group values, averaged over the whole energy spectrum. This equation applies to all isotopes present in the fresh fuel or produced by fission and those originating from neutron absorption or decay of other nuclides. Since the nuclei of the primary fission products generally have a considerable excess of neutrons, they are unstable and decay often via complicated chains (Massimo 1976). The number of isotopes to be treated in a burnup calculation is on the order of 100. They include heavy nuclides, fission products, and the isotopes derived from them by neutron absorption and decay.

## 5.2. Deep-Burn and Transmutation

HTGR has several characteristics that enable transmutation of transuranic elements from commercial spent nuclear fuel. t has a higher epithermal component than LWRs, allowing for better plutonium utilization (neutron energies that produce plutonium fission).The first advantage is the high thermal flux and large mean free path of neutrons in the graphite. The second is the high thermal conductivity of the graphite matrix. In a LWR, the reactivity of fuel is directly linked to the moderator density of the nearby coolant. U-238 free fuels with high plutonium concentrations are generally known to give a large thermal flux depression resulting in a power spike within a fuel lattice. This power spike can potentially increase the heat flux past the critical heat flux in an LWR because the water coolant is a two-phase fluid. However, this is not an issue for an HTGR because the helium coolant is a single phase fluid (Bays et.al. 2006).

Gas-cooled reactors, either prismatic or pebble bed, because of the high burnup capability of TRISO particles provide an attractive means to deal with the current and future used fuel from nuclear reactors as an efficient and cost-effective alternate to disposal in a permanent repository. Because of the high level of fissile fuel utilization and transmutation of fertile TRU, the reactor has been called the Deep-Burn High Temperature reactor, DB-HTR (Rodriguez et.al. 2003).

In the deep burn concept, unburned uranium in the LWR waste is separated and recycled. The transuranic actinides left behind are packaged and loaded as HTGR fuel. Some transuranic actinides are fissionable and become driver fuel in the DB-HTR. Others are not, but behave as fertile fuel, becoming fissionable by neutron capture and decay, thus sustaining the nuclear reaction in the deep burn HTGR, and eventually being fissioned. Some of the fission products and the small amount of transuranic actinides that are still left after the deep burn reactor cycle could further be destroyed or stabilized in an accelerator driven deep burn system. Either way, deep burn reduces the toxicity levels and the duration of that toxicity in the LWR waste repository (Rodriguez et.al. 2003, Venneri and Hamilton 2010).

The transmutation of the TRU materials from LWR spent fuel can be extended by reprocessing the deep-burn spent fuel and using it as fuel for fast spectrum reactors to reduce the long-lived material to near zero. After burning in the DB-HTR, the fuel compacts are removed from the irradiated graphite fuel blocks and packaged for disposal. The Deep Burn program is also investigating the options for reuse of graphite fuel blocks or reducing their contamination sufficiently to allow disposal as lower level waste (Versluis et.al. 2008).

The key advantages to HTGR fuel are higher power density, higher burnup, higher temperature and temperature gradients, and allowing higher neutron fluence, compared to conventional LWRs (General Atomics 1996).The advantages are primarily because of the higher thermal conductivity and melting temperature of the graphite matrix. The hexagonal structure (for the prismatic core) of the HTGR assembly allows for symmetric loading of different fuel types into the lattice. The thermal mean free path of graphite is large enough that there is negligible spectrum variation between neighboring fuel compacts of different isotopic composition. Graphite being a moderator causes gradual thermalization. Each collision of a neutron with a carbon atom in the graphite slows down the neutron much less than a collision with a hydrogen atom in the case of a water reactor. As a result, there is an abundance of opportunities for neutrons in the HTGR to interact with non-fissile actinides that exhibit large resonance capture cross sections in the thermal and epithermal energy range.

The large values of resonance capture cross sections for the non-fissile LWR waste transuranics, coupled with the fact that these actinides are fertile (they become fissile upon absorption of neutrons), makes it possible to use them as (1) burnable poisons - to compensate for excess reactivity of the fissile elements, and (2) resonance absorbers - to provide prompt negative feedback (Rodriguez et.al. 2003, Venneri and Hamilton 2010). In nuclear reactors, these roles are traditionally taken by parasitic absorbers such as boron or erbium. However, in deep burn, they are taken by the fertile transuranics, which are transmuted to fissionable isotopes in the process (Rodriguez et.al. 2003, Venneri and Hamilton 2010, Versluis et.al. 2008).

Neutron capture in the fuel leads to creation of plutonium, minor actinides, including americium, curium and fission products. Americium is the second most abundant transuranic element after plutonium in the transuranic vector. The unique characteristic of Am-241 is that, unlike the odd numbered uranium and plutonium isotopes, Am-241 is fertile but it transmutes into the fissile isotope Am-242 (Bays et.al. 2006). Transmutation effect of thermal neutron capture in neptunium and americium results in fertile plutonium isotopes. This accumulation of fertile plutonium reduces the attractiveness of the waste for weapons usage (exhibiting high proliferation resistance).

The fuel in the reactor is used until the Uranium-235 content becomes too low to sustain the chain reaction before being moved to the spent nuclear fuel pool. Nuclear reactor waste is significantly more hazardous than natural uranium ore, even after a million years. But treating or transmuting(uranium separated from the waste and plutonium and minor actinides transmuted) could add significant value to the disposal process, as small thermal cross sections of minor actinides make them less likely to fission, and plutonium has tremendous energy, which is a proliferation concern. Transmutation could lead to reduction of waste volumes into more stable forms that are less toxic and not attractive for use in nuclear weapons.

Cesium and strontium have been a concern of the nuclear industry because of their long half-lives (about 30 years) which make them hard to transmute. They must therefore be isolated and contained until it is almost all decayed. Technetium and iodine isotopes of concern are also very long lived isotopes but could be converted into stable isotopes if enough neutrons are available, which is similar to transmuting plutonium and the minor actinides (Rodriguez and Baxter 2001). Rodriguez and Baxter (2001) have also shown the potential impact of removing and transmuting the plutonium and actinide wastes in Figure 14.

Neutrons can fission atoms of the irradiated materials, creating atoms of lower atomic weight that are generally more stable and, when absorbed by the irradiated atoms, can form heavier atomst hat are likely to fission when hit by other neutrons. Both capture and fission cross sections are an order of magnitude higher for thermal neutrons than for the fast neutron spectrum, meaning they could be effectively utilized to essentially destroy all of the proliferation concern plutonium isotopes (Pu-239 and Pu-241).

In fast reactors, the fission cross-sections in fast neutron region are smaller than in the thermal region; however fission-to-absorption ratios are higher in the fast neutron region. Minor actinides are very hard to fission in any energy level because of their small cross sections. Still, their relative destruction rates are much better than in thermal reactors, provided they are exposed to high neutron flux.

The benefits of transmutation could be utilized by transmuting plutonium in thermal spectrum and further utilizing fast reactors for destroying minor actinides thereby leading to a stable nuclear waste form.

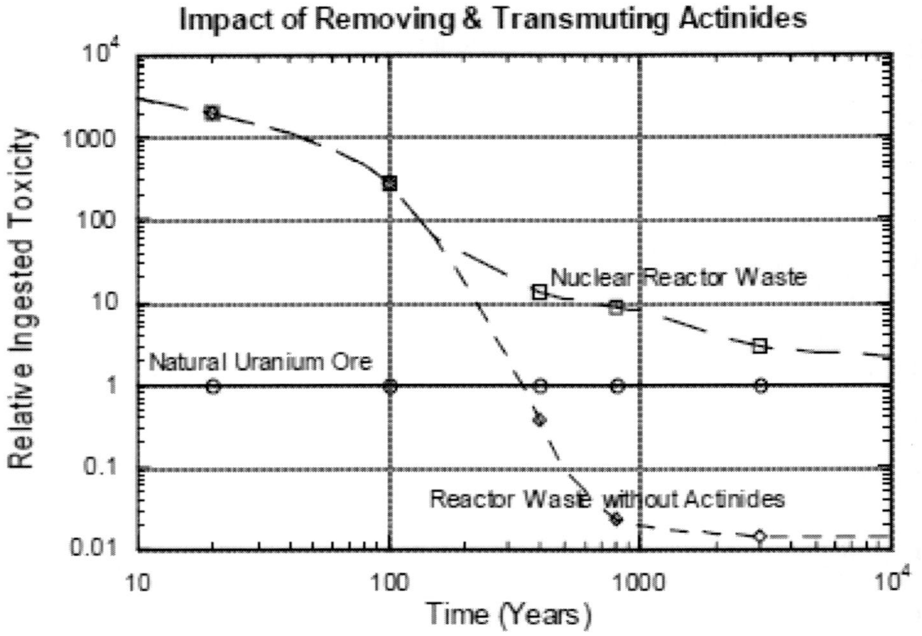


Figure 14. Impact of removing and transmuting actinides (Rodriguez and Baxter 2001).

## 6. Process Heat Applications

The strategic goal of the HTGR is to broaden the environmental and economic benefits of nuclear energy in the United States from power production to meet the energy needs and also by demonstrating its applicability to market sectors not being served by LWRs. A large number of industrial applications exist that can be envisioned with a ROT of 750°C. These are referred to as near-term applications, where as applications requiring higher ROTs than 700°C are referred to as longer-term applications.

Process industrial applications (such as hydrogen production via steam methane reforming of natural gas and high temperature steam electrolysis, substitute natural gas production, oil sands recovery via steam assisted gravity drainage, coal to liquid production, natural gas to liquids production, methanol to gasoline production, ammonia production, ex situ oil shale, and in situ oil shale) will have large process energy requirements and, in the future, represent potential markets for HTGRs. Long term (>700°C) and near term (<700°C) goals for process applications that the HTGR could support are shown in Figure 15.

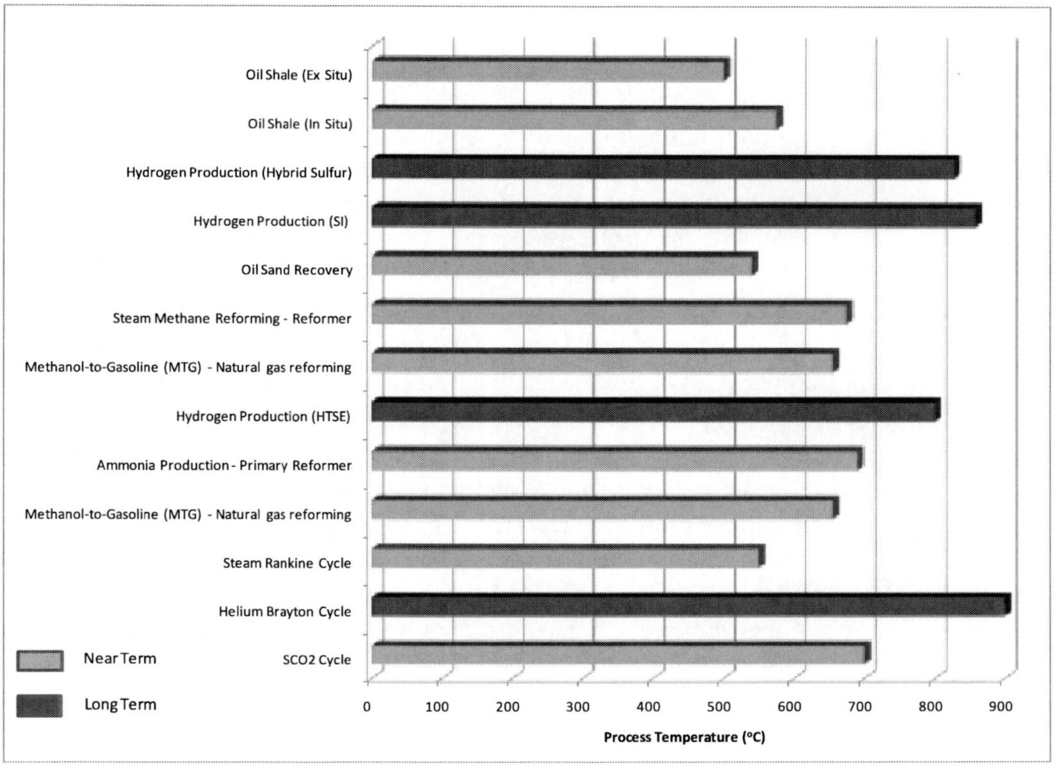


Figure 15. Required Temperature for Process Heat Application.

# 7. Challenges with HTGR

Currently following challenges are identified with HTGR:

- Material challenges at higher ROTs.
- Closed Helium Brayton power conversion system
- Graphite dust generation and oxidation issues.
- Tritium permeation and contamination of integrated system and products—tritium is mainly produced by neutron capture reaction in Boron from graphite impurities and neutron activation of He-3.
- Long schedules requirements before an HTGR could be licensed and deployed, and inability to predict future requirements, leading to decreased private sector interest.

# Discussion and Conclusion

The design of the modular HTGR is to a large extent governed by the requirement of safe decay heat removal. The power density of the core is chosen such that for all operation and accident conditions the after heat can be transferred to the environment by heat conduction and natural convection for pebble bed core and the after heat removal of the reactor vessel is by free convection air cooling for prismatic core without exceeding maximum fuel temperature (Kroeger 1988).He coolant is inert and remain single phase under all conditions. He coolant flows downward through coolant channels in the graphite fuel elements in the reactor core and then further through the cross vessel to the power conversion system. The gas coolant (He) is neutral with respect to reactivity. With the change in coolant temperature there is no direct reactivity effect, only an indirect effect due to temperature change in the moderator which is very slow due to large thermal inertia (Dam, 1992). The use of coated particles enables fuel to withstand very high temperatures (1600°C) before failure and exhibit high retention of fission products for long periods of time. Graphite core provides high heat capacity leading to very slow thermal response (heatup or cooldown), and structural stability at very high temperatures (LaBar, 2002), leading to substantial time on the order of days compared to minutes for other reactors, with reactor operators to take the required actions in order to restore normal operating conditions. This inherent feature makes the high temperature gas cooled reactor very attractive and will certainly be of tremendous assurance for reactor safety, when it goes through licensing. The HTGR will also produce less heavy metal radioactive waste because of high thermal efficiency and high fuel burnup. The HTGR fuel has high proliferation resistance because of low volume fractions of fissile (plutonium and uranium) fuel and the refractory characteristics of TRISO particles which make it very hard to retrieve fissile materials. The HTGR is therefore highly proliferation resistant (LaBar 2002).

Deep Burn HTGR is a robust and intrinsically safe way to burn transuranic actinides from LWR waste. However, beyond robustness and safety, deep burn is an effective way to (1) reduce toxicity of the waste in a repository to levels below those of natural uranium in uranium mine, while (2) converting waste energy into useful electric energy. Deep Burn could provide profound benefits in terms of better use of uranium resources, reduction in long term nuclear materials proliferation risks and high level storage requirements.

In an accident scenario such as loss of coolant accident, heat production will exceed the heat removal rate. In order to limit this rise in temperature, a high thermal inertia reactor such as an HTGR is very desirable. With the increase in core temperature the inherent negative reactivity due to fuel heat-up will drive the reactor towards the shut down stage, causing the reactor to reach decay power levels after a few minutes. The HTGR design has two active systems that can be used to remove decay heat, as explained by LaBar (2002): the power conversion system and the shutdown cooling system. In the event the active system fails, the RCCS surrounding the reactor vessel will be able to remove the decay heat and keep the reactor safe through conduction (decay heat to pressure vessel) followed by transfer of heat through radiation (pressure vessel to RCCS) and natural convection in the RCCS. Having such a feature is not an add-on but an essential requirement.

The refractory nature of the graphite and its enormous thermal capacity coupled with negative reactivity temperature coefficients makes the HTGR particularly insensitive to power excursion and loss of coolant accidents. The combination of solid moderator, completely ceramic core, chemically and neutronically inert coolant, high thermal capacity, and good fission product retention by the coated particles make the HTGR a very safe and reliable reactor to meet future demands.

## ACKNOWLEDGMENTS

References and findings herein to any specific commercial product, process, or service by trade name, trademark, manufacturer, or otherwise, does not necessarily constitute or imply its endorsement, recommendation, or favoring by the U.S. Government, any agency thereof, or any company affiliated with Idaho National Laboratory.

## REFERENCES

Angelino G., *Perspectives for the Liquid Phase Compression Gas Turbine*, Journal of Engineering for Power, Trans. ASME, Vol. 89, No. 2, pp. 229-237, 1967.

AREVA NP Inc., *NGNP with Hydrogen Production Preconceptual Design Studies Report*, Executive Summary; 12-9052076-000; June 2007.

Bays, S., Sabharwall, P., Herring, S., and Weaver, K., *Thermal Analysis for a Heterogeneous VHTR Transmutation Fuel Block as a Function of Burnup*, Proceedings of ICAAP'06, June 4-8, 2006.

Beck, J.M., Garcia, C.B., and Pincock, L.F., *High Temperature Gas cooled Reactors Lessons Learned Applicable to the Next Generation Nuclear Plant*, INL/EXT-10-19329, 2010.

Collins, J.W., *Next Generation Nuclear Plant Project Technology Development Roadmaps: The Technical Path Forward for 750-800°C Reactor Outlet Temperature*, INL/EXT-09-16598, 2009.

Crozier J., General Atomics, *Engineering Services for the Next Generation Nuclear Plant (NGNP) with Hydrogen Production Test Plan-Steam Generator for 750°C Reactor Outlet Helium Temperature*; 911174, Revision 0; December 16, 2008.

Dam V. H., *Physics of Nuclear Reactor Safety*, Rep. Prog. Phys, 11, 2025-2071, 1992.

Dewson, S. J., and Thonon, B., *The Development of High Efficiency Heat Exchangers for Helium Gas Cooled Reactors*, Report # 3213, International Conference on Advanced Nuclear Power Plants (ICAP), Cordoba, Spain, 2003.

DOE, *A Technology Roadmap for Generation-IV Nuclear Energy Systems*, GIF-002-00, U.S.DOE Nuclear Energy Research Advisory Committee and the Generation IV International Forum, December 2002.

DOE, *DOE Fundamentals Handbook – Thermodynamics/Heat Transfer/Fluid Flow*, Volume 1of3, DOE-HDBK-1012/1-92, 1992.

DOE, *Probabilistic Risk Assessment for the Standard Modular High Temperature Gas-cooled Reactor*, DOE HTGR-8601, Rev. 5, General Atomics, April, 1988.

Dostal, V., Driscoll, M.J., and Hejzlar, *A Supercritical Carbon Dioxide Cycle for Next Generation Nuclear Reactors*, MIT-ANP-TR-100, 2004.

Gougar, H. D., and Schultz, R., *Next Generation Nuclear Plant Methods Technical Program Plan*, Document ID: PLN-2498, 2010.

*GT-MHR Conceptual Design Description Report*, General Atomics, Project No. 7658, 1996.

*Heatric™ Homepage*, available online at http://www.heatric.com accessed on 20th January, 2011.

Hosegood, *An independent engineer's view of the modular HTGR*, Nuclear Engineering and Design, Vol. 109, pp. 221-226, 1988.

Idaho National Laboratory, *HTGR Mechanistic Source Term White Paper*, INL/EXT-10-17997.

Koekemoer W., Westinghouse Electric Company LLC, *Next Generation Nuclear Plant Report onUpdate of Technology Development Roadmaps for NGNP Steam Production at 750°C-800°C*;NGNP-TDI-TDR-RPT-G- 00003 Revision 0; April 2009.

Kroeger P G, *Roc. Int. Top. Meeting Safety of Next Generation Power Reactors* (Seattle) pp. 206-20, 1988.

LaBar, M.P., *The Gas Turbine – Modular Helium Rector: A Promising Option for Near Term Deployment*, GA-A23952, 1-8, 2002.

Lohnert, *The consequences of water ingress into the primary circuit of an HTR-Module – from design basis accident to hypothetical postulates,* Nuclear Engineering and Design, Vol. 134, pp. 159-176, 1992.

MacDonald, P.E. et. al., *NGNP Preliminary Point Design-Results of the Initial Neutronics and Thermal Hydraulic Assessments*, INEEL/EXT-03-00870, 2003.

MacDonald, P.E., *NGNP Preliminary Point Design – Results of the Initial Neutronics and Thermal-Hydraulic Assessments*, INEEL/EXT-03-00870, 2003.

Massimo, L., *Physics of High Temperature Reactors*, Pergamon Press, 1976.

McKellar, M.G., Boardman, R.D., Patterson, M.W., *Power Cycles for the Generation of Electricity from a Next Generation Nuclear Plant*, Idaho National Laboratory, TEV-694, 2009.

Melese, G. and Katz, R., *Thermal and Flow Design of Helium Cooled Reactor*, American Nuclear Society, ISBN: 0-89448-027-8, 1985.

Moorman, R., *A Safety Re-evaluation of the AVR pebble bed reactor operation and its consequences for future HTR concept*, Berichte des Forschungszentrums Jülich; 4275, ISSN 0944-2952, 2008.

NIST, *NIST Chemistry Webbook*, webbook.nist.gov.

Oh, C. H., Davis, C., Siefken, L., Moore, R., NO, H. C., Kim, J. , Park, G. C., Lee, J. C. and Martin, W. R., 2006, *Development of Safety Analysis Codes and Experimental Validation for a Very High Temperature Gas-Cooled Reactor,* Final Report, Idaho National Laboratory, INL/EXT-06-01362, March 2006.

Oh, C.H., Kim, E.S., No, H.C., Cho, N.Z., *Experimental Validation of Stratified Flow Phenomena, Graphite Oxidation, and Mitigation Strategies of Air-ingress Accident*, INL/EXT-10-20759, 2010.

Pappano, P., *Nuclear Graphite For High Temperature Reactors*, Modular HTGR Technology Course(PowerPoint presentation), Atlanta, 2009.

Rodriguez C., Baxter, A., McEachern, D., Fikani, M., Venneri, F., *Deep-Burn: making nuclear waste transmutation practical*, Nuclear Engineering and Design 2805, 1-19, 2003.

Rodriguez, C., and Baxter, A., *The Value of Helium-Cooled Reactor Technologies for Transmutation of Nuclear Waste*, 9th International Conference on Nuclear Engineering, April 8-12, 2001.

Schultz et al., *Next Generation Nuclear Plant Methods Technical Program Plan*, INL/EXT-06-11804, Rev 0.26, September 2006.

Shah, R. K., and Sekulic S.P., *Fundamentals of Heat Exchanger Design,* John Wiley and Sons, 2003.

Sherman, S. R., and Chen Y., *Heat Exchanger Testing Requirements and Facility Needs for the NHI/NGNP Project, WSRC-STI-2008-00152*, April 2008.

Stone & Webster Engineering Corp., *Preliminary safety information document for the standard MHTGR*, HTGR-86024, Rev. 13, September 1992.

Takeda, T. and Hishida, M., *Studies on Molecular Diffusion and Natural Convection in a Multicomponent Gas System*, International Journal of Heat and Mass Transfer, Vol. 39, No. 3, pp. 527–536, 1996.

Venneri, F., and Hamilton, C., *Fuel Cycle Analysis of Deep Burn HTRs used for LWR TRU Management*, Proceedings of HTR 2010, Prague, Czech Republic, October 18-20, 2010.

Versluis, R.M., Venneri, F., Petti, D., Snead, L., McEachern, D., "*Project Deep Burn: Development of Transuranic Fuel for High-Temperature Helium Cooled Reactors*," Proceedings of the 4th International Topical Meeting on High Temperature Reactor Technology, Washington, D. C. September 28 – October 1, 2008.

In: Energy-Environment-Economics
Editor: Riad Benelmir
ISBN: 978-1-62948-862-2


*Chapter 7*

# MATERIALS AND COMPONENTS FROM NPP DECOMMISSIONING: OPPORTUNITIES OF INVESTIGATION BY NEUTRON TECHNIQUES

***Massimo Rogante****
Rogante Engineering Office, Italy

## ABSTRACT

Lifetime prolongation and safety are key issues in planning and management of Nuclear Power Plants (NPPs), as knowledge of materials' conditions and related ageing processes is necessary either to ensure high levels of dependability or to deal with severe natural and plant–centred events. Conventional investigations, nevertheless, can present a lack of information, thus the data obtained require to be complemented.

Decommissioning is not only a mere activity of installation's conclusion, but also an important opportunity to get materials, parts and worked systems submitted for decades to degradation and ageing. In dismantling a NPP, so, new diagnostic activities and measures for the installation of new components can be suggested to enhance safety and reliability of NPPs.

Neutron investigations have lately become a progressively more considerable probe across a wide range of disciplines and they can reveal important properties about materials. Neutrons are becoming ever more helpful in the non–destructive characterisation of industrial materials and components of nuclear/traditional interests: applications of neutron–based methods, moreover, are being developed in a range of new sectors [1, 2].

In this paper, the following neutron techniques are briefly described, which can be exploited in the NPP field: small angle neutron scattering (SANS) for the micro- and nano-structural characterization as well as the radiation damage assessment; neutron diffraction (ND) for the determination of internal and sub–surface residual stresses (RS); prompt gamma activation analysis (PGAA) for qualitative and quantitative analyses of the material's constitutive main and trace elements. Some applications are also described. The available results, together with those from comparisons between naturally aged, new (non-aged) components and reference basic materials, can permit investigating in a

* Phone + 39 0733 775248; fax + 39 0733 1941315; email: main@roganteengineering.it.

supplementary way NPPs parts, helping to increase safety levels and service life of nuclear installations, with remarkable positive impact by either environmental or economical point of view.

## INTRODUCTION

Decommissioning of NPPs or nuclear installations means the termination of operations and the entire removal of the plant from service (the so called "green field") or its transformation into an out–of–service condition. The involved phases include planning, post operation, characterisation, decontamination, dismantling and safe storage. In dismantling, in particular, equipment and structures with varied dimensions and sizes and made of diverse materials are cut and segmented. This is a unique chance to perform innovative analyses of really aged materials and components, such as pipes, plates, welding, irradiated materials of the reactor vessel and cements. The outcome of these diagnostics can help explaining different aspects connected with ageing and degradation – e.g., embrittlement, thermal treatment (HT), radiations and chemical attacks produced by corrosion, stress corrosion cracking (SCC) – and generating effects towards the progress of future NPP installations. Neutron radiation damage and thermal ageing are fundamental factors determining the reactor vessel lifetime. Neutron radiation associated with elevated operative temperatures, e.g., is able to produce important nano– and micro–structural changes in metallic components, deteriorating their mechanical characteristics and favouring creep, embrittlement, hardening and swelling.

Traditional investigation methods involved by the existing norms for the NPP sector include, e.g., mechanical tests (they can measure only the macroscopic properties' changes), X–rays and mechanized ultrasonic investigations concerning the pre–service inspection and phased array ultrasonic testing [3]. The obtainable information, however, are not exhaustive and require to be completed. Resistance to ageing and creep depends notably on nano–and micro–structure (e.g., pores, precipitates and dislocation groups) formed and evolved in the metals: non–uniform degradation can accelerate crack's propagation, critically reducing the service life of components and systems [4]. Neutron techniques present significant advantages, in comparison with the classic methods of materials' analysis and fractography, contributing to solve essential problems associated with the methodological limitations of these methods.

Materials and parts after NPP dismantling (e.g., concrete as a shield of the reactor, inside components of pumps, piping, pressurizer, pumps of the primary circuit, vapour generator and welds) can be investigated by using the following techniques, with possible consequent indications for the manufacture of new components:

- *Nano- and micro–structural characterization by SANS.* Fundamental information on chemical or magnetic inhomogeneities such as matrix heterogeneities (gas bubbles, pores and precipitates) can be achieved - considering also the eventual local deviations in the composition and in the technology – supplying indications on the ageing and degradation levels of the studied materials and components
- *Internal and sub–surface RS assessment by ND.* The results, besides the evaluation of the tensile or compressive condition, consent studying the efficiency of post-welding

relaxation treatments or the effect of further heat treatments, the RS relaxation fraction at the work temperatures and after a long time and the RS concentration related to notches and cracks. A peak profile analysis can be performed, moreover, to reveal plasticity by the broadening/increase in the full width at half maximum (FWHM) of the diffraction peaks, as a measure of sub-structural evolution with plastic strain.

– Qualitative and quantitative analyses of the material's constitutive elements by PGAA.

A comparison can be performed, by adopting these neutron techniques, between components submitted to natural ageing, new components (non–aged) and reference basic materials.

# METHODS

## Small Angle Neutron Scattering

SANS is a non-destructive technique that accurately and completely characterises materials at the micro- and nano-levels, providing statistically precise information averaged over a macroscopic volume. The following parameters relative to the scattering objects (defects), in particular, can be monitored: size, concentration, volume fraction and area of interface. The theoretical bases of the SANS technique can be found in various references [1, 2, 5–8]. The same specimen can be studied or measured any number of times after either further usage or treatment. This technique, averaging over a macroscopic sample volume, provides information with high statistical precision.

Figure 1 shows, e.g., a scheme related to a SANS instrumentation, in this case that of the Budapest Research Reactor. The neutrons wavelength of this facility can be changed in the range 3-25 Å and monochromatic wavelengths can be selected by a multidisc type velocity selector. Neutrons are guided by super–mirrors to the investigated object and the scattered neutrons are detected by a $BF_3$ filled, two dimensional position sensitive 64×64 pixels detector. The scattered angle covers a momentum transfer 0.003-0.5 $Å^{-1}$ *q*–range, allowing density, composition and magnetization fluctuations in materials under test to be measured on a length scale in the range 5-1400 Å.

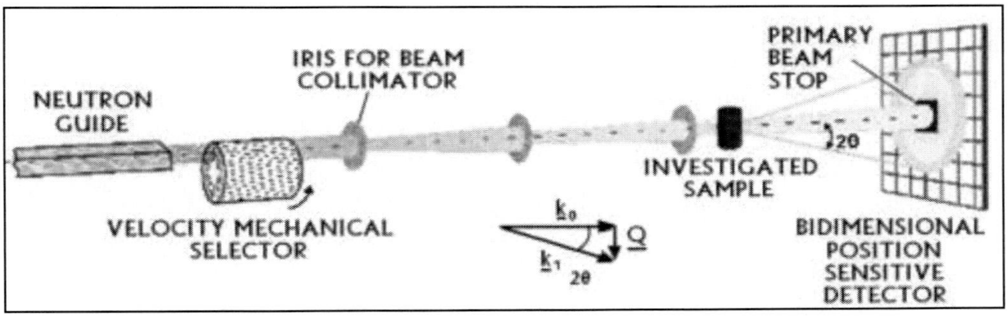


Figure 1. Scheme of a small angle neutron scattering diffractometer.

The scattered intensity at small angles is proportional to the square of the contrast, i.e., the difference between the scattering length densities of the so called matrix and the nanometric sized objects located inside the matrix [7, 8]. When the monochromatic plane wave illuminates the sample, scattered waves of variable intensities are emitted in all directions of space. The incident and scattered wave at an angle of $2\theta$ can be described by their wave vectors and $\bar{k}_1$, which defines the scattering vector $\overline{Q}$ :

$$\overline{Q} = \bar{k}_1 - \bar{k}_0 \tag{1}$$

In the case of elastic scattering:

$$|Q| = \frac{4\pi \sin\theta}{\lambda} \tag{2}$$

Where $\lambda$ is the wavelength. The measured intensity at the detector can be expressed as:

$$I(\lambda,\theta) = I_0(\lambda)\Delta\Omega\eta(\lambda)TV\frac{\partial\sigma}{\partial\Omega}(Q) \tag{3}$$

where $I_0$ is the incident flux, $\eta$ is the detector efficiency, $T$ is the sample transmission, $V$ is the sample volume and $(d\sigma / d\Omega)(Q)$ is a function known as the microscopic differential cross–section. $(d\sigma / d\Omega)(Q)$ has the dimension of $(\text{length})^{-1}$ and it is usually expressed in units of $cm^{-1}$. A Guinier fit is made subsequent to a proper calibration and radial averaging. In the Guinier region, the expression:

$$I(\lambda,\theta) = const(\lambda)\exp(-Q^2 R_g^2/3) \tag{4}$$

describes the behaviour of the scattering intensity.

In (4), $R_g$ is the average radius of gyration and $I$ the measured intensity, containing the different constant values characteristic of the instrumentation and the analyzed sample such as transmission, detector pixel size and sample volume. The Guinier model allows determining the gyration radius (that gives information on the scattering objects) and the measured intensity values versus $Q$ ($Å^{-1}$). SANS has revealed his helpfulness to investigate micro– and nano–structural defects in materials submitted to irradiation or thermo–mechanical treatments (e.g., creep, fatigue and ageing).

## Neutron Diffraction

Knowledge of directional and spatial RS distribution is necessary to understand material's behaviour, being a decisive factor for safety, quality and service life assessment. Stress concentration should be avoided in order to increase fatigue life and keep away from

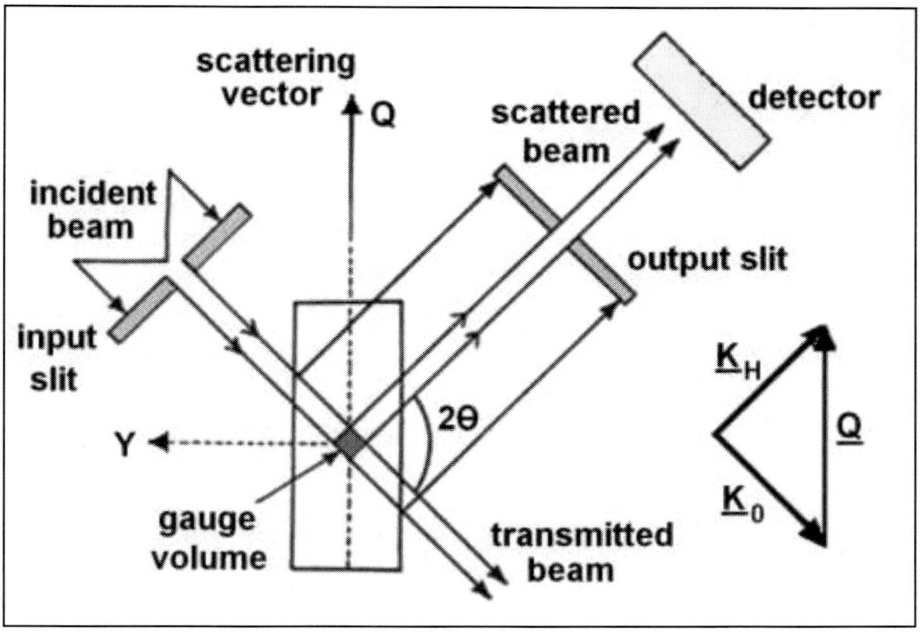


Figure 2. Scheme of a strain measurement by neutron diffraction.

material's ageing related phenomena. ND represents almost certainly the most complete method for non–destructive strain/stress measurement [1, 2, 9 and 10]; this technique is able to provide data also including the effects of undesirable thermal variations, which are difficult to be studied only theoretically. In the ND strain measurement, a collimated neutron beam having a wavelength $\lambda$ is diffracted by the investigated sample, then it passes through a second collimator and reaches the detector. Both collimators slits define the investigated volume (see Figure. 2).

The Bragg's law:

$$n\lambda = 2d_{hkl}\sin\theta \tag{5}$$

(where the integer $n$ is the diffraction order, $2\theta$ is the ample take–off angle related to the maximum of the Bragg diffracted intensity peak and $hkl$ are the Miller indices of the studied lattice planes) allows determining the interplanar distance $d_{hkl}$. Only those grains in the investigated specimen whose lattice planes are oriented to allow Bragg's reflection contribute to the scattered beam. Bragg's reflection happens as the normal to the reflecting planes bisects the incident and scattered neutron beams, hence being parallel to the variation in scattering vector $\overline{Q}$. The main axes of the strain tensor can be assessed by rotating the specimen in the instrumentation and measuring $d_{hkl}$ shift as a function of orientation. The lattice strain is calculated by using the relation:

$$\varepsilon_{hkl} = \frac{d_{hkl} - d_0}{d_0} \tag{6}$$

where $d_0$ is the $hkl$ interplanar distance in a stress–free reference material.

The stress values can be obtained, therefore, by using the relations:

$$\sigma_{ii} = \varepsilon_{ii}\left(\frac{E(1-\nu)}{(1+\nu)(1-2\nu)}\right)_{hkl} + (\varepsilon_{jj} + \varepsilon_{kk})\left(\frac{\nu E}{(1+\nu)(1-2\nu)}\right)_{hkl} \tag{7}$$

where: *i, j, k* = 1, 2, 3 and E, ν are the elastic constants of the considered material.

The FWHM of the Bragg's peak, finally, can be related to the number of dislocations and hence to the plastic strain of the investigated material. Any alteration in the FWHM from the reference distribution (on an unstressed sample) may be correlated to the enhance of plastic deformation, so the plastic zone may be assessed as the area where the FWHM is over that for the unstressed material.

## PROMPT GAMMA ACTIVATION ANALYSIS

PGAA is based on the detection of characteristic prompt gamma photons originating in (n,γ) nuclear reactions. Every atomic nucleus, apart from $^4$He, may undergo a (n,γ) reaction with diverse probabilities and the energies of the emitted gamma photons are characteristic for every given isotope. The gamma peaks intensities are proportional to the amount of a given isotope and this phenomenon consents to use a quantitative elemental (isotopic) analysis method known as PGAA or PGNAA (prompt gamma neutron activation analysis) [1, 11].

The prompt gamma spectra are recorded using a high purity Germanium detector surrounded with Bismuth Germanate scintillators dedicated to perform the Compton–suppressed measurement mode. The signals coming from the detectors are treated using a multichannel analyser and the spectra are evaluated using a Hypermet–PC; the element identification is based on a prompt–gamma library [12].

The quantitative analysis is based on the following considerations. The detected γ–ray intensity $A_E$ is directly proportional to the mass $m$ of a given element, the analytical sensitivity $S$ and the measurement time $t$, such that $A_E = m \cdot S \cdot t$ . The detected count rate in a certain γ peak is proportional to the number of nuclei emitting the γ photons of a given energy and it can be determined using the equation:

$$\frac{dN_P}{dt} = \int_V \int_{E_n=0}^{\infty} \frac{\mu(r)N_{Av}}{M} \sigma_\gamma(E_n)\Phi'(E_n, r)\varepsilon'(E_\gamma, r)dE_n dr \tag{8}$$

where $N_P$ is the peak area, $\mu(r)$ is the density of the element of interest in the point r, $N_{Av}$ is the Avogadro number, $M$ is the element's atomic mass, is the local neutron flux, is the detector efficiency and $\sigma_\gamma$ is the partial γ–ray production cross section, defined as:

$$\sigma_y = \sigma_0 \cdot I_y \cdot \theta \tag{9}$$

where $\sigma_0$ is the thermal neutron absorption cross section, $I_\gamma$ is the probability of γ–ray emission and $\theta$ is the isotopic abundance. In the case of a small, thin and homogeneous sample and of detector efficiency independent of the sample position, the thermal equivalent flux $\Phi_0$ is defined in order to have the same reaction rate, as follows:

$$R \equiv N \int_{E_n=0}^{\infty} \Phi(E_n) \cdot \sigma(E_n)\ dE_n = N \int_{E_n=0}^{\infty} v \cdot n(E_n) \cdot \frac{\sigma_0 v_0}{v}\ dE_n = \\ = N \cdot \sigma_0 \int_{E_n=0}^{\infty} v_0 \cdot n(E_n)\ dE_n = N \cdot \sigma_0 \cdot \Phi_0 \tag{10}$$

The peak area, thus, can be calculated from the equation:

$$N_P = \rho \cdot t_m = m \cdot S \cdot t_m = \frac{m N_{Av}}{M} \Phi_0 \sigma_\gamma \varepsilon(E_\gamma) t_m \tag{11}$$

and the analytical sensitivity can be obtained from the equation:

$$S = \frac{N_{Av}}{M} \cdot \theta \cdot \sigma_0 \cdot I_\gamma \cdot \Phi_0 \cdot \varepsilon(E_\gamma) \tag{12}$$

According to (10), lower energy (cold) neutrons mean higher thermal equivalent flux that, according to (12), means better sensitivities. The mass ratios, or equivalently the weight–percentage ratios of arbitrary elements “*X*” and “*Y*”, are independent of both the actual amount of each sample and of the exact neutron flux and they can be assessed from peak area ratios and sensitivity ratios as follows:

$$\frac{w_X(\%)}{w_Y(\%)} = \frac{m_X}{m_Y} = \frac{A_X}{A_Y} \cdot \frac{S_{\gamma,Y}}{S_{\gamma,X}} \tag{13}$$

The sensitivities are determined by standardisation or comparator measurements and they are collected in a γ–ray spectrum catalogue for PGAA [12]. The detection limits depend on the composition of the investigated specimen and they can be enhanced by increasing the acquisition time.

PGAA, due to the high penetrability of neutrons, is able to provide an average composition of the inside material, i.e., of the component in its whole. This technique can be employed, e.g., to examine whole specimens or their fragments. A principal component analysis (PCA) would evidence similes and differences between the investigated samples.

## EXPERIMENTAL AND RESULTS

Welding, as a major source of RS, frequently generates large tensile stresses whose maximum value is roughly equivalent to the material's yield strength, balanced by minor compressive RS wherever else in the considered component. Contraction of molten weld metal all through solidification is opposed by surrounding colder metal; thus, RS are created. The most significant problems linked with welding processes comprise also the formation of heterogeneities that induce, ordinarily, a higher density of mechanical defects and a lower chemical stability. The lifetime of components submitted to cyclic loadings, high temperature and/or erosion, consequently, is reduced, as creep, fatigue and wear damages in welded zones result amplified. Intercrystalline and inter–granular SCC, in addition, can be associated to narrow welded zones (e.g., in pipes and vessels) and this is due to the RS generated by the production procedure and to the eventual presence of aggressive elements [13].

The global state–of–the–art of welding project processes related to nuclear/traditional industrial applications consists practically in numerical models reproducing welding procedures and in primary interests in welding investigations to enhance welds qualities. An experimental programme was planned on a new welding project method, based not only on FEM calibrated against conventional experimental results, but also on fundamental real parameters directly related with microstructural investigations of different welded joints (various metallic materials and welding processes). The considered methods include neutron techniques. The characterisation of different types of welded joints by neutron–based methods also represents a helpful database, developing the nanoscopic safety criterion to forecast possible fracture processes in welds. The results of this project can be used in a wide range of industrial applications, signifying a satisfactory and inclusive capability of welding qualification, ready to be used in the design guidance for NPPs related welded joints [14].

Several characterization studies of welds have been carried out, till now, showing the advantages of adopting neutron techniques to obtain essential data not available by using other means and the results compared with FEM calculations. The trends in ND and SANS applications for material science are already declared to investigate monolithic and bi–metallic welds, joints based on friction welding, repair welding, effects of radiation on RS and defect formation, TIG welds. Neutron techniques make possible also:

- to investigate the factors that control the fatigue behaviour of welded components
- to use analytical methods to assess the total fatigue life of welds subjected to variable–amplitude loading histories and surface treatments, in order to find some possible routes for improving the fatigue strength
- to apply the probabilistic methods to estimate possible nucleation of micro– and nano–cracks within a joint metal (fatigue damage processes) [14].

The RS levels assessment by ND allows revealing the unknown cause–effect relationships between the current condition of a given component's material under study and its potential future failure modes under operating conditions. The understanding of such relationships consents predicting which types of failure modes are possible to occur and the material resistance to crack propagation under operating conditions.

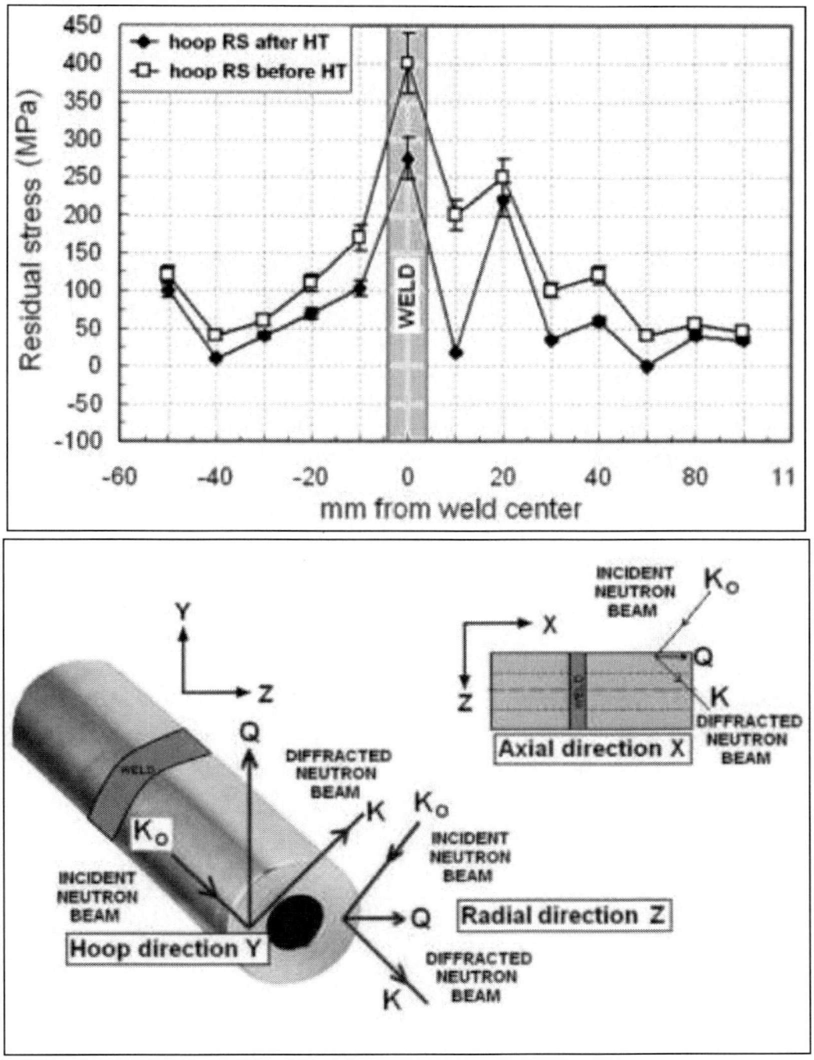


Figure 3. Hoop residual stresses (5 mm depth) determined by neutron diffraction in a 2.25 Cr1Mo ferritic arc welded pipe before and after heat treatment, with representative geometry of diffraction.

To build a suitable failure predictive model, e.g., requires the involvement of valid methods for specimen's collection, cataloguing and data analysis. An example of RS determination by ND in pipes is shown in Figure 3, which reports hoop RS (5 mm depth) assessed by ND in a 2.25 Cr1Mo ferritic arc welded pipe before and after relaxation heat treatment [15]. Neutron-based investigations of NPPs related welding can be performed with reference to dismantled parts and new welded joints made of the same constitutive materials. Welds in austenitic stainless steel pipes of NPPs can experience delayed reheat cracking in the heat affected zone and repair of existing welds is frequently necessary once major cracking arises. Knowledge of the influence of repair welds on the RS status in welded component is crucial, as well as the efficacy of the repair and the possibility to forecast the kinetics subsequent creep damage: ND technique, in this case, is capable to advance the information on the consequences of RS alterations linked with the repair action. Welded metals, whose fracture risk appears more significant than that in the base metal, involve the key problems of materials' strength. The higher risk of joints' fracture is due to mechanical stresses (non–uniform) and other factors inducing ageing (e.g., HT consequent thermal-related stress, fatigue, corrosion). Existing concepts of metal fracture, together with welded joints, are essentially based on the optical and electron microscopy data of surface structures or defects in thin slices. A number of researches from the fractography sector have been directed towards the scanning electron microscopy (SEM) to quantify the features in the non–planar fracture surface [16–17]. The main characteristics that release these quantitative data are the true fracture surface areas and the fractal properties of profiles and surfaces, which have been investigated as a part of various fracture studies. The notion of fractals was developed by Mandelbrot, who underlined the different natural self–similar structures [18]. Two most important classes of fractal structures exist, i.e., aggregated clusters of small particles (volume fractals) and systems with irregular interfaces and grain boundaries (surface fractals). SANS measurements consent to verify the fractal nature of structures, as well as to assess the fractal dimensions [19]. This technique has proved to provide crucial information in studying turbo-machinery parts providing a unique possibility of non–destructive characterization of precipitates [20-21]. For instance, a study has been performed to recognize the structural features of Udimet 520 and Udimet 720 alloys as dependent on temperature (800-900°C) and ageing time (250-6000 h) in spatial scale ~ $10^1$-$10^2$ nm, parameters responsible manly for the functional properties of these materials.

Two Inconel 738 turbine blades were investigated by SANS, one in the as-fabricated state and the other one after 25,000 hours of service. Nine positions along the blade edge (see Fig. 4) were tested by adopting the neutron beam cross-section of (3×15) $mm^2$ and using the neutron wavelength $\lambda$=2.1 Å. Figure 4 reports the volume distributions of precipitates in dependence on their size, for positions 1 and 9 of both blades.

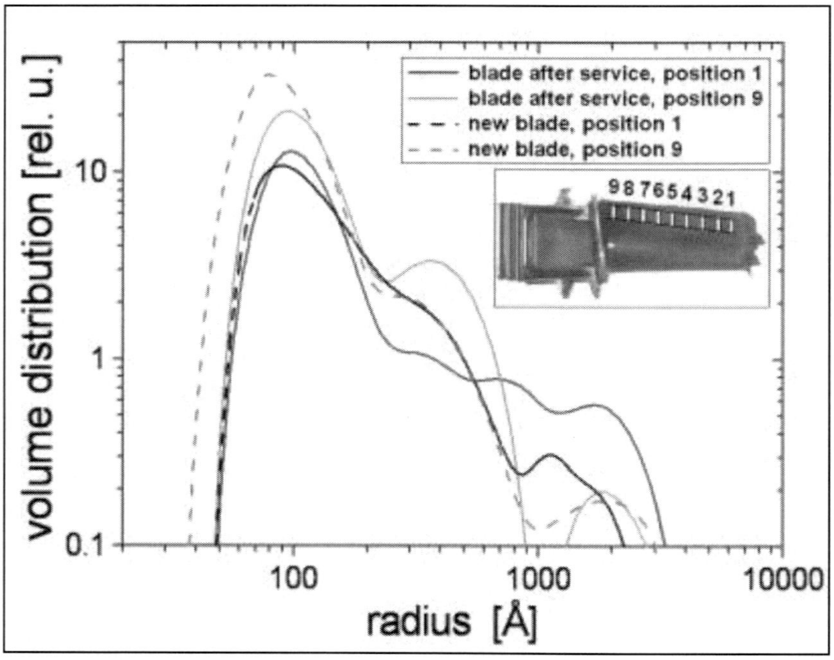


Figure 4. Volume distributions of precipitates in dependence on their size, for positions 1 and 9 of two Inconel 738 turbine blades.

Three populations of precipitates differing by their characteristic size are shown and the differences along the scan appear larger in the blade after service.

Figure 5 shows the parameter $I_o/R_G^3 \sim \varphi$ being proportional to volume fractions for small, middle (1, 2) (a) and large (3) (b) structural elements in Udimet 520 vs. ageing time (800°C), where $I_o$ is the intensity scattered from the particles fraction, $R_{Gi}$ the gyration radius and $\varphi$ the volume fraction of particles [22].

In the reactors of working NPPs, materials are submitted to elevated neutron radiation and severe thermal conditions. To trace the micro- and nano-structure alterations disjointedly due to high temperature and radiation, the structure and precipitate morphology of industrially prepared reactor vessel steel materials have been investigated by SANS, especially concerning thermal ageing. Samples cut from 15Kh3MFA steel cracked specimens prepared from the Charpy impact test and belonging to standard VVER–440 reactor materials were investigated, after a thermal ageing process at 295°C for 360 days. An anisotropic precipitate structure formed at the nanoscale level was revealed (due to the macroscopic texture), together with the inhomogeneity of the vessel raw material [23].

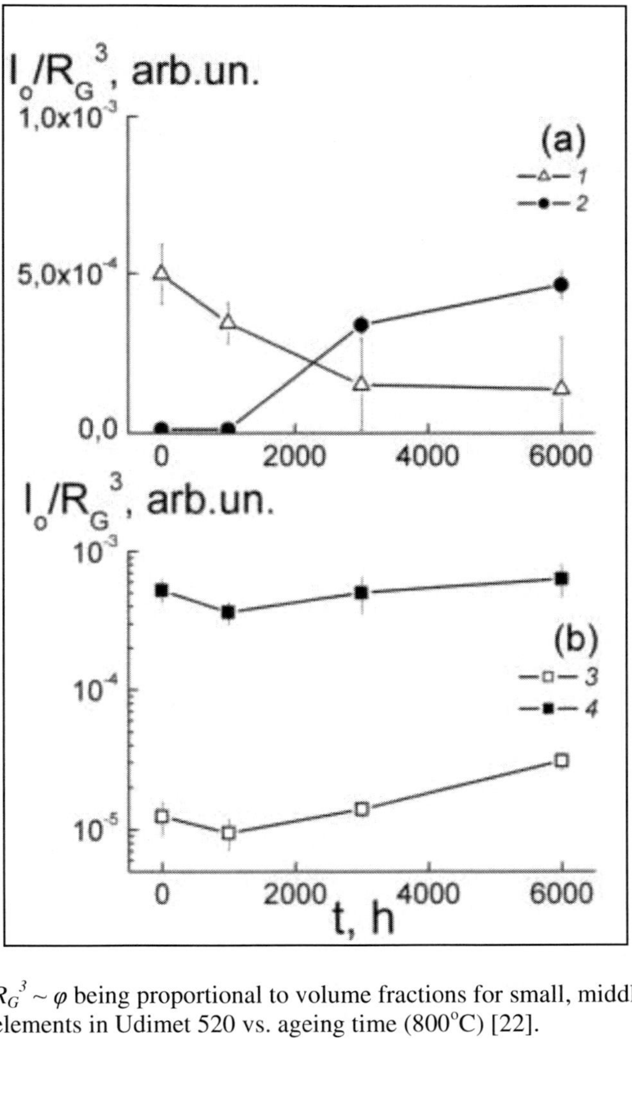


Figure 5. Parameter $I_o/R_G^3 \sim \varphi$ being proportional to volume fractions for small, middle (1, 2) (a) and large (3) (b) structural elements in Udimet 520 vs. ageing time (800°C) [22].

Thermo–mechanical stresses related with the pulse nature of thermonuclear fusion reactor can be simulated, in order to evaluate these stresses. An AISI 316L steel specimen, for example, was submitted to cyclic surface heating and simultaneously cooled by low pressure water flowing through channels across the specimen. Finite element methods and experimental (ND) RS values were assessed as a function of depth along a line from surface to central hole axis [24].

A 100 mm thickness sample cut out from a 08H18N10T austenite stainless steel plate, wire welded in electric arc, was investigated by SANS. Base metal and welded joint were thermally treated, cooling down in water from the initial temperature of 1050°C. Welded joint was scanned by a thin neutron beam to obtain SANS data for base and welded metal. The cross sections showed three fractions of particles with gyration radii: $R_{g1}$~20 Å, $R_{g2}$~9 Å and $R_{g3}$≤1 Å (point like defects). The amount of these defects in base metal was larger by factor 5, as compared to welded metal [14].

A feasibility study for a SANS investigation of P91martensitic steel (9Cr1MoVNb) specimens obtained by cutting 24" pipes (thickness=14 mm) has been carried out. The test system contains longitudinal welds in straight parts and double axial welds in the connections. The system is submitted to 3000–8000 hours accelerated creep tests, i.e., to 7,000 start cycles after 100,000 hours (max. gradient <10°C/min.). The test temperature ranges from ~545°C to ~625°C, while the test pressure is ~50bar. The HT produces the growth of some inclusions (precipitates), allowing to obtain also their characteristics (number and size distribution) by knowing the chemical nature of the precipitates (e.g., $Cr_6C_{23}$ carbides). It has been planned the investigation of samples from the original material - including welds -, as well as those related to after the test, with the aim to check microstructural changes (e.g., nano-defects and voids). The comparison of SANS cross sections for fresh prepared and aged samples is able to provide data on size concentration and distribution of new defects induced by the mechanical treatment. The main purpose of this creep tests is to determine, before reaching the 30,000 operative hours, the accelerated data that facilitate to predict the stress/strain level at 100,000 hours. A second aim is to contribute to the elaboration of constitutive equations able to perform interpretative analyses of effects (e.g., temperature, stress and strain), which is a correct step for considering parameters such as temperature, hold-time and relations with primary/secondary phases of creep. These equations would represent the strain as a function of temperature and stresses, also taking into account the voids' changes after HT [19].

Dismantling activities of NPPs can involve both mechanical and hot cutting processes such as plasma torch and oxyacetylene arc metal. Analyses performed on pipes and plates exposed to the reactor coolant steam (by use contaminated components below the free release level of $1 \cdot 10^4$ Bq/m$^2$, not yet considering radioactivity) allowed accurate estimations of the chemical and physical characteristics of the emissions produced during hot cutting tests, as well as technical parameters such as cutting time and cutting rate vs. pipe diameter or plate thickness. The results highlighted the kind of troubles that can emerge in these cutting processes, in particular focusing on a comparison of the effects of plasma and oxyacetylene cutting processes and the chemical composition of the dusts collected by filtering the gaseous emission [25–28]. In this case, SANS can be exploited to obtain useful complementary information. An investigation test of filters made of hydrophobic membrane bonded to high density polyethylene support filters and used to capture the gaseous emissions from hot cutting tests of pipes was carried out by Rogante Engineering Office using the SANS

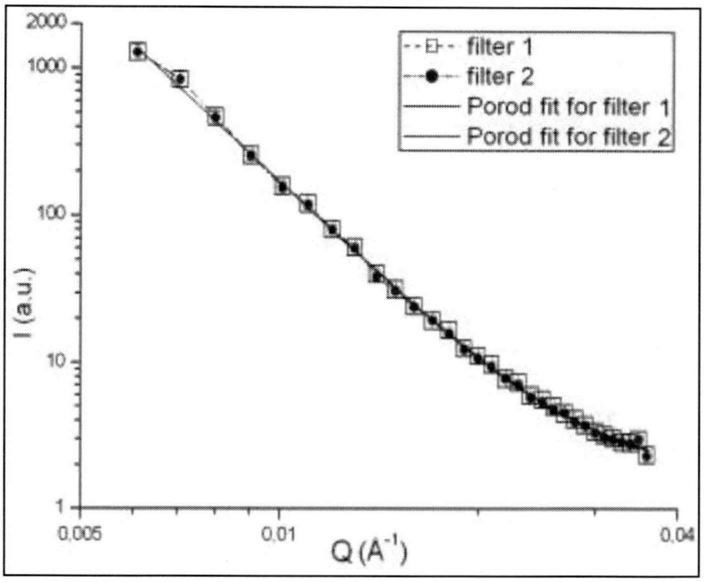


Figure 6. SANS graphs (Porod fits) of two membrane filters used to capture the gaseous emissions from hot cutting tests of pipes.

instrumentation of the BRR. Since the inhomogeneities forms distribution, with reference to this SANS facility, can be evaluated in the range $10^0$-$10^4$ Å, SANS data on the captured micronic dusts can be considered complementary to those obtained by transmission electron microscopy (TEM). While SANS data are obtained from micro– and nano–structural characteristics and they are averaged on a volume in the range between 1 $mm^3$ and 1 $cm^3$, in fact, TEM is referred to very limited surface. The sample–to–detector distance fixed for the measurement was $d$=5.6 m and the mean neutron wavelength was $\lambda$=11.22 Å. Figure 6 reports the SANS graphs related to the analysis of two filters with very similar concentrations of the captured particles.

In the same graphs, $Q$ is the momentum transfer (scattering vector) and $I$ is the intensity measured at the detector. Figure 5 shows that scattering occurs satisfactory: the Porod exponent is close to 4, which means that the scattering objects have sharp surfaces.

PGAA also can be considered to analyse filtering supports (new and after service), used for physico–chemical analyses of the gaseous emissions produced during hot cutting. This technique would consent to get information on the diverse particles typologies in the collected dusts and also to verify the eventual presence of uranic and trans–uranic elements and isotopes. PGAA spectra, in which the evident peaks are characteristic of the chemical elements present (according to the energies of their respective γ–rays characteristic peaks) would provide a qualitative analysis. Values of the elements' concentrations (expressed in %mass or in μg/g units and taking into account the exact determination of the γ–peak intensities) would give a quantitative analysis. The investigation could involve additionally metallic cut slags and HEPA filter fragments taken from the filtering system.

A final consideration concerns the analysis of cements by SANS. Study of dense cements, ceramics and related materials (e.g., cement stones) is significant to create novel advanced components for nuclear–safety related structures, with high functional properties (such as ageing resistance, cracks' formation, hardness, durability, stability of mechanical modules and ecological criteria). These materials can be developed to be considered also as corrosion resistive coatings for nuclear waste containers. Knowledge of the thermo–physical properties of high–performance concretes in broad temperature and moisture ranges is key information for realistic modelling of potential accidents in NPPs [29]. High quality quartzitic aggregates with especially low porosity and remarkable mechanical properties, for instance, have shown better performances than more porous siliceous and calcareous aggregates [30]. Conventional investigations of cements, including those for nuclear industry, use standard methodologies to get information on aspects primarily related to micro-structure, fatigue behaviour and computational mechanics. These investigations rely essentially on standard tests, theoretical studies and simulation-based analyses. Porosity, e.g., is a main parameter for cement, affecting mechanical and heat-insulation features. This parameter is generally assessed through gravimetric techniques, gas and liquid porosimetry (e.g., mercury intrusion porosimetry) and acoustic methods [31]. The central questions that can be solved by using SANS technique include the following aspects:

- researches on hydration processes (contrast variation, by using light and heavy water to comprehend the mechanisms of water molecules association and their structuring in ceramics' matrix) in the formation process of cement stones as dependent on the addition of special entities (e.g., nano–particles, polymers, low molecular components varying hydrogen bonds, such as salts, acids, bases);
- observation of kinetics of water reactions and nanoscale migration (localization) for short (seconds) and long (hours, days) times in association with strength and hardness at microscopic level of the cement stone;
- study of the influence of disperse mineral fillers to cement compositions, in the case of cement mixtures [32];
- study of porosity and of the influence of disperse mineral fillers to cement compositions, assessment of fractal dimension, size distribution and other characteristics (open, closed pores, cracks) at the scale from 1 Å to 100 Å and at large distances (from 102 Å to 103 Å, by using ultra–SANS).

A feasibility study has been carried out for a full SANS investigation of cement samples, e.g., of a ordinary Portland cement mixed with Rhoximat® PAV 22 (Vinyl acetate and vinyl versatate) copolymer resin and nano $Al_2O_3$ (γ type), with different $\gamma Al_2O_3$/PAV22 ratios, the water/cement ratio remaining stable [31]. SANS will supply, in this case, significant data on basic parameters connected with degradation, fracture and other phenomena, allowing also a more reliable lifetime assessments.

## Conclusion

Development and testing of prototype NNPs components should be connected with a study in depth of defectoscopy and microstructural characteristics of the involved constitutive materials, in order to understand more deeply their behaviour and assess their suitability. The dismantling activity of a NPP decommissioning is a real opportunity to carry out important comparative studies of fresh prepared samples and aged parts, able to supply key data.

Neutron techniques represent a valuable advance, contributing to better study the ageing process and to develop the knowledge of various safety–related properties and other important characteristics of the involved materials.

## REFERENCES

[1] Rogante, M. In: *Proceedings of 1st Italian Workshop for Industry on Industrial Applications of Neutron Techniques*; Rogante, M.; Ed; Rogante Engineering: Civitanova Marche, IT, 2008; pp 40–120.

[2] Rogante, M.; Rosta L. *SPIE Proc*. 2005, *5824*, 294–305.

[3] Fischer, E. In *Encyclopedia of Materials: Science and Technology*; *Buschow, K. H. J.; Cahn, R. W.; Flemings, M. C.; Ilschner, B.; Kramer, E.; Mahajan, S.; Veyssiere, P.; Eds.;* Elsevier Science: Oxford, UK, 2001; pp 5979–5983.

[4] Rogante, M.; Lebedev, V. T. In *Proceedings of the 31st ESReDA Seminar*; Petterson, L.; Cojazzi, G. M.; Eds.; EUR 22887 EN; JRC European Commission: Ispra, IT, 2007; pp 58–68.

[5] Glatter, O.; Kratky, O. *Small Angle X–ray Scattering*; Academic Press: London, UK, 1982.

[6] Williams, C.; May, R.P.; Guinier, A. In: *Characterisation of Materials*; Lifshin E.; Ed.; VCH Verlagsgesellschaft: Weinheim, DE, 1994; Vol. 2B of *Materials Science and Technology*, pp 611–656.

[7] Rietveld, H. M. *Acta Crystallogr*. 1967, *22*, 151–152.

[8] Rietveld, H. M. *J. Appl. Crystallogr*. 1969, *2*, 65–71.

[9] Noyan, I. C.; Cohen, J. B. *Residual Stress – Measurement by Diffraction and Interpretation*; Ilschner B.; Grant N. J.; Eds.; Springer–Verlag: New York, US, 1987; p 276.

[10] Rogante, M. *Caratterizzazione, mediante scattering neutronico, di materiali e componenti per l'impiantistica nucleare ed industriale*; Ph. D. thesis, University of Bologna, IT, 1999.

[11] *Révay, Zs*.; Belgya, T.; Kasztovszky, *Zs*.; Weil, J. L.; Molnár, G. L. *Nucl Instrum Meth B. 2004, 213, 385–388.*

[12] Révay, Zs.; Molnár, G. L.; Belgya, T.; Kasztovszky, Zs.; Firestone, R. B. *J Radioanal Nucl Chem*. 2001, *248*, 395–399.

[13] Rogante, M.; Battistella, P.; Cesari, F. *Int. J. Hydr Energy*. 2006, *31*, 597–601.

[14] Rogante, M.; Lebedev, V. T.; Kralj, S.; Rosta, L.; Török, Gy. *Multidiscipline Modeling in Mat and Str*. 2006, *2*, 419–433.

[15] Genta, P.; Rogante, M. *TOPEJ*. 2009, *2*, 12–16.

[16] Wright, K.; Karlsson, B. *J. Micr. 1983*, *130*, *37–51*.

[17] Exner, H. E.; Fripan, M. *J. Micr. 1985*, *138*, 161–178.

[18] Mandelbrot, B. B. *The Fractal Geometry of Nature*; Freeman: San Francisco, US, 1982.

[19] Rogante, M.; Cesari F. G.; Lebedev, V. T. In: *Hydrogen Materials Science and Chemistry of Carbon Nanomaterials*; Veziroglu, T. N.; Zaginaichenko, S. Y.; Schur, D. V.; Baranowski, B.; Shpak, A. P.; Skorokhod V. V.; Kale A.; Eds.; NATO Security

through Science Series – A: Chemistry and Biology; Springer: Dordrecht, NL, 2007; pp 135–144.

[20] Rogante, M.; Šaroun, J.; Strunz, P.; Ceschini, G. F.; Ryukhtin, V.; Lukáš, P.; Marinčák, V. *Kovove Mater.* 2005, *43*, 371–381.

[21] Rogante, M.; Lebedev, V. T. *Mat. Design* 2008, 29, 1060–1065.

[22] Rogante, M.; Lebedev, V. T. Small angle neutron scattering comparative investigation of Udimet 520 and Udimet 720 samples submitted to different ageing treatments, submitted to Int. Journal of Alloys and Compounds.

[23] Rétfalvi, E.; Török, Gy.; Rosta, L. *Appl Phys A.* 2002, *74 [Suppl.1]*, s1415–s1417.

[24] Albertini, G.; Bruno, G.; Fiori, F.; Rogante, M.; Rustichelli, F. In: *Modelling of steel microstructural evolution during thermomechanical treatment*; Tomellini, R.; Ed.; Steel Research EUR17585; European Commission: Luxembourg, LU, 1997; pp 185–195.

[25] Cesari, F.; Rogante, M.; Giostri, A. *Nucl.Eng. Design.* 2008, *238*, 2801–2810.

[26] Cesari, F. G.; Rogante, M.; Giostri, A. In: *Transactions, 19th International Conference on Structural Mechanics in Reactor Technology SMiRT 19*; International Association for Structural Mechanics in Reactor Technology: Raleigh, US, 2007; paper # W01/4.

[27] Cesari, F. G.; Rogante, M.; Giostri, A.; Conforti, G. In: *Proc. Int. Conf. on Nuclear Engineering ICONE14*; American Society of Mechanical Engineering: New York, US, 2006; paper 89451.

[28] Cesari, F.; Giostri, A.; Rogante, M.; Sirito, E.; Sirito, M. In: *Proc. 11th Int. Scientific Conf. on the Contemporary Achievements in Mechanics, Manufacturing and Materials Science CAM3S`2005*; Dobrzański, L. A.; Ed.; International Organising Committee of the Scientific Conferences AMME World Press: Gliwice, PL, 2005; pp 1111–1116.

[29] Vodák, F.; Černý, R.; Drchalová, J.; Hošková, Š.; Kapičková, O.; Michalko, O.; Semerák, P.; Toman, J. *Cem Concr Res.* 1997, *27*, 415–426.

[30] Černý, R.; Drchalová, J.; Rovnaníková, P. *Cem Concr Res.* 2001, *31*, 1129–1140.

[31] Rogante, M.; Selezneva, A. In: *Proceedings 2nd International Conference Mechanical Technologies and Structural Materials MTMS 2011*; Živkovic, D.; Ed.; Croatian Society for Mechanical Technologies: Split, HR, 2011; pp 39-45.

[32] Domanskaya, I.K. *Sroykompleks Srednego Urala.* 2002, *12*, 42.

In: Energy-Environment-Economics
Editor: Riad Benelmir
ISBN: 978-1-62948-862-2


*Chapter 8*

# KNOWLEDGE MANAGEMENT OF MAINTENANCE ACTIVITIES FOR POTABLE WATER DISTRIBUTION BUSINESS

***Mohammad D. Al-Tahat*** [*] ***and Hasan Al-Shoubaki***
Industrial Engineering Department, University of Jordan, Amman, Jordan

## ABSTRACT

The purpose of this paper is to develop and to present a maintenance knowledge management system that can be used to enhance management efficiency of maintenance and operations activities at water distribution companies, a big Jordanian water distribution company labeled as Muna is used as a case study. A maintenance knowledge management system that depends basically on the failure tree analysis is designed. All failures, symptoms, and causes as well as all maintenance procedures and follow up actions are diagnosed and mapped. Basic Design Principles of Expert Systems (DPES) is pursued. Consequently new knowledge management system composed of five main component is developed, these components are; Knowledge database, reporting and history, relationships, rule-based expert system, and links and interfaces. It has been found that the developed maintenance knowledge management system has a high potential for maintaining Muna's assets by increasing mean time to fail and decreasing mean time to repair.

The system has been used effectively to manage and support the operations and maintenance departments in Muna Company. The developed system has a positive impact to Muna specially when comparing it with the existing used one. It is built based on day-by-day manual data collecting, knowledge base is constructed by a knowledgeable maintenance engineer corporate with the maintenance experts; it is recommended for future similar work to collect the maintenance knowledge using questionnaires or an automated system like Supervisory Control and Data Acquisition (SCADA) system. This work adds to Muna and to similar business systems more valuable knowledge utilization system instead of the old restricted one. That helps the maintenance crew to follow up the assets in all company locations rapidly and efficiently.

---

* Corresponding author: Tel.: +962 6 5355000, Ext. 22933. Fax: +962 6 535 5588; altahat@ju.edu.jo.

**Keywords:** Knowledge management, Knowledge and information, Knowledge use/utilization, Maintenance planning, Management of potable water

## INTRODUCTION AND LITERATURE REVIEW

Improving water delivery is a pressing societal goal, and Knowledge Management (KM) has a positive role in effecting change when providing this service. Information science research has led to understanding theories and practices of knowledge use within the informing professions, also KM has the potential to influence and enhance other professional disciplines. KM is a beneficial framework to help water distribution companies manage their practices and ultimately administer quality service to their customers.

The main focus of this paper is to develop and to present a maintenance knowledge management system that can be used to enhance the efficiency of maintenance and operations activities at water distribution companies, a case study to achieve this purpose will be employed. Developing of such maintenance knowledge system is required for effective maintenance and operations practice of water providing systems that will require a substantial investment. The return on this investment is depending on the appropriateness of the developed system, the effectiveness of its implementation and the personnel commitment to it. To achieve the objectives of this paper Design Principles of Expert Systems (DPES) is followed. Many research works have been conducted to apply DPES to manage knowledge of different companies such as Su et al. (2001), Su et al. (2006), Emmanouilidis et al. (2006), Patel et al. (1996), Kerr (2007), Choi et al. (2006), Thomas et al. (2004), Reilly (2001) and many others.

Su et al. (2001) pointed that the nature and quality of users' interaction with expert systems has been an increasing concern. The design of the user interface is influenced by many factors, amongst these the mental model of the user's thinking processes, aspects of usability and the explanation capabilities of the knowledge-based expert system. Another study by Su et al. (2006) pointed that the Taiwan motorcycle industry got its start in 1952 when gas engines were installed on bicycles to create the island's first motorbikes. The diagnostic errors often result in wasting customers' time and money or producing new failure mechanisms.

These phenomena indicated that the main problem is lack of knowledge sharing and consistent direction that is referable. Flexible software solution has been developed by Emmanouilidis et al. (2006) to facilitate a wide range of monitoring scenarios, where monitoring, novelty identification and machinery diagnostics, can easily be customized. The objective of Emmanouilidis work was to develop a system capable of performing novelty detection and failure diagnosis, even in cases when very little a priori knowledge exists about the monitoring task.

Another study by Kerr (2007) pointed that the city of Lima Ohio introduced a special software as same as Computerized Maintenance Management System (CMMS) to nearly 100 personnel in its utilities department to facilitate its customer service, water treatment, water delivery, sewer collection, and wastewater treatment operations. The software replaces multiple systems and databases to provide a single, easy to-use system for computerized maintenance of its water and wastewater operations.

# DEVELOPING THE REQUIRED KNOWLEDGE MAINTENANCE SYSTEM

The proposed Knowledge Management System is structured as shown in figure 1, it is composed of the following main parts:

1. Knowledge database. Knowledge is collected from many fields and departments of the company. It includes - As shown in the upper left part of figure 1- various types of information regarding, technical details, sites coordinates, equipments, maintenance staff, and many other maintenance-related details. Database is used to create status and/or history reports about any asset. And it is used to formulate different maintenance algorithms that may use for labor efficiency calculations and designing maintenance schedules.
2. Reporting, history and calculations. This part is concerning of importing and exporting data, generating historical maintenance reports through retrieving of information resources available in the database, updating and modifying database.
3. Rule-based Expert System (ES). It is the core of the developed maintenance management system, composed as shown in figure 2, of *knowledge base* (the rules and the facts), *inference engine*, and *user interface*. It determines the symptoms, causes, and the solutions for every probable failure, and clarifies all the needed explanations regarding maintenance procedures, maintenance equipments and tools, manpower capacity, and the standard time of maintenance action. Knowledge base incorporates all maintenance knowledge of Muna Company. Maintenance procedures are decoded from maintenance experts into rules and facts by knowledge engineer. Rule -based technique is used in the inference engine which acquired the rules and facts from the knowledge base to generate decision regarding root cause of maintenance problems and their solutions, as well as to propose predictive maintenance procedures. The nature and quality of users' interaction with expert systems has a big concern. Understanding of the user's cognitive structure and the users' task is of a big concern when designing user interface. A good interface should enhance user/expert system interaction and task performance.
4. Maintenance and operations relationships. Different maintenance activities and functions should be performed according to maintenance objectives. These activities are classified into managerial maintenance activities and technical ones. Managerial activities include, planning, scheduling, supervising, controlling etc. Planning is the setup of objectives and the determination of ways to achieve them. It is defining which job to do and how to do it, Scheduling is defined as deciding when to do the job. Finally, controlling is the assurance that activity plans are executed properly. These activities are executed by maintenance manager and by heads of departments. On the other hand, technical activities include preventive maintenance activities; diagnostics and repair of failures, calibration, construction, in addition to quality control are executed by maintenance technicians and supervisors.

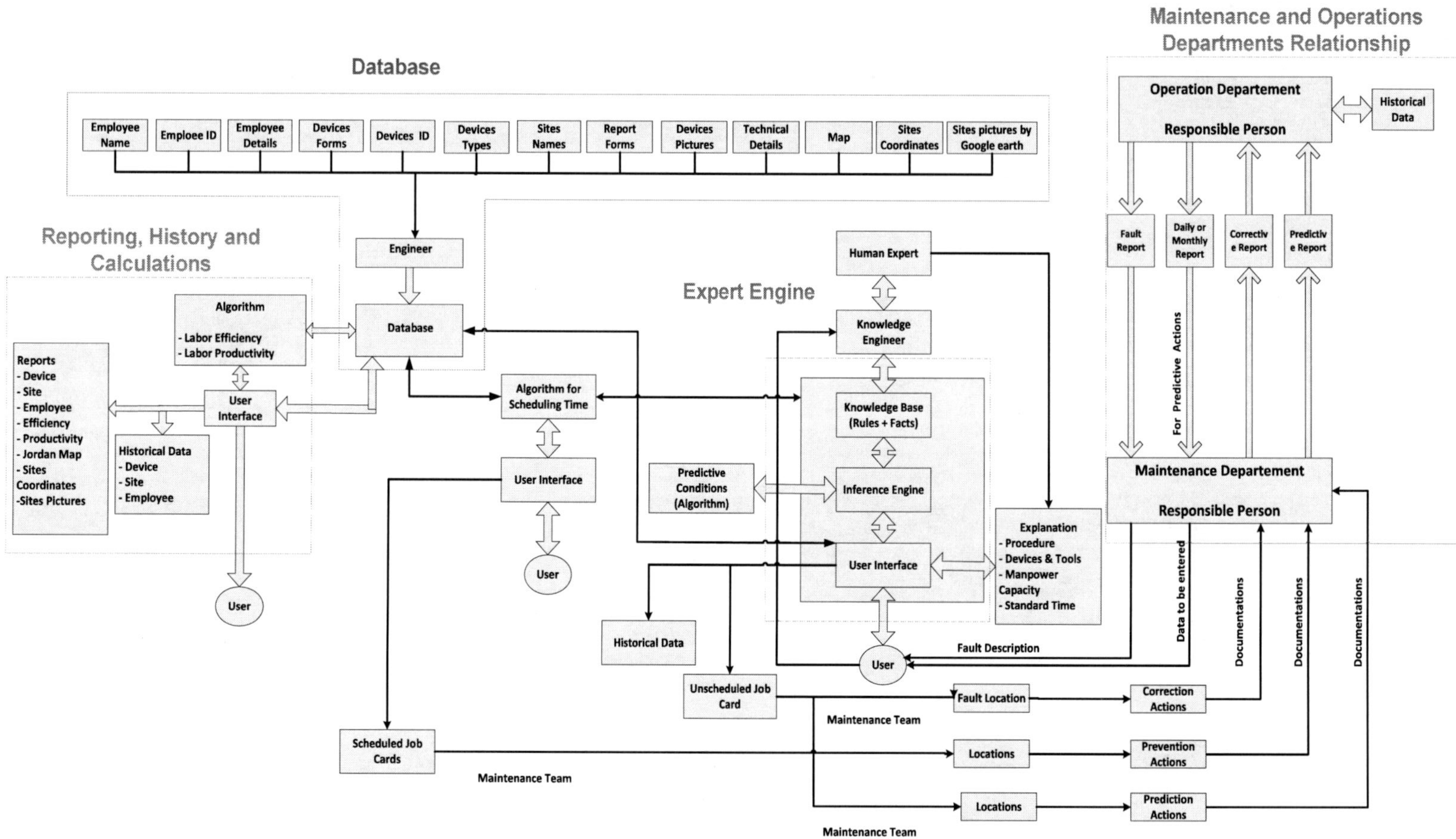


Figure 1. Main structure of the developed maintenance knowledge management system.

Operations department performs two types of reports, first, following up the failure daily reports that sent to the maintenance department to take an action to solve problems (failure), second, statistical data regarding daily and/or monthly reports are analyzed by maintenance management to create prediction maintenance programs.

5. The flask that holds all connections, interfaces, and links between the main parts of the proposed knowledge management system. Visual Basic VB programming language is used to build such flask. Visual Basic (VB) is the third-generation event-driven programming language and Integrated Development Environment (IDE) from Microsoft for its Component Object Model (COM) programming model. VB is considered a relatively easy to learn and use programming language, because of its graphical development features and BASIC heritage, that enables rapid application development (RAD) of Graphical User Interface (GUI) applications, access to databases using data access objects, remote data objects, and creation of ActiveX controls and objects. The developed Visual Basic program enables to, run maintenance algorithms on the computing agent like the computers, assists in making more effective use of Microsoft Excel sheet, provides a suitable user interface in our worksheet, and expresses and manipulate complex logic formulas when needed. Figure 3 presents the main window of the system flask coded by Visual Basic VB, the figure also shows the included icons and their explanation.

Three main types of maintenance schedules are generated out of this part of the system, these are; preventive maintenance schedule, corrective maintenance schedule, and predictive maintenance schedule.

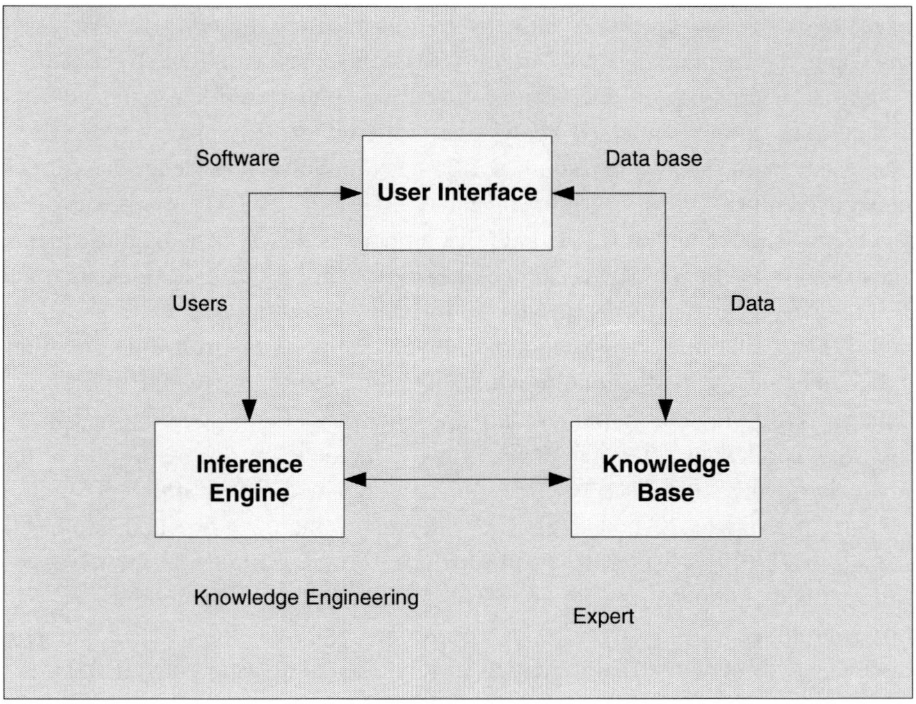


Figure 2. The building blocks of the expert system.

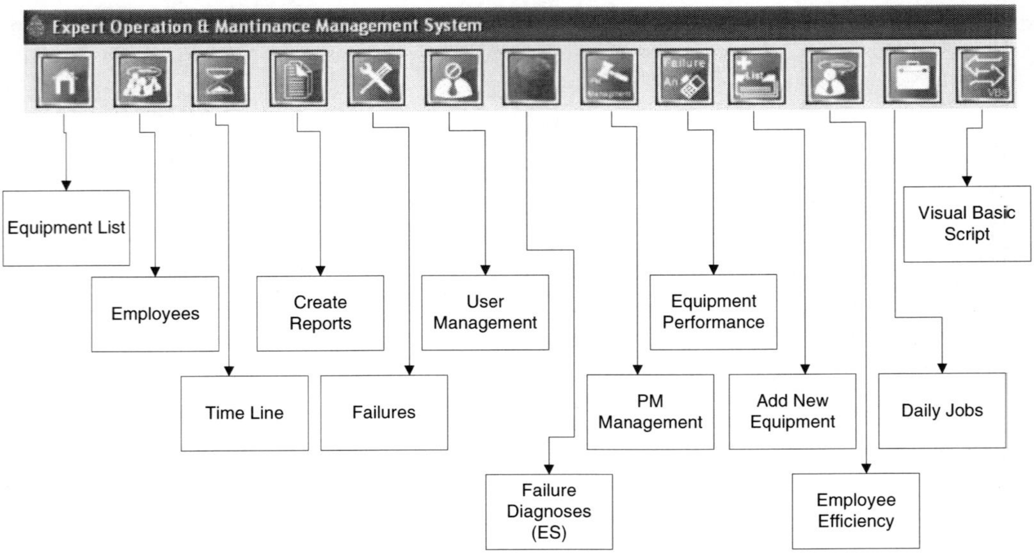


Figure 3. Icons and their explanations in the main window of system flask coded by VB.

## KNOWLEDGE ACQUISITION AND PROBLEM SOLVING

Liebowitz (1997) defines knowledge acquisition as "The process of extracting, structuring, and organizing knowledge from several knowledge sources, so that the problem-solving expertise can be captured and transformed into a computer-readable form". The knowledge base of the proposed knowledge management system involves knowledge acquisition and representation. Knowledge to be represented is extracted from the sources where it exists. These include documented rules, procedures, written text, and contents of existing databases or undocumented knowledge in the minds of human experts (Kerr, 1991). Figure 4 presents some knowledge sources used when building knowledge-based, this model is presented by Turban (1988).

Knowledge representation is a multidisciplinary subject that applies theories and techniques from other fields (Sowa, 2000), like, logic that provides the formal structure and rules of inference. Knowledge representation formalizes and organizes the knowledge. One widely used representation is the production rule, or simply rule. A rule consists of an IF part and a THEN part (also called a condition and an action, see Figure 5). "IF" part lists a set of conditions in some logical combination. The piece of knowledge represented by the production rule is relevant to the line of reasoning being developed if the IF part of the rule is satisfied; consequently, the "THEN" part can be concluded, or its problem-solving action taken. Expert System whose knowledge is represented in rule form is called rule-based expert system (Rich et al., 1991). Each rule specifies what to do, or what conclusion to draw, under a set of well-defined circumstances. Some sets of typical rules are; rules to represent facts, rules to identify symptoms, rules to suggest repair procedure, and rules to prioritize maintenance.

Maintenance Knowledge representation depends on troubleshooting of failures through diagnostics of symptoms and their root causes. A failure accompanies a number of symptoms that assigned to a number of causes, called causes of symptoms or causes of the failure. Thus a failure can be represented simply as: *Failure= {Symptoms, Causes of Symptoms}*.

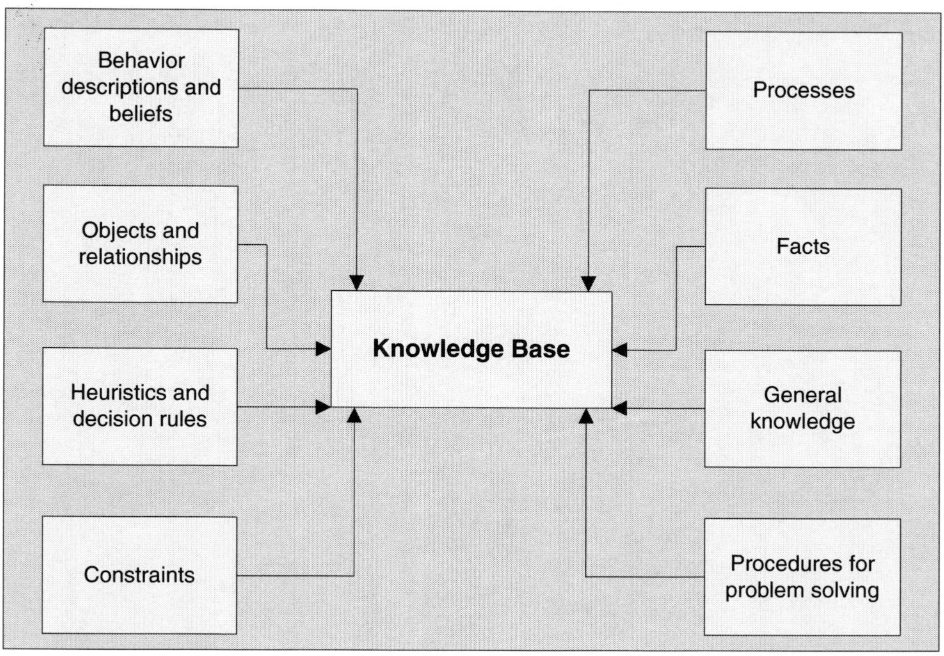


Figure 4. Knowledge sources and representation.

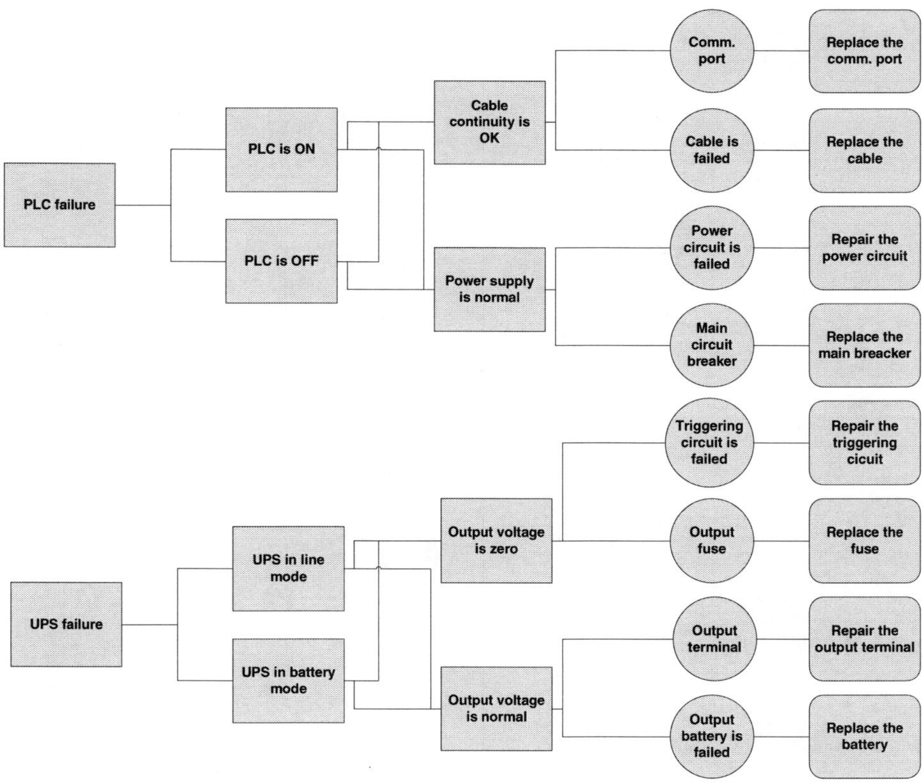


Figure 5. The rule-based technique: Decision making.

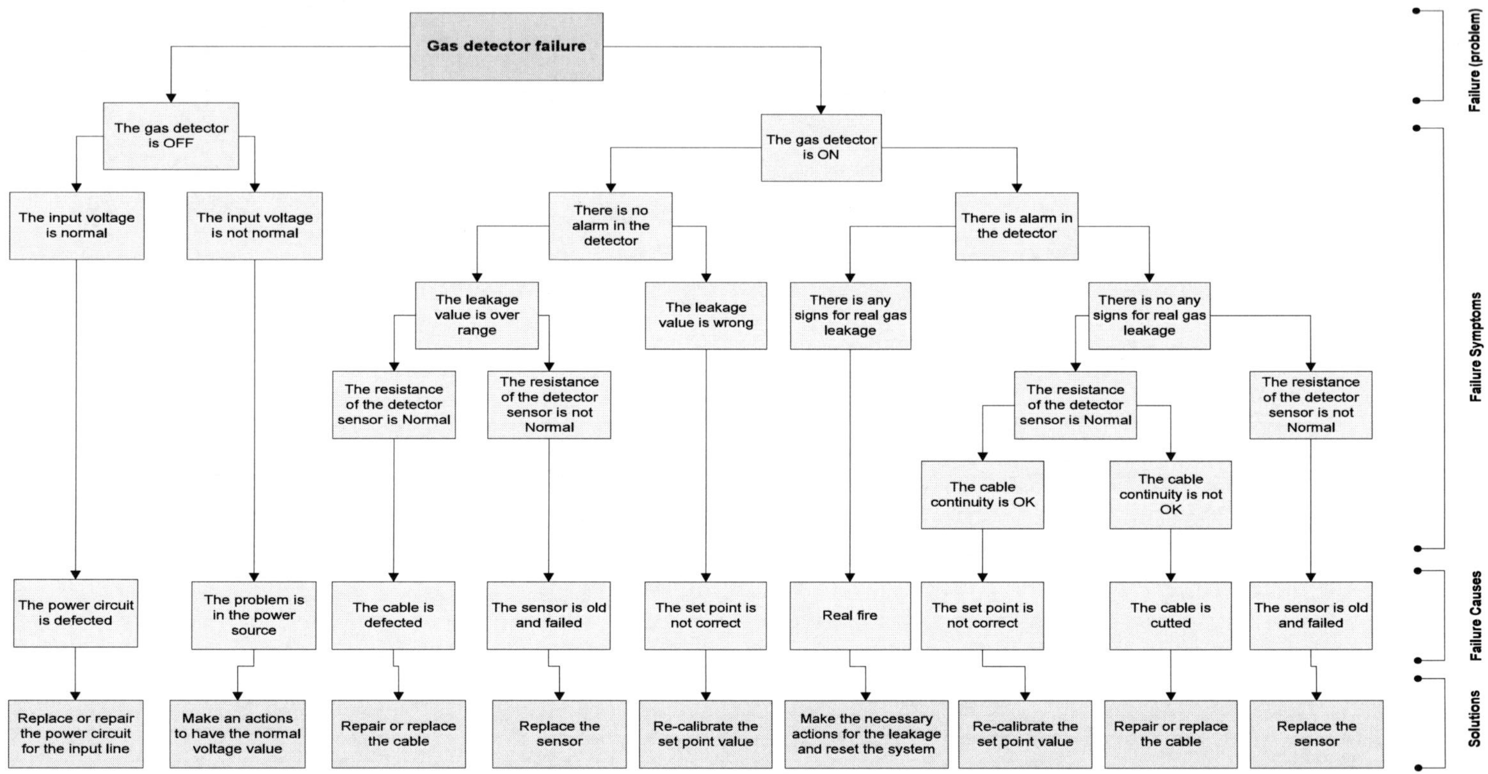


Figure 6. Symptoms, causes and solutions of gas detector failure.

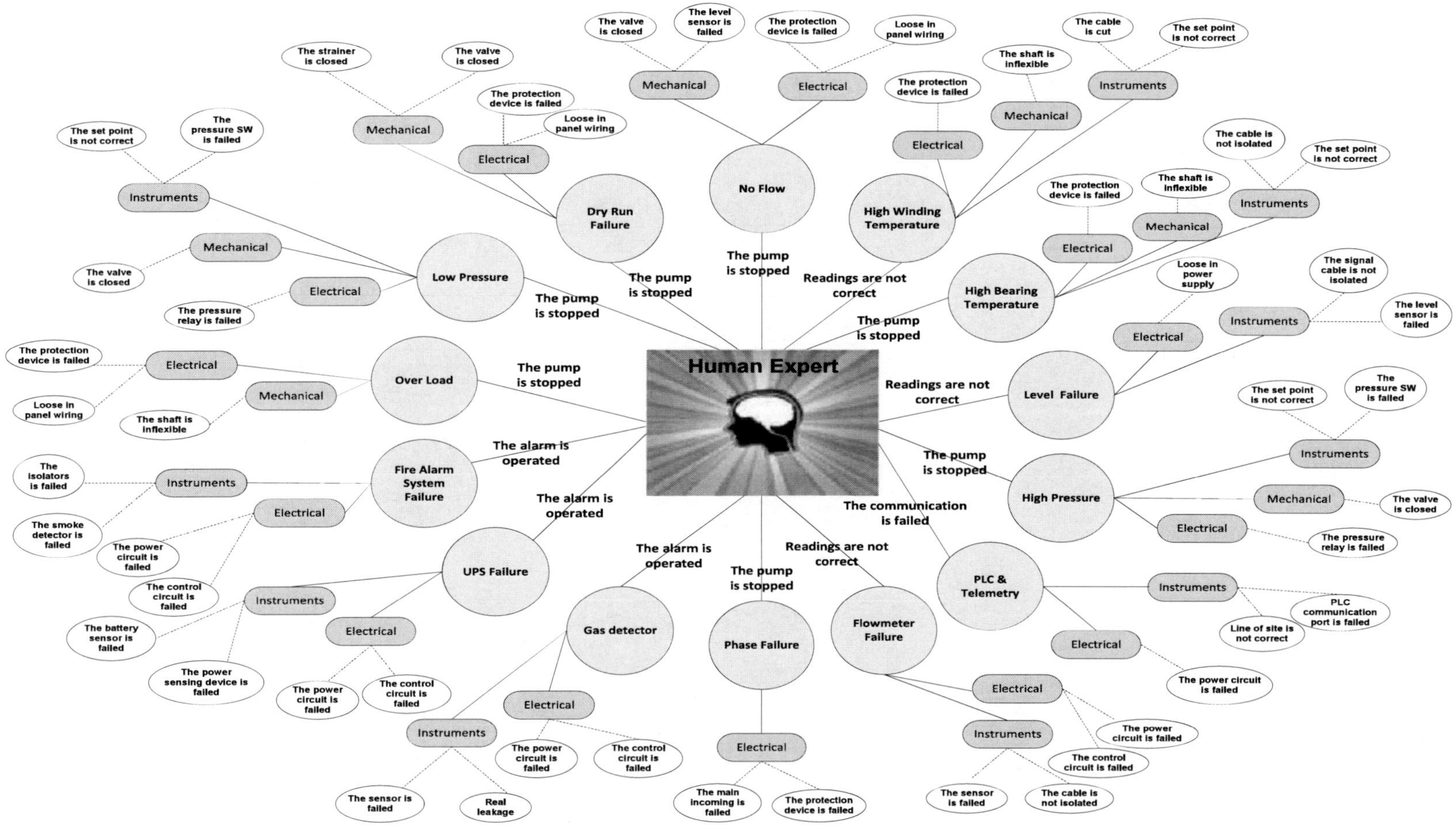


Figure 7. Failures mapping as well as their root causes and symptoms.

Using this pattern of representation gas detector failure which is explained in details by figure 6 can be described as:

$$Gas\ detector\ \ failure = \begin{Bmatrix} Real\ \ fire & Gas\ leakage \\ Not\ normal\ resistance & Old\ and\ \ failed\ \ sensor \\ Gas\ Leakage & Defected\ \ Cable \\ Not\ Normal\ Voltage & Error\ Power\ \ Source \end{Bmatrix}$$

This problem solving process is including; 1) Problem diagnostics which means the definition of both the failure symptoms and causes, 2) Problem analysis, implies the suggestion of possible solutions, and 3) Decision making, where the appropriate solution, action or decision concerning the problem failure is chosen. Consequently, solved failure with all details is stored in the database to be used later. This problem-solving method is built into the inference engine based on forward and sometimes backward chaining as a control method to manipulate and to use knowledge to form a line of reasoning. In Muna Company sixteen different failure types were observed, these types as well as their root causes and symptoms are mapped as presented in Figure 7.

**Table 1. Comparison between the developed maintenance knowledge system and the existing used one**

| Dimension | The developed system | The existing used system |
|---|---|---|
| Programming language | Visual Basic | Oracle |
| Searching type | By ID, location, and equipment type | By ID |
| Format of Equipment History reports | Detailed table | Descriptive text |
| User | Multi users | Single user |
| Maintenance Schedules | Corrective, Preventive, and Predictive | Corrective and preventive |
| Calculations | Equipment efficiency, labor efficiency, availability, and failure rate | Equipment efficiency |
| Data Equipment Format | A special form for each equipment type | There is no any form for the equipments |
| Diagnostics and Solving of Failures | Through rule-based technique and the knowledge base expert system with the explanations for each question and solution | Through human experts guessing from the last experiences |

## CONCLUSION

A real maintenance knowledge management system for potable water distribution Company labeled as Muna is developed. The system enables following up works, actions and failures of systems in maintenance departments of the company effectively. Failure tree analysis technique is used successfully to evaluate failures for the Company's assets. Basic design principles of expert systems are followed. The developed system composed of; knowledge database, reporting and history, relationships, rule-based expert system, and links and interfaces. One main advantage of the developed system is that it increasing the availability of the equipments and decreasing the mean time to repair. Table 1 illustrates the major differences between the developed system and the existing used one. Connection between the developed maintenance management system and inventory management is recommended as an extension to this work.

## REFERENCES


Choi, K. Wing, K. and Janet (2006), *Prioritization of Planned Maintenance Works in Public Hospitals in Hong Kong,* Doctoral Dissertation, The faculty of architecture, The University of Hong Kong, Hong Kong.

Emmanouilidis, C., Jantunen, E., and Macintyre, J. (2006), Flexible software for condition monitoring, incorporating novelty detection and diagnostics. *Computers in Industry*, Volume 57, Issue 6, Pages 516-527.

Kerr, R. (1991). *Knowledge-Based Manufacturing Management: Applications of Artificial Intelligence to the Effective Management of Manufacturing Companies.* Addison-Wesley Publishing Company, Singapore.

Kerr, S. (2007), *Cityworks Advances Preventive Maintenance for Treatment Plant Assets at Lima and Ohio*, VELOCITIE Partner News, Ohio, USA.

Krajewski, Ritzman, and Malhotra (2007), *Operations Management Processes and Value chains*, 8th edition, New Jersey: Pearson-Prentice Hall.

Liebowitz, J. (1997), *The Hand Book on Expert Systems.* West University of Timisoara, Romania.

Patel, S.A. and Kamrani, A.K. (1996), Intelligent Decision Support System for Diagnosis and Maintenance of Automated Systems. *Computers and Industrial Engineering*, Volume 30, Issue 2, Pages 297-319.

Reilly, M. (2001), Design of Computerized Maintenance Management System for Radionuclide Monitoring, Doctoral Dissertation, School of Engineering and Applied Science, University of Virginia, Virginia state USA.

Rich, E., and Knight, K. (1991). *Artificial Intelligence*, 2nd edition. McGraw-Hill Book Company, New York.

Sowa, J. (2000). *Knowledge Representation: Philosophical and Computational Foundations.* Books Cole Publishing Co., Pacific Grove.

Su, K.-L., Hwang, S.-L., and Chou, Y.-F. (2006), Applying knowledge structure to the usable failure diagnosis assistance system. Expert systems with applications, Volume 31, Issue 2, Pages 370-382.

Su, K.-W., Liu, T.-H., and Hwang, S.-L. (2001), A developed model of expert system interface (DMESI). *Expert systems with applications*, Volume 20, Issue 4, Pages 337-346.

Thomas, L. (2004), Intelligence moves plant from preventive to predictive maintenance, *the global magazine for automation and control*. InTech, July 2004, fosteReprints 866-879-9144.

Turban, E. (1988). *Decision Support and Expert Systems: Managerial Perspective*. Macmillan Publishing Company, New York.

In: Energy-Environment-Economics
Editor: Riad Benelmir

ISBN: 978-1-62948-862-2


*Chapter 9*

# RENEWABILITY ANALYSIS OF MICROALGAE BIODIESEL USING DIFFERENT TRANSESTERIFICATION PROCESSES

***Ehiaze Ehimen, Zhifa Sun* and Gerry Carrington***
Physics Department, University of Otago,
Dunedin, New Zealand

## ABSTRACT

An assessment of the renewability of microalgae derived biodiesel produced using different transesterification schemes is carried out and the in-situ and conventional transesterification processes are compared. The microalgae biomass cultivation and the biodiesel production process renewability are assessed by comparing the minimum work required to restore the non-renewable resources degraded in the process with the useful work available from the main process products. If the maximum work obtained from the products is larger than the restoration work, the process is considered to be renewable. In a present day scenario, where fossil fuel sources are used for the production of the raw process materials, all the transesterification processes are shown to be non-renewable. The influence of the electricity generation scheme, raw material source and the heating fuel type on the process renewability is also examined. The renewability of the process of in-situ transesterification of microalgae lipids to biodiesel is found to improve significantly when renewable electricity is used, the reacting alcohols are obtained from biomass fermentation, and heat pump technology is used for drying and heating.

## INTRODUCTION

The potential for the use of microalgae biomass for biodiesel production has been examined in the literature (i.e., Nagle and Lemke, 1990; Miao and Wu, 2006; Chisti, 2007),

* Corresponding author. Tel: +64 3 479 7812, Email: zhifa@physics.otago.ac.nz.

and the advantages of this feedstock relative to common terrestrial oil sources (i.e., soybeans and rapeseed oils) have been highlighted.

Because the cost of the biomass feedstock oil contributes 60-75% of the final biodiesel cost (Mata et al., 2010), the relatively higher cost of microalgae oil (compared to traditional oil sources) is considered to be a factor potentially limiting its use (Chisti, 2007). The high cost associated with microalgae production and conversion is considered to be mainly due to the process energy requirements (Chisti, 2007). The production economics of microalgae derived biofuels is, however, still under dispute, with some investigations indicating that microalgae oil can compete on cost with current conventional oil sources and petro-diesel (Huntley and Redalje, 2007).

To promote the use and competiveness of microalgae biomass for biodiesel production, Chisti (2007) put forward a research agenda to improve the energetics, economics and practicality of biofuel production from microalgae. This included enhancing the algal biology at the genetic level, improved bioreactor engineering, and optimisation of the biofuel production processes. This paper does not deal with either the optimisation of the microalgae biomass or oil productivity, or the photo-bioreactor technology. This study focusses on the third approach, exploring how modifications to the microalgae oil extraction and transesterification process could influence the overall energetics of biodiesel production using microalgae biomass.

The production of biodiesel from microalgae has been largely investigated using the conventional route, as demonstrated in Nagle and Lemke (1990) and Miao and Wu (2006). This involves the extraction of the lipids from the microalgae biomass followed by its conversion to alkyl esters and glycerol. The methanolysis studies of Nagle and Lemke (1990) indicated that the most important variable in the microalgae lipid conversion process is the type of catalyst used. The use of acid catalysts (HCl) results in a higher fatty acid methyl ester (FAME) yield per microalgae lipid input compared with the alkaline catalyst (NaOH) under the same reaction conditions. This difference is attributed to the characteristic high free fatty acid (FFA) content of microalgae lipids (Miao and Wu, 2006), which tend to reduce the conversion efficiency with alkaline catalysts.

The use of enzyme catalysed and supercritical transesterification methods, as alternatives to the conventional method for the production of microalgae biodiesel, have also been investigated (i.e., Li et al., 2007; Demirbaş, 2009). The supercritical method (based on the reacting alcohol properties) involves the use of high reaction temperatures and pressures i.e., > 240°C and >8 MPa respectively, without catalysts (Saka and Kusdiana, 2001).

For the large-scale production of microalgae biodiesel, the possibility of using cheap and energetically efficient conversion technologies to overcome the disadvantageous costs of microalgae biomass production is considered in this paper. The relatively high costs of the lipase catalysis (Fukuda et al., 2001), and the prohibitive operational and energetic costs of the supercritical method (Marchetti and Errazu, 2008), when compared with the inorganic catalysed routes, limits the short-medium term commercialisation of those methods. These processes are not considered further in this paper. In addition, safety concerns about the use of high temperature and pressure supercritical transesterification may further hinder its use for microalgae biodiesel production (Saka, 2006). Accordingly, only inorganic catalysis is considered for the transesterification process in this study.

An alternative to the conventional lipid extraction and transesterification process, the "in-situ" transesterification method and its modifications, are the main focus of this paper, as potential tools for improving the energetics of microalgae biodiesel production. The in-situ process facilitates the conversion of the biomass oil to fatty acid alkyl esters (FAAE) directly from the oil bearing biomass, thereby eliminating the solvent extraction step used to obtain the oil feedstock in the conventional method. The one-step in-situ transesterification method is considered by Haas et al. (2007) to have the potential to reduce the energetic, raw material and economic requirements for biodiesel production, compared with the conventional oil extraction and transesterification processes. This method may be especially advantageous for use with microalgae, since the extraction of microalgae lipids is usually accomplished by solvent extraction, rather than cheaper physical extraction methods (for example, expellers), as utilised in conventional oil crops (Johnson and Wen, 2009).

The influence of important reaction parameters (i.e., temperature, time, biomass moisture content and stirring) on FAME production from microalgae biomass by the in-situ transesterification process has been reported by Ehimen et al. (2010). The results of that study show that increases in the reaction temperature and alcohol volume favour the production of FAME. However, within the experimental conditions used, an increase in the reacting molar methanol to oil ratio > 315:1 does not show significant changes in the percentage mass conversion of the oil to FAME. In that study dried *Chlorella* biomass (with an oil content of 0.28 g/ g dry microalgae) was used as the feedstock (Ehimen et al., 2010).

The percentage mass content of the FAME in the biodiesel improves significantly with more stirring in the in-situ process (Ehimen et al., 2010). Biomass drying also plays an important role, with an increase in the moisture content of the biomass resulting in significant reductions of the equilibrium FAME yields expressed as g FAME/g biodiesel (Johnson and Wen, 2009; Ehimen et al., 2010). Extensive biomass drying may therefore be needed prior to biodiesel production by the in-situ process. To further reduce the large reacting alcohol requirements for the in-situ process and improve the FAME yields, the use of ultrasound stirring, as well as the integration of co-solvents (i.e., diethyl ether and pentane) in the transesterification process, has also been demonstrated in the literature (i.e., Xu and Mi, 2011; Ehimen et al., 2011).

The aim of this study is to evaluate the potential energetic advantages associated with the in-situ process and its modifications (i.e., the use of ultrasound agitation and the use of co-solvents) for microalgae biodiesel production, compared with the conventional process. To achieve this goal an adequate quantification technique is required to compare the renewability of different processes. The renewability analysis in this paper is based on an assessment of the net gains (or losses) obtained from the use of microalgae for biodiesel production. This involves an examination of the fossil fuel inputs into the microalgae biomass and fuel production process, as well as the useful work obtainable from the fuel products from the different transesterification processes. Schemes which could be potentially applied to improve the renewability of the use of microalgae for biodiesel, are also highlighted.

The results obtained from the process model, as presented in Ehimen et al. (2011), are used to estimate the energy and raw material requirements of the various in-situ and conventional transesterification processes. These requirements are then used to make a simplified assessment of the renewability of the different microalgae biodiesel production options and energy recovery routes.

## RENEWABILITY ASSESSMENT OF TRANSESTERIFICATION PROCESSES

The renewability analysis in this study is based on an assessment that quantifies the net gains (or losses) obtained from the use of microalgae for biodiesel production. The renewability evaluation is based on the process energy availability, or exergy (*B*), which is the work that a system can produce if it is brought to physical, thermal and chemical equilibrium with the environment in a reversible manner. This corresponds to the maximum work extractable from a given system (Ayres et al., 2006). The main objective of this paper is to use the exergy approach to establish a suitable indicator for the renewability of different biodiesel production processes. This indicator is presented in the following section. The system boundaries for the renewability evaluation are then defined, and the renewability of the different biodiesel production processes evaluated.

## QUANTIFICATION OF THE RENEWABILITY INDICATOR

A resource can be regarded as renewable when regeneration mechanisms exist to return the resource to its original state without depleting other resources i.e., a cyclic transformation is facilitated. The production of biodiesel from microalgae, and its subsequent combustion in diesel engines, could be considered as a truly renewable resource when the ideal process cycle shown in Figure 1 is feasible. In the ideal process solar energy is harnessed by the thermochemical cycle which involves the following step-wise carbon transformations (Figure 1):

- Microalgae biomass is naturally cultivated using water, atmospheric $CO_2$, and nutrients, the photosynthetic process being driven by solar energy.
- The lipid fraction of the microalgae biomass is obtained and converted to biodiesel without the need for fossil fuels.
- The biodiesel produced is combusted in diesel engines to produce useful work (Wp), with $CO_2$ and water as the main process by-products.

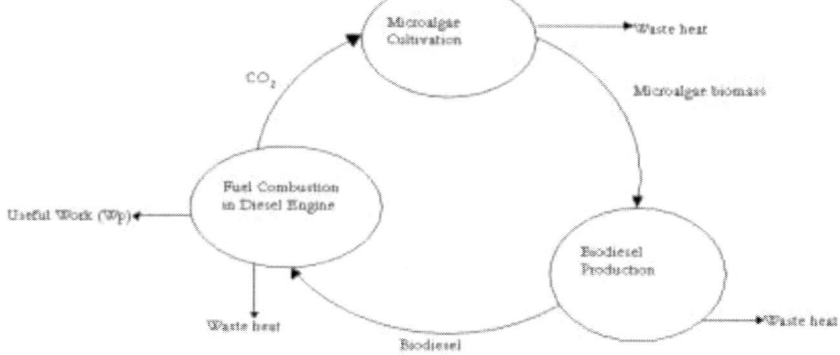


Figure 1. The ideal microalgae-biodiesel cycle.

It must be mentioned that the cycle presented above is only a simplified one and does not take into account other factors such as nitrogen oxides and particulate emissions from the system.

For the ideal process, it is assumed that all of the chemical reactants and by-products are recyclable, with only low quality heat rejected by the microalgae-biodiesel cycle to the environment. Furthermore, all of the $CO_2$ released from the fuel combustion process, and the organic wastes produced by the biodiesel production process, are assumed to be recycled into microalgae production. The renewability of the microalgae biodiesel production process would, however, be compromised when non-renewable resources are consumed in the biofuel processing units or in restoring the environment to its initial state. Resources classified as non-renewable resources (NRRs) are those which are consumed at a faster rate than they are regenerated naturally. For NRRs the work required for the restoration of the environment to its initial state is greater than that produced from these resources (Berthiaume et al., 2001). Fossil fuels such as petroleum and coal are NRRs. The direct use of fossil fuels and their by-products (i.e., fertilisers) in the microalgae biodiesel production process, although increasing the productivity of the process, could potentially reduce its overall renewability.

The NRRs consumed during the biomass and the biodiesel production processes must be accounted for to determine the renewability of biodiesel production. The process exergy consumption approach, as described by Berthiaume et al. (2001) and Szargut et al. (1987), is used in the renewability assessments in this study. Following this method, this paper compares the maximum work (Wp) obtainable from the final products of the biodiesel production process with the minimum work (Wr) required for restoration of the non-renewable resources consumed in that process to their initial environmental state (Berthiaume et al., 2001).

The renewability indicator (Ir) used in this study was defined by Berthiaume et al. (2001) using Eq. 1:

$$Ir = \left. (Wp - Wr) \middle/ Wp \right. \qquad (1)$$

The process renewability is evaluated on the basis of the Ir estimated for the different processes:

Ir = 1, indicates a fully renewable system, with Wr = 0, as in the ideal microalgae-biodiesel system.

- $0 < Ir < 1$, for a partially renewable system.
- Ir = 0, indicates a system in which the work produced by the biofuel and the restoration work required are equal. In such cases, the non-renewable resources could be considered for use directly instead of the biofuel.
- $Ir < 0$, for a process which consumes more restoration work than it produces as output work. Such processes are considered as non-renewable, and are less efficient than similar processes where the non-renewable resources are used directly for the intended biofuel application.

The renewability indicator, Ir, is applied in this paper to assess the microalgae transesterification processes examined and to compare the different processes studied.

## The System Boundaries Used in This Study

As thermodynamically required, the boundaries for the microalgae biodiesel production systems examined here start with the cultivation and harvesting of the microalgae biomass and end after the biodiesel and glycerol upgrading processes.

The system inputs include microalgae cultures, $CO_2$, nutrients, electricity, fossil fuels and process chemicals. The system outputs are biodiesel, glycerol, calcium sulphate ($CaSO_4$), microalgae biomass residues, and process heat and material wastes. Figure 2 shows the boundaries of the system and the various material and energy flows considered in this paper.

The material and energy inputs for the transesterification processes are considered on the basis of the requirements for the conversion of 1 t dry microalgae biomass. With the process assumed to be in steady state, and the energy input for the recovery of the excess methanol and solvents accounted for, only the additional raw materials needed to make up the required reacting quantities are used as process inputs.

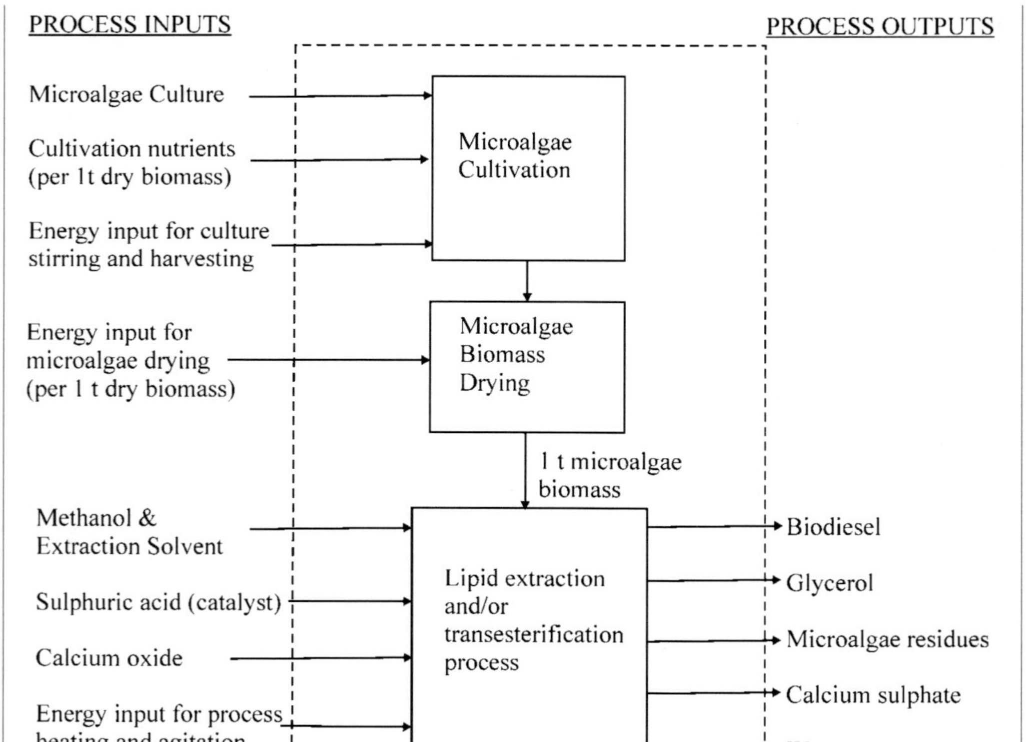


Figure 2. The system boundary (represented by the dashed lines) and material and energy flows considered in this study.

# Estimation of the Process Wr

## Microalgae Biomass Production and Harvesting

The minimum work inputs into the microalgae production and harvesting stages are still contentious, as shown in Chisti (2008) and Reijnders (2008). The Wr of the biomass production process depends on the specific microalgae species cultivated, the type of bioreactor used, the microalgae culture, and the climatic conditions available. The differences in the published reports are mainly due to the developmental nature of microalgae technology for low-cost biofuel production.

For this study it is assumed that the *Chlorella* biomass has a lipid content of 27.6 % (w/w, on a dry biomass basis) and that the biomass productivity of the cultivation and harvesting systems is 20 g dry microalgae/m$^2$/day, as presented in Ehimen et al. (2010) and Goldman (1980).

The mean values for the major elements found in microalgae (expressed as a percentage of ash free dry weight) are reported by Benefield and Randall (1980) to be: Carbon (C), 52%; Hydrogen (H), 8.0%; Oxygen (O), 31.0%; Nitrogen (N), 8.0%; and Phosphorus (P), 1.0%. This elemental composition is similar to that obtained for the *Chlorella* samples by Ehimen et al. (2010), which is used as the basis for most of the study assumptions. These nutrients must be supplied to the microalgae cultures in adequate amounts to facilitate biomass synthesis and maintain productivity levels (Becker, 1994). Atmospheric $CO_2$ is considered to meet the culture carbon requirement, with the biomass H and O fractions obtained from the culture water. Only the major nutrients, N and P, are considered in this study as process fertilisers for biomass cultivation. Ammonia ($NH_3$) and urea ($CO[NH_2]_2$) are commonly used as N sources in microalgae cultivation systems (Becker, 1994). Only the use of $NH_3$ is considered in this paper, since it is relatively cheaper and is reported to be easier to handle than urea for application in microalgae cultivation systems (Neenan et al., 1986). Triple super-phosphate (TSP) is used as the reference P fertilizer due to its widespread use in agricultural applications (Patzek, 2004). The areal $NH_3$ and TSP requirements for the microalgae biomass cultivation process are the same as those used by Ehimen (2010).

Mixing of the microalgae cultures, pumping the water and microalgae cultures, as well as the harvesting and concentration of the algal biomass (i.e., via centrifugation) are some of the processes which require electrical input (Becker, 1994; Becker and Venkatraraman, 1980; Neenan et al., 1986). The process electricity demand provided by Chisti (2008) is used in this study to obtain the biomass production electricity requirement. For biomass cultivation and harvesting, this is 8.77 and 0.30 MJ/kg microalgae oil respectively (Chisti, 2008). This electrical energy demand, which is nine times less than that obtained by Becker and Venkatraraman (1980) for similar raceway cultivation systems, is taken to represent the process electrical requirement in a present day scenario. With an estimated extracted lipid content of 0.276 g lipids/g *Chlorella* biomass (Ehimen et al., 2010), this represents an electrical energy input of 2448.9 MJ/t dry *Chlorella* biomass produced for the microalgae cultivation and harvesting processes.

The renewability assessment carried out is initially based on electricity generated from typical fossil fuel generation systems. An analysis of the influence of different electricity sources (i.e., the use of hydro-electricity) on the overall process renewability is conducted later in the sensitivity analysis section.

## Microalgae Biomass Drying

Although detailed reports of the energy requirement for the microalgae drying have not been found in the literature, this processing step is considered to require a significant energy input (Chen et al., 2009). The energetic demand for the microalgae biomass drying is included in this study by considering two drying technologies, (a) a conventional hot air dryer and (b) a heat pump dryer, for reducing the moisture content of the harvested biomass to the levels suited for the transesterification process. A post-harvest microalgae moisture content of 72.5% (w/w, on the basis of the dried biomass), as described in Ehimen et al. (2010), is used in this study.

The specific moisture extraction rate (SMER), which is the mass of water extracted from the product per unit energy consumed in the drying process (kg/kWh), is used to define the energy efficiency of the biomass drying processes. During drying the SMER reduces with time as moisture removal from the biomass becomes more difficult, due to hygroscopic retention of moisture in the product. This effect can be observed for wood drying using dehumidifiers, as demonstrated by Sun et al. (1999), where the reported process SMER clearly decreases towards the end of the drying process. Although factors such as the thickness of biomass layers, the dryer air state and the drying time also influence the SMER of the drying process, only reported average SMERs are used for the preliminary estimation of the biomass drying energy requirement. A maximum SMER value of 1.55 kg moisture/kWh, based on the latent heat of water vaporisation at 100°C (Raghavan et al., 2005) is used to estimate the conventional dryer energy demand. This is used to represent a best-case scenario for conventional dryers (Raghavan et al., 2005). For the removal of 725 kg of moisture from the post harvested microalgae, this relates to an energy input of 467.74 kWh for the recovery of 1t dry biomass.

For biomass drying using a heat pump dryer, a SMER of 4 kg moisture/kWh electricity (Carrington, 2010) is used in this study.

For the biomass drying stage, it is initially considered that the combustion of fossil fuels is used to generate the heat required for the drying process. The use of a heat pump biomass dryer, and its influence on the renewability of the over-all process, is explored in the section: Renewability Sensitivity Analysis. In the fossil fuel scenario, it is initially assumed that diesel is used as the heating fuel. This is assigned a limiting first law thermal efficiency of 100% for the combustion process, bearing in mind that best operating efficiencies of ≈ 80% are typical (Carrington, 1978). The analysis presented in the section: Estimation of the Process Wr, is initially based on the use of separate heat and power (SHP) installations (i.e., electrical generators and/or on-site boilers) for providing the process electrical and heating requirements. The use of combined heat and power (CHP) systems as an efficient approach for power and thermal energy generation from a single fuel source has been discussed widely in the literature (i.e., Pilavachi, 2000; Graus et al., 2007). However, the use of CHP systems for the process electricity and heat generation, and its influence on the process renewability, is not considered in this study due to the assumption of 100% operating efficiency for heating fuel combustion in the SHP installations.

In assessing different transesterification processes, the same biomass cultivation and drying processes are assumed to be used. Consequently the same energy inputs are used, per unit of microalgae biomass, for the biomass production, harvesting and drying steps for the transesterification processes considered.

## Microalgae Biodiesel Production

The microalgae lipids are extracted and transesterified (conventional transesterification) or transesterified directly (in-situ transesterification) from the dried biomass. This involves the use of methanol as the reaction alcohol, and $H_2SO_4$ as the process catalyst. To estimate the process renewability of biodiesel production by the transesterification processes considered, the minimum work (Wr) required for the restoration of the non-renewable resources consumed is initially estimated. The maximum useful work available from the process products (Wp) is then determined. The values of Wp and Wr are then used to obtain the renewability indicator (Ir), using Eq.(1).

## Mechanically Stirred *In-Situ* Transesterification

### *Cumulative Exergy Consumed for Production of NRRs Consumed in the Mechanically Stirred In-Situ Process*

Using the process material and minimum heat demands, the cumulative net exergy (CNEx) needed to restore the NRRs used in the mechanically stirred in-situ process is shown in Table 1. The process inputs (i.e., methanol, catalyst and CaO) and outputs (i.e., biodiesel, glycerol and $CaSO_4$) are based on the equivalent requirements for the in-situ transesterification of 1 t dried microalgae biomass, as obtained from the process models presented in Ehimen et al. (2011). For the stirring of the in-situ transesterification reactors, a mixing electrical energy requirement of 1.42 MJ/t biodiesel produced (Sorguven and Özilgen, 2010) is used in this study. A total hourly mass flow of biodiesel of 0.281 t is obtained for the mechanically stirred in-situ transesterification process after the conversion of 1 t dry *Chlorella*. The stirring electrical energy requirement per hour is therefore 0.8 MJ for the two transesterification reactors used. It is initially assumed that the process heat demands are supplied by diesel fuel combustion. A maximum useful work (exergy) for the diesel heating fuel of 44.4 MJ/kg (Szargut and Morris, 1987) is used in this analysis. The quantity of heating fuel (kg/t dry microalgae biomass consumed) required to meet the minimum heat demand (after process integration) of the different transesterification processes is estimated on the assumption of a 100% conversion of the fuel to heat. The influence of using other common boiler fuels (i.e., coal, natural gas and wood pellets) on the process renewability is examined subsequently in the section: Renewability Sensitivity Analysis. The use of heat pump drying is also explored in that section. The cumulative exergy required to restore the NRRs used to supply the transesterification raw materials, the heating fuel, and the electricity requirements

(CNEx), considered to be representative of a present day scenario, are largely obtained from Szargut and Morris (1987). This data has been widely used in recent literature (i.e., Berthiaume et al., 2001; Yang et al., 2009; Sorguven and Özilgen, 2010) to determine the exergy cost for the production of NRRs in a present day scenario.

***Useful Work (Wp) Available from the Products of the Mechanical Stirred In-Situ Transesterification Process***

The exergy content of the biodiesel and glycerol products is used to determine the maximum useful work (Wp) obtainable from the transesterification process. The sum of the maximum work available from the transesterification products is used to estimate the process Wp. The maximum useful work of the FAME and unreacted oil content of the biodiesel product are treated separately to quantify the work available from the transesterification product. The total useful work (Wp) available from the products derived from the in-situ transesterification (with mechanical agitation) of 1 t dry microalgae is shown in Table 2.

## *In-Situ* Transesterification with Ultrasound Agitation

For the low frequency ultrasonication of the in-situ mixture, the assessment in this section is based on the process electrical energy estimates obtained by Tyrrell (2010). Two 4 kW, 20 kHz ultrasonicator attached to the reactors are considered for the estimation of the agitation electrical energy requirement in this section. This is due to the fact that this commercially available ultrasonicator size as reported to be suited for biodiesel reactors with a flow rate of 1-3 $m^3$ transesterification mixture/h (Tyrrell, 2010). This ultrasonicator size is also reported to be adequate for microalgae cell disruption for processes with a biomass flow rate of 0.1-0.8 $m^3$ microalgae/h (Tyrrell, 2010). It therefore seems to be adequate for the mass flow rates involved for the in-situ transesterification of 1 t/h of dry microalgae biomass using a molar methanol to oil ratio of 105:1. The minimum work required to restore the NRRs depleted in the ultrasound agitated in-situ transesterification of 1 t of *Chlorella* biomass is given in Table 1. The excess methanol recovered using distillation methods is re-used in the transesterification process. Furthermore, the mass flow of the main products following the application of the ultrasound assisted in-situ transesterification process for the conversion of 1 t of *Chlorella* is used to estimate the Wp obtainable from the process. The sum of the work available from the products (Wp) for the transesterification processes considered is shown in Table 2.

## *In-Situ* Transesterification with Diethyl Ether Used as a Process Co-Solvent

The cumulative exergy required to produce the materials and energy inputs for the mechanically stirred in-situ transesterification process (per ton of microalgae biomass), incorporating diethyl ether (DEE) as a co-solvent, are shown in Table 1. The energetic requirement for stirring the transesterification reactors is assumed to be same as that for the mechanically stirred in-situ process in the section above: Mechanically Stirred *In-Situ*

**Table 1. Minimum restoration work required for the NRRs depleted for the transesterification processes examined per ton of *Chlorella* biomass**

| | | In-situ (mechanical stirring) | | In-situ (ultrasound) | | In-situ with DEE (mechanical stirring) | | In-situ with DEE (ultrasound) | | Conventional process (mechanical stirring) | |
|---|---|---|---|---|---|---|---|---|---|---|---|
| NRRs (unit) | CNEx (MJ/unit) | NRR quantity (kg) | Wr (MJ) | NRR quantity (kg) | Wr (MJ) | NRR quantity (kg) | Wr (MJ) | NRR quantity (kg) | Wr (MJ) | NRR quantity (kg) | Wr (MJ) |
| Microalgae biomass production | | | | | | | | | | | |
| Ammonia (kg) | 30.9[1] | 80.00 | 2472.00 | 80.00 | 2472.00 | 80.00 | 2472.00 | 80.00 | 2472.00 | 80.00 | 2472.00 |
| $P_2O_5$ (kg) | 7.52[2] | 20.00 | 150.4 | 20.00 | 150.4 | 20.00 | 150.4 | 20.00 | 150.4 | 20.00 | 150.4 |
| Electricity (MJ) | 4.17[3] | 2448.90 | 10211.91 | 2448.90 | 10211.91 | 2448.90 | 10211.91 | 2448.90 | 10211.91 | 2448.90 | 10211.91 |
| Microalgae drying-diesel (kg) | 53.20 | 37.93 | 2017.88 | 37.93 | 2017.88 | 37.93 | 2017.88 | 37.93 | 2017.88 | 37.93 | 2017.88 |
| Biodiesel Production | | | | | | | | | | | |
| Methanol (kg) | 73.08[4] | 104.20 | 7614.94 | 56.64 | 4139.25 | 46.59 | 3404.80 | 36.54 | 2670.34 | 30.02 | 2193.86 |
| $H_2SO_4$ (kg) | 9.10[5] | 270.00 | 2457.00 | 279.00 | 2538.9 | 270.00 | 2457.00 | 279 | 2538.90 | 260.00 | 2366.00 |
| Butanol (kg) | | | | | | | | | | 2.00 | 120.26 |
| CaO (kg) | 10.05[6] | 154.37 | 1551.42 | 159.52 | 1603.18 | 154.37 | 1551.42 | 159.52 | 1603.18 | 148.67 | 1494.13 |
| Reactor stirring (MJ) | 4.17 | 0.82 | 3.42 | 28.80 | 120.10 | 0.82 | 3.42 | 28.8 | 120.10 | 1.23 | 5.13 |
| Process heating-diesel (kg) | 53.20[7] | 92.86 | 4940.15 | 37.20 | 1979.04 | 18.82 | 1001.22 | 10.46 | 556.47 | 48.40 | 2574.88 |
| Total Wr (MJ) | | | 31419.28 | | 25232.66 | | 23270.05 | | 22341.18 | | 23606.45 |

[1] CNEx values based on ammonia gas production via steam reforming of natural gas (Szargut & Morris, 1987).

[2] CNEx value for $P_2O_5$ obtained from Berthiaume et al. (2001) based on the production energy values demonstrated in Wittmus et al. (1975).

[3] Typical CNEx value for electricity generated from fossil based fuels (Szargut & Morris, 1987).

[4] Methanol produced from crude oil (Szargut & Morris, 1987).

[5] CNEx of sulphuric acid based on production via the application of the Frasch process and sulphur combustion using ground sulphur (Szargut et al., 1988).

[6] CNEx of calcium oxide based on CaO production via the mining, crushing and calcining of limestone as the process feedstock (Szargut et al., 1988).

[7] Typical diesel fuel from crude oil CNEx value (Szargut & Morris, 1987).

**Table 2. Useful work available from the in-situ transesterification products (per ton of microalgae biomass converted)**

| | In-situ (mechanical stirring) | | In-situ (ultrasound) | | In-situ with DEE (mechanical stirring) | | In-situ with DEE (ultrasound) | | Conventional process (mechanical stirring) | |
|---|---|---|---|---|---|---|---|---|---|---|
| Product | Product quantity (kg) | Wp (MJ) | Product quantity (kg) | Wp (MJ) | Product quantity (kg) | Wp (MJ) | Product quantity (kg) | Wp (MJ) | Product quantity (kg) | Wp (MJ) |
| FAME | 281.88 | 11472.52 | 295.58 | 12030.11 | 264.24 | 10754.57 | 284.73 | 11588.51 | 239.85 | 9761.89 |
| Algae oil | 6.76 | 271.41 | 1.00 | 40.15 | 23.77 | 954.37 | 10.83 | 434.82 | 8.17 | 328.03 |
| Glycerol | 27.69 | 512.82 | 29.01 | 537.26 | 25.86 | 478.93 | 27.95 | 517.63 | 23.52 | 435.59 |
| $CaSO_4$ | 374.73 | 22.48 | 387.22 | 23.23 | 374.73 | 22.48 | 387.22 | 23.23 | 360.90 | 21.65 |
| Total Wp (MJ) | | 12278.71 | | 12630.75 | | 12210.34 | | 12564.19 | | 10547.16 |

Transesterification. Since the process is in steady state, and the diethyl ether solvent is assumed to be fully recovered from the transesterification product stream, the diethyl ether is not one of the NRRs consumed, although the heat energy required for co-solvent recovery is included in the process energy assessment.

The total useful work available from the main products of the mechanically stirred in-situ transesterification of 1 t *Chlorella* biomass, using a reacting methanol to oil ratio of 79:1 and diethyl-ether as a co-solvent, is given in Table 2.

## *In-Situ* Transesterification of Microalgae with Co-Solvent (Diethyl Ether) and Ultrasound Agitation

The minimum work required to restore the degraded NRRs consumed in the in-situ transesterification of 1 t of *Chlorella* biomass, involving diethyl ether as co-solvent and ultrasonic agitation, to their initial states is presented in Table 1.

Similarly, as presented earlier, the sum of the product chemical exergies for the transesterification of 1 t *Chlorella* biomass, with a molar ratio of reacting methanol to oil of 26:1, is given in Table 2.

## Conventional Extraction and Transesterification of the Microalgae Biomass Lipids

For the conventional process, the material inputs include the extraction solvents required for the initial stripping of the oil feedstocks from the microalgae. The conventional transesterification process is assumed to be conducted with a reacting methanol to oil molar ratio of 50:1, as described by Ehimen (2010a). The process heat input is based on the minimum hot utility requirement after process integration. The conventional transesterification process is assessed on the basis of the exergy needed to restore the NRRs consumed in the extraction and transesterification of 1 t of dried *Chlorella* biomass to their initial environmental state. This is given in Table 1.

To estimate the cumulative exergy consumed for the butanol production, it is initially assumed that the process butanol is synthesised by acid-catalysed hydration of propylene derived from the autothermic cracking of crude oil. This section uses an exergy cost of 60.13 MJ/kg butanol produced (Szargut et al. 1987), which is similar to the value for crude oil derived ethanol put forward by Sorguven and Özilgen (2010). The influence of the use of butanol derived from fermentation processes on the transesterification process renewability is examined later in the sensitivity analysis section.

The sum of the useful work (Wp) available from the main products of the conventional lipid extraction and transesterification process is presented in Table 2.

## Renewability Evaluation of the Different Transesterification Processes

Using the renewability indicator (Ir) from Eq. 1, Figure 3 compares the calculated Ir for the different transesterification processes assessed in the above sections: Mechanically Stirred *In-Situ* Transesterification to Conventional Extraction and Transesterification of the Microalgae Biomass Lipids. It must be noted that process heating is assumed to be supplied by diesel combustion and electricity is generated using fossil fuel based sources. The renewability assessment provided here thus represents a present day production scenario.

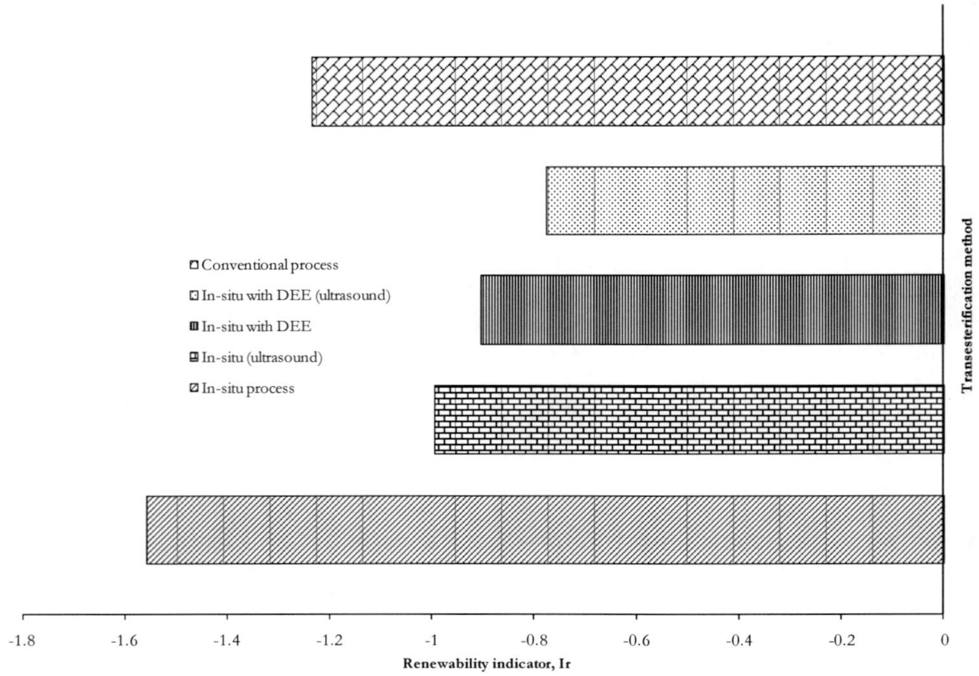


Figure 3. Renewability indicators for the different transesterification processes in a fossil fuel based scenario (per ton microalgae biomass converted). Here DEE refers to the diethyl ether used as a reaction co-solvent.

From Figure 3 it can be seen that for all the processes examined the estimated Ir is negative. This means that more work is required for the restoration of the non-renewable resources consumed in the transesterification process to their initial environmental states than is produced by the main products of the transesterification process. Nevertheless, the modifications to the in-situ transesterification process considered in this study are observed to improve its renewability. The modified in-situ processes involving low frequency ultrasound and co-solvents are seen to be more renewable than the conventional method for microalgae FAME production under the process conditions of a present day scenario.

The use and recovery of large reacting alcohol quantities in the in-situ transesterification process appear to affect the overall renewability of this method negatively. The use of a reacting molar ratio of methanol to oil of 315:1 for the mechanically stirred in-situ process (including the heat requirement for the alcohol recovery) contributes ≈40% of the overall Wr calculated for this process. The in-situ method involving the use of ultrasound and co-solvent helps to reduce the Wr, thus improving the transesterification process renewability. However that method is also seen to have negative renewability (Ir) values, mainly due to the primary resources used for electricity production and raw materials recovery. It is initially considered in this study that the resources consumed in the microalgae production and conversion are derived from fossil fuel sources (i.e., methanol from crude oil). This assumption is based on a present day scenario where most of the industrial inputs are derived directly or indirectly from such sources. The influence of the source of these raw materials on the process renewability is further explored in the following section. Changes in the electricity generation scheme and the choice of the fuel source for the process heat generation are also examined in the following section, since they also influence the process renewability.

## Renewability Sensitivity Analysis

### *Electricity Generation Scheme*

The cumulative exergy consumed for the production of the process electricity inputs used in the renewability assessment in the above sections: Mechanically Stirred *In-Situ* Transesterification to Conventional Extraction and Transesterification of the Microalgae Biomass Lipids, is based on a value of 4.17 MJ fossil fuel consumed per MJ electricity generated, as determined by Szargut and Morris (1987). Compared with other process inputs, the electricity requirement for the different biomass production and transesterification processes (from fossil fuel sources) is seen to be the largest single contributor to the overall process Wr (≈30-46% of the total Wr). In scenarios where electricity is generated from other sources, the exergy required for the process electricity would differ. In locations with hydro-generated electricity, the minimum restoration work required for the electricity production is 0.006 MJ/MJ generated hydro-electricity (Berthiaume et al., 2001). To assess the effect of this change, the cumulative exergy cost of 4.17 MJ/MJ electricity for fossil fuel electricity generation is replaced with 0.006 MJ/MJ in Table 1, and the values of Wr and Ir for the different processes recalculated. Assuming the biomass production and transesterification electricity requirements are met using hydro-electricity, the recalculated values of Ir are given in Table 3. DEE in Table 3 refers to the diethyl ether co-solvent.

Compared with fossil fuel generated electricity, the use of hydro-electricity improves the process renewability for all processes, the total restoration work, Wr, being reduced by 32, 40, 43, 46 and 43 % for the in-situ (mechanical stirred), the in-situ (ultrasonicated), the in-situ (with diethyl ether), the ultrasound agitated in-situ with diethyl ether, and the conventional processes respectively. The improved renewability due to the change in electricity source is illustrated by the Ir column in Table 3.

By comparison with Figure 3 this shows that a renewable electricity source makes a critical contribution to the renewability of the biomass production and transesterification processes examined. The use of other renewable electricity generation sources, such as wind and solar, are not considered, since it is expected that CNEx values similar to those for hydro-electricity would be obtained.

**Table 3. Process renewability with the use of hydro-electricity**

| Process | Wr | Wp | Ir |
|---|---|---|---|
| In-situ (mechanical stirring) | 21220 | 12280 | -0.73 |
| In-situ (ultrasound) | 14920 | 12630 | -0.18 |
| In-situ with DEE (mechanical stirring) | 13070 | 12210 | -0.07 |
| In-situ with DEE (ultrasound) | 12020 | 12560 | 0.05 |
| Conventional process (mechanical stirring) | 13400 | 10550 | -0.27 |

***Source of Process Raw Materials (Inputs)***

The analysis presented in the above sections: Mechanically Stirred *In-Situ* Transesterification to Conventional Extraction and Transesterification of the Microalgae Biomass Lipids, is based on the assumption that the process inputs are derived from fossil based sources. Under that scenario the cumulative exergy consumed for the production of the process raw materials influences the process renewability significantly. For example, the exergy consumed for the methanol used for the different transesterification process contributes 10-25% of the total estimated Wr. To assess the influence of using alternatively sourced raw materials for the transesterification processes, the impact of producing the reacting alcohols by biomass fermentation is examined here.

The large scale production of ethanol by biomass fermentation has been widely detailed in the literature (i.e., Lin and Tanaka, 2006; Hattori and Morita, 2010). The use of ethanol as a reacting alcohol in the transesterification process has also been demonstrated in the literature (i.e., Al-Widyan and Al-Shyoukh, 2002). In this section the use of ethanol is examined in order to investigate the impact of fermentation sourced alcohols on process renewability. In this analysis the cumulative exergy cost (CNEx) required for the production of ethanol via the fermentation of sugars is 26.40 MJ/kg ethanol, following Sorguven and Özilgen (2010). The same assumption of using biomass derived ethanol is also applied to the conventional biodiesel production process to provide an even comparison with the in-situ processes.

Butanol is used as the extraction solvent for the conventional process in this study because it can be produced in large quantities by fermentation with various biomass materials, such as starch, sugars and non-cellulosic substrates – including algae biomass, as feedstocks (Jones and Woods, 1986).

Instead of the use of butanol derived from propylene, as previously evaluated in the section: Conventional Extraction and Transesterification of the Microalgae Biomass Lipids, the effect of using butanol derived from fermentation processes on the conventional extraction and transesterification process is examined. The process butanol solvent is assumed to be produced using the acetone-butanol-ethanol (ABE) fermentation process. A conservative exergy cost of 26.40 MJ/kg butanol produced is used in this section, based on the cumulative exergy consumed for the production of fermentation derived alcohols, as presented by Sorguven and Özilgen (2010).

The minimum restoration work required for the production of the reacting alcohol, as presented in Table 1, is replaced with the values adopted in this section, and the Ir of the different processes recalculated. In this re-evaluation the process electricity requirement is assumed to be met by hydro-electricity. The influence of the use of the alcohols obtained from biomass fermentation on the Wr of the processes and the corresponding Ir is shown in

Figure 4. The sum of the useful work (Wp) available from the main products of the different transesterification processes is also presented.

Figure 4 indicates that all the modified in-situ transesterification processes (i.e., using ultrasound agitation and/or co-solvents) can be deemed renewable when fermentation derived alcohols and hydro-electricity are used for microalgae production and conversion to biodiesel. The ultrasound agitated in-situ process using diethyl ether as a co-solvent is seen to be superior to the other transesterification processes with regard to the process renewability.

The mechanically stirred in-situ transesterification process is seen to still be non-renewable, despite the use of hydro-electricity and fermentation alcohols for the conversion process. The overall process heat demands (for the biomass drying and recovery of the stoichiometrically excess methanol) are seen to contribute significantly to the minimum restoration work calculated for the process. With the use of diesel fuel to meet the process heating needs, the exergy cost associated with the process heating contributes 42.64% of the total process Wr.

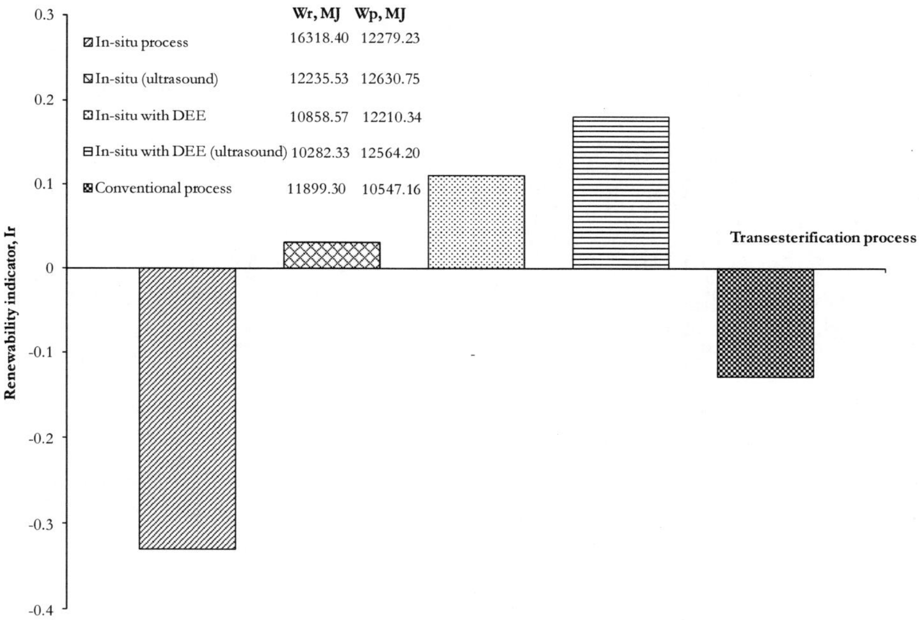


Figure 4. Result of using fermentation alcohols and hydro-electricity on the microalgae biodiesel renewability. DEE refers to the reactions using diethyl ether as a co-solvent.

The influence of the choice of the process heating fuel is therefore carried out in the following section to determine if any further improvement in the process renewability can be achieved. The use of electrically driven heat pumps to provide the heat requirement of the different processes is also considered.

## *Choice of Process Heating Fuels or Technology*

The use of petro-diesel fuels for boilers has been considered to supply the minimum process heating needs. The influence of the boiler fuel on the process renewability is examined in this section, by considering the effect of using other common industrial boiler fuels, coal and natural gas, on the estimated Ir. The use of wood pellets is also explored.

The effect of supplying the process heat using electric heat pumps is also explored. This technology is well established for heating water and other fluids (i.e., Carrington and Knopp, 1983; Anderson et al., 1985). The influence of coal and natural gas on the process renewability is investigated because of their widespread use for industrial heat generation. In this section the values of useful work available from coal and natural gas are 29.00 and 50.75 MJ/kg respectively, as established by Szargut and Morris (1987). The quantity of heating fuel (kg/t microalgae biomass consumed) required to meet the minimum heat demand (after process integration) of the different transesterification processes is estimated on the assumption of a 100% conversion of the fuel to heat. Although this assumption is not realistic, it provides the most optimistic conversion scenarios for the use of such fuel sources. The cumulative exergy cost associated with the production of the heating fuels is 30.44 and 57.87 MJ/kg consumed, for coal and natural gas respectively (Szargut and Morris, 1987). For wood pellets the average useful work available (Wp) is 20.0 MJ/kg (Bilgen et al., 2004). A conservative value of the exergy cost for the pellet production of 6.4 MJ/kg wood pellets, is used for the process renewability evaluation, after Patzek and Pimentel (2005) and Theka and Obernbergera (2004). This incorporates the exergy costs for the production of the NRRs consumed in the biomass cultivation, wood felling, collection, transportation, chipping and drying process. The wood drying process is considered to be performed using best available technology tube bundle dryers with wood residues used as the process heating fuel (Theka and Obernbergera, 2004).

For the heat pump process heating, a conservative first law thermal conversion efficiency of 40% (Carrington, 2010) is used to estimate the electricity inputs for this method. As earlier highlighted, the use of isothermal heat pump drying (Catton et al., 2010) operated using hydro-electricity, which reduces the exergy cost for the restoration of the NRRs degraded in the biomass drying process, is considered in this section. This is carried out using a SMER of 4 kg moisture/kWh (Carrington, 2010).

The estimated Ir obtained by the use of different process heating fuels and technologies (including biomass drying) is shown in Figure 5.

The process renewabilities presented in Figure 5 are estimated on the basis of the use of hydro-generated electricity and fermentation derived reacting alcohol for the biomass and biofuel production processes. The heating fuels considered in this section are used, where required, to meet the demands of both the biomass drying process and process heating.

The results in Figure 5 show that the types of heating fuel and technology used for biomass drying and process heating significantly affect the renewability of the process. The use of heat pumps for biomass drying and process heating is seen to result in the best improvements in the renewability indicators (Ir) of all the transesterification processes, compared with the Ir obtained using other fossil based heating fuels or wood pellets.

The use of wood pellets is shown to reduce the total Wr of all the transesterification processes considered by 8-16%, with the process operated with hydro-electricity, compared with the use of fossil fuels. The modified ultrasound agitated in-situ process using diethyl ether is seen to be the most renewable of the transesterification processes considered. The mechanically stirred in-situ process, which was previously considered as non-renewable with fossil based fuels, is seen to be renewable when heat pumps are used for drying and heating, as well as when wood pellets are used for process heating.

Although the use of heat pump heating and drying is not as widely applied commercially at present compared with conventional boiler and drying systems, this technology appears to offer a lot of promise for the viability of biodiesel production from microalgae as a renewable technology.

Comparing the process Ir of the different fossil fuel derived heating fuels, the use of coal is seen to lead to increased renewabilities when compared with diesel fuel and natural gas. This is attributed to the current ease of coal extraction and processing compared with the other fossil fuels.

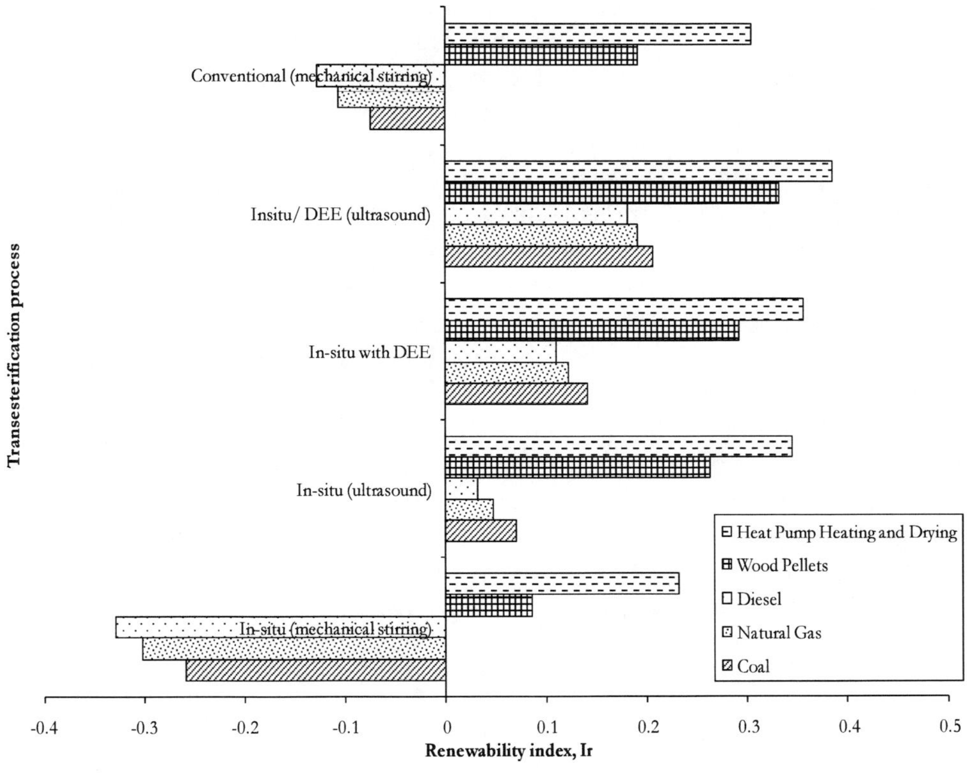


Figure 5. Influence of choice of heating fuel on the renewability of the different microalgae biodiesel processes.

# CONCLUSION

The process renewability is adjudged by comparing the minimum work consumed for the restoration of the non-renewable resources depleted by the conversion process with the maximum work obtained from the main process products. The results obtained show that in a present-day fossil fuel scenario, all the processes are non-renewable. Here, the ultrasound agitated in-situ transesterification process with diethyl ether as co-solvent is shown to be less non-renewable than the other conversion processes.

Generally, the modifications applied to the in-situ transesterification process (i.e., use of low frequency ultrasound and co-solvents) exhibit significant improvements in process renewability compared with the mechanically stirred in-situ process. The results of this study show promise for the use of the modified in-situ processes for microalgae biodiesel production.

Altering the electricity generation scheme from a fossil-based source to hydro-electricity (or wind generation) for the biomass production and transesterification methods leads to an improvement in the renewability of all the routes investigated. Nevertheless, all of the processes considered remain non-renewable, with the exception of the ultrasound agitated stirred in-situ process using diethyl ether as co-solvent.

A significant improvement in the process renewability is obtained by the use of alcohols sourced from biomass fermentation, with all the modified in-situ processes considered exhibiting positive Ir values. The use of heat pumps to supply the process heat and to provide biomass drying is shown to increase the process renewability compared with commonly used fossil heating fuels and wood pellets. The mechanically stirred in-situ process is shown to be renewable with the use of fermentation derived alcohols for the transesterification process, and the process heat and biomass drying supplied by either heat pumps or wood pellets.

To improve the renewability of the biomass production and transesterification processes, energy generation from the post transesterified microalgae residues i.e., Ehimen et al. (2009, 2011a) could also be examined. The use of the digestion process for $CH_4$ recovery and waste treatment could provide a significant increase in the useful work available from the process products, primarily due to the $CH_4$ produced. In addition, recycling the nutrient-rich digestate of the anaerobic digestion process into the microalgae cultivation unit could further increase the renewability of microalgae biodiesel production using the in-situ process. The cultivation nutrients could also be supplied by anaerobically digested manure or organic wastes, as has been demonstrated by Levine et al. (2010) for *Neochloris oleoabundans* cultivation for biodiesel production. The digestate could thus be considered as the major nutrient supply, without the need for in-organic fertilisers, as is assumed in this paper. Post-treated organic industrial and municipal wastes could also be considered for use in the microalgae cultivation step.

## REFERENCES

Al-Widyan, M.I., Al-Shyoukh, A.O. 2002. *Bioresour. Technol.* 85, 253-256.

Anderson, J.A., Bradford, R.A., Carrington, C.G. 1985. *Int. J. Energ. Res.* 9, 65-89.

Ayres, R.U., Ayres, L.W., Masini A. 2006. In: *Sustainable Metals Management*. Eds. A. von Gleich, R.U. Ayres, S. Göß ling-Reisemann. Springer Publishing, Dordrecht, the Netherlands. pp 141-194.

Becker, E.W. 1994. *Microalgae: Biotechnology and Microbiology*. Cambridge Studies in Biotechnology. Cambridge University Press, Cambridge.

Becker, E.W., Venkataraman, L.V. 1980. In: *Algae Biomass*. Eds. G. Shelef, C.J. Soeder. Elsevier/North-Holland Biomedical Press. pp 35-50.

Benefield, L.D., Randall, C.W. 1980. *Biological Process Design for Wastewater Treatment*. Prentice-Hall Inc, Englewood Cliffs, New Jersey.

Berthiaume, R., Bouchard, C., Rosen, M.A. 2001. *Exerg. Int. J.* 1, 256-268.

Carrington, C.G. 1978. *Ind. Shellman* 41, 19-21.

Carrington, C.G. 2010. *Personal Communication*. 18 October 2010.
Carrington, C.G., Knopp T.C. 1983. *Int. J. Energ*. Res. 7, 255-267.
Catton, W., Carrington, C.G., Sun, Z.F. 2010. *Int. J. Energ. Res*. DOI: 10.1002/ er.1704.
Chen, P., Min, M., Chen, Y., Wang, L., Li, Y., Chen, Q., Wang, C., Wan, Y., Wang, X., Cheng, Y., Deng, S., Hennessy, K., Lin, X., Liu, Y., Wang, Y., Martinez, B., Ruan, R. 2009. *Int. J. Agric. Biolog. Eng*. 2, 1-30.
Chisti, Y. 2007. *Biotechnol.Adv*. 25, 294-306.
Chisti, Y. 2008. *Trends. Biotechnol*. 26, 351-352.
Demirbaş, A. 2009. *Energ. Sources A*. 31, 163-168.
Ehimen, E.A. 2010a. PhD Thesis. University of Otago, Dunedin, New Zealand.
Ehimen, E.A. 2010. *Energ. Sources A*. 32: 1111-1120.
Ehimen, E.A., Connaughton, S., Sun, Z., Carrington, C.G. 2009. GCB Bioenerg. 1, 371-381.
Ehimen, E.A., Sun, Z.F., Carrington, C.G., 2010. *Fuel* 89, 677-684.
Ehimen, E.A., Sun, Z.F., Carrington, C.G., 2011(in press). *Procedia Environ. Sci.*
Ehimen, E.A., Sun, Z.F., Carrington, C.G., Birch, E.J., Eaton-Rye J.J. 2011a. *Appl. Energ*. 88, 3454-3463.
Fukuda, H., Kondo, A., Noda, H. 2001. *J. Biosci. Bioeng*. 92, 405-416.
Goldman, J.C. 1980. In: *Algae Biomass Eds*. G. Shelef, C.J. Soeder. Elsevier/North-Holland Biomedical Press. pp 343-359.
Graus, W.H.J., Voogt, M., Worrell, E. 2007. *Energ. Policy* 35, 3936-3951.
Haas, M.J., Scott, K.M., Foglia, T.A., Marmer, W.N. 2007. *J. Am. Oil. Chem. Soc*. 84, 963-970.
Hattori, T., Morita, S. 2010. *Plant Prod. Sci*. 13, 221-234.
Huntley, M.E., Redalje, D.G. 2007. *Mitigation and Adaptation Strategies for Global Change* 12, 573-608.
Johnson, M.B., Wen, Z. 2009. *Energ. Fuels*. 23, 5179-5183.
Jones, D.T., Woods, D.R. 1986. *Microbiological Reviews* 50, 484-524.
Levine, R.B., Constanza-Robinson, M.S., Spatafora, G.A. 2010. *Biomass. Bioenerg*. DOI: 10.1016/j.biombioe.2010.08.035
Li, X., Xu, H., Wu Q. 2007. *Biotechnol. Bioeng*. 98, 764-771.
Lin, Y., Tanaka, S. 2006. *Appl. Microbiol. Biotechnol*. 69, 627-642.
Marchetti, J.M., Errazu, A.F. 2008. *Energ. Convers. Management* 49, 2160-2164.
Mata, T.M., Martins, A.A., Caetano, C.S. 2010. *Renew. Sustain. Energ. Rev*. 14, 217-232.
Miao, X., Wu, Q. 2006. *Bioresour. Technol*. 97, 841-846.
Nagle, N., Lemke, P. 1990. *Appl. Biochem.Biotechnol*. 24/25, 355–361.
Neenan, B., Feinberg, D., Hill, A., McIntosh, R., Kerry, K. 1986. Fuels from Microalgae: Technology Status, Potential and Research Requirements. *Solar Energy Research* Institute. Golden, Colorado. Task No. 4513.20.
Patzek, T.W., Pimentel, D. 2005. *Critical Rev. Plant Sci*. 24, 327-364.
Pilavachi, P.A. 2000. *Appl. Thermal Eng*. 20, 1421-1429.
Raghavan, G.S.V., Rennie, T.J., Sunjka, P.S., Orsat, V., Phaphuangwittayakul, W., Terdtoon, P. 2005. *Brazilian J. Chem. Eng*. 22, 195-201.
Reijnders, L. 2008. *Trends. Biotechnol*. 26, 349-350.
Saka, S. 2006. *The 2nd Joint International Conference on "Sustainable Energy and Environment (SEE 2006)"*. 21-23 November 2006, Bangkok, Thailand. Paper C-043(O).
Saka, S., Kusdiana, D. 2001. *Fuel* 80, 225-251.

Sorguven, E., Özilgen, M. 2010. *Renew. Energ*. 35, 1956-1966.

Sun, Z.F., Carrington, C.G., Bannister, P. 1999. *Drying Technol* 17, 731-743.

Szargut, J., Morris, D.R. 1987. *Energ. Res*. 11, 245-261.

Szargut, J., Morris, D.R., Steward, F.R. 1988. *Exergy Analysis of Thermal, Chemical, and Metallurgical Processes.* Hemisphere Publishing Corporation. New York, US.

Theka, G., Obernbergera, I. 2004. *Biomass. Bioenerg*. 27, 671-693.

Tyrrell, J. 2010. *Personal communication with Dr. John Tyrrell*, DKSH NZ, Auckland.

Xu, R., Mi, Y., 2011. *J. Am. Oil. Chem. Soc*. DOI: 10.1007/s11746-010-1653-3

Yang, Q., Chen, B., Ji, X., He, Y.F., Chen, G.Q. 2009. *Comm. Nonlinear. Sci. Num. Simul.* 14, 2450-2461.

In: Energy-Environment-Economics
Editor: Riad Benelmir
ISBN: 978-1-62948-862-2


*Chapter 10*

# A Critical Review of Organizational Structure and Reforms of the Indian Electricity Sector

***S. C. Tripathy*** [*] ***and Tripta Thakur***
Ghanashyam Hemalata Institute of Technology and Management,
Bhuan, Chhaitana, Puri, Orissa, India

## 1. Introduction

Ever since its independence in 1947, India has taken rapid strides in the Power Sector both in terms of enhancing power generation and in making available the power to widely distributed geographical boundaries. The installed generation capacity in the utility sector has increased from a mere 1,362 MW in 1947 to about 1, 27,056 MW in September 2006 (MoP, 2007). The power transmission and distribution network has also grown substantially. The per capita consumption of electricity in the country has increased from 15 KWh in 1950 to about 606 KWh per annum in 2004-05 as worked out based on UN guidelines and this is projected to grow to about 1000 KWh per annum in 2012 (Ministry of Power, MOP).

At the time of independence, only about 1500 villages of the country had access to electricity and about 6430 pump sets were energized. The scenario has changed significantly over the years. Till July 2006, 14842467 pump sets were energized (Central Electricity Authority (CEA)) and it has been possible to extend electricity to about 438934 villages, thereby resulting in electrification of 74% villages (CEA, 2006). As per rough estimates, out of this about 18,000 villages are located in remote and difficult areas and it is not possible to extend power supply to these villages through the existing power grid. Electrification of these villages, therefore, is proposed to be achieved through various sources of distributed generation including non-conventional sources of energy.

Baring above areas, the entire country has been provided electricity mainly through the State Electricity Boards (SEBs). Historically the SEBs have not adequately been able to

[*] E.Mail : sarattripathy1@gmail.com.

provide substantive generation expansion with corresponding expansion in transmission and distribution networks in order to meet the ever-growing demand. This resulted into an era of growing power shortages. This fact was apprehended as far back as mid seventies when it was felt that the Central Government should come forward to take up generation and transmission projects under the Central Sector to assist and augment the efforts of States for improving the power supply position. Agencies like National thermal power corporation (NTPC), National hydropower corporation (NHPC) and Power Grid Corporation of India Limited (PGCIL) were accordingly set up in a phased manner.

The supplementary efforts by Central Sector Utilities in adding substantial generation and implementing new transmission projects have helped in improving the overall power scenario in the country. During the early nineties, generation was opened to Private Sector with a view to mobilize additional resources. Despite these efforts, the condition of the power supply industry has not been up to mark (Thakur et al., 2005) due to compounding effects of numerous attributes that owe their existence to the organizational structure of the power sector, and especially to the maladies afflicting distribution sector.

This paper evaluates the Indian power sector, explores its evolution and structure, and examines the electricity reforms in India with special emphasis on the recently enacted electricity Act 2003 that has major sector revamping.

## 2. Organization Structure of Electricity Supply Industry in India

The power sector in India has been regulated and owned for many years by various government agencies and organization. The subject of electricity is covered under the concurrent list in the Constitution of India, implying that both the Central and State government have the power to legislate on matters concerning the sector. The Indian power sector is divided into five regions: Northern, Western, Southern, eastern, and the North-Eastern Region: and almost every state in India has its own state owned electric utility (SOEU). There exist a multitude of institutions that govern and constitute the Indian electricity sector as illustrated in Figure 1. The Central Electricity Authority (CEA) is responsible for power planning at the national level. CEA advise the Ministry of Power (MoP) on matter concerning the national power policy, and national power planning. The Thermal Power Corporation (NTPC) is the largest thermal power generating company in India, which was incorporated in 1975. The National Hydro Power Corporation (NHPC) was also incorporated in 1975 to harness the vast hydro, tidal and wind potentials of the country. The power finance corporation (PFC) was incorporated in 1986 to function as the prime financial institution dedicated to the growth and the overall development of the Indian power sector. PFC provides funds to SOEUs and private power utilities. The power Grid Corporation of India Limited (PGCIL) was established in 1989 to operate regional and national power grids and facilities the power transmission reliability, security and economy. The Power Trading Corporation of India Limited (PTC) was constituted in 1999 to purchase and sell electric power, to plan and promote an efficient and reliable power trading and distribution system, with majority equity participation by the PGCIL, NTPC, PFC, and other financial institutions. Regulatory matters in the Center are taken care by Central Electricity

Regulatory Commission (CERC) and the respective State Electricity Regulatory Commissions (SERC) takes care of corresponding regulatory issues at the state level.

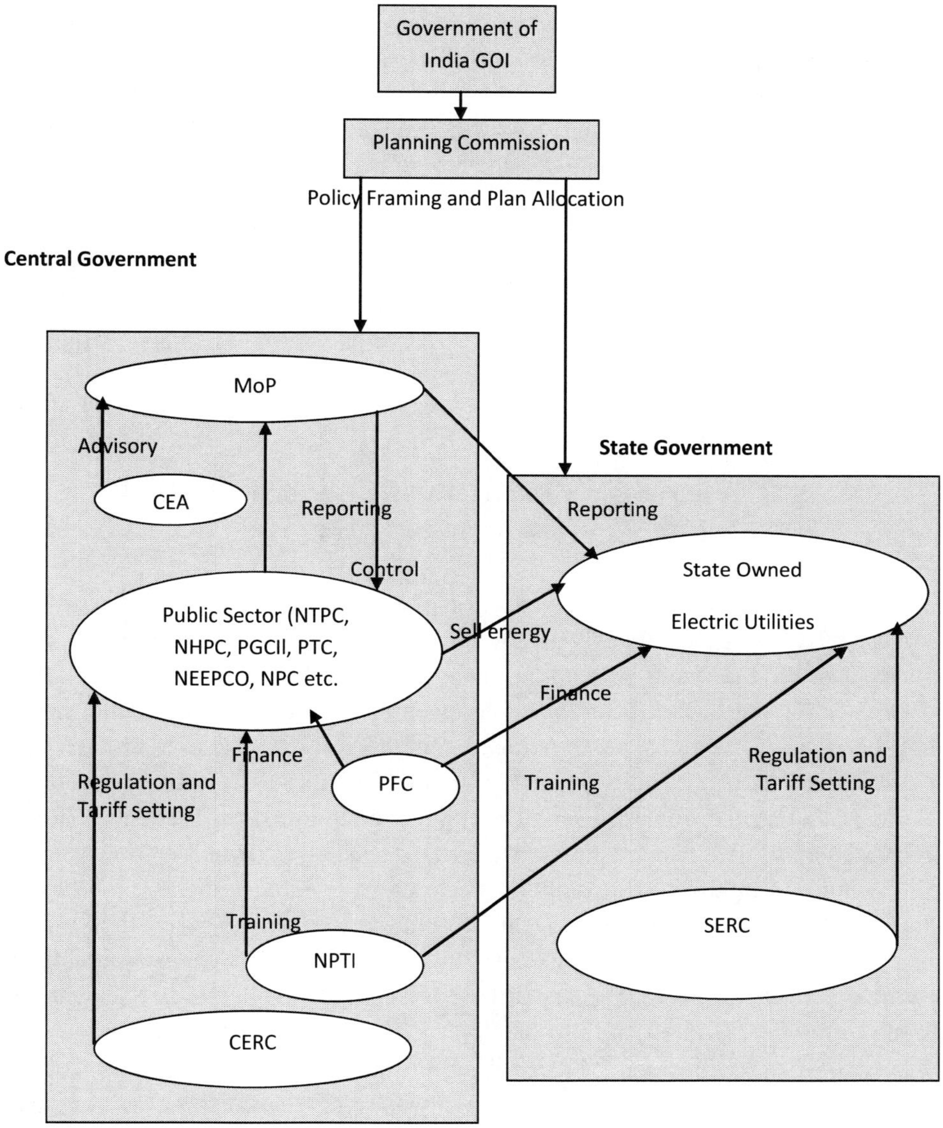


Figure 1. A Schematic View of The Indian Power Sector.

The National Power Training Institute (NPTI) acts as the national apex body for the human resources development of the power sector in India, while the Energy Management Center (EMC) was set up in 1989 to strengthen energy management capabilities in the country through workshops.

Trainings, seminars, multimedia, awareness campaign, etc. pending the power sector revamping as envisaged in the Electricity Act 2003, the major players in Indian sector continue to be the SOEUs. To arrest the deteriorating state of the sector, the Government of India (GoI) and a number of SOEUs have begun developing and implementing strategies suggested by the Ministry of power and the world bank strategic studies such as those reported by Kessides (2004), and Bacon and Jones (2002). Most of these suggestions rivet on drive to introduce corporate governance and promote private sector participation both (foreign and domestic). Improvement of government structure and commercialization with privatization has been limited to only two states; Orissa and Delhi.

As on 31 st March 2004, there were 13 SOEUs, 7 state electricity departments and 6 union territories electricity departments existing in the country (CEA, 2005). Haryana, Uttar Pradesh, Andhra Pradesh, Karnataka, Rajasthan, Madhya Pradhesh and West Bengal were the first Indian States that have unbundled their activities and separated generation, distribution and transmission functions. These states have commercialized their activities but they are still in effective government control. Some other states like Maharashtra, Gujarat, Uttar Pradesh and West Bengal have private licensees (including 18 cooperative societies) and 3 Muncipal licenses at the end of year 2002-0. (CEA, 2005). As of Nov 2005, there are 18 Licensees for interstate trading including both public and private traders.

## 3. Review of Performance of the Indian Electricity Sector

The Indian Electricity sector remained a complete State monopoly with social objectives till the year 1991. The social objectives did help serve the underprivileged domestic and the agricultural class but the results were undesirable in the long run as continued cross subsidies spelled disaster for the overall economy. The electricity prices rose in the industrial sector that bore the brunt of the subsidies provided to the domestic and agriculture sector and by 1999-2000 the tariffs for the industrial sector became 15 times that in the agriculture sectors and 2.1 times that in the domestic segment (Planning Commission (b), 2002), forcing the industry to step up its own captive power plants. This resulted in dwindling power sales to the industrial sector: the industrial consumption going down from 67% in 1960 to 40% by 1991 (Tata Energy Research Institute, New Delhi, 1993) and to nearly 30% by the year 1998-99 (Planning commission, 2002). The result was that the Electricity Boards accumulated huge losses as the recovery of average financial cost of supply through average revenue realized has declined from 76.7% in 1996-97 to 68.58% in 2001-02. In 1995-1996 while 9 of the 19 SEBs incurred losses, by the year 2000-2001, all of them were in the red (IEA, 2002). SEBs were increasingly unable to pay for the electricity they purchased from the central public-sector power companies, or from independent power producers (IPPs).

This coupled with the policies such as unmetered changes in agriculture (flat rates charged based on pump capacities) and often in the domestics sectors coupled with large scale thefts, resulted in deteriorating O and M status reflected in mounting T and D losses that became very high. The power supply position as on September 2006, indicated a peak deficit of 12.8% and energy deficit of 8.3% (MoP, 2007) at the All India level as against a peak deficit of 18% and energy deficit of 11.5% during 1996-97 while the gap between average financial cost of supply and average revenue realized increased from 50 cents/Kwh in 1996-97 to 110 cents/Kwh in 20001-02. (Planning Commission, 2002).

**Note: subsequent the year 2002 several of the SEBs have unbundled their operations, so that further data for SEBs is not available.*

# 4. Background of the Electricity Reforms

The reasons for reforms differ between countries- especially between industrial and developing countries. In mature industrial countries pressure for change has grown with the emergence of excess capacity and from disillusionment with capital-intensive generation projects triggered by the oil crises of he 1970s (Kessides, 2004). In developing and transition countries reforms have been driven by the poor operating and financial performance of state-owned electricity systems (with low labour productivity, poor service quality, and high system losses), lack of public funds for badly needed investments, unavailability of service for large portions of the population, and government desires to rise revenue through privatization (IEA, 1999; Bacon and Besant-Jones, 2001).

It is for these reasons that the recent years have witnessed wide and rapid acceptance of power Industry restructuring in several developing countries. This has also been necessitated by the pressing need for improvements in the existing power services in most of these countries, many of which are plagued by sub-optimal sector performance.

Huge demand-supply gap is often a universal problem in developing countries and the distribution sectors are frequently financially crippled. Serious cash flow constraints result in palpable curtailment of much needed investment in expansion and maintenance of services and reflect in poor sector performance. High distribution losses, poor management, low market densities, poor metering and billing practices and weak institutions are some of the common problems besieging the developing nations (Alberto, 2004), and India has been no exception.

## 4.1. The Case of India

The Indian Electricity sector remained the State monopoly with overriding social objectives till the year 1991. Reforms introduced since 1991, did not result in significant improvement in the financial credit worthiness of the SEB's and could not induce capacity addition in the sector. For example, there were significant capacity slippages form the 5 year Plan targets both in VIII and IX Plans (to the extent of 46% and 53% respectively)

This was despite the fact that reforms have been very comprehensive and have achieved significant progress with respect to the following aspects.

- Separation of policy, regulatory and operational aspects including set up of independent Regulatory Commission at the Central and the State level
- Increase commercial autonomy to distribution companies
- Unbundling for focused benchmarking across entities and transparent administration of unbundled entities.
- Institutional reforms and capacity building initiatives including development of organizational capabilities
- Transparent governance of the sector including better subsidy administration and well defined enabling legal, policy, regulatory and commercial frameworks.

**Table 1. Plan-wise Composition of Installed Capacity**

| Plan | Target (MW) | | | | Achievement (MW) | | | |
|---|---|---|---|---|---|---|---|---|
| | Central | Sate | Private | Total | Central | Sate | Private | Total |
| Ninth Plan 1997-2002 | 11909 | 10748 | 17588 | 40245 | 4504 | 9450 | 5061 | 19015 |
| Eight Plan 1992-1997 | 12858 | 14858 | 2810 | 30538 | 8157 | 68352 | 14304 | 16422.6 |

Source : Planning Commission (2002).

The issue before the government was not the achievements per se, but concern with respect to the pace of reforms and whether they fulfilled stakeholder's expectations. The government was therefore, inclined to take steps regarding framing of appropriate legislation and policy framework that would expedite the reforms and would revamp and restructure the power industry.

## Major Legislative Initiatives in India

The power sector in India has witnessed four important Legislations:

1. The Indian Electricity Act, 1910,
2. The Electricity (Supply) Act, 1948,
3. The ERC Act, 1998, and
4. The Electricity Act, 2003.

The key feature of these acts have been summarized as follows:

*1. The Indian Electricity Act, 1910*

- Provided basic framework for electric supply industry in India;
- Growth through private licensees with license provided by state government;
- Provision for license for supply of electricity in a specified area;
- Legal framework for laying down of wires and other works; and
- Provision for laying down relationship between licensee and consumer.

*2. The Electricity (Supply) Act, 1948*

- Mandated creation of SEBs; and
- Need for the state to step in (through SEBs) to extend electrification (so far limited to urban centers) all across the country.

*3. The ERC Act, 1998*

- Provision for setting up of CERC/SERC with powers to determine tariffs;
- Constitution of SERC optional for states; and
- Distancing of government from tariff setting process;
- Rational for change in legislative framework.

4. The Electricity Act 2003

Despite the above legislation in place, the need was still felt for devising a new legislation governing the Indian power sector to take into account the changed condition under the reforms initiated since 1991, and also due to the following;

Requirement for harmonizing and rationalizing provisions in the existing laws.

- To create a competitive environment which would result in enhancing quality and reliability of supply to consumers.
- Distance regulatory responsibilities of the government.
- To obviate the need for individual states to enact their own reform laws.
- To introduce modern concepts like power trading, open access, appellate tribunal etc.
- To provide special provisions for rural areas.

This new legislative called the electricity Act, 2003 formulates a comprehensive legislation imparting renewed thrust to coordinated development of power sector in India.

## 5. The Electricity Act 2003: What It Entails

The electricity act, 2003 provides a comprehensive yet flexible legislative framework for power development and envisions a sector characterized by a competitive market in power where the regulators and the power utilities play increasingly significant roles. The act aims to provide the paradigm shift by progressive introduction to competition and choice. The intent of the Act is to provide complete commercial autonomy to buy and sell power. The provisions of the act have come into force since June 10, 2003 and the Act consolidates and augments the previous Electricity Acts.

The Act is move towards creating a market-based regime in the Indian power sector and consolidates the laws relating to generation, transmission, distribution, trading and use of electricity. It generally takes measures conductive to development of electricity industry, thereby promoting completion, protecting interest of consumers and supply of electricity to all areas.

It also takes care of retionalization of electricity tariff ensuring transparent policies regarding subsides, promotion of efficient environmentally begin policies, constitutions of a Central Electricity Authority and regulatory commissions, establishment of Appellate tribunals, etc.,. Steps that shall entail impacts on every segment of the power sector, be it generation, transmission or distribution, and shall introduce large scale changes in the current power industry structure (Figure 2). The main feature of the Act are summarized in Table 2.

Provision in the Act will finally change the present Single-Buyer Model to Multi-Buyer model. There would be several players operating at the different stages of the power industry: generation, transmission and distribution. Open access to transmission and distribution segments resulting in strengthing of infrastructure, which is critical for the Indian power sector.

In the medium to long term, competition created among various categories of players will hopefully, lead to improved efficiency and better customer service standards. Wholesale and eventually retail markets shall provide choices for end user in terms of supplier, reliability of supply, and competitive tariffs. This will also lead to a situation where monopolies enjoyed SEBs for buying/selling power will cease to exist, leading to market-determined tariff structures.

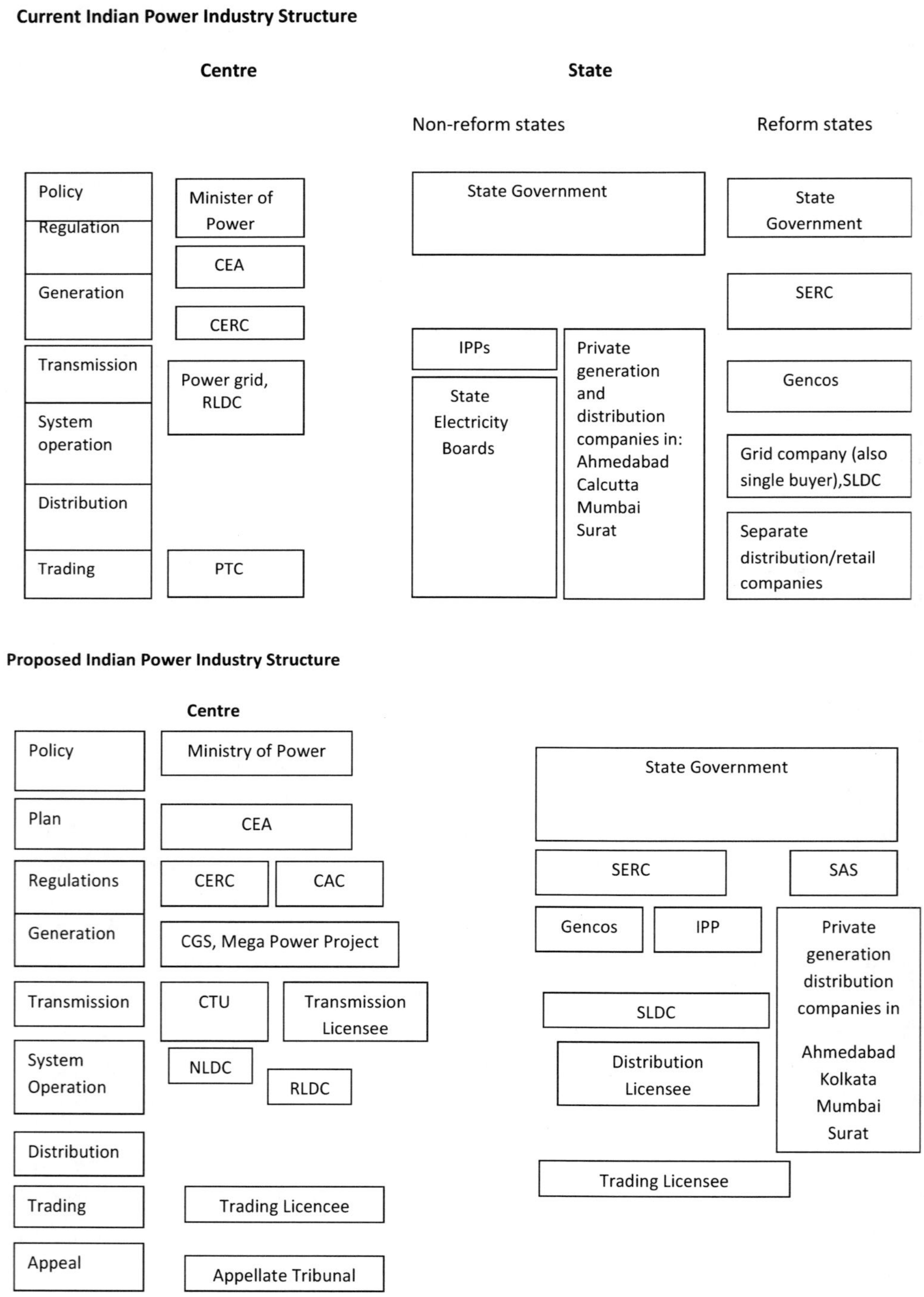


Figure 2. Current Versus Proposed Indian Power Industry Structure.

**Table 2. A summary of the Major Features of the Electricity Act 2003**

| Sl. No. | Category | Key Feature |
|---|---|---|
| 1 | Objectives | Encouraging autonomous regulation with the separation of policy regulation and operational aspects.<br>Rationalizing tariff and lowering the cross-subsidization levels<br>Creating competition in the industry<br>Ensuring supply of electricity to all areas<br>Protecting consumers interests |
| 2 | Policy | A National Electricity Plan shall be prepared in accordance with National Electricity Policy every 5 years<br>National policy on stands alone systems for rural areas and Non-conventional energy systems.<br>National policy on electrification and local distribution in rural areas. |
| 3 | Restructuring | Vertical integration instead of horizontal unbundling of State Electricity Boards (SEBs) to make them financially strong<br>State governments will have the freedom to decide the sequence and phases of restructuring, and also retain the integrated structure of the SEB for a limited period<br>Introduction of the concept of power trading as a distinct activity, and the introduction of a spot market for bulk electricity. |
| 4 | Generation | Removal of captive power plants from the ambit of licensing and other permissions<br>Generators can contract directly with DISCOMs<br>DISCOMs can have embedded generation<br>Captive generation allowed freely-can supply to associates<br>Elimination of Licensing requirement and techno-economic clearances for generation projects except hydel projects |
| 5 | T and D | Provision for Private participation in distribution<br>Surcharge for open access to meet current cross-subsidy burden (except for CPP's)<br>Dedicated transmission lines allowed (not regulated)<br>Central and State transmission Utilities (CTU and STUs) not permitted to trade Transmission licensees allowed<br>Multi Year Tariff (MYT) recommended<br>Bidding allowed<br>Provision of Non-discriminatory open access mandatory, Open access in distribution is to be introduced in phases, SERCs will frame regulations within 1 years regarding phasing-in of open access, open access only when allowed by regulator<br>Fixed charges for third party sales for service obligation of incumbent<br>Non-exclusive licensing.<br>Parallel networks explicitly allowed<br>Mandatory metering within 2 years (extendable by GoI) |
| 6 | Regulation | Gradual (progressive) reduction and ultimate elimination of Cross-subsidization<br>The formation of a State Electricity Regulatory Commission (SERC) in every state to be mandatory |
| 7 | Legislation | Provision for appropriate legislation to check power theft<br>Provision of Appellate Tribunal |
| 8 | Consumer protection | Consumer protection against failure to meet the standards of performance<br>Mechanism for redressal of consumer grievances |

## 5.1. Impact Analysis of the Electricity Act 2003

### *Impact on Generation*

Despite undertaking massive expansion projects the Indian Power Sector has not been able to much the rapidly growing demand for reliable and cost-effective supply. The demand has continued to grow at a compound annual rate of growth of nearly 8% and has outstripped the supply leading to a winding gap. To bridge its future demand supply gap, India would need capacity addition of nearly 100,000 MW in the coming 10-12 yeas. The Ministry of power (Mop) estimates that the additional capacity requirement to meet these storages is about 10,000 MW every year. This translate into an investment of about US$10 billion per annum. In this scenario to fulfill the aim of Indian Government to provide power to all by 2010 appears ambitious and practically difficult to achieve. To bridge the gap between investment requirements and available public resources, the active and large-scale participation of the private sector is necessary. In 1991, Indian power sector was opened for impendent power producers (IPPs), but the results were not up to expectation-by march 2006, only 12% of the installed capacity belonged to the private sector (Table 2). in view of the demand-supply gap, capacity slippages and to ensure greater private sector participation, the Electricity Act 2003 envisages new policy measures in the generation sector.

The act will free entry into generation thereby delicensing generation. However, hydro electronic projects will require usual clearance from the CEA, besides those from environment ministry and the state government. Any generating company may establish, operate and maintain a generating station and have grid connectivity if it complies the technical standards, laid by CEA. The IPPs companies can access transmission lines without any discrimination or can construct heir dedicated lines. IPPs or any generating company would have complete commercial autonomy to sell its power to any entity (However, as per the Act Transmission Licensee, National and Regional Load Dispatch Centers and Central and State Transmission Utilities cannot engage in trading business).

### *Importance* of Captive Power Generation f*or India*

Another area in which the Act is likely to have an impact is the captive power generation. The installed captive power plant (CPP) capacity in India as in July 2002 was 24,000 MW i.e., about one-fourth of the total installed capacity. Despite restriction, captive power growth at over 8 percent rate has been higher than the 5 percent growth in total installed capacity (Infraline.com (a), 2003). Before electricity act 2003 the policies for setting up CPPs were not uniform, varying from state to state, and substantial policy differences existed in term of wheeling and banking charges, third party sales, fuel use and size restrictions on CPPs amongst various states, so that the overall scenario was quite confusing for an investor.

**Table 3. Percentage Contribution of the Government and the Private Sector in Power Sector (2006)**

| Central Sector State Sector Private Sector | | | |
|---|---|---|---|
| Generation | 32.5 | 55.5 | 12 |
| Transmission | 15 | 84.6 | 0.4 |
| Distribution | Nil | 88 | 12 |

Source: Mop, 2007.

The Electricity Act 2003 endeavors to simplify these procedural hassles. Captive plants will be freely permitted henceforth, not only for captive use in the promoter's own plants but also for the use of groups of industries, thus imparting new perspectives to the current policies on captive power generation in India. The act effective dose away with any permission required form the SEBs or CEA for setting up the captive power plant and transmission networks, only the relevant technical details need be provided to the state Government so that it may assess the demand supply position., long term impact of such capacities and formulate policies accordingly. CPPs are also free from surcharge due to the cross subsidy. Any surplus power generated by the plants can be offloaded into the grid. Through, the entry barriers to setting up captive power plants have been eased, the supply of electricity from captive generating plants through the grid shall be regulated in the same manner as the generating stations of a generating company.

**Impact on Transmission**

So far the Indian Transmission system has been taken care of by the Power Grid Corporation of India Limited (PGCIL). The ministry of power (MoP) had notified the PGCIL as he Central Transmission Utility (CTU) after the Electricity Laws (Amendment) Acts, 1998. The PGCIL operates over 39,000 Circuit Km of EHV transmissions network across the country and has the distinction of being ranked among the top six transmission for utilities in the world . For making region-wise demarcation for efficient, economical and integrated transmission and supply, Government of India (GOI) has also constituted Regional Load Dispatch Centers (RLDCs).

Although the Electricity laws (Amendment) Act, 1998 had allowed independent transmission service providers to set up transmission service providers to set up transmission lines for interstate and intra-state transactions, but under the direction, control and supervision of the Central Transmission Utility (CTU). Section 27(A) of Part IIA of the Electricity Laws (Amendment) Act,1998 empowered the CTU to exercise supervision and control over the interstate transmission system. The Regional Load Dispatch Centers (RLDCs), which are mandated to coordinate the integrated operation of the power system in a particular region, are required by the law to be operated by the CTU. Also the section 27C (4) of the Electricity Laws (Amendment) Act, 1998 further stated that the applicant first obtains the approval of the CTU. Thus CTU has been given extra-ordinary power by the Electricity Laws (Amendment) Act,1998. This was justified on the basis of the fact that in any retail competition regime the role of transmission licensee is that of being the wires owner. Because of its integrated nature, this business has to be a regulated business, and transmission capacity can be procured on a competitive basis from multiple players by any centralized agency CTU or STU as the case may be. Grid planning function has to remain with CTU or any other system operator for safety, security and reliability reasons.

The Electricity Act 2003 permits multiple licenses in parallel transmission and distribution lines; thereby making a provision for private transmission licenses and in the process provides a fillip to power trading. Thus Act will introduce non-discriminatory open access to network. The CTU and STU are entrusted with the task of undertaking transmission of electricity and planning and coordinating the inter-state and intra-state transmission network in the country. CUT and STU will have right to provide direction, control and supervision of private transmission service providers. For the smooth operation of Transmission sector the Central Government has been entitled to establish a center at the national level, to be known as the National Load Dispatch (NLDC) and in state the State Load Dispatch Centers (SLDCs) for optimum scheduling and dispatch of electricity among the Regional Load Dispatch Centers (RLDCs). NLDCs shall be operated by Government but shall not engage in the business of trading in electricity. Similarly RLDCs and SLDCs will take care of Regional dispatch and intrastate dispatch respectively; they are also governed by Government and shall not engage in the business of trading in electricity. Similarly RLDCs and SLDCs will take care of Regional dispatch respectively; they are also governed by Government and shall not be engaged in the business of trading of electricity. Pending notification by the governments, the CTU, and STU shall operate the RLDC and SLDCs respectively. RLDCs/SLDCs shall comply with principles, guidelines and methodological as specified by the CERC/SERCs in the Grid Code.

### Impact on Utilities Operations Due to New Transmission System

Bulk supply business will cease to exist with the State Transmission Utility (STU) as per the provisions of the ACT. If the Transmission Company (Transco) intends to continue in trading, the contracts may have to be transferred to a separate trading entity owned by the Government. The Transmission Companies (Transco) would need to commerce a two-way/three way separation of business.

In the new regime, the STUs have been entrusted with critical planning function to ensure that congestion does not occur. Separation of responsibilities between STUs and System Operator (SO) in terms of functions would need to emerge especially with regard to the planning and implementation parts. In any case implementation of open access conditions would be complex. The SO will need to implement the rules of open access but the ownership could be a contentious issue. The implementation responsibilities and powers not clearly outlined in the Act and would need to emerge through the regulator process. Also the statement arrangements would increasingly become sophisticated and therefore, SOs needs to start preparing for the emerging scenario.

The merit Order dispatch would also need particular attention. The SO will be the primary agency to ensure that there is no abuse of market power. Ideally the market mechanisms should result in an automatic development of Merit Order conditions. However that may be long way away considering the realities of the present Indian electricity market. What needs to be especially noticed and taken care of is the fact that the available transmission capacity currently is almost fully committed to existing generators and there is little surplus capacity. Unless present lines are modified, little electricity will flow directly to buld users. This may entail considerable financial requirement, planning and commitment.

### Impact on Distribution

Many states are adopting reforms in distribution sector by adopting private participation in this sector. The Act provides solutions to the root problems of the SEBs, regarding their poor financial health due to continued subsidies (currently 1.5% of GDP) resulting in yawning gap between the national average cost of generation of Rs.3.50 per unit and the average realization of Rs.2.50 per unit. In India the Industrial and commercial sectors have been bearing the brunt of the subsidies provided to the domestic and agriculture sector. Despite continuing reforms cross subsidies reforms cross subsidies have continued to thrive mainly because of political compulsions and as a rational for serving the underprivileged through universal provisioning. The distortions in the tariff rates are severe: by 1999-2000 the

Tariffs for the industrial sector became 15 times that in the agriculture sector and 2.1 times that in the domestics agent (calculations from raw data in Planning Commission .

At the international level it is now a well recognized fact that cross-subsidy regimes prove counter-productive in the long run and that such regimes are sub-optimal. For example, Chattopadhyay (2004), assert from a study conducted in the Indian background that demonstrated cross-subsidies lead to wastage of resources, and rising price elasticity of electricity demand in industry, forcing the industry to seek energy alternatives leadings to declining revenues from the industry reflected in the share of industry in the total energy sales. In 2001-02, it was 29% as against 33% in 1996-97 (Planning Commission (a), 2002). This apparently forced the government to seek changes in the distribution sector in the Electricity Act 2003.

Following are the important features of the Act for the Distribution sector:

- Non-exclusive licensing. Parallel Networks explicitly allowed.
- Mandatory metering within two years (extendable by Electrical Regulating Commission (ERC)) from date appointed by Government. CEA may direct installation of meters at stages/locations deemed proper.
- Compensate consumer for failure to meet performance standards
- Non discriminating open access to be introduced in phases.
- Such open access may be allowed before the cross subsidies are eliminated on payment of a surcharge in addition to the charges for wheeling as may be determined by the state Commission. This surcharge shall be utilized to meet the requirements of current level of cross subsidy within the area of supply of the distribution licensee. However the surcharge and cross subsidies shall be progressively reduced and eliminated (as specified by the State Commission). Such surcharge shall however, not be livable in case open access is provided to a person who has established a CPP for carrying electricity to the destination of his own use.
- It shall be the "duty" of any distribution licensee to supply on request within a month (period extendable by the regulating commission for certain cases), subject to agreeing to pay the price.
- Where open access is allowed, consumers may enter into an agreement with any person for supply or purchase of electricity on such terms and conditions (including tariff) as may be agreed upon by them.
- Any distribution licensee can enter into generation business.
- Subsidies, if any, would be paid in advance, as prescribed by the regulatory commission. This is a positive development and an assurance to the Distribution Companies (DISCOMs), yet it may not really take away the entire risk that private DISCOMs would be exposed to, since it would depend on the state government's willingness and ability to service the subsidy requirement.

## Implications of the Act for Power Distribution

The Act allows distribution to be separated from supply. Supply circles limited even up to the level of substations could be given to private parties, thus effectively doing away with the requirement of large utilities and business parties. A much wider choice of parties could thus be tapped for privatization of distribution than has been possible in the Orissa and Delhi models.

In the new framework, innovation at various levels will be required to keep supply costs down. These could include adoption of alternative technologies like distributed generation, or alternate arrangements made for high cost energy (example, non-conventional energy being traded through Power Corporation (PTC).

Isolated grids would need active consideration (as also mandated by the law if enacted in current form) and demand side management (DSM) will become an even more important imperative for maintaining supply standards. The new environment would invariable require effective regulatory management to minimize business risks.

Due to the positive environment provided by the Act competition in the Distribution sector will increase in the medium to long term, especially due to open access condition. The larger consumers will have access to various suppliers. Hence, DISCOMs will have to work on strategies to retain consumers. The players will also have to focus on controlling the costs, and hence will have to concentrate on improving the existing infrastructure.

### Implementation of Open Access: Core Issues

Open Access implies that a number of licensees can reach a consumer through a network operated by a DISCOM. The access charges (called the wheeling charges) to be paid by the licensee shall be determined by the regulatory commission. Such non-discriminatory open access is a prerequisite to competition. The rules regarding open access are decided by CERC, to provide well-defined framework for the implementation of the open access system would further depend on the proper implementation of the merit order dispatch system. There might also be considerable time lag amongst various States to implement the open access system, as the State Electricity Regulatory Commissions (SERCs) have the discretion to allow open access to retail consumers in the time frame that suits them. Therefore, there needs to be a binding framework for introducing open access to retail consumers with a definite timetable synchronized with cross subsidy elimination path.

The open access system would also call for greater infrastructure investments, as its implementation requires excess available transmission capacity in the system. However, presently several sectors have current transmission capacities fully committed to the existing generators with little surplus. Another important pre-requisite for implementing the open access systems is electronic and real time metering at all levels of the transmission system. All this would invariably result in additional investment requirement.

Eventually open access may lead to declining electricity process in line with the international experiences. For example, electricity prices in UK have fallen by over 30% over the last decade and similar trends have been observed in USA and European countries (Haldea, 2003). In India also there is every chance that prices may eventually fall (after an initial upsurge to adjust to the realistic costs), once private investors start producing more power and able to sell it directly to the consumers. All this however, may take a long time spanning more than a decade. The open access system is likely to become fully operation in phases gradually over a long time once operational, the bulk buyers may be the ones who may benefit in terms of the choice of suppliers as well as the competitive prices. The implications of the policy change in the form of open access would be large- scale, interesting shall unfold only in the various empowered state be greatly influenced by the policy frameworks that are adopted by the various empowered state commissions.

## Trading: A New Policy Perspective

Subsequent to the Electricity Law (Amendment) Act, 1998, the power Trading company (PTC) was established with majority equity participation by the PGCIL along with the NTPC, PFC, and other Financial institutions. This structural arrangement was not a very healthy sing for an evolving competitive market, since trading is a service wherein competition is natural and should be provided. Thus this service needed to be opened to more players. The electricity Act, 2003 allow third party sales, and introduces the concept of trading of bulk electricity. The trading activity will be recognized as a distinct licensed activity. The basic intent is to introduce competitive market in India ; there are no restrictions on the number of players in the business of electricity trading. Whosoever qualifies the eligibility criteria (technical) to be set by CERC or relevant serc, would be eligible for norm specified by the commission. Distribution licensee and generation companies can engage in power trading. However, NLDC, RLDC, SLDC, CTU and STU and transmission licensees have not been allowed to trade in power. This is again to avoid an undue advantage to these players. ERCs will have the right to fix a ceiling on trading margins in intra-state trading, if necessary.

## Role of the Regulator

Role of Regulator shall be very crucial as happens in any free and open power market structure. The role will be important especially for promoting competition, fixing reasonable charges for transmission, generating tariffs, fixing wheeling and cross-subsidy charges, and in protecting the consumers from the rising prices of electricity, more so in time of shortages. In the regulated markets the role of the regulator shifts from setting prices to keeping vigilance that the markets remains competitive and non-monopolistic and to ensure market development. Table.2 illustrates the role of the regulatory commissions (Central Electricity Regulatory Commission (CERC) and the State Electricity Regulatory Commissions (SERCs) in new regime that the Electricity Act 2003 ushers.

The testing times for the regulator shall be to see through the transitional period that may take a few years. The smoothness with which the transition period is navigated would depend on the promptness of the regulator and their jurisdiction, which has not yet been completely clarified. Also, for any policy to materialize it is advisable to set time limit for its provisions to come into effect. In this case, however, the Act does not set time frame for One of the major function of regulatory Commission shall be to advise the government in matters pertaining to "promotion of competition, efficiency and economy in activities of the electricity industry" (The Electricity Act, 2003).

The transfer of all functions listed in the Act from federal state governments to respective regulators. The act leaves a lot of improvement for selection of regulators; the criteria for the qualification for the selection of regulator are very general and leave undefined much that is desirable in terms of proven technical credentials. Furthermore, the regulators shall be appointed initially for the period of three years and their term can be extended by the government for another three years. The implications of all of these are significant and far-reaching; the Act makes electricity regulators subservient to the governments that appoints, renews their term and terminates them, making them vulnerable to political pressures.

The Electricity Act 2003 provides for the establishment of an Appellate Tribunal by notification of the Central Government for the purpose of hearing appeals against the orders of adjudicating officer or the Regulatory Commissions. For the purpose of discharging its functions the Tribunal will have the same powers as a civil court and its orders shall be equivalent to the decree of a civil court. Chairperson of the Appellate Tribunal shall be equivalent to the decree of a civil court. Chairperson of the Appellate Tribunal shall exercise general power of superintendence and control over the Regulatory Commissions. The Act also defines the Terms of the Office and terms and conditions of the service for the members and Chairperson of the Tribunal. Thus it is hoped that the regulator shall be regulated by an independent forum that shall redress the ultimate grievances.

**Table 4. Role of Regulatory Bodies in the New Regime**

| | Aspect | Earlier Role | Role Subsequent To Act |
|---|---|---|---|
| CERC | Generation Tariff | Regulating tariffs for Central and Inter-State Generating stations | Regulating tariffs for Central and Inter-State generating stations |
| | Transmission Tariff | Transmission tariff setting for inter-state/region transmission Facilities | Determine tariffs, charges for intervening lines in case of disagreement |
| | Open Access | No defined role | Facilitate open access. Determine Surcharge for access by consumers |
| | Cross-subsidy | No role | As above |
| | Trading tariffs | Concept did not exist | Unregulated. However regulator may fix trading margins. Bidding allowed |
| | Load dispatch | No role defined, except for system Security aspects | Dispute resolution powers provided to CERC |
| | Licensing | For transmission facilities | For transmission and trading. License conditions to be determined within Lyr. |
| | Promoting Competition | State objective-however Mechanisms not define | Market development identified as key objective. Caps can be fixed |
| | Captives/NCE | No role | Dispute resolution powers on access to transmission system. |
| SERC | Generation/Bulk Tariff | 1. Approving generation tariff. 2. Regulating Bulk Supply tariff | 1.Approving generation tariff. 2.regulating Bulk Supply tariff function is now replaced by trading function in the Act for which margins will be determined by the appropriate commission. |
| | Transmission Tariff | Determining transmission tariff on cost-plus basis (sixth schedule) | Specify only if Licensees cannot agree upon tariffs. Determine surcharge |
| | Open Access | No defined role | Promote open access with surcharge as deemed necessary. |
| | Cross-subsidy | Objective defined-specifics on | Cross-subsidy mechanism absent |
| | Trading tariffs | Concept did not exist | Unregulated. However regulator may fix trading margins. |
| | Load dispatch | No role defined except that the Procurement should be efficient | Dispute resolution powers envisaged |
| | Licensing | Issuing powers for transmission, Bulk supply distribution and retail | Trading included. License conditions to be determined within 1 yr |

## The Electricity Act and Rural Electrification

Indian economy is predominately agrarian and as much as 72.2% of the population resides in the rural areas (Census, 2001). Going by the current rate of electrification nationally about 1 million new connections are added each year, whereas the annual growth in households is 1.85 million. Approximately 78 of the 138 million rural households do not have access to electricity services. With the number of households likely to grow to 157 million by 2012, in order to achieve the goal of electrifying all rural households would imply that about 97 million households would have to be electrified by 2012 or about 10 million households would have to be electrified by 2012 or about 10 million household per year (based on 2002 figures). This implies higher rate of electrification coupled with massive investments with minimal returns in terms of revenues. It is not very likely that the private sector will find the task attractive and the only hope would be the international aid in the form of loans etc. with Gol promoting a number of schemes and programmes to accelerate the pace of village electrification, much-needed impetus has in the past, come from the international arena.

The Electricity Act 2003 envisages a National Policy (to be prepared by central Government in consultation with the state Government) on stand- alone systems (including those based on renewable sources of energy and other non-conventional sources of energy) for rural areas. Hopefully, this would make the process of electrification self-sustaining in the rural areas by making it a commercial process.

This would also pave the way for the non-conventional producers of power, reinforced by the fact that the Act allows a person (notified by the State Government) to generate and distribute electricity in rural areas without getting a license from the Commission. The new policy will help attract necessary investment in generation as well as distribution from small and medium private players and also help the rural areas grow economically and socially.

The Act also proposes for formulation of a National policy for rural electrification and local distribution in rural areas. This policy would exclusively deal with bulk purchase of power and management of local distribution in rural areas through local Institutions such as the Panchayats (elected village committees), users' associations, co-operative societies, non-governmental organizations or franchisees.

The policy is to be formulated by the Central Government, in consultation with the State Government and the State Commissions. Under the Act, it will be obligation of the Appropriate Government to supply electricity to all areas including villages and hamlets. This would also help expedite rural electrification programme.

Overall the policy perspective is likely to benefit the smaller private sector enterprises that deal in renewable and non-conventional energy sources. An attempt has certainly been made to introduce private sector participation in rural electrification, as no license would be required for generation and distribution in rural areas.

This is a positive development even from the perspective of strategic investors in the DISCOMs. Not only is there greater scope for smaller players in power business now but also there is hope to light the life of every individual in the country as electrification of remote villages has been an unenviable task so far.

## The National Electricity Plan-2004

Section 3 (1) of the Electricity Act 2003 has mandated for the Central Government to prepare the National Electricity Policy and Tariff Policy in consultation with the State Governments and CEA for development of power System based on optimal utilization of resources. Accordingly, CEA has been paying special attention to the preparation of a National Power Plan[9]. Since the inception of the Electricity Act 2003, radical structural. Changes are being implemented in the power supply industry, and these have a significant bearing on the concept and procedures in the domain of power system planning. The recently announced National Electricity Plan 2004 has been prepared taking into account these developments, as also to provide a blue-print for development of the Power Sector. This plan is in accordance with National Electricity Policy and is expected to function as a facilitating lever for implementation of the Electricity Act 2003. As per the Policy, this Plan is for the ongoing $10^{th}$ Plan periods and includes perspective plan for $12^{th}$ Plan period.

The plan shall serve as a roadmap towards optimum growth of the Power Sector and shall have the following objectives:

- Universal access to electricity- available for all households in nest five years.
- Availability of Power-Demand to be fully met by 2012. Energy and peak shortages to be overcome and spinning reserve to be made available.
- Supply of Reliable and Quality power of international standard at reasonable rates by 2012.
- Per Capita consumption of electricity to be increased to over 1000 units by 2012.
- Minimum lifeline consumption of 1 unit/household/day.
- Protection of consumers' interests.

Further, the National Electricity plan, 2004 would be for short-term framework of five years while giving a 15 years perspective and would include:

- Short-term and long-term demand forecast for different regions;
- Suggested areas/locations for capacity additions keeping in view the economics of generation and transmission, losses in the system, load requirements, grid stability security of supply, quality of power including voltage profile etc, and environmental considerations including rehabilitation and resettlement;
- Integration of such possible locations with transmission system and development of national grid including type of transmission systems and requirement of redundancies;
- Exploration of different technologies available for efficient generation, transmission and distribution; and
- An addressal of issues concerning Generation, Transmission, Distribution, R&D, Power trading and Market development, Private Sector participation, Training and HRD, Energy Conservation, Cogeneration and Renewable and Non-Conventional Energy Sources, Rural
- Electrification, and Consumer Protection and Dispute Resolution.

## Consumer Protection

The entire Indian power sector has so far been driven by populist measures that have cultivated subsidy regimes. In turn these policies have promoted system inefficiencies in the form of poor power quality and poor service qualities. It is generally perceived in India that both of these will improve in the new restructured set-up. However, Sharma (2003) refers the studies by Colette (200) and paddon and small (1999) to drive home the fact that consumer interests in the privatized and restructured regimes may actually not be looked after so well as in generally perceived. Perhaps in line with such apprehensions, the Electricity Act 2003 has spelled "protecting interest of consumers and supply of electricity to all areas", in its preamble as specific policy direction to ensure consumer protection.

The Act provides protection to consumers with reference to standards of performance. If a licensee fails to meet the specified standards then disregarding any penalty that may be imposed or prosecution pronounced, the regulatory commissions will have powers to impose appropriate compensation to be paid to the affected person within the time limit of 90 days. In addition, every licensee will have to furnish information about the level of performance achieved by it to the commission and the ERCs shall arrange for publication of such information at least once every year. Beside, State advisory committees constituted by respective SERCs will have consume protection as a declared objective.

The Act also provides for establishment of a forum by the distribution licensee, within six months from the appointed date or date of grant of license, whichever is earlier, for redressal of grievance of the consumer in accordance with the guidelines specified by SERC. To protect honest consumers who pay, the Act also lays out provisions for specific penalties and punishments such as imprisonment for electricity theft or theft of electrical lines, tampering of meters, or deliberate or negligent wastage of electrical energy.

The Act has also made it the policy of the government to endeavor to supply electricity to all areas. Distribution licensee has to supply electricity within one month from the date of request for supply, except where capital works are required for connectivity. Failure of distribution licensees to supply within the said period would attract penalty. In the Appropriate Government to review the quality of power supply and consumer satisfaction. Separate polices for the rural areas where majority (72.2%) of consumers reside, have been envisaged thus giving priority to the social objective of power supply to all.

## Environmental Concerns

Environmental impact of the reforms is an area that is of vital concern; however linkages between reforms and mental concerns are not yet established. Khanna and Zilberman (1997) demonstrated the potential of institutional and economic policy reforms that provide incentives for adoption of efficiency-enhancing production practices to reduce carbon emissions while increasing net electricity generation. Similarly Kulshreshtha (2000) argues that coal using technologies might improve (fluidized bed combustion etc.) resulting in better utilization of poor quality coal, and alternatively, utilities can opt for coal beneficiation or might simply import good quality coal, thereby decreasing emissions. Fuel substitution is another way out for reducing emissions; countries like china are, in fact, actively promoting the policy of replacing coal by other environment benign fuels such as oil, renewable energy and Gas. Other studies subscribe to a conflicting viewpoint. For example, field data from Australia suggest that the reforms have actually led to increasing emissions (ABARE, 2000). Sharma (2003). Also demonstrates that the trends indicate increasing carbon dioxide contributions from the power sector ever since the introduction of the competitive markets. This might possibly be an outcome of benchmarking criteria that often employ short-run marginal costs environment friendly fuels and technologies with the relatively cheaper but not so benign fuel and technologies might tax the environment, but do not form part of the traditional benchmarking process, and hence the potential power producers may not have much incentive (unless forced by regulatory measures) to reduce carbon emissions.

Environmental impacts assume significance for India particularly because India, the second most populated country, contributes nearly 5% of global $CO_2$ emissions and there are possibilities of increasing this share in future. This increasing trend, combined with industrialized countries success in stabilizing their $CO_2$ emissions, may make its global share even larger (Reddy and Balachandra, 2003). The power sector is amongst the greatest contributors to the emissions. India is the world's third largest producer of coal and this coal has high ash content and low calorific value. With coal contributing as much as 59% of the fuel type in the total installed capacity as on June 2002 (MoP, 2002 (a) and the fact that increased coal consumption over the past 4 decades has led to a nine fold increase in energy related carbon emissions , the implications for the environment might actually be far reaching.

Conspicuously, the Electricity Act 2003 is largely, and almost completely, silent on matters concerning the environment. As a matter of policy, it does however recognize "promotion of efficient and environmentally benign policies" as one of the objectives right in its preamble, nut does not spell out detailed policy directives in this regard. It is completely left to the individual regulatory commissions to enact measures that further the said objective and to the usual government clearances for large thermal and hydel projects. Under such a scenario, it is likely that environmental concerns may take a back-seat, more so because the government is preoccupied with the challenge of providing reliable and affordable power supply to all as a matter of priority.

## Shortfalls and Limitations of Electricity Act 2003

It is expected that the Electricity Act 2003 will bring about revolutionary changes in the Indian Power Industry and would bring around a transformation that would act as model policy reforms for other developing countries. The Act however, does leave certain policy issues unresolved which have to be decided by regulators in consultation with appropriate government (Table-2). These issues include:

- Network pricing complexities- resulting in capacity excess or shortage
- Mechanism for pricing supply till open access is allowed.
- Settlement complexities in context of present cash shortage in system.
- Transmission bottlenecks negating open access.
- Environmental Concerns have not been addressed with reference to the new reform and Restructuring polices.

This is yet another category of Issues that have been addressed only partly in the Act. These issues include:

- Market dominance of few dominant generators that may result in increased prices.

**Table 5. Major Policy Issue of Concern In the Act, 2003**

| Sl. No | Category | Policy Issues |
|---|---|---|
| 1. | Generation | • Will existing Generation Companies directly contract with DISCOM ?<br>• How will the Genco contracts be allocated to DISCOMs?<br>• Will there be a government intermediary (trader) to take over existing contracts.<br>• Can existing Licensees with direct transmission links function without a transmission license?<br>• (But with a transmission license they cannot trade/resell)<br>• What is he voltage definition of transmission?<br>• How is capacity to be scheduled and dispatched ? What are the Merit Order Implications?<br>• Can Gencos trade among each other? Is reselling to Gencos by traders allowed?<br>• Is group capacity going to lead to inefficient generation?<br>• Who sets the detailed rules for group captives?<br>• Can existing contracts be opened up by virtue of generation being a regulated activity? |
| 2. | Transmission | • How would CTU redistribute cross subsidy surcharge<br>• Would new consumers bear cross-subsidy burden for open access?<br>• Is merit-order dispatch a role of the LDC anymore? Or do the Discoms/Customers/Traders need to schedule their own generation?<br>• Institutional capacity in the STUs for transmission constrained scheduling and settlements lacking |
| 3. | Distribution | • How quickly will cross subsidies and the cross subsidy charges be eliminated?<br>• Will open access and free captive generation lead to rapid erosion of paying consumer base of the DISCOMs?<br>• Will parallel networks lead to cherry picking?<br>• So regulators fix license areas or intending licensees define it?<br>• How will the DISCOMs gear up to metering challenges?<br>• How is network access to be priced (form of MYT)? |

Mechanism for preventing sharp increase in trading prices- The market may not send appropriate price signal (or security) for new generation capacity.

There are other open ends that might cause concern:

- The Act has not specified the time frame for the regulators to allow open access system and for the Government to transfer all the functions (specified in the Act) to the regulator.
- The Act, while recognizing the fact that cross-subsides should be eliminated, does not specify any concrete time frame for elimination of the cross-subsidies.

The timeframe and adherence to a strict timetable for implementation of the various provisions of the various provisions of the Act will hold tremendous importance for the policy initiatives to succeed.

**Absence of Information Dissemination and Performance Evaluation**

One of the major areas of concern is the absence of stress on developing specific monitoring processes that would incorporate building an effective data reporting system leading to periodic review of performances, and steps that would lead to dissemination of information across various stakeholders, including the common people. Researchers like Bajaj and De (2004) have pointed out the importance of review of sector performance and dissemination of information as "an important component of succeeding in building acceptability of reforms" and towards the fact that as the sector moves towards broader reforms, it would be necessary for the regulatory bodies to develop strong and structured information dissemination and communications programme.

The documents of the government of India also lay stress on fostering efficiencies in the operation of the electricity services. For example, the roadmap for the Tenth Plan has the following reform objectives (Tenth Plan,2002); "Improving efficiency in all the three segments viz generation, transmission and distribution, either by creating separate profit centers with full accountability within the vertically integrated structure, or unbundling SEBs into generation, transmission and distribution entities or through other models of reform depending on the choice of the state government". Also there is stress on encouraging competition and private participation in "each element of the electricity value chain" (Tenth Plan, 2002). However, despite this stress on fostering efficiencies and competition in the sector, there has been little stress on measuring sector performances and inducting efficiencies in the operations of the electricity utilities-steps that are pre-determinants of restructuring, regulation and commercialization (including possible privatization) of services.

The Electricity Act, 2003, has left it to the discretion of individual regulatory commissions to implement performance measurements and no consistent policy with regard to the collection of information, formation of data bank, evolving frameworks and methodologies for performance appraisals, and dissemination of information etc. has been framed for uniform implementation across the states. The role of the CEA and CERC under such circumstances assumes large significance[10].

## CONCLUSION

The Electricity Act 2003 has spelled a distinct paradigm shift in the electricity sector in India. The Act encompasses wide reforms in electricity generation, transmission and distribution. However, distribution remains the weakest link, and presently represents the core central issue in the entire gamut of reforms in the electricity sector. The shift of focus towards distribution sector is a natural corollary to the efforts at making the sector financially viable since revenue is ultimately earned through the consumer who is positioned at the distribution network. The fundamental goal of distribution reforms is the creation of a structure that encourage competition, and provides efficient coverage to currently unserved and under-served areas. Therefore, we focus exclusively on the Indian electricity distribution sector, and analyses the related reforms in detail, suggesting the reorganization of the sector, and advocating alternative institutional arrangement for ensuring greater sector efficiency.

## REFERENCE

MOP 2007 Ministry of power, Government of India, available on line, http://powermin.nic.in/jsp-sevlets.

CEA,2006: Progress report on pump set energisation as on 31 july,2006, Central Electricity Authority-Distribution planning and Development Div, New Delhi.

THE ELECTRICITY ACT 2003: The Gazette of India, Extra ordinary, 2003, Part II, Section 1, No-39, Ministry of Law and Justice, Government of India, New Delhi.

Tripta thakur, S.G.Deshmukh , S.C.Kaushik and Mukul Kulshrestha, "Impact assessment of the Electricity Act, 2003 on the Indian Power Sector, Energy Policy, Vol.33(9),2005,pp.1187-1198.

N.Kessides, N.Ioannis, 2004: *Reforming Infrastructure Privatization, Regulation and Competition* "A world Bank Policy Research report".

R.W.Bacon, and J.E.Besant Jones, 2001. "Global Electric Power Reform, Privatization, and liberalization of the Electric Power Industry in Developing Countries"; *Annual Review of Energy and Environment*, Vol.26, PP.331-359.

In: Energy-Environment-Economics
Editor: Riad Benelmir

ISBN: 978-1-62948-862-2


*Chapter 11*

# ENERGY VERSUS ENVIRONMENT

*Michel Feidt*
LEMTA-Université de Lorraine, France

## ABSTRACT

The keynote lecture objective to this Conference is to examine the new paradigm that has become crucial for the humanity at the beginning of the new century, where the Economy (and especially the Finance sector) prevails.

Even though attempts have been made in order to understand the necessity to include the humanity in his natural and immediate environment and generally its evolution in the Universe, it seems that the need for a Sustainable Development doesn't represent a priority for the moment, maybe because of the lack of adapted criteria to measure this Development (Evolution for the thermodynamicist). At the same time, there is another very important factor that generates this situation, the current global economic crisis, which leads to cutting subsidies for developing clean technologies and therefore efficient technologies. We propose here to see how it's possible to extend the usual criteria for efficiency in the world of physics (especially thermodynamics) to the technical (and technological), then the economical and environmental (Ecology) sectors: from the efficiency notion to that of the quality factor. We can draw consequences in technical terms, but also economical and societal: the need to pass from the Sustainable Development to the Finite Development (Registered trademark). Some usual applications from the Energy sector will be given as examples, to illustrate this idea: combined heat and power plants, heat pumps, combined systems.

**Keywords:** Exergy, energy, thermodynamics

## 1. INTRODUCTION

The stakes for optimizing energy are they included in the paradigm Energy vs. Environment? We propose to analyze it here in an historical manner according to science and technology. The industrial revolution which dates from two centuries has "released" the man by offering him the power of machines, based initially on the use of a solid fuel, the coal. And then, the more recent development of transport (especially the automobiles, aircraft, etc.) has allowed, by an increased mobility, the commercial and human trade as we have known since approximately 1 century: the oil, liquid fuel, is the main vector.

Finally, the gas (gaseous fuel) comes reinforce the two precedents effects, especially in the industrial and domestic world. But in the rush and carelessness, only the quantitative aspects related to the energy were accounted for (Extensity, how would say the thermodynamicists), without take care that it was from sources stored, therefore finite and exhaustible, and passing almost entirely by the heat and combustion.

The brutal awareness of this reality with the occasion, especially, of the oil shocks, at the end of the 20th century, obliges us to reconsider the way we use energy, not only of the quantitative point of view (Extensity), but also from the qualitative point of view (intensity): the ubiquitous use of the electricity (electric-power) is a proof.

But also, and especially the relations and consequences for the Environment. The latter are emerged first in the form of acid rain, and it/has led us to deal with the presence of sulfur in fuels; the event the most noticed was the London fog. And then there was the hole in the ozone layer, which has led in the 1980s to a reconsideration, mainly because of the use of CFCS, then HCFCs and finally HFCs.

This has not been without drastic consequences on the refrigeration industry (freezing, under-freezing, air conditioning, air treatment) that concerned our societies in their daily lives.

Finally after a few tens of years, the global warming has become a major concern, with the evolution of the rate of $CO_2$ in the atmosphere. The result is an index well known to all, in the form of the GWP (Global Warming Potential). It should be noted on this occasion that this index represents in fact a Global Warming Potential equivalent in terms of $CO_2$ (the reference), but that it integrates other compounds (including methane, the refrigerants, ... ).

Knowing that energy and the environment are interdependent, what relevant criterion, the science (THERMODYNAMICS) can propose to allow the engineers to control the energy technologies necessary for the fulfillment of human needs and our societies, without endangering our environment? The preservation of our needs and our environment can pass through the exergy concept which dates from 1889.

Section 2 is dedicated to a brief history of the concept of energy/Exergy and its implementation in perspective until this day. Section 3 shows the complementarity and the differences as compared to other methods more or less similar existing already. Section 4 attempted to justify the choice of the energy/exergetic analysis to deal with the problems of energy-environment.

Section 5 shows the current status of the applications of this analysis. Section 6 discusses the current deficiencies in the view of the results, and allows for placing in perspective by proposals for extensions considered desirable.

## 2. Put into Perspective the Concept of Exergy

According to our bibliographic sources and after recent exchange with Gaggioli (2.11 ), it appears that the concept of exergy was invented by Gouy (1889) in the form o "usable energy" in 1889. At the same time Stodola (1898) ran an article on same vocation. But it will have to wait until 1956 for that RANT (1956) introduced the word "Exergy" for " technical disponible work".

Since things have evolved slowly, with a starting acceleration since the years of 2000 under the impulse of the physicists and engineers. To illustrate this fact, we will give here a succinct vision and selective resulting from the choice of the author (it is hoped that this choice is relevant but it remains open to the discussion). And with the years 2000 has emerged a tribune dedicated to the exergy in the form of the international journal called ''Exergy, (2004)''. An article by Rosen (2002), shows the importance of education in thermodynamics through the concept of exergy which is not yet unanimously transferred and accepted as a basic tool.

There have been books specifically devoted to this method that is the EXERGETIC Analysis, built on the concept of exergy. Among these books, one of the most recent is due to Szargut (2005) and includes as captioning: Technical and ecological applications.

It is therefore at the heart of the subject and the author is one of the inescapable references in this area at the international level. The French community is not without action, both in the past (Marchal, 1956; Le Goff, 1979) to the present (Queiros-Conde et al., 2011).

To show and confirm this rise in power of the concept of exergy, the reader is referred to a recent article of Sciubba and Wall (2007) who wants to be a brief and commented history on the exergy since his birth in 2004.

Even if you may not share the entire document, the latter is very comprehensive and it is done state on the date of issuance of more than 2600 references, which are accessible by a link on the site of the journal (http://www.icatweb.org/vol10/10.1/sciubba-wall.pdf). We can only be convinced, given the amount of work on the concept of exergy and its applications, the relevance and usefulness of the latter.

# 3. THE EXERGETIC ANALYSIS

## 3.1. The Definition of the Exergy

The most recent definition of the exergy can be put in the following form: theoretical maximum of noble energy usable for a system (or process) S evolving (or carried) to its thermodynamic equilibrium with its immediate environment according to a process in which S interacts only with its immediate environment (Benelmir, 1989).

Note: the designation "Exergy concept" is used in Europe, whereas it is rather "Available Energy" in the United States (Usable Energy).

## 3.2. The Forms of the Exergy and the State of Reference

In each form of energy (Cf Feidt, 2006, chapter 3) correspond a form of the exergy (Sciubba-Wall, 2007, p. 3). In summary, the noble energies correspond to Exergies associated (often reduced to the unit of mass when it is possible: specific mass energy).

It should be noted simply that the chemical energy is not fully convertible into another form of energy (for example mechanical or electric); for a pure substance the difference in concentration between the system S(c) and the environment (c0) occurs, along with the potential for chemical reference μ0. It sees here reveal the importance of the state of reference.

For more in the case of the radiative electromagnetic energy, unanimity is not gained and the discussions are continuing (Pons, 2011).

Finally we note that the reference mass has no meaning for the radiative energy, as for the electrical energy.

Contrary to all the noble previous energies (coherent), thermal energy (non-coherent) is an energy which workable part is dependent of a factor of Carnot θ, which is worth:

$$\theta = 1 - \frac{T_0}{T_Q} \tag{1}$$

$T_0$ is the reference temperature, that of the environment generally assume isothermal

$T_Q$ is the temperature representative of the system S, and this must be precised.

For the configuration very common of an open Thermo mechanical system, it comes classically per unit of mass for a pure body:

$$\theta_x = h - h_0 - T_0(s - s_0) + \frac{v^2 - v_0^2}{2} + g(z - z_0) + \mu - \mu_0 + RT_0 \ln(\frac{c}{c_0}) \tag{2}$$

The first 3 terms represent the physical Exergy.

The 2 following terms represent the mechanical kinetic and potential exergy.

The last 2 terms represent the chemical exergy of the pure body.

The Exergy concept is a function of state (as compared to the baseline status of the chosen environment). It is remembered that the Exergy, E is retained in accordance with the following relationship:

$$E_x = E - An \tag{3}$$

An- represents the ANERGY associated with a processing whatsoever. It corresponds to the destruction of exergy in any real transformation (therefore irreversible) such as:

$$\Delta An = T_0 \Delta S_i \tag{4}$$

ΔSi is the entropy created in the thermodynamic transformation.

Note: The environment may be subject to of developments (for example the atmosphere can vary in temperature, pressure, humidity, according to various time scales seasonal, daily, instantaneous (meteorology) and the exergy varies in consequences. This may be considered and has been long recognized (Gögus).

The relationship (4) is very important because it indicates the fundamental link existing between the exergetic analysis and the entropy analysis, introduced and developed under the impetus of A. Bejan (1982) for the coupled transfers from heat and matter. It is to be noted that the exergetic analysis applied to the energy systems date from the same period (Ahern, 1980).

The entropy analysis is independent of any environmental reference; it therefore appears as a method more absolute, a method of physicist. The exergetic analysis located the system (or process) in its environment; due to this fact it is a method of engineers, and as we are concerned more today.

### 3.3. Efficiency and Optimization

All of the existing studies since the birth of thermodynamics (Brodiansky, 2006) show the concept of efficiency as a key concept, linking the consumption D to a useful effect U. The simplest case, introduced by Carnot, led to the concept of energy efficiency ηe, of a Thermomechanical engine:

$$\eta_E = \frac{EU}{UD} \tag{5}$$

This relationship illustrates well the fact that this efficiency involves only the quantities of energy, regardless of their quality. A discussion therefore exists as to the extension of this concept of efficiency, which then allows him to integrate the extensive and intensive aspect (Feidt, 2009).

To this day, according to Sciubba and Wall (2007), three definitions of the Effectiveness emerge from this discussion:

- an exergetic efficiency also called efficiency within the meaning of the second law:

$$\eta_{Ex} = \frac{E_x u}{E_x D} \tag{6}$$

With Exu, the useful exergy and ExD, the expended exergy

The reversibility coefficient:

$$\Psi = \frac{Exergiedesproduits}{\sum Exergied'entrée} \tag{7}$$

The exergy destruction coefficient:

$$\xi = \frac{Exergiedétruite}{\sum Exergied'entrée} = \frac{T_0 \Delta S_i}{\sum Exergied'entrée} \tag{8}$$

The definition (6) that characterizes for us exclusively the exergetic efficiency (to the exclusion of the efficiency within the meaning of the second law, Cf Feidt, 2009) is very useful. On the contrary the relationship (7) which is an efficiency within the meaning of GRASSMANN (Feidt, 2002) do not seem to have a lot of usefulness

The ratio of destruction of exergy (8) appears more interesting because it connects to the exergy of constant entry, the exergetic optimization to the entropy optimization. It is here the minimization of dissipation. This minimization has been particularly studied by A. Bejan (1982) for the coupled transfer of material and heat, in the form of a method known as EGM (entropy Generation minimization). It appears to the author that this vision is reductive as it is explained in the paragraph that follows.

## 3.4. Stakes of the Exergetic Optimization

Sciubba and Wall (2007) emit the idea that except in Thermo-economy, "from 1970 to 2000, all of the work is relative to the procedures of optimization: the challenge is to define the objective function the most appropriate to maximize the exergetic efficiency of a system (or process) for an entry of resources imposed ". These authors summarize the work made in this direction, under a number of topics:

- theoretical developments
- applications to the conservation of energy
- Improvement of the Efficiency
- progress in chemical processes
- development of design tools
- study of the properties of raw materials and of standard states of reference.
- work to educational and informative vocation

### *3.4.1. Energy Efficiency*

It is proposed here to pose the problem more precisely in terms of optimization in targeting on energy efficiency in a first time (Feidt, 2009).

It is certain that the concept of efficiency is a difficult concept and that it must be used with caution. We have shown that the efficiency within the meaning of the second law ($\eta_{II}$) that we will call rather Fq (quality factor) is not independent of the system environment, and the constraints associated with it. For example in the case of Carnot engine (well known), it comes:

In energy expenditure imposed, $Q_0$

$$F_q(Q_0) = 1 - \frac{\Delta S_i}{(T_{SH} - T_{SC}) Q_0} \tag{9}$$

TSH and TSC, temperatures respectively of the hot thermostat and the cold thermostat

To useful effect imposed, W0

$$F_q(W_0) = 1 / \left[1 + \frac{T_{SC} \cdot \Delta S_i}{W_0}\right] \tag{10}$$

The relations (9) and (10) clearly show that under certain very specific conditions (constraints), the optimization of the quality factor corresponds to the minimization of the entropy created by the dissipation. But this optimization rest energy, although in relationship with the environment of the system (here source and sinks).

### *3.4.2. Exergetic Efficiency*

The system is here deliberately considered in its reference environment. The application to a Carnot machine for reverse cycle (Feidt, 2009) led to the following results:

- for the machine, the exergetic efficiency is:

$$\eta_{Ex}(\text{machine}) = \frac{COP_{\text{réel}}}{COP(T_0, T_C)} \quad (11)$$

This efficiency depends on the temperature $T_C$ of the cold fluid cycle, and $T_0$, the temperature of the ambient thermostat.

- for the system (machine, hot and cold thermostats, environment to $T_0$), the exergetic efficiency becomes:

$$\eta_{Ex}(\text{système}) = \frac{COP_{\text{réel}}}{COP(T_0, T_{SC})} \quad (12)$$

This efficiency then depends on $T_0$ and Tsc, the temperature of the cold thermostat. The immediate consequence is the difference between the relative exergetic efficiencies and the quality factor: they are identically in the particular case $T_0 = T_{SH}$, therefore the need to situate the machine or the system (process) in its environment.

The expansion to other references that thermal ($T_0$) appears as a challenge for the environmental studies and the criteria to associate (Feidt, 2009).

### *3.4.3. Optimization Energetic And Constraints*

Energy optimization in which we limit ourselves now, shows the main objectives functions following:

with $Max(Eff.) = Max\left|\frac{UE}{EC}\right|$ (13)

We see from (13) that Max (Eff.) differs from Max (EU) and Min (EC). A differentiated optimization is therefore needed with any constraints to identify. The minimization of rejects R is clearly an environmental optimization. It is concluded that in general there is a matrix of optimization for objectives constraints functions or non-constraints. Table 1 gives a statement of these from an energy approach.

The transposition to an exergetic approach is immediate. It should be noted that the extension to the economy is well covered, while that on the environment still deserves a careful review. The application to engines was recently illustrated (Feidt, 2009). The same approach has been applied for machines to reversed cycles (cold machines, heat pumps).

### 3.5. Conclusion - Prospects

The EXERGETIC optimization appears as a natural crucial extension; therefore, the use of energy becomes intimately linked to the environment. The most common practice presupposes an environment represented by its reference temperature T0 (constant). This is happy because the intensive variable temperature is connected to the variable s extensive entropy, so that the entropy analysis or the exergetic analysis joins together.

**Table 1. Matrix of optimization of a system (or process) forced or non-forced**

| $F_0$ \ Constraints | | E.U. | D.E. | Efficiency | $S_T$ | Finite Dimensions | Costs | Others |
|---|---|---|---|---|---|---|---|---|
| E.U. | MAX | | MAX | MAX | MAX | MAX | MAX | MAX |
| D.E. | Min | | | Min | Min | Min | Min | Min |
| Efficiency | MAX | MAX | MAX | | MAX | MAX | MAX | MAX |
| $S_T$ | Min | Min | Min | Min | | Min | Min | Min |
| Finite Dimensions | Min | Min | Min | Min | Min | | Min | Min |
| Coûts | Min | Min | Min | Min | Min | Min | | Min |
| Others (Environment) | | | | | | | | |

Still a variable reference temperature complicates the problem. A reference in multiple variables (pressure, concentrations, chemical potential …) must also be developed, if to take into account environmental concern becomes more important. Results exist in this direction, but are to develop.Finally, the THERMODYNAMICS optimization shows that the potential objective functions are numerous, as well as the constraints. Also the development of a search is often multi-criteria (the example of the efficiency criterion is a very good illustration). It appears to us that the development of an Optimal THERMODYNAMICS in Finite Dimensions is a subject that is very current, in whose development we work (FDOT, Finite Dimensions Optimal THERMODYNAMICS).

## 4. Applications of the Exergetic Analysis

### 4.1. Technical Applications

The primary application of the exergetic analysis is without doubt in the world of engineering (Gool, 1992; Hammond, 2007), mainly from energy conversion systems, but also the chemical processes. In the fields of mechanical engineering, a lot of work related to the production of mechanical power (electric) and the associated components: steam turbines; combustion turbine; systems for renewable energy (since the 1980's); other systems (cogeneration, three and polygeneration; Fuel Cells; engines of all types, including in the field of aeronautics and space).

The heat exchangers and the associated networks have been the object of particular attention, given the importance of these basic components, in the system and energy processes and the existence of a compromise between the irreversibilities of heat transfer and matter (A. Bejan, 1982). In the areas of engineering for processes (chemical engineering), the cryogenics has been driving during the 1950s; and then the extension was made to all types of reverse cycle machines, progressively up to now (Sankaranarayanan, et al., 2010). In parallel, the exergetic analysis has been applied to chemical reactions, the combustion. For many specific applications energetic analysis have been developed, including the drying, the uptake of CO2, but also in the areas of the distillation, desalination, and agri-food industries. It should be noted that for the distillation, entropy analysis remains very used (Kjelstrup et al., 2010).

It can be concluded here, that despite the study of all of these works the subject remains not exhausted, and there is still much to be done in term of hybridization and the integration of system.

## 4.2. The Environmental Applications

As it was stated in paragraph 3 above, the exergy is an indicator of the interaction of the system with its immediate environment; the exergy is therefore intrinsically linked to the environmental impact. This is confirmed and developed in the work and recent book of Szargut, (2005).But there exists to this day numerous variants included: (Frangopoulos et al., 1992) the environmental analysis and its relationship to the economy (Frangopoulos et al., 1997); the Exergetic analysis applied to the environmental impact and to the emissions of rejects (Rosen-Dincer, 1997, 1999); the exergy, a measure of the environmental impact (Sciubla, 1999); etc.

Sciuba noted that the difficulty of "Measuring" the "ecological cost of what is commonly called "pollution" is not simple and clear, because it can be quantified in monetary (economic) or exergetic terms ". So appears a new relationship to the economy, which will be resumed later in this referenced article.

The measure of the environmental impact need to clarify the attached criteria. The approaches are trying to develop the ecological criteria. Angulo Brown (1991) and Yan have proposed an ecological function based on thermodynamics, but different from the exergy:

$$\dot{E} = \dot{W} - T_0 \cdot \dot{S}_I \quad (14)$$

More recently $U_{st}$ (2006) introduced the concept of ecological coefficient of performance (ECOP):

$$ECOP = \dot{W}/(T_0 \cdot \dot{S}_I) \quad (15)$$

These criteria seem not fulfilling all the conditions necessary for taking into account of the environmental impact. The progress remains to do, to connect and extend various definitions in use for example in the field of combustions: relationship of ecological coefficient ε, with the indicator of pollution.

$$\pi_q = (CO_2)_{cq}/PCI \tag{16}$$

This is to get closer to the indicators GWP (greenhouse Warning Potential) and TEWI which are commonly used in the field of cold (Feidt et al., 2009).

Other methodologies have also been proposed, such as the Emergy method (Odum, 1995), or the method TERE (Total Equivalent Resource) is called exergy by physicists) (Reistad, Zhang, 1998). The management of resources also involves the recycling of materials and the reuse of materials in parts (there is necessity of differentiate), these two operations limiting the impact on the environment (by minimization of rejects)

The Life Cycle Analysis (LCA) is practiced since the years 1997-1998, but need as the emergy method, enrichment and a validation of databases.

# 5. Discussion-Conclusion

## 5.1. CEC, Contents of Cumulative Exergy

In fact the various attempts of extension mentioned in paragraph 4.2 , were preceded by the proposal due to Szargut, regarding the method of the content of cumulative exergy (CEC, Cumulative Exergy Content). This encryption method of exergetic cost is constructed on the additivity of the exergy: each process adds (Algebraically) of the exergy in input and output, destroyed of the exergy by irreversibilites: the exergy of the useful effect is therefore a cumulative exergy along the process chain. Valero (1986) formalized mathematically the Exegetic Cost of a product $E_o$, as a linear function of entry exergies $E_i$:

$$E_0 = \pi\{E_i\} \tag{17}$$

$\pi$, matrix of cost transfer of the chain process.

As a result of this fact the Exegetic cost method of Valero represents the linearized approach of the method CEC attributed to Szargut. These methods are methods of physicists.

## 5.2. Thermoeconomy

As has been seen, the engineer is always faced with the economy: the cost concept, or benefit is always present (Valero, 1986).

Taking into account the economic aspects dated from the 1960s; it has been developed in a concomitant way in the United States under the impulsion of Evans and Tribus, with the application thermo-Economics, and in Europe under the impulsion of Szargut, with the name of Exergo-Economics.

It appears to us that the Thermo Economic Efficiency (ETE) can be estimated on 3 bases, either the Cost of Investment (CI), or the cost of operation (CF; Operating and Maintenance Cost), either the overall cost, sum of the previous (Total Cost). The result is a multiplicity of points from the individual to the collective, to which the reader should pay attention, because influencing fundamentally on the result of optimization.

Even more, it seems to us an essential difference in the general area of the Thermo Economy; the difference between the Energo Economy and the Exergo Economy. If the first is well established and used in the world of the engineer, with many energo-economic correlations (Le Goff , 1979), this is not the same for the Exergo Economy which still requires a large work: the imprecision of language appears for example in (Tsatsaronis, 1993) regarding the optimization of the of energetic systems.

The thermo economic analysis proposed is of type Exergo economy. This article is interesting because it must appear the optimization; in effect an economic optimum in overall cost exists inherently as a compromise between the cost of investment and the cost of operation, the optimum on energy basis differs from that on the exergetic basis.

There are differences regarding the exergo-economy optimum according to the methods; this is reflected in the "dead loop" of a cogeneration system using combustion turbine (Tsatsaronis-Pisa, 1994).

## 5.3. Extended Exergetic Method (EEA, Extended Exergy Accounting)

This method is intended to express the energy costs, material, on the exergetic basis, but also the environmental costs, which are still at the beginning.

Sciubba (2001) proposed in the extensions of the method 5 factors of cost associated with the capital C, the mass M, the energy E, the work L, the environment O (method of CMELO); this method which looks for a scale of unique values is the continuation of MELTIDE + K method proposed by Le Goff (1979).

The work is to continue deepening the connections with the life cycle analysis, to account for the finite time; relationship has thermodynamics in finite time, and has the concept of Sustainable Development (Meyer, et al., 2009; Wall, 2010). It seems that the development must be more finite than sustainable.

## 5.4. Thermodynamics and Optimization

The whole of the previous methods leads to the optimization of systems and processes (Feidt, 1996), both from a design view but also diagnosed (static optimization, generally nominal regime), and of a control command point of view (dynamic optimization).

But the variety of objectives (energy, economic, environmental, entropy…) led in a multiobjective optimization, whose methods are a subject of active research (Martínez-Eliceche, 2009). Two big classes of problems result: the non-constraint optimization (natural), the constraint optimization which is without doubt the most common. We advocate and work to the development of Thermodynamics in Finite Dimensions (physical), (FDOT, Finite Dimensions Optimal THERMODYNAMICS) (Feidt, 2009).

## 5.5. Extensions of the Application Areas

### *5.5.1. Complex Systems*

The progresses achieved and future of exposed methods will help to the quantification of more and more fine efficiency for complex systems at various scales. A local analysis supported by the CFD is developed on entropy basis.

### *5.5.2. Biological Systems*

Application to biological systems is still in a preliminary stage, as was shown by the recent symposium organized by SFT (2011).

The corresponding approach is not unrelated with the Emergy concept which remains an approximate concept, according the rigor of "thermodynamics".

### *5.5.3. Societal Systems*

There is an exergetic balance at the level of a society.

The first study in this area is due to Reichstad (1975) in a United States exergetic balance report. It was followed by Nakicenovic (1996), Ersterag (2000-2001), Hammond (2001) studies for other countries, attempting to make comparison. Accordingly this becomes a useful decision tool for the politicians, policy makers, and the agencies.

## 5.6. Conclusion-Perspectives

Exergy concept is old and has given rise to quite a number of monographs (including in France), also this review articles (see in the reference list). This concept has remained too long the preserve of the academic world.

The relationship with the Economy and the Environment makes a candidate of choice for a general criterion of efficiency. But the work is not yet completed.

From this or these criteria, it is then possible to develop an Optimal THERMODYNAMICS in Finite Dimensions (FDOT), which characterizes the world in which we live (progressing according to the thermodynamics). The works are still in this meaning and must be open to other fields than the hard sciences given the generality of the proposed approaches.

## REFERENCES

Angulo Brown, J, 1991, *J. Appl. Phys.*, 69, 7465-7469.

Bejan, A., 1982, "*Entropy generation trough heat and fluid flow*", Wiley Interscience, New York, US.

Benelmir, R., 1989, Second Law Analysis of a Cogénération Cycle, PhD Thesis,Georgia Tech, US.

Brodiansky, V.M., 2006, *"Sadi Carnot 1796-1832"*, Presses Universitaires de Perpignan.

Erstenag, I, et al., (2000), *Energy,* 25 (b), 957-973.

Erstenag, I, et al., (2000), *Energy*, 26 (3), 253-273.

Exergy 2004, www.inderscience.com

Feidt M., 1996, "Thermodynamique et Optimisation energetique des systemes et Procedes", *TEC et DOC*, Paris, France (2e edition).

Feidt M., 2009, *Termotehnica,* 2, 26-33,

Feidt M., et al., 2009, "Communications to IIR-IIF workshop", Antony, France (Accepted for publication in Int. J. of Refrigeration).

Feidt M., Lang S., 2002, *Entropie,* 242, 2-11.

Feidt, M, 2006, "*Energetique*", Dunod, France.

Frangopoulos, C, 1992, Proc. of Int. Symp. on Efficiency Costs Optimization and Simulation of energy systems, 231-239.

Frangopoulos, C, Caralis, Y., 1997, Energy Conversion Management, 38(15-17), 1751-1763.

Gaggioli, R., 2011, "*Communication personnelle*".

Gogus, Y., Communication personnelle, Ahern, J.E., 1980, "*The exergy method of energy systems analysis*", Wiley Interscience, New York, US.

Gool V., 1992, *Energy*, 17(8), 791-803.

Gouy, M., 1889, "*J. de Phys.* 2e serie, t. VIII, 501-518 (in French)

Hammond, G.P., 2007, *Applied Energy*, 84(7-8), 675-700.

Hammond, G.P., Stapleton, A.J., *Journal of Power and Energy*, 285 (2), 141-162.

Journees SFT Paris 2011;

Kjelstrup, S. et al., 2010, "*Non-equilibrium Thermodynamics for Engineers*" World Scientific Publishing, Singapore.

Le Goff, P. 1979, *Energetique Industrielle*, TEC et DOC, Paris, France.

Le Goff, P., 1979, "*Energetique Industrielle, Technique et Docummentation*", Paris, France.

Marchal, R., 1956, "*La thermodynamique et le theoreme de l'energie utilisable*", Dunod, Paris, France.

Martinez, P., Elicheche, A.M., 2009, *Computer Aided Chemical Eng*, 27 (c), 1869-1874.

Meyer, L., (2009), *Energy,* 34(1), 75-80.

Nakicenovic, N, et al., 1996, *Energy,* 21(3), 223-237.

Odum, 1995, "*Maximum Power*", University Press of Colorado, US, (chapter 28).

Pons, M., 2011.

Queiros Conde D., et al. 2011.

Rant, Z., 1956, *Forsch Ing_Wes.*, 22(1), 36-37 (in German).

Reichstad, R.M., Zhang M, 1998, HVAC Research, 4(1), 45-65.

Reistad, G.M., 1975, *Journal of energy for Power Trans of ASME*, 97(3), 429-434.

Rosen, M.A. Dincer, I, 1997, *International Journal of Energy Research*, 21(7), 643-654.

Rosen, M.A. Dincer, I, 1999, *International Journal of Energy Research*, 23(13), 1153-1163.

Rosen, M.A., 2002, "*Exergy", an International Journal*, 2,207-210.

Sankaranarayanam, K., et al., 2010, "Efficiency and Sustainability in the Energy Chemical Industries" (second edition), CRC Press (Taylor and Francis Group), Boca Raton (Fl.), US.

Sciubba, E, (2001), *Int. J. Exergy*, 2, 68-84.

Sciubba, E. 1999, *Proc. of the ASME division*, AES vol. 39, 573-581.

Sciubba, E., Wall, G. 2007, *Int. Journal of Thermodynamics*, 10(1), 1-26.

Stodola, A., 1898, *Zeitschrift der VDI*, 23 (38), 1086-1091, (in German)

Szargut, J, 2005, "*Exergy Method-Technical and Ecological applications*", WIT Press, Southampton, UK.

Szargut, J, 2005, "Exergy Method-Technical and Ecological applications", WIT Press, Southampton, UK.

Tsatsaronis, G, 1993, Prog. *Energy Combust.* J, 19, 227-257.

Tsatsaronis, G, Pisa, 1994, J., *Energy*, 19(3), 287-321.

Ust, Y., 2006, *J. of The Energy Institute*, 79(1), 47-52.

Valero, A., et al., 1986, *ASME, AES*, vol. 2-1 et 3, 17 et43.

Wall, G. 2010, "*The Open Env. Eng, J*", 3, 21-32.

In: Energy-Environment-Economics
Editor: Riad Benelmir
ISBN: 978-1-62948-862-2


*Chapter 12*

# STIRLING REFRIGERATING MACHINE: CONFRONTATION OF DIRECT AND FINITE PHYSICAL DIMENSIONS THERMODYNAMICS METHODS TO EXPERIMENTS

***Lavinia Grosu*[*,1], *Stoian Petrescu*[2], *Catalina Dobre*[2], *and Pierre Rochelle*[1]**
[1]Université Paris Ouest Nanterre La Défense, Laboratoire Energétique, Mécanique et Electromagnétisme, France
[2]Politehnica University of Bucharest, Department of Engineering Thermodynamics, Splaiul Independentei, Bucharest, Romania

## ABSTRACT

The paper presents a theoretical and experimental study of the reverse Stirling cycle applied to a β type machine. It is based on analytical models developed for refrigerating machines by using a direct method (DM) and a Finite Physical Dimensions Thermodynamics (FPDT). The theoretical results are compared to experimental ones. The effect of the piston speed on the COP is emphasized. Although the heat losses due to imperfect regeneration are differently evaluated, the two methods present similar performance variation.

**Keywords:** Reverse Stirling cycle; Finite Physical Dimensions Thermodynamics; Direct Method, irreversibility causes

---

[*] E-mail: mgrosu@u-paris10.fr.

# NOMENCLATURE

$A$ heat exchange surface, $m^2$
$b$ distance between the regenerator matrix wire, $m$
$COP$ coefficient of performance, -
$c$ specific heat, $Jkg^{-1}K^{-1}$
$D$ diameter, $m$
$h$ convective heat transfer coefficient, $Wm^{-2}K^{-1}$
$K$ heat exchanger conductance, $WK^{-1}$
$k$ losses factor in regenerator, -
$m$ mass, $kg$
$N$ number of screens of the regenerator
$n$ engine speed, $rot.s^{-1}$
$p$ pressure, $Pa$
$Pr$ Prandl number, -
$\dot{Q}$ heat flow, $W$
$r$ gas constant, $Jkg^{-1}K^{-1}$
$S$ stroke of the piston, $m$
$T$ temperature, $K$
$V$ volume, $m^3$
$\dot{W}$ mechanical power, $W$
$X$ regenerative losses coefficient, -
$y$, $z$ adjustment parameters, -

## Greek Symbols

$w$ speed of engine piston, $ms^{-1}$
$w_{SL}$ speed of the sound at $T_L$ temperature, $ms^{-1}$
$\tau$ extreme temperatures ratio for the circulated fluid, -
$\varepsilon$ volumetric compression ratio ($V_{max}/V_{min}$), -
$\nu$ kinematics viscosity, $m^2s^{-1}$

## Subscripts

$CC$ Carnot cycle
$\varepsilon$ depending on $\varepsilon$
$H$ hot on heat sink side
$h$ hot on working gas side
$g$ gas
$in$ intake
$irrev$ irreversible
$l$ low on working gas side

*L* low on heat source side
*MF* refrigerating machine
*m* average value
*max* maximum
*min* minimum
*out* exit
*p* constant pressure (specific heat)
*rev* reversible
*R* regenerator
*v* constant volume (specific heat)

# 1. Introduction

Direct Method - DM (PETRESCU S., HARMAN C., 1994; PETRESCU S. et al., 2003), consists of the study and evaluation of the irreversibilities of thermal machines by analyzing the cycle, step by step, and directly integrating the equation of the First Law combined with Second Law of Thermodynamics with finite speed, for each process of the cycle. It is there obtained in a straightforward manner, the expression of efficiency or coefficient of performance depending on the speed of the processes and geometric and functional parameters. So far the Direct Method was used to study Stirling engines, classic and solar machines (PETRESCU S. et al.. 2003; 2010), refrigeration machines and Stirling heat pumps (PETRESCU S. et al., 2001; 2009), Otto engines (PETRESCU S. et al., 2008), Diesel engines PETRESCU S. et al., 2009), and also for the study of the Carnot Cycle, (PETRESCU S. et al., 2009), (CHEN L.G. and al., 2011).

The Finite Physical Dimensions Thermodynamics method (FPDT) (FEIDT M., 1996, 2009; GROSU L., 2000; ROCHELLE P., 2005), is a method which takes into account the finite time, finite speed and finite geometric dimensions. This method introduces exo-ireversibilities due to the finite heat transfer, between heat source/sink and working fluid, and also takes into account the constraints which the engineer is confronted with in practice. P. Rochelle and L. Grosu (GROSU L. et al., 2008), (MARTAJ et al., 2007) have shown that machines operating such similar Carnot cycles must be described using physical parameters as maximum pressure, maximum volume, rather than gas mass. It is considered essential to use the rotational speed as the main variable, since heat and mass transfer are dependent in a direct manner with speed.

In this article, we propose to confront the theoretical and experimental studies of the reversed Stirling cycle applied to a special type β refrigeration machine. This comparison will allow us to validate the two methods, for the thermal operating parameters of the studied machine.

# 2. The Experimental Device

The analyzed installation is composed of a β type thermal machine, characterized by an arrangement of piston engine, displacer and heat exchangers in a single cylinder. It can operate as an engine by using an electrical resistance located at the top of the cylinder, or as a refrigerating machine by driving the shaft with an electric motor.

This thermal machine is equipped with pressure sensor, piston instantaneous position sensor, thermocouples, ampermeter, voltmeter and with a device composed of photodiodes equipped with a drilled disc, which allows measurement of the speed of rotation.

The working fluid is air, which is assumed to behave as a perfect gas.

## 3. Direct Method Algorithm Applied to the Reversed Cycle

The present study analyzes the Stirling refrigerating machine performances by introducing the external irreversibilities of heat transfer with finite temperature difference. Calculation takes into account also the internal irreversibilities due to pressure losses generated by the finite speed and incomplete regeneration of heat in the regenerator, which are all related and determined function of the finite speed of the process cycle (piston speed or engine speed).

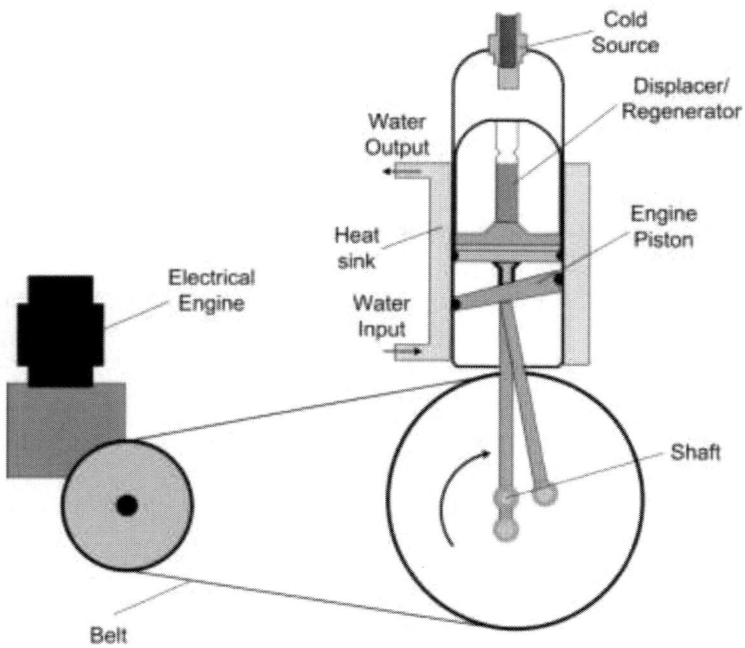


Figure 1. The experimental device.

Using the Direct Method, (PETRESCU S., HARMAN C., 1994; PETRESCU S. et al., 2003), the coefficient of performance of refrigerating machines, COP is expressed as a product between the Carnot cycle COP and the Second Law Efficiency, an idea used frequently in the literature in order to take into account the losses caused by the irreversibilities of the cycle (PETRESCU S. et al., 2001; 2009). It was shown that the expression of the performance coefficient is obtained as follows:

$$COP_{MF} = \underbrace{\frac{1}{\frac{T_h}{T_l} - 1}}_{COP_{MF,CC}} \cdot \underbrace{\left[1 - \frac{X(T_h/T_l - 1)}{r/c_v(T_m)\ln\varepsilon}\right]}_{\eta_{II,irrev,X}} \cdot \underbrace{\frac{1}{1 + \frac{3\left(1 - \frac{1}{\varepsilon}\right)\sum \Delta p_i}{p_{min}(T_h/T_l - 1)\ln\varepsilon}}}_{\eta_{II,irrev,\sum \Delta p_i}} \quad (1)$$

with:

$COP_{MF,CC}$ – coefficient of performance of reversible Carnot refrigerating machine, operating at $T_h$ hot gas temperature, respectively $T_l$ cold gas temperature.

$\eta_{II,irrev,X}$ – the Second Law Efficiency (PETRESCU S. et al., 2001; 2003; 2010; Feidt M., 1996) which takes into account the losses due to imperfect heat regeneration.

$\eta_{II,irev,\sum \Delta pi}$ – the Second Law Efficiency (PETRESCU S. et al., 2001; 2003; 2010), which takes into account the irreversibility due to pressure losses generated by the finite speed of processes, the gas flow in the regenerator and the two heat exchangers of the machine.

The irreversible power required by the refrigerating machine can be expressed with a unique equation, also resulting from the application of Direct Method in the studied reversed cycle.

The irreversible power required by the refrigerating machine can be expressed with a unique equation, also resulting from the application of Direct Method in the studied reversed cycle.

$$\dot{W}_{MF,irrev} = z\dot{W}_{rev} + \dot{W}_{loss,\Sigma\Delta p_i}$$
$$= \left[zm_g r(T_h - T_l)\ln\varepsilon + 3\left(1 - \frac{1}{\varepsilon}\right) - V_{min}(\Sigma\Delta p_i)\right] \cdot \frac{w}{2S} \quad (2)$$

The adjustment parameter, $z$, is deduced from the validation calculation scheme, by confronting the experimental results of the actual machine with theoretical ones ($z = 0.80$), and the piston speed is $w = 2nS_p$.

The study of the coefficient of regenerative losses (PETRESCU S. et al., 2008) and of the pressure losses in the machine due to the finite speed process is based on the studies of Stirling engines functioning with finite speed and imperfect regeneration.

The detailed analysis of the processes with finite speed in the regenerative, coupled with the First Law of Thermodynamic carry out to an analytical expression for the regenerative losses coefficient, $X$, which has been validated on 12 real engines and 16 operating regimes of these engines (PETRESCU S. et al., 2008).

Given that the Stirling engines have the same main geometric features both as an engine in operation, and as a refrigerating machine, it is considered that the coefficient of losses resulting from the irreversibilities in the regenerator has a similar expression to the one in the engine.

$$X = X_1 y + X_2(1 - y) \quad (3)$$

where the terms involved in equation (3) are presented in equations (4) and (5):

$$X_1 = \frac{1+2M+e^{-B}}{2(1+M)} \tag{4}$$

$$X_2 = \frac{M+e^{-B}}{1+M} \tag{5}$$

$$M = \frac{m_g c_{v,g}}{m_R c_R} \tag{6}$$

$$B = (1+M)\frac{h A_R}{m_g c_{v,g}} \cdot \frac{S}{w} \tag{7}$$

with $y$ an adjustment coefficient ($y=0.43$).

Convective heat transfer coefficient in the regenerator is:

$$h = \frac{0.395(4p_m / rT_l)w_g^{0.424} \cdot c_p(T_m) \cdot \nu(T_m)^{0.576}}{(1+\tau)\left[1 - \frac{\pi}{4[(b/D_w)+1]}\right] D_R^{0.576} \cdot Pr^{2/3}} \tag{8}$$

The regenerator has the diameter $D_R$, is composed of $N$ metal screens, having components wires whose diameter is $D_w$, and distance between two wires is $b$.

The other values involved in the above relations are evaluated, taking into account the average values of gas pressure and temperature in the cycle.

$$\frac{\Sigma \Delta p_i}{p_{min}} = \frac{1}{3\left(1-\frac{1}{\varepsilon}\right)} \left[ \left(\frac{w}{w_{SL}}\right) \gamma \left(1+\sqrt{\tau}\right) ln\varepsilon + 5\left(\frac{w_g}{w_{SL}}\right)^2 N + F \right] \tag{9}$$

with: $F = \frac{3(0.94+0.045w)10^5}{4p_{min}}$. (10)

## 4. Finite Physical Dimensions Thermodynamics (FPDT) Method Applied to Reversed Exo-Irreversible Stirling Cycles with Imperfect Regeneration

In this work FPDT method is applied to an exo-irreversible reversed Stirling cycle with imperfect regeneration, represented in Figure 2.

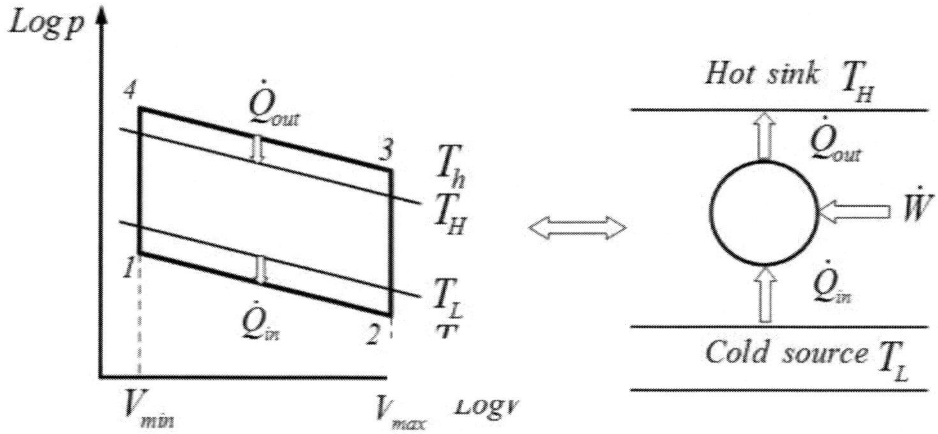


a) b)

Figure 2. Reversed exo-irreversible Stirling cycle (a), energy balance scheme (b).

Transferred energies are given by following relations:

• heat provided to the hot sink, by the working gas at temperature $T_h$, in the case of perfect regeneration:

$$Q_{inrev} = Q_{12} = p_{max} V_{max} \frac{ln\varepsilon}{\varepsilon} \frac{T_l}{T_h} = E_\varepsilon \frac{T_l}{T_h} = E_\varepsilon \frac{T_l}{T_h} \tag{11}$$

• heat transferred in a perfect regenerator (heat stored and restored by the working gas) during an isochoric transformation:

$$Q_R = m_g c_v (T_h - T_l) = \frac{m_g r T_h}{(\gamma - 1)} \left( 1 - \frac{T_l}{T_h} \right) \tag{12}$$

If the regenerator is imperfect, its efficiency is described by the relationship:

$$\eta_R = \frac{Q_R - Q_{R+}}{Q_R} \tag{13}$$

where $Q_{R+}$ is the heat that would be extracted from the one produced on the cold source and to the one provided to the hot sink ($Q_{R+}>0$). It can be expressed as:

$$Q_{R+} = (1 - \eta_R) Q_R = \frac{m_g r T_h}{(\gamma - 1)} \left( 1 - \frac{T_l}{T_h} \right) (1 - \eta_R) \tag{14}$$

The gas mass is considered to be fully transferred from the hot volume to the cold volume and vice-versa (neglecting the dead volume).

Replacing the $m_g r T_h$ product with $p_{max} V_{min} = \frac{p_{max} V_{max}}{\varepsilon}$, one obtain:

$$Q_{R+} = \frac{p_{max}V_{min}}{\gamma - 1}\left(1 - \frac{T_l}{T_h}\right)(1 - \eta_R) = \frac{p_{max}V_{max}}{\gamma - 1}\frac{1}{\varepsilon}\left(1 - \frac{T_l}{T_h}\right)(1 - \eta_R) \quad (15)$$

$$Q_{R+} = E_{\varepsilon} k\left(1 - \frac{T_l}{T_h}\right) \quad (16)$$

Notation $k$ is used to define the losses factor in the regenerator:

$$k = \frac{(1 - \eta_R)}{\ln\varepsilon(\gamma - 1)} \quad (17)$$

• heat exchanged in case of imperfect regeneration:

$$|Q_{out}| = |Q_{34}| - Q_{R+} = |Q_{outrev}| - Q_{R+} = E_{\varepsilon}\left[1 - k\left(1 - \frac{T_l}{T_h}\right)\right] \quad (18)$$

$$Q_{in} = Q_{12} - Q_{R+} = Q_{inrev} - Q_{R+} = E_{\varepsilon}\left[\frac{T_l}{T_h} - k\left(1 - \frac{T_l}{T_h}\right)\right] \quad (19)$$

It follows from this, mechanical work consumed per cycle, in absolute value:

$$|W| = |Q_{out}| - Q_{in} \quad (20)$$

This mechanical work is independent of regeneration efficiency. Heat flow transferred on the hot sink and cold source can be obtained also, taken into account the engine speed, $n$:

$$\dot{Q}_{out} = n|Q_{out}| = nE_{\varepsilon}\left[1 - k\left(1 - \frac{T_l}{T_h}\right)\right] = K_h(T_h - T_H) \quad (21)$$

$$\dot{Q}_{in} = nQ_{in} = nE_{\varepsilon}\left[\frac{T_l}{T_h} - k\left(1 - \frac{T_l}{T_h}\right)\right] = K_l(T_L - T_l) \quad (22)$$

where $K_l$ and $K_h$ are the cold source respectively hot sink conductance.

The coefficient of performance of refrigerating machine, $COP$, using Stirling cycle can be determined by the relation:

$$COP = \frac{Q_{in}}{W} \quad (23)$$

# 5. Confrontation of the Results

The main hypothesis of this analysis is that the expansion and the compression spaces are isothermals. It is also considered that the working gas is a perfect gas and its total mass remains constant throughout the experiment (closed thermodynamic system). Using geometric and functional parameters, measured or determined by the acquisition program (CassyLab) and using the calculation algorithm of the two studied thermodynamic methods, the following developments are obtained depending on the finite speed of processes, which is considered to be the common variable. (Table 1 and Figures 3-7)

**Table 1. Confrontation of The Performances (B) Obtained By Different Methods for The Same Initial Conditions (A)**

*a)*

| $\dot{Q}_{in}$ [W] | $n$ [rot/s] | $T_H$ [°C] | $T_L$ [°C] | $p_{min}$ [Pa] | $p_{max}$ [Pa] | $K_l$ [W/K] |
|---|---|---|---|---|---|---|
| 16.72 | 2.65 | 24.3 | 5.85 | 99780 | 211600 | 0.9 |

b)

| COP | | |
|---|---|---|
| Exp. | FPDT | DM |
| 1.03 | 1.17 | 1.2 |

The Direct Method introduces an efficiency that takes into account the heat exchange in the regenerator (Figure 3). Introducing this term tends to approach the experimental values to the analytical results, which validates the calculation model proposed by The Direct Method. The FPDT method takes into account heat exchanger conductances, which are identified by experiment. The evolution of the FPDT algorithm calculated factors that have a strong influence on the performance of Stirling refrigerating machine, is shown in Figure 5-7. For an overview of the losses of the machine, the results of the study by confrontation with experimental results are presented. Calculation results highlight the effect of irreversibilities on the performance of each machine and also the influence of finite speed processes. Thus, this study shows a significant increase of the real COP coefficient of performance of the studied refrigeration machine, with the increase of piston speed. (Figure 4).

The variations are similar for the two analyzed methods with an overestimation at low speed (Figure 7) even if the evaluation of heat losses due to imperfect regeneration is done by different models (due to introduction of different physical quantities, $k$ respectively $X$). There are showed also evolutions of experimental COP and Carnot cycle COP.

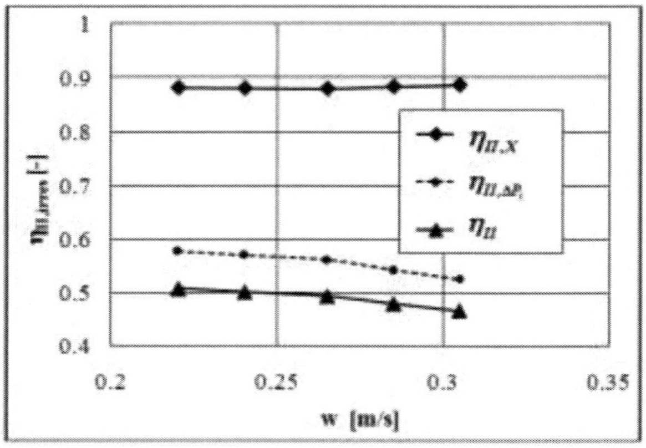


Figure 3. Second Law efficiencies variation that characterize the losses due to the finite speed of Stirling refrigerating machine.

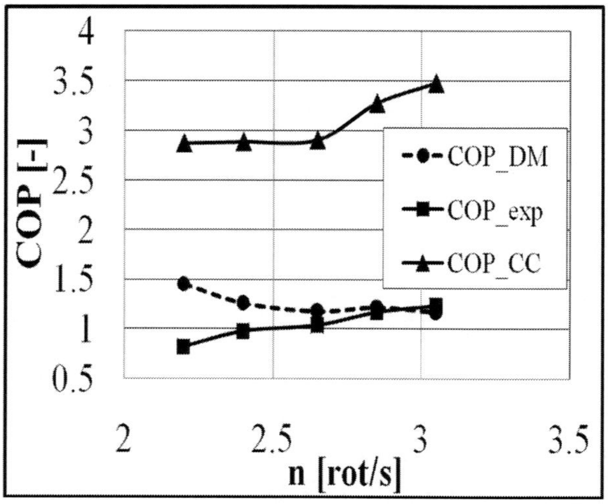


Figure 4. COP of Stirling refrigerating machine versus piston speed, using the direct method.

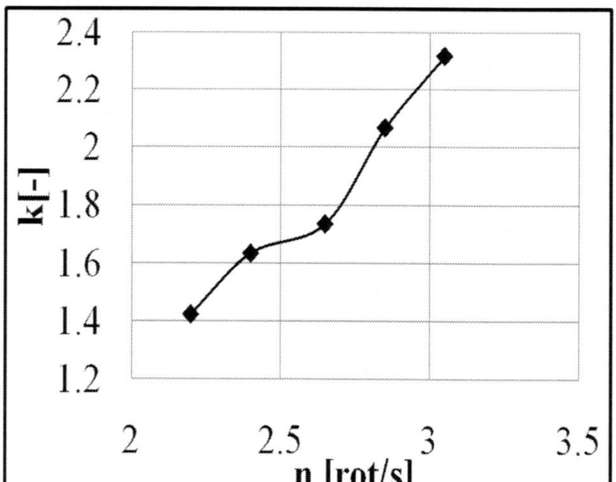


Figure 5. Losses regenerator factor versus rotation speed of the shaft.

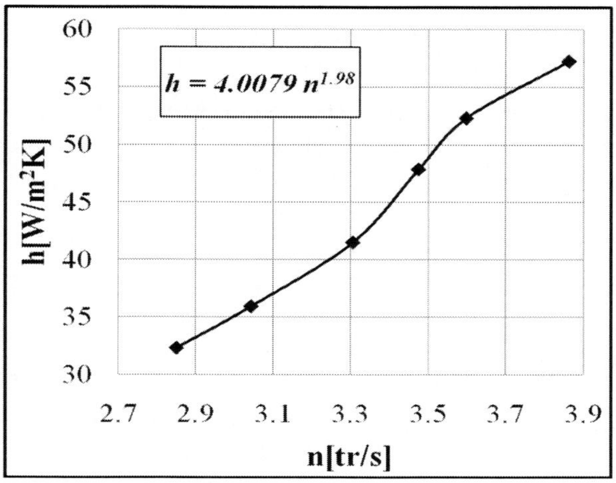


Figure 6. Heat transfer coefficient versus rotation speed of the shaft.

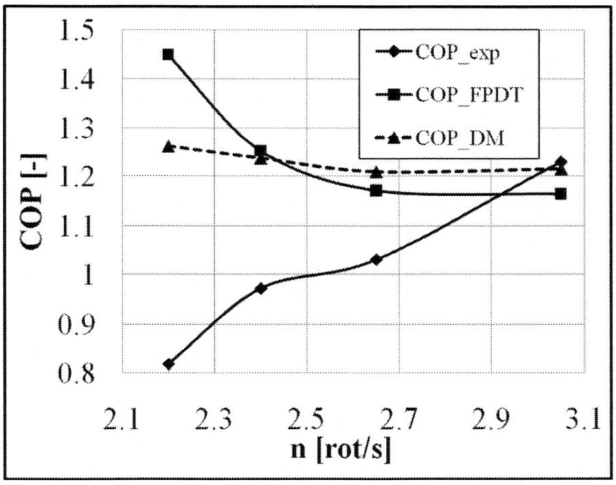


Figure 7. COP values obtained with different methods versus rotation speed of the shaft.

## CONCLUSION

This paper presents a study of a Stirling refrigeration machine, using two thermodynamic analysis methods: DM and FPDT. Stirling cycles have been studied, taking into account the finite speed of the process, but also other physical parameters such as $p_{max}$, $V_{max}$, $T_h$ and $T_l$. Reversed cycle performances were evaluated considering internal and external irreversibility. For different attempts, we conducted several measurements obtaining values for temperatures, pressures and instant gas volumes. The results of the two thermodynamic models have been presented by comparing them with experimental results. That led to validation of the two calculating methods for operating refrigeration parameters of the considered Stirling machine.

The purpose of this study was to compare the two methods of thermodynamic analysis and to determine which of the two is closer to experimental reality. The result is surprising because the two methods give very similar results, with the observation that the direct method is more accurate for low speeds but at higher speeds systematic error from experiment for the two cases.

## ACKNOWLEDGMENTS

The work has been funded by the Sectorial Operational Programme Human Resources Development 2007-2013 of the Romanian Ministry of Labour, Family and Social Protection through the Financial Agreement POSDRU/88/1.5/S/60203.

## REFERENCES

Chen L.G., Feng H.J., Sun F.R. (2011), "Optimal piston speed ratio analyses for irreversible Carnot refrigerator and heat pump using finite time thermodynamics, finite speed thermodynamics and direct method" *Journal of Energy Institute* Vol.84, N°2, pp. 105-112.

Feidt M. (1996), "Optimisation énergétique des systèmes et procédés", 2nd ed., *Techand Doc*, Paris.

Feidt M. (2009), "Optimal thermodynamics- new upperbounds", Entropy, 11, pp. 529-547.

Feidt M. (2011), "Thermodynamics applied to reverse cycle machines", *Int. J. Refrig*, 33, pp. 1327-1342.

Grosu L. (2000), "Contribution à l'optimisation Thermodynamique et Économique des Machines à cycle inverse à deux et trois réservoirs de chaleur". PhD thesis, Institut National Polytechnique de Lorraine.

Grosu L., Rochelle P., Martaj N. (2008), "Thermodynamique à échelle finie: optimisation du cycle moteur de Stirling pour ingénieur", 11-13 juin 2008, *COFRET*, Nantes-France.

Martaj N., Grosu L., P. Rochelle, «Thermodynamic study of a low temperature difference Stirling engine at stady state operation», *Int. J. of Thermodynamic*, Vol. 10, No.4, pp. 165 – 176, 2007.

Petrescu S., Harman C. (1994), "The Connection between the First and Second Law of Thermodynamics for Processes with Finite Speed. A Direct Method for Approaching and Optimization of Irreversible Processes", *J. of The Heat Transfer Society of Japan*, 33, *128*, pp.60-67.

Petrescu S. et al. (2001), "Les cycles des machines à froid et des pompes à chaleur à vitesse finie", *Rev. Entropie,* 232, 48-54.

Petrescu S. et al. (2003), "Application of the Direct Method to Irreversible Stirling Cycles with Finite Speed", *Int. J. of Energy Research*, 26, pp. 589-609.

Petrescu S. et al. (2003), "A Scheme of Computation, Analysis, Design and Optimization of Solar Stirling Engines", Proc. of the 16$^{th}$ Int. Conf. on Effic., Cost, Optimiz., Simul. and Enviromental Impact of Energy Systems – ECOS'03, Copenhagen, Denmark, I, pp. 1255-1262.

Petrescu S. et al. (2003), "Thermodynamics with Finite Speed versus Thermodynamics in Finite Time in the Optimization of Carnot Cycle", Proc. of the 6$^{th}$ *ASME-JSME* Therm. Engineer. Joint Conf., Hawaii, USA, CD.

Petrescu S. et al. (2008), "Optimization of the Irreversible Otto Cycle using Finite Speed Thermodynamics and the Direct Method", Proc. of the 10th WSEAS Int. Conf. on Math. and Comput. Methods in Science and Engineering (MACMESE'08), Bucureşti, Romania, II, pp. 51-56.

Petrescu S. et al. (2009), "Optimization of the Irreversible Diesel Cycle using Finite Speed Thermodynamics and the Direct Method", Bul. of Transilv. Univ. of Braşov, Romania, 2(51) – Series I, *1*, vol.1, pp. 87-94.

Petrescu S. et al. (2009), "Irreversibility Generation Analysis of Reversed Cycle Carnot Machine by using the Finite Speed Thermodynamics", Termotehnica, Ed. *AGIR*, Bucarest, XIII, *1*, pp. 43-48.

Petrescu S. et al. (2010), "A Methodology of Computation, Design and Optimization of Solar Stirling Power Plant using Hydrogen/Oxygen Fuel Cells", *Energy*, 35, *2*, pp. 729-739.

Rochelle P. (2005), "LDT Stirling engine simulation and optimization using finite dimension thermodynamics", In Thermo- and GFD Modelling of Stirling Machines, pp. 358-3.

In: Energy-Environment-Economics
Editor: Riad Benelmir

ISBN: 978-1-62948-862-2


*Chapter 13*

# ON THE PERTINENCE OF A BASED THERMODYNAMICS LAWS MODEL FOR THE SIMULATION OF A BETA TYPE REFRIGERATING STIRLING MACHINE

***Nadia Martaj[1], Lavinia Grosu[1], Pierre Rochelle[1], Stephan Savarese[2] and Rachid Bennacer[3]***

[1]Laboratoire LEME, Université Paris Ouest Nanterre-La Défense, Ville d'Avray
[2]Comsol, France
[3]Dpt GC/ LMT-Ecole Normale Supérieure – Cachan, Cachan Cedex, France

## ABSTRACT

The problems involved in the destruction of the ozone layer, due to the use of organo-chlorinated compounds describing an inverse cycle, lead us to take into account the development of new alternatives. Thus, refrigerating Stirling or Ericsson cycles, using air or another inert gas, might become beneficial alternatives. This study relates to a Beta type refrigerating Stirling machine, comparing experimental measurements with results of two numerical simulations, first one in 0D and second one in 2D (CFD).

## NOMENCLATURE

$A$ heat transfer area, [m$^2$]
$c_p$ specific heat at constant pressure, [J/kg K]
$C$ stroke, [m]
$D$ diameter, [m]
$k$ heat exchange coefficient, [W/(m$^2$.K)]
$m$ mass, [kg]
$N$ engine speed, [rev/min]
$p$ pressure, [Pa]

$\dot{Q}$ heat flux, [W]

$Q$ heat [J]

$r$ gas constant, [J/(kg K)]

$T$ temperature, [K]

$t$ time, [s]

V volume, [$m^3$]

$W$ work, [J]

$\dot{W}$ mechanical power, [W]

## Greek Symbols

$\varepsilon$ porosity , -

$\phi$ phase lag, [°]

## Subscripts

c compression

d displacer

e expansion

g gas

h hot/high

l low

p working piston

r regenerator

t total

w wall

## 1. Introduction

The use of the refrigerating Stirling machine, using air or another inert gas as working fluid, represents an interesting alternative to mitigate the problem of the ozone layer destruction by organo-chlorinated refrigerants [1-4]. The theoretical cycle is composed of two isothermal processes within heat exchangers, and two isochoric processes associated to the working gas flow through the machine regenerator. This cycle has, theoretically, a performance coefficient equal to that of Carnot cycle, assuming a perfect regenerator (Figure 1).

This cycle could be studied using finite time, finite speed or more generally finite physical dimensions thermodynamic method [5-8]. The engine have three heat exchangers and the working gas is everywhere within the engine; temperature depends not only on time but also on the considered position in the engine (different parts). Thus, it is useless to compare gas evolution in the Stirling engine with a theoretical thermodynamic cycle because of this space-time no homogeneity of the working gas.

In 1871, Schmidt [9] proposed a simplified analysis, known under the name of "*ideal analysis*", which inspired several models developed later [10-12]. In order to take into account the local phenomena, it is necessary to develop coupled analysis, more complete and more realistic than the previous analyses by simulating the thermomechanical processes and also fluid dynamics. One of the limitation of such analysis improvement is the increase in computing time [13-16].

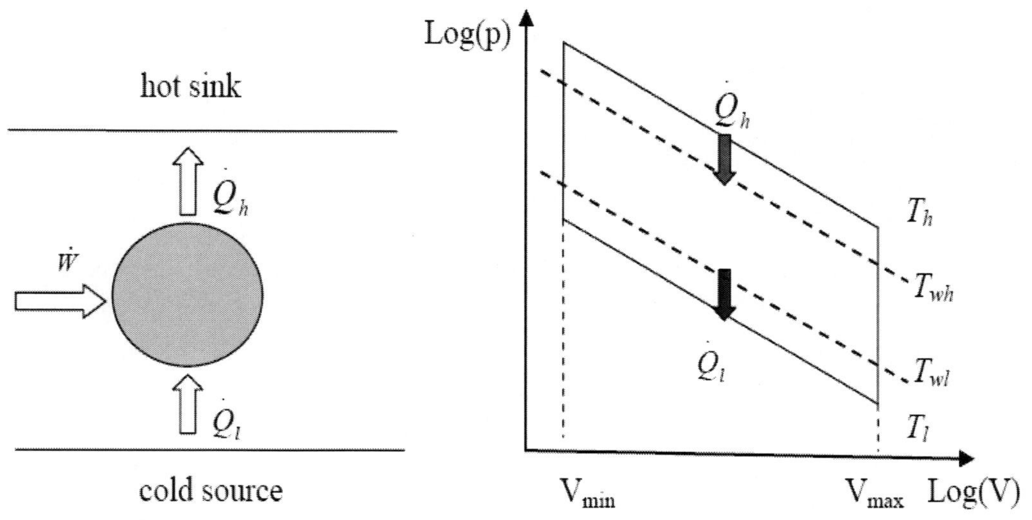


Figure 1. Inverse exo-irreversible Stirling cycle.

## 2. Studied Refrigerating Machine

The experimental device (Figure 2) is composed of a cylinder of highly resilient glass inside which two pistons undergo a reciprocating alternative movement with a phase lag of 110°. This cylinder is surrounded by a double-wall sleeve inside which a water flow constitutes the hot source. On the top part of the cylinder, a thermocouple measures the temperature of the cold space.

The displacer forces the passage of the gas from the bottom part to the top part of the cylinder and conversely. It also holds a porous material which is used for the heat storage/release, thus playing the regenerator role. Such porous domain is heated when the fluid pass from the hot source to cold source. The stored sensitive energy is restored when the flow reverse occur from the cold source to the hot one. Such gas exchanges on the external surfaces in contract with the hot source (or the cold source) and the efficiency of such exchange will control and define upper system limit.

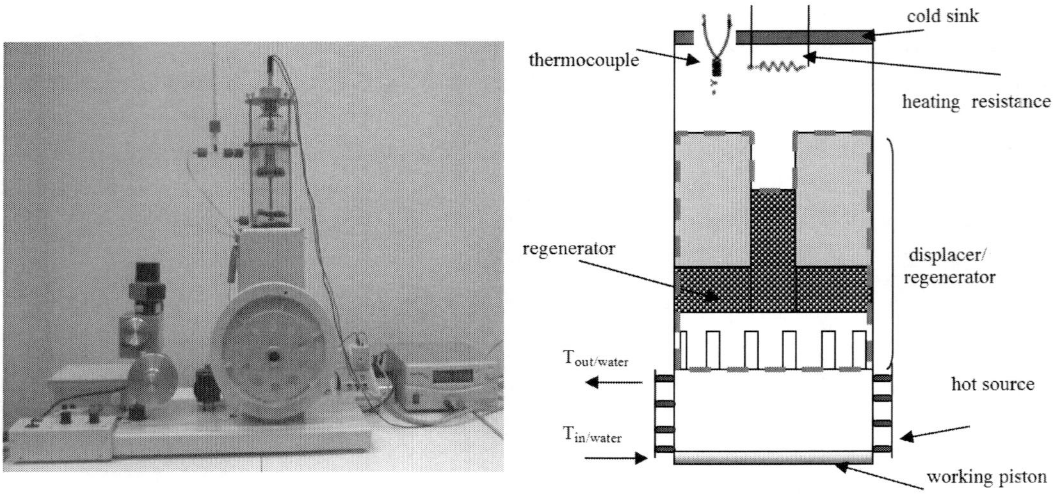


Figure 2. Stirling Beta machine a) Image of the bench test b) Scheme of the transparent cylinder of the refrigerating.

# 3. Modelling

Two models were carried out, one named "0-D" based on the thermodynamics transformation laws, and a second, named "2-D", more elaborated, based on the numerical solution obtained after solving the coupled mass, momentum and energy conservation equations, applied to the working gas in the regenerator (porous media), compression and expansion spaces.

## 3.1. 0-D Modelling

This analysis relies on the division of the refrigerating machine in 3 spaces: expansion, compression and regeneration spaces (Figure 3).

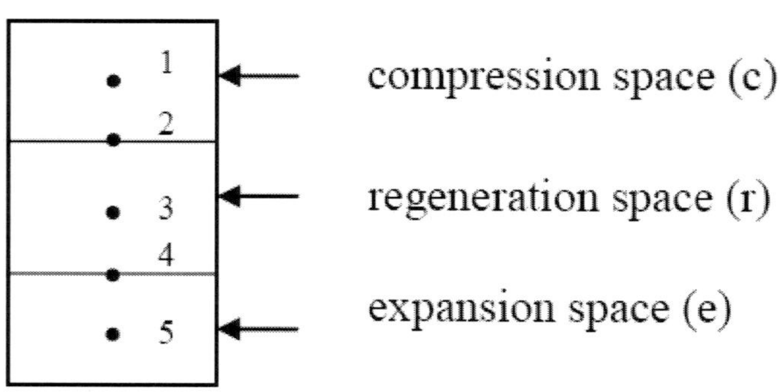


Figure 3. Representation of the machine 3 spaces with their boundaries.

The working gas is supposed as perfect gas. We also assume constant average temperatures inside the compression ($V_c$ ) and expansion spaces ($V_e$), in contact with the heat/sink sources. The regenerator temperature, $T_r = T_3$, is the logarithmic average of these two neighbour spaces temperatures.

The gas mass contained in each volume is estimated using the temperature and pressure variables with the perfect gas law:

$$m_c = \frac{p_c V_c}{rT_1} \; ; \; m_r = \frac{p_r V_r}{rT_3} \; ; \; m_e = \frac{p_e V_e}{rT_5} \tag{1}$$

The total gas mass in the machine (cylinder) is given by $m = m_c + m_r + m_e$ .

Thus, the instantaneous pressure, assumed to be uniform in the whole engine, so eq. 1 becomes

$$m = \frac{p}{r}\left(\frac{V_c}{T_1} + \frac{V_r}{T_3} + \frac{V_e}{T_5}\right)$$

and the homogeneous pressure is given by the following relation:

$$p = m\ r \Big/ \left(\frac{V_c}{T_h} + \frac{V_r}{T_r} + \frac{V_e}{T_l}\right) \tag{2}$$

The elementary masses in each volume are calculated by using equations (1):

$$\begin{cases} dm_c = dm_1 = \dfrac{pdV_c + V_c dp}{rT_1} \\ dm_e = dm_5 = \dfrac{pdV_e + V_e dp}{rT_5} \\ dm_r = dm_3 = m_r \dfrac{dp}{p} \end{cases} \tag{3}$$

The elementary mass flows through the interfaces and the associated temperatures are given by the following equations:

$dm_2 = dm_c$ , if up flow $dm_2 < 0,\ T_2 = T_3$ if not $T_2 = T_1$

$dm_4 = -dm_e$ , if up flow $dm_4 < 0,\ T_4 = T_3$ if not $T_4 = T_5$

The pressure differential $dp$ is given by using equation (2):

$$dp=-p(\frac{dV_c}{T_1}+\frac{dV_e}{T_5})/\left(\frac{V_c}{T_1}+\frac{V_r}{T_3}+\frac{V_e}{T_5}\right) \tag{4}$$

The elementary heat transfers in these 3 spaces are obtained using the energy conservation equation applied to each volume:

$$\delta Q_c=\left(\frac{c_v}{r}+1\right)p\,dV_c+\frac{c_v}{r}V_c\,dp+c_pT_2\,dm_2$$

$$\delta Q_r= V_r\frac{c_v}{r}dp+c_p\,(T_4\,dm_4-T_2\,dm_2\,) \tag{5}$$

$$\delta Q_e=\left(\frac{c_v}{r}+1\right)p\,dV_e+\frac{c_v}{r}V_e\,dp-c_pT_4\,dm_4$$

Elementary work in compression space, $\delta W_c=-p\ dV_c$ , and expansion space, $\delta W_e=-p\ dV_e$ , allows to determine, after integration, the work used during a cycle.

$$W=W_e+W_c \tag{6}$$

The temperatures of these two spaces are given by the Fourier-Newton law, using the heat transfer coefficients obtained by experimental tests.

$$\left|\dot{Q}_h\right|=k_h\,A_h\,(T_h-T_{wh}\,)\rightarrow T_h=T_{wh}+\frac{\left|\dot{Q}_h\right|}{k_h\,A_h} \tag{7}$$

$$\dot{Q}_l\ =k_l\,A_l\,(T_{wl}-T_l)\qquad\rightarrow T_l=T_{wl}-\frac{\dot{Q}_l}{k_l\,A_l} \tag{8}$$

The equations of this 0-D model are solved using the Simulink tool.
The different parts

- Mask of the Stirling refrigerating machine
- The pressure and its differential are drawn below (Figure 4):

The results obtained from the established dynamic simulation allows the temperature, pressure and work/energy versus time.

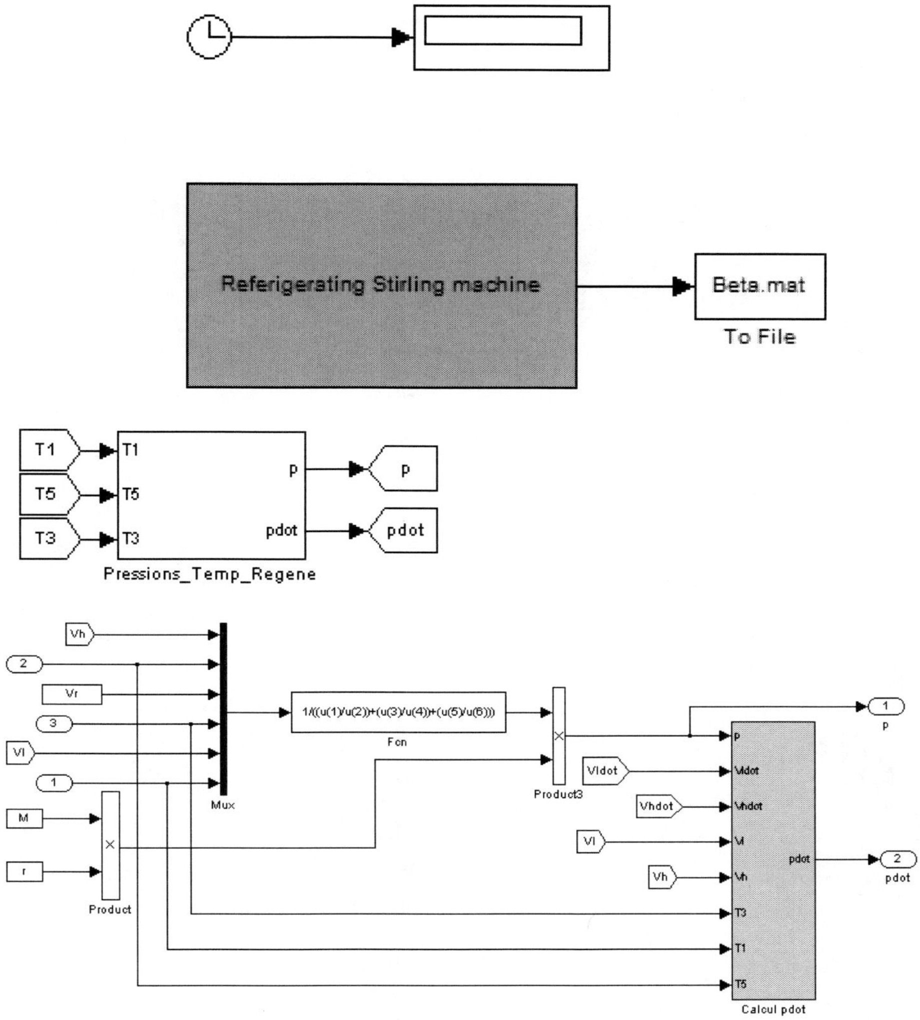


Figure 4. Block Simulink of a) refrigerator stirling and b) pressure model.

## 3.2. 2-D Modelling

The assumptions considered for the equations used in 2D modelling are as follows:

- the gas flow is turbulent,
- the gas thermophysical properties are time dependant (function of the instantaneous local temperature),
- the physical properties of the porous material (regenerator) are uniform and constant,
- energy transport is mainly controlled by conduction and convection where radiation exchange is neglected based of moderate temperature machine.

The Navier Stokes equations are numerically solved under the finite element approximation by using the commercial software COMSOL Multiphysics.

## 4. Results and Discussions

The controlling used data for the 2 modelling methods, are summarized in the Table 1.

**Table 1. Relative data of the simulated case**

| $T_{wl}$ | $T_{wh}$ | N | $D_m$ | $D_d$ | $\phi$ | C |
|---|---|---|---|---|---|---|
| K | K | rev/s | m | m | ° | m |
| 278.85 | 297.30 | 2.65 | 0.065 | 0.065 | 110 | 0.04 |

The results obtained by the 0-D model are the instantaneous variables: temperature, pressure, volume and mass. For the 2D model, in addition to these variables, we also obtain the instantaneous gas velocity, the temperature and pressure fields. An example of instantaneous temperature and pressure files are illustrated in figure 5.

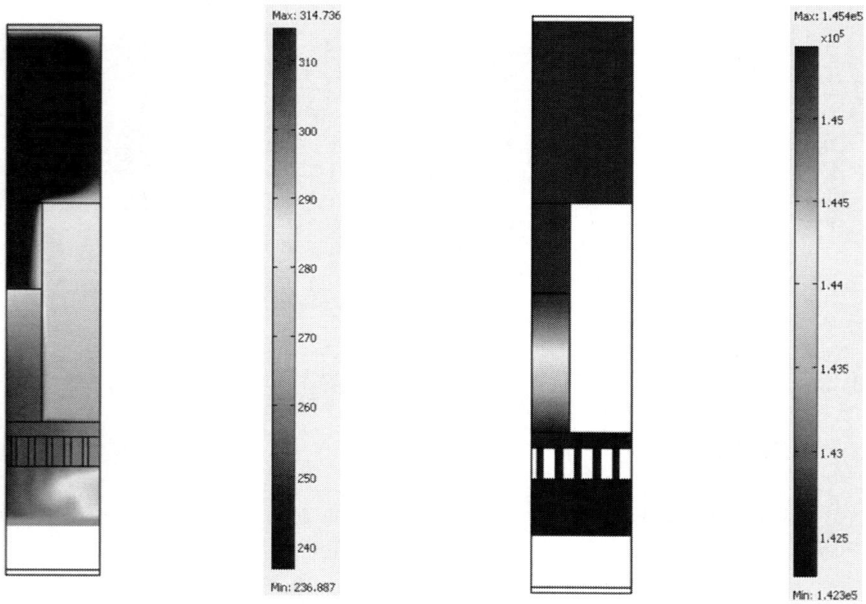


a) b)

Figure 5. Temperature (a) and pressure (b) fields for the displacer at the bottom dead center point and the working piston with phase lag of 110°: and for a porosity of 0.6.

The 2-D results illustrate clearly that the overall pressure in the compression and expansion domain is almost constant (fig. 5-b) and the complex flow modify mainly the obtained friction throw the regenerator and is main key for the local heat transfer as represented by the heterogeneous temperature field in the bottom domain (Figure 5-a)

The cycles obtained with these two models are presented on figure 6. The indicated work is calculated by integration on a cycle. The numerical results are presented in table 2 and compared to the experimental one.

**Table 2. Comparison of the obtained work by the different approaches**

| | 0D | 2D | experiment |
|---|---|---|---|
| W by cycle [J] | 5.9 | 9.2 | 5.7 |

One remarks that despite the 0-D Model seems relatively simple it permits to estimate with good precision the delivered work providing the use of parameters obtained by experiments. In our case, we use experimental heat exchange coefficients estimate on our bench test, which can explain such agreement.

For the 2-D model, we have considered the regenerator as a porous matrix with a given porosity (0.6 in our case). A significant difference between experiment and numerical simulation can be observed and explained by the value of the porosity which has been approximated.

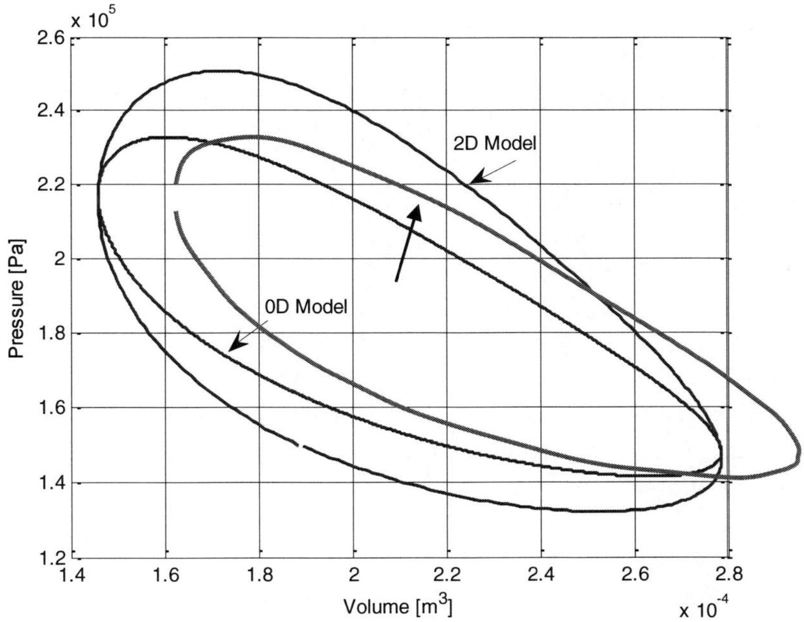


Figure 6. (p, V) diagram obtained by 2-D and 0-D numerical simulation.

## Conclusion

This work presents an experimental and numerical study of a beta type refrigerating Stirling machine.

The confrontation of the results permits to validate 0-D isothermal model for known heat transfer coefficient. We have to underline that it is difficult to get such value with precision and must be indentified by experimental approach.

So, the *a priori* known exchange value (it became as controlling parameters) and under the constant temperature and pressure on expansion and compression domain we converge to equivalent given work by 0-D and experimental approach.

Concerning the 2-D model, it allows to validate some used assumptions in 0-D and also to define the limit of validity. The CFD work allows also the heat transfer coefficient identification.

In a future work, we will be able to change the characteristics of the regenerator (porosity, permeability, etc.) and to carry out several experimental tests in order to get the optimal working conditions. We will also be able to find heat exchange coefficient variation laws, which was considered as constant in our models 0-D model.

These cross considerations could improve and validate the 0-D and 2-D models, and eventually improve the next experimental Set-up.

## REFERENCES


[1] G. T. Reader, C. Hooper, "Les machines Stirling", Edit. E and F.N. SPON, *New Fetter Lane*, London, (1983).

[2] Ph. Nika, F. Lanzetta, "Développement d'une machine frigorifique Stirling de petite taille, adaptée à des niveaux thermiques modérés", *Journal de physique*, vol. 5, N°6, (1995), 835-861.

[3] Ercan Ataera, H. Karabulut, "Thermodynamic analysis of the V-type Stirling-cycle refrigerator", *International Journal of Refrigeration* 28, (2005), 183–189.

[4] N. S. Petrescu, L. Grosu, M. Costea, P. Rochelle, C. Dobre, C. Petre : "Analyse théorique et expérimentale d'une machine à froid de Stirling", Colloque francophone COFRET 2010, *IASI,* Roumanie, Vol. LVI, Fasc.3a, (4 – 7 mai 2010), 441 – 453.

[5] L.G. Chen, H.J. Feng and F.R. Sun, "Optimal piston speed ratio analyses for irreversible Carnot refrigerator and heat pump using finite time thermodynamics, finite speed thermodynamics and direct method", *Journal of Energy Institute* Vol.84, N°2, (2011), pp. 105-112.

[6] S. Petrescu et al., "Les cycles des machines à froid et des pompes à chaleur à vitesse finie", *Rev. Entropie,* 232, (2001), 48-54.

[7] M. Feidt, "Optimisation énergétique des systèmes et procédés", 2nd ed., *TechandDoc,* Paris (1996).

[8] L. Grosu, "Contribution à l'optimisation Thermodynamique et Économique des Machines à cycle inverse à deux et trois réservoirs de chaleur". PhD thesis, *Institut National Polytechnique de Lorraine*, (2000).

[9] G. Schmidt. "Theorie der Lehmann'schen calorischen Maschine. "Z. Des Ver. deutsher Ingenieue, XV(2), (1871) 98-112.

[10] S. Bonnet. "Moteurs Thermiques À Apport de Chaleur Externe: Étude d'un Moteur Stirling et d'un Moteur Ericsson", PhD thesis, Université de Pau et des Pays de l'Adour, (2005).

[11] N. Martaj, L. Grosu, P. Rochelle, "Thermodynamic study of a low temperature difference Stirling engine at steady state operation", *Int. J. of Thermodynamic*, Vol. 10, No.4, (2007), 165 – 176.

[12] L. Grosu, P. Rochelle, "Application de la méthode de Schmidt avec régénération imparfaite aux 3 types de moteur Stirling-nouvelles solutions analytiques", Congrès SFT, Vannes, Tome 2, (26-29 mai 2009), 895-900.

[13] N. Martaj, P. Rochelle, L. Grosu, R. Bennacer and S. Savarese, "Moteur Stirling à faible différence de températures (LTD): confrontation simulations numériques et expérimentation", *Congrès SFT, Toulouse*, (3-6 juin 2008), 729-735.

[14] N. Martaj, P.Rochelle, L.Grosu, R.Bennacer, S.Savarese, "Multiphysics CFD simulation of an Stirling engine with a porous regenerator", Diffusion in Solids and Liquids DSL, Paris, (05-07 july 2010).

[15] H.Z. Hassan, A.A. Mohamad, R. Bennacer, "Simulation of an adsorption solar cooling system", *Original Research Article Energy*, Volume 36, Issue 1, (January 2011), 530-537.

[16] N. Martaj, R.Bennacer, P.Rochelle, L.Grosu, "Simulation dynamique d'un moteur: cas du Stirling de type gamma", *Mécanique and Industries* 11, (2010), 69–74.

In: Energy-Environment-Economics
Editor: Riad Benelmir
ISBN: 978-1-62948-862-2


*Chapter 14*

# OVERVIEW ON GAS TURBINES AND COMBINED CYCLES WITH $NO_X$ REDUCTION CONSIDERATION

***V. Cenusa, Riad Benelmir* and A. Merabtine***
University of Lorraine, France

## ABSTRACT

This paper intends to present an overview of gas turbine and combined gas and steam turbine cycles with an emphasis on combustion chambers with $NO_X$ reduction and Heat Recovery Generators. It is based on a literature survey issued mostly from manufacturers data as General Electric or Alstom. The objective is highlight the need of technico-economic optimisation in the design and implementation on the installations.

## 1. INTRODUCTION

Technology of Gas Turbine (GT) Plants is one of the most dynamic in the world. In this field, improvements where realised mostly in special materials and coating resisting at high temperature conditions [1] as Directionally Solidified (DS) materials [2] and thermal Barrier Coating (TBC), in burners with low $NO_X$ emissions as EnVironmental (EV) burners of Alstom ([3-6]), Dry Low $NO_X$ (DLN) burners of General Electric [7, 8], Dry Low Emissions (DLE) burners of Westinghouse-Rolls Royce, Hybrid Burner (HB) of Siemens KWU [9, 10] and PreMix (PM) burners of Mitsubishi Heavy Industries [2], and also in cooling technologies of high temperatures parts.

GT plants cover a large domain of electrical power production (3 to 265 MW) and reach up to 40 % efficiency. Most of these gas turbines are aeroderivative machines, as the LM6000 gas turbines of Nuovo Pignone and Motoren und Turbinen-Union, the LM6000 Sprint gas turbines of GE S and S Energy Products, the TRENT and WR21 gas turbines of Rolls-Royce and the Mercury50 gas turbine of Solar Turbines Inc. Combines cycles with GT and Steam

* E-mail: riad.benelmir@unniv-lorraine.fr.

Turbines (ST) reach up to 57% efficiency. Gas turbines used in these combined cycles are "heavy-study" machines as the GT24 and GT26 gas turbine of Alstom Power, the MS9001FA gas turbine of Nuovo Pignone and General Electric Power Systems, the V94.3A and W501G gas turbines of Siemens AG Power Generation Group KWU. In order to fit the existing conception of turbines operating at 40 Hz frequency to the frequency of 50 Hz, manufacturers used with success and low risks the method of aerodynamic homothertia as for the case of General Electric MS9001E and MS9001F adapted gas turbines based on the initial MS9001E and MS9001F gas turbines, where this procedure was applied for almost all the components of the gas turbine installation except for the combustion chamber [11].

**Table 1. Performances of the MS9001E and MS9001F gas turbines (open cycle) [8]**

| Model TG | Inlet temperature | Exhaust gases | | Gross power | Net efficiency |
|---|---|---|---|---|---|
| | | Flow rate | temperature | | |
| | °C | kg/s | °C | MW | (on LHV) % |
| MS 9001E | 1124 | 409.6 | 537 | 123.4 | 33.8 |
| MS9001F | 1260 | 612.2 | 583 | 212.2 | 34.2 |
| Natural gas – *base load – ISO conditions*<br>*Gross power : at the terminal of alternator*<br>*Net efficiency : at the terminal of transformer* | | | | | |

GT plants are compact installations with an investment cost lower than for ST plants and allow lower noise levels through the introduction of a "Noise Reducer" in the path of the combustion products, between the exhaust of the gas turbine and the inlet of the stack, and which resist to mechanical and thermal stress [12]. For the case of combined GT and ST plants, the Noise Reducer is within the Heat Recovery Steam Generator (HRSG).

In order to increase power production in a combined cycle, additional combustion is conducted in the HRSG due to sufficient excess air in the combustion gases issue from the gas turbine. The efficiency of the combined cycle is reduced since heat of combustion obtained in the HRSG is used to produce power only in the steam cycle. Due to this handicap, additional combustion is used usually for temporary increase of the power of the combined cycle (peak load).

## 2. Aeroderivatives and Heavy-Duty Gas Turbines

GT plants are classified in two categories following the technology of construction [13]: (a) Aero-derivative machines fitted to ground operation (from few MW of tens of MW) which are derived from the aviation turbo-reactor with important adaptation due to high velocity of turbine exhaust gases and from the naval turbo-propeller with minor adaptation due to a lower velocity; (b) Heavy-Duty machines which are developed especially for power production (from tens to hundreds of MW). Aerderivative GT in aviation engines have been designed for maximum thermal efficiency (minimum fuel consumption) since the aircraft had a limited quantity of fuel to cover large distances. Produced power is usually lower than 50 MW and the combustion gases have low thermal potential with respect to combined cycle applications but sufficient enough for cogeneration applications. The pressure ratio varies between 27 and

30. Heavy-Duty GT and Aeroderivative GT with turbo-propulsion have been designed for maximum specific power production. In Heavy-Duty GT, the flue-gas, with a mass flow rate between 350 t/h and 560 t/h, have a temperature ranging from 550 to 590 °C. This thermal potential, greater than in Aeroderivative machines, is appropriate for combined cycles applications. The pressure ratio varies between 11 and 17. A particular case of Heavy-Duty machines is sequential combustion GT where the pressure ratio can reach values comparable with the Aeroderivative machines. For example the GT24 and GT26 gas turbines (*figure 1*) have a pressure ratio of 30. These turbines have also an important net efficiency (approximately 58 %) in combined cycle (*figure 2*).

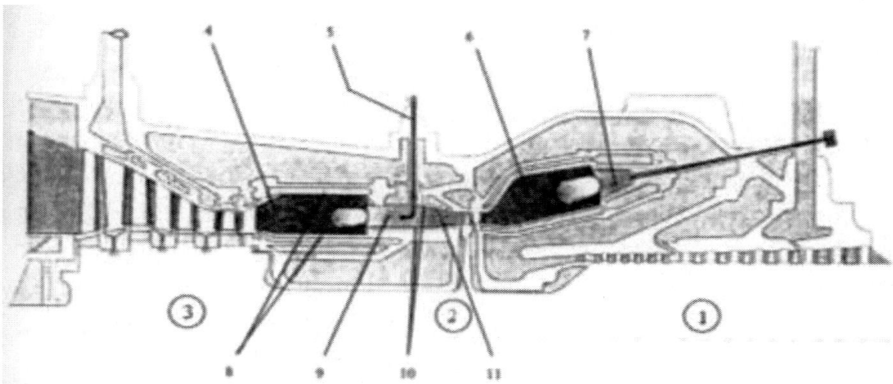


1 compressor
2 high pressure stage turbine
3 low pressure stage turbine
4 **SEV** burners
5 fuel injection
6 **EV** comb. chambers
7 **EV** burners
8 convection cooling
9 mix zone
10 vortex generator
11 **SEV** burners with effusion cooling

Figure 1. GT24 and GT26 gas turbines cross section [3].

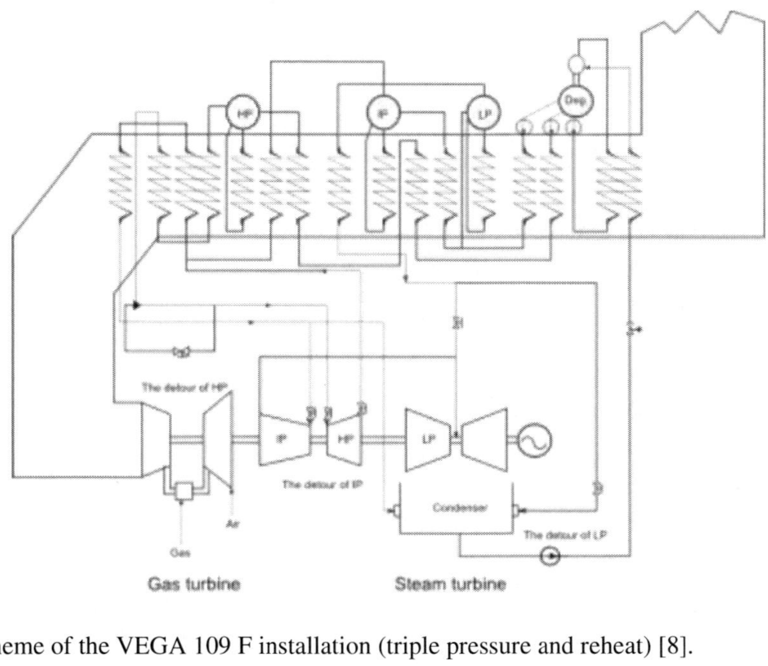


Figure 2. Scheme of the VEGA 109 F installation (triple pressure and reheat) [8].

## 3. Combustion Chambers and $NO_X$ Reduction

For an increase of power and efficiency of gas turbines, manufacturers had to design turbines with higher inlet temperatures and pressure ratio [22] leading an increase of the combustion temperature and consequently of $NO_X$ emissions. In order to maintain $NO_X$ emissions within acceptable limits, gas turbines manufacturers have worked on improvement of combustion chambers. Water or steam injection [14] in the combustion chamber is the most usual process in $NO_X$ reduction. However, (liquid) water injection under the constraint of a constant combustion temperature, needs an increase of fuel consumption (in order to handle water vaporisation) leading a lower efficiency of the installation despite a slight increase of the produced power due to a higher specific heat of steam present in the combustion gases. Steam injection could be arranged by using the sensible heat recuperated from turbine exhaust gases leading a slight increase of fuel consumption (since only minor superheating of the steam up to the combustion temperature is needed) and of power (for the same reasons mentioned in liquid water injection). Hence, the efficiency is almost unchanged.

Air cooling of the combustion reduces the combustion air leading an increase of $NO_X$ emission. In order to overcome this inconvenient steam cooling is preferred to air cooling as the "Steam-cold Combustor" [2] of Mitsubishi Heavy for the M501G turbines.

In general, modern combustion chambers are the annular type and offer numerous advantages [6] as a uniform distribution of the temperatures. The DLN burners of General Electric, of dry type in $NO_X$ reduction, have a lower flame temperature due to a poor oxygen combustion [8]. In Integrated Gasification Combined-Cycle (IGCC) a technology called "Clean Coal" is commonly used [15] as in the combined cycle based on the MS7001FA and V94.3 heavy-duty gas turbines [15-16]. The combustion chamber in the GT24 and GT26 gas turbines of Alstom Power use dry type burners (without water or steam injection) in $NO_X$ reduction: the EV burner for the first combustion chamber and the SEV (Sequential EV) burner for the second combustion chamber [4-6]. The configuration of this burners [5, 6] allows a perfect centrifugal mixing of fuel and air [6].

## 4. The Heat Recovery Steam Generator (HRSG)

The Heat Recovery Steam Generator (HRSG) is the interface between the gas cycle and the steam cycle where up to 80 % of sensible heat is recuperated. The recuperation factor is represented by $\beta$ (1):

$$\beta = \frac{\Delta T_{HRSG}}{\Delta T_{\max}} \tag{1}$$

where $\Delta T_{HRSG}$ is the temperature variation of the flue gases and is the maximum temperatures pinch between the flue gases and the steam.

For a combined cycle with heavy-duty gas turbine, the electric power produced in the bottom cycle (after the recuperated heat) represents approximately half the electric power produced by the top gas turbine cycle (table 4). In cogeneration applications (combined electricity and heat production) the global thermal efficiency is very high as in the VEGA109F combined cycle based on the MS9001F gas turbine where the global thermal efficiency could reach 89 % for a large domain of production [8].

Important exergy destructions occur in the HRSG due to important temperature differences between the combustion gases and the water/steam. In order to reduce these irreversibilities (exergy destructions) and still recuperate efficiently the sensible heat of combustion gases multi-pressure levels HRSG are used. From the thermodynamic view, thermal efficiency increases with the number of pressure levels, however the limiting constraints on the investment cost limit drastically this number of pressure levels, requiring a technico-economical optimisation with respect to the pressure ranges and the number of pressure levels. HRSG in combined cycles of high power use two or three pressure levels of steam production (table 2, table 4, table 5) which reheating of steam between the high pressure and intermediate pressure stage of the steam turbine (Figure 2 and 3, table 5).

The lower temperature difference between the combustion gases and the water/steam is called the pinch point (in the water side, this point is at the junction of the economiser and the vaporiser).

Exergy destruction increases with the pinch point. In order to reduce the pinch point, we have to increase heat transfer surfaces which will end up with an increase of investment cost requiring again a technico-economical optimisation.

In the VEGA109F combined cycle based on the MS9001F gas turbine (Figure 2) the temperature pinch is about 9°C [8]. For a fixed combustion gas temperature at the inlet of a single steam pressure level HRSG the combustion gas temperature at the outlet is a function of the steam pressure level and the temperature pinch.

A decrease of the (production) steam pressure would lead an increase of irreversibilities due to increasing temperature differences between the combustion gas and the water/steam. On the other hand, an increase of steam pressure at constant steam turbine inlet could have negative effects on the turbine [13]. Optimal pressure for steam production for some combined cycle installations performances are presented in tables 2 and 5.

**Table 2. Performances of the VEGA 109F combined cycle (MS9001F gas turbine) [8]**

| | Inlet of steam turbine | | Gross power | Gross efficiency |
|---|---|---|---|---|
| | Pressure | Temperature | | |
| | Bar | °C | MW | % |
| 109 F<br>one steam pressure level | 65 | 540 | 329.9 | 51.9 |
| 109 F<br>two steam pressure level | 100/4.6 | 540/190 | 340.0 | 53.4 |
| 109 F<br>two pressure level and reheat | 110<br>28/4.6 | 540<br>540/265 | 343.8 | 54.0 |
| 109 F<br>three steam pressure level and reheat | 110<br>28/4.6 | 540<br>540/265 | 347.2 | 54.5 |
| *Natural gas – air temperature: 8°C condenser pressure: 25 mbar* | | | | |

**Table 3. Technical data of the GT24 and GT26 gas turbines (basic cycle- fuel: methane) [3, 23]**

| | | GT24 | GT26 |
|---|---|---|---|
| Fuel | | Natural Gas | |
| Frequency | Hz | 60 | 50 |
| Electrical power output | MW | 183 | 265 |
| Electrical efficiency (LHV) | % | 38.3 | 38.5 |
| Heat rate (LHV) | kJ/kWh | 9400 | 9350 |
| Compressor pressure ratio | - | 30 | 30 |
| Exhaust gas flow rate | kg/s | 391 | 562 |
| Exhaust gas temperature | °C | 640 | 640 |
| Turbine speed | rpm | 3600 | 3000 |
| $NO_X$ emissions | ppm | < 25 | < 25 |
| Number of compressor stages | - | 22 | 22 |
| Number of turbine stages | - | 5 | 5 |
| Number and type of combustion (annular) | - | 1 EV | 1 EV |
| | | 1 SEV | 1 SEV |
| Number of burners (fuel injection) EV/SEV | - | 30/24 | 30/24 |

**Table 4. Technical data of the KA24-1 ICS and KA26-1 combined cycles installation [18, 23]**

| | KA24-1 | KA26-1 |
|---|---|---|
| Electrical net power output | 271 MW | 393 MW |
| Electrical net efficiency (LHV) | 57.6 % | 57.9 % |
| Compressor pressure ratio (GT24) | 30 : 1 | |
| Burners type | EV/SEV, dry low - $NO_X$ | |
| Combustible | natural gas, fuel N° 2 | |
| Frequency | 60 Hz | 50 Hz |
| Electrical gas turbine output | 183 MW | 265 MW |
| Electrical steam turbine output | 97 MW | |
| Water/Steam cycle | Two steam pressure level and reheat | |
| HRSG | Horizontal | |
| HP steam | 565 °C, 160 bar | 565 °C, 118 bar, 300 t/h |
| Steam turbine construction | Double stage | |
| Alternator cooling | With air | |
| $NO_X$ | ≤ 25/24 vppm (dry gas/fuel with water injection) | |
| fuel : methane – ISO conditions – wet cooling tower – $p_{condenser}$ = 70 mbar | | |

## 5. Sequential Combustion

Sequential combustion allows an increase of (electric) power production, hence efficiency, without the need to increase (gas) turbine inlet temperature comparatively to standard gas turbines. However, In order to face with downstream steam cycle needs (in combined cycle) this process is constrained by an optimum gas turbine exit temperature requiring almost to double the pressure ratio [4].

In gas turbines with sequential combustion as the GT24 and GT26 (Advanced Cycle System [17]) the fuel is introduced sequentially in two combustion chambers separated by a zone for (partial) expansion of a combustion gases.

The primary fuel introduced in the first combustion chamber (EV burner) is mixed with the compressed air and the combustion gases are expanded in the first stage of the turbine. Hence, these gases, still rich in air, enter the second combustion chamber (SEV burner) where secondary duel is injected.

The combustion gases are then expanded in the four stages of the turbine to a pressure still greater than atmospheric pressure. At base load, the combustion gases temperature is almost of same magnitude at the exit of the first and second combustion chambers [3] (figure 1).

## 6. The KA24-ICS Combined Cycle Installation

The main components arrangement of the combined cycle installation KA24-ICS ([17] to [19]) on the power line (figure 3) allows a high operating flexibility since the Self Shifting and Synchronising (SSS) device delimits the turbo-generator from the steam turbine which could operate alone.

Since the efficiency of a turbine is proportional to the shaft speed (which is limited by the airfoil length), the shaft speed of the high pressure stage of the steam turbine (about 9000 rpm) could be higher than the shaft speed of the lower (and intermediate) shaft speed by means of a gearing device [18].

The cooling system is either direct cooling within the rotor (General Electric) [11] or indirect cooling by means of a heat exchanger for the preliminary air cooling [18] (Siemens, Alstom) or by injection of water in the compressed air [19].

The advantages of direct cooling are the simplicity of the mechanism and low thermal gradient [11] (the material of the airfoil resists better to thermal fatigue).

However it needs higher cooling air flow requirements and leads lower combustion air. The use of a heat exchanger may achieve intense cooling and needs less air flow (or a reduction if the flow sections) due to the lower temperature of the cooling air and/or more combustion air with a possibility of heat recuperation for the steam cycle. This cooling system leads higher thermal gradients and investment cost.

The cooling of the electric generator can be achieved with hydrogen or air [20, 21]. Air cooling (GT24 and GT26 gas turbine generator) offers higher reliability than with hydrogen cooling while the investment and maintenance costs are lower but the efficiency is lower due ventilation losses [20].

The HRSG of the KA24-ICS combined cycle (figure 3) is characterized by [18,19] (a) two steam pressure levels for production and reheating (LP: 7 bar and 320 °C; HP:160 bar and 565°C); (b) heat recovery from the airfoil, used partially for the production of high pressure steam (increasing therefore the combined cycle efficiency); (c) high combustion gas temperature at the inlet of the HRSG (around 640 °C, making its efficiency comparable with the efficiency of standard HRSG with three steam pressure levels).

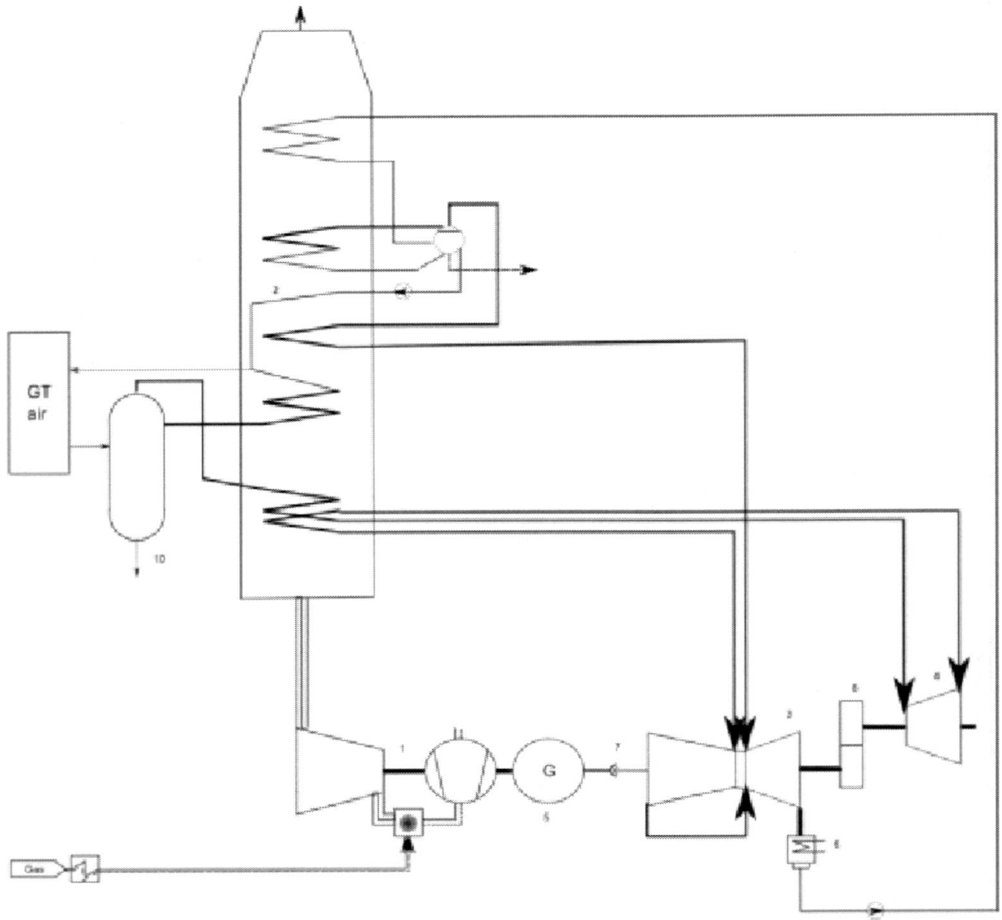


1 Gas Turbine (GT24)
2 HRSG (dual pressure and reheat)
3 Steam Turbine (IP and LP)
4 Steam Turbine (HP)
5 Alternator with air cooling
6 Condenser
7 Self-Shifting and Synchronizing
8 Gearing
9 Branching to the GT24 cooling
10 Separator with purge

Figure 3. Scheme of KA24-ICS combined cycle installation [18, 19].

## CONCLUSION

The use of sequential combustion in gas turbines seems to be a reasonable compromise for $NO_X$ reduction but care should be made regarding the cost since higher pressure ratio may lead to higher investment costs.

On the other hand, the use of multi-pressure heat recovery generators raises also the need for a technico-economic optimisation with respect to the number of pressure levels and the optimal value of these pressures. Another question regards the part of maintenance costs in configurations with Self Shifting and Synchronizing devices.

## REFERENCES


[1] Toulios M., McKenzie D., L'utilisation performance des matériaux par l'analyse inélastique avancée. Application aux turbines à gaz, Revue Technique *GEC ALSTHOM,* 17, (1995).

[2] Murase T., Tsukuda Y., Akita E. Iwasaki Y., *Current* operating status of M501G gas turbine, *ECOS 2000*, Part 4, (2000), 1857-1868.

[3] Joos F., Brunner P., Werning B.S., Syed K., Développement du système de combustion séquentielle pour la famille des turbines à gaz GT24/GT26, *Revue ABB*, 4, (1998), 4 – 16.

[4] Neuhoff H., Thorén K., Les nouvelles turbines à gaz GT24 et GT26 – rendements élevés, grâce à la combustion séquentielle, *Revue ABB,* 5, (1994), 4 – 7.

[5] Vogel M., La turbine à gaz – le composant principal de la technique moderne des centrales thermiques, *Revue ABB,* 5, (1994), 11 – 15.

[6] Joos F., Brunner P., Stalder M., Tschirren S., Expériences pratiques avec le système de combustion séquentielle de la famille des turbines à gas GT24/GT26, *Revue ABB*, 5, (1998), 12 – 20.

[7] Cannon M.F., Molière M., Depietro S.M., Perkavec M., Les technologies « bas $NO_X$ » pour les turbines à gaz. Expérience et références d'EGT, *Revue technique GEC ALSTHOM*, 19, (1996), 1 – 16.

[8] Paren J., Parietti C., Centrales à cycles combinés VEGA 109F 3 pressions resurchauffé, *Revue Technique GEC ALSTHOM*, 4, (1991), 15 – 26.

[9] Krebs W.¸ Walterskötter R., Flohr P., Walz G., Hoffmann S., Effect of burner design parameters on thermoacoustic stability of annular gas turbine combustors, *ECOS2000,* Part 4, (2000), 2061 – 2077.

[10] Hermann J., Orthmann A., Hoffmann S., Berenbrink P., Active and passive control of combustion oscillations in a 260 MW heavy-Duty gas turbine, *ECOS 2000*, Part 4, (2000), 2089 – 2102.

[11] Brandt D.E., Colas M., Une nouvelle technique de pointe turbine à gas 50 Hz, *Revue Technique GEC ALSTHOM*, 4, (1991), 3 – 14.

[12] Rouget F., Michelat L., Babaud G., Analyse thermomecanique transitoire appliquée aux turbines à gaz «heavy-duty. L'échappement de la MS 9001F de Gennevilliers», *Revue Technique GEC ALSTHOM*, 16, (1995), 33 – 48.

[13] Alexe Fl., Cenusa V., Marculescu C., Ene M., Optimizarea centralelor electrice de medie putere cu ciclu combinat gaze – abur, Produducerera, transportul si distributia energiei electrice si termice, 10, (2000), 16 – 23.

[14] Saad M. A., Cheng D. Y., Steam injection technology for utility-size gas turbines, *ECOS'98*, Volume II, (1998), 725 – 732.

[15] Pilleul R., Intégration du cycle combine VEGA 109FA dans les procédés de gazéification, *Revue Technique GEC ALSTHOM*, 14, (1994), 1 – 16.

[16] Altafini C. R., Mirandola A., Stoppato A., Analysis and simulation of an integrated gasification combined cycle power plant, *ECOS'98*, Volume II, (1998), 641 – 647.

[17] Frutschi U. H., Les nouvelles turbines à gaz GT24 et GT26- l'arrière plan historique de l' "advanced Cycle System", *Revue ABB*, 1, (1994), 20 – 25.

[18] Bachmann R., Schultz R., KA24-1 ICS™ – succés sur le marché des centrales énergétiques standards, *Revue ABB*, 3, (1999), 4 – 11.

[19] Svoboda R., Sviluppo di cicli combinati ad elevate efficienza, La Termotecnica, Novembre, *Anno LIII* – n. 9, (1999), 83 – 88.

[20] Stephan C. E., Baer J., Ziemmermann H., Neidhöfer G., Egli R., Nouveau turboalternateur refroidi à l'air de la classe de 300 MVA, *Revue ABB*, 1, (1996), 20 – 28.

[21] Buck D., Le système électrique des centrales à cycle combiné ABB, *Revue ABB*, 2, (1995), 15 – 23.

In: Energy-Environment-Economics
Editor: Riad Benelmir

ISBN: 978-1-62948-862-2


*Chapter 15*

# ONE STEP REACTION RATE MODEL FOR APPLICATION OF HCCI COMBUSTION ON ENERGY EFFICIENT AND ENVIRONMENT FRIENDLY THERMAL ENGINES

***Fadila Maroteaux*** * ***and Jean-Baptiste Millet***
Université de Versailles Saint Quentin en Yvelines,
Laboratoire d'Ingénierie des Systèmes de Versailles (LISV), Vélizy, France

## ABSTRACT

To reduce the number of tests during engine tuning phases, a model with a very short computational time to simulate Diesel homogeneous combustion (Homogeneous Charge Compression Ignition HCCI) is needed. Therefore, the goal of this study is to provide the engine manufacturers with a simple physical combustion model to assist engine tuning and engine management system optimization, with the aim of predicting in-cylinder pressure evolutions and mean effective pressure (IMEP). A reduced model driven by to two state variables: the temperature and the mass fraction of burning fuel in the combustion chamber, is described in this paper.

The chemistry is modeled by a global degradation reaction where the reaction rate coefficient, usually modeled through the Arrhenius law is driven in this approach by a global function Ω(T). This global function takes into account the slow dynamic of the cool flame and the fast dynamic of the main ignition and the transition between the two stages. A drawback of the global approach is the introduction of some new parameters which need to be correlated to give reliable results; this has required an important and large parametric study to calibrate the reaction rate coefficient. The proposed model validated for an engine running with premium diesel fuel is then autonomous, meaning that the model parameters are function of the engine operating conditions only.

The results show that this type of model can be useful to describe the ignition delay time of HCCI combustion and the rate of heat release, with very short computational

* Corresponding author: E-mail address: fadila.maroteaux@iut-velizy.uvsq.fr.

times around two seconds. Model against experimental data shows overall satisfactory agreements in terms of in-cylinder pressure and of mean effective pressure (IMEP).

**Keywords:** HCCI combustion, one step reaction, ignition, global reaction rate

## LIST OF SYMBOLS

$A_{LT}$ : Model parameter

$A_{HT}$ : Model parameter

$C_{LT}$ : Model parameter

$C_{HT}$ : Model parameter

$c_v$ : Specific heat at constant volume $[J.K^{-1}.kg^{-1}]$

$h$ : Convective heat transfer coefficient $[J.K^{-1}.m^{-2}]$

$m$ : Total mass $[kg]$

M : Molecular weight

N : Engine speed

$P$ : Pressure $[Pa]$

$P_a$ : Standard atmospheric pressure $[Pa]$

$P_{in}$ : Inlet pressure $[Pa]$

$\dot{Q}_{comb}$ : Combustion rate of heat release $[J.CA^{-1}]$

$\dot{Q}_w$ : Heat transfer rate to the walls $[J.CA^{-1}]$

$Q_{LHV}$ : Lower heating value of fuel $[J.kg^{-1}]$

$S$ : Surface of the combustion chamber $[m^2]$

$T$ : Temperature of the system $[K]$

$T_a$ : Standard atmospheric temperature $[K]$

$T_{LT}$ : Low temperature regime model

$T_{in}$ : Inlet temperature $[K]$

$T_{HT}$ : High temperature regime model

$T_W$ : Walls temperature $[K]$

$u$ : Specific internal energy $[J.kg^{-1}]$

$V$ : Volume $[m^3]$

$\Phi$ : Equivalence ratio

$X_{C_7H_{16}}$ : Initial molar fraction of n-heptane

$X_{EGR}$ : Initial molar fraction of EGR

$Y_{C_7H_{16}}$ : Mass fraction of n-heptane

$Y_f$ : Mass fraction of fuel

$\theta$ : Crank angle (CAD)

$\gamma$ : Ratio of constant pressure specific heat to constant volume specific heat

$\Omega$ : Global function or global reaction velocity

$\Omega_{LT}$ : Low temperature regime function or low temperature reaction velocity

$\Omega_{HT}$ : High temperature regime function or high temperature reaction velocity

TDC : Top Dead Center

BTDC : Before Top Dead Center

# 1. INTRODUCTION

Homogeneous Charge Compression Ignition (HCCI) offers a multitude of advantages, which may lead to improved emissions while maintaining high thermal efficiency [1-3]. However, this combustion process is mainly controlled by chemical kinetics [4-5]. Therefore, there is no direct control method as in spark ignition engine or conventional diesel engine. In recent years, this combustion mode has been the subject of numerous experimental and theoretical investigations. Numerical investigations to model the HCCI process can be summarized in three approaches: single-zone models [1, 2, 6], multi-zone models [4, 5, 7-10] and multi-dimensional CFD models [11-13]. Single and multi-zone models concentrate on accurate modeling of chemical kinetics and require computational times around minutes or hours. Multi-dimensional CFD calculations with detailed chemistry require weeks, depending on the number of species included in the mechanism, and 3D simulations become realizable with reasonable computation times (days) when a reduced chemical mechanism (number of species lower than 30) is used. However, these computation times are not suited to assist the engine tuning. Car manufacturers need models with very short computational time (around few seconds) to simulate the engine in the whole operating range; this step is very important for the engine tuning to reduce the number of engine experiments on the engine test bench, which are very expensive. In case of HCCI combustion (or premixed combustion) the lower computational cost of a simpler chemistry is preferred.

The objective of this study is to provide the engine manufacturers with a simple physical combustion model to assist engine tuning and engine management system optimization, with the aim of predicting in-cylinder pressure evolutions and mean effective pressure (IMEP). The model has to reproduce qualitatively the behavior of HCCI combustion: the cool flame and the main ignition with the transition between both stages.

This paper explores the suitability of one-step function to describe the two stages ignition during homogeneous combustion process. The approach used here is different from the approach used to model combustion chemistry. Consequently, instead of reducing chemical reactions mechanism, we have opted for a simple zero-dimensional model with one step reaction (i.e., one step function), keeping as much as possible the behavior of HCCI process. The proposed model is reduced to two state variables: the temperature and the mass fraction of burning fuel in the combustion chamber. The model is driven by a function (one step reaction), which takes into account the slow dynamic of the first stage oxidation and the fast dynamic of the main ignition as well as the transition between both stages. The proposed model excludes detailed chemistry and consequently is not able to predict NOx emissions.

The modeling methodology adopted in this work is described in the next section. The complete model is reported in section 3 of this paper, following by the model calibration in section 4. The results are presented and discussed in the fifth section and the conclusions are reported in the last section.

## 2. Conservation Equations

As mentioned before the HCCI process proceeds in two stages and these oxidation steps are observed for most fuels. The first step is associated to the low temperature regime, and the time delay between the first step and the main ignition (second step) is due to the negative temperature coefficient regime NTC region.

The objective of the reduced model is to reproduce qualitatively these trends. The main idea of the reduction methodology is to find a global function able to model these two oxidation stages and the transition between them, while maintaining the description of the system with a minimum number of state variables.

As a reminder, when the inlet and exhaust valves are closed the mixture is treated as a closed system with no mass crossing the boundary (one zone approach), so the total mass (m) of the mixture is constant:

$$m = \sum_{i=1}^{K} m_i \qquad \dot{m} = 0 \tag{1}$$

where: $m_i$ is the mass of the $i^{th}$ species and K the total number of species in the mixture. The dots denote differentiation with respect to time.

The individual species are produced or destroyed according to:

$$\dot{m}_i = V\dot{\omega}_i M_i \tag{2}$$

where: $\dot{\omega}_i$ is the molar production rate of the $i^{th}$ species by elementary reaction, $M_i$ is the molecular weight of the $i^{th}$ species and V is the volume of the combustion chamber function of time.

The generalized energy equation for a closed thermodynamic system with the zero-dimensional assumption may be written as:

$$m\dot{u} = -P\dot{V} - \dot{Q}_w - \dot{Q}_{comb} \tag{3}$$

where:

$\dot{Q}_w$ is the total heat transfer rate to the walls.

$\dot{Q}_{comb}$ is the energy source term due to the reactions.

$p\dot{V}$ is the displacement work.

With the assumption that the heat transfer mechanism is due only to forced convection. Equation (3) is developed to obtain a differential equation for the change in temperature of the thermodynamic system as (see reference 15):

$$mc_v\dot{T} = -P\dot{V} - hS(T - T_w) - V\sum_{i=1}^{K}\dot{\omega}_i M_i u_i \quad (4)$$

where: h is convective heat transfer coefficient and S is the cylinder head, piston and liner surface area function of time. and $u_i$ are respectively the walls temperature and the internal energy of the $i^{th}$ species.

The system of equations (2) and (4) represents the model based on a closed system first law analysis (i.e., single zone model) with the possibility of including several species, the dimension of the system is then equal to (K+1). In this study, the first objective is to start from minimum state variables model, meaning that the dimension of the system is reduced to its minimum value. Consequently, the natural variable to describe the combustion progress (i.e., fuel consumption) is the mass of burning fuel in the combustion chamber. Besides, in order to take into account the heat release induced by fuel consumption, the energy balance equation must be kept (i.e., equation (4)). Consequently, two dimensions are necessary in this case; we have kept the temperature and the mass of burning fuel in the combustion chamber. With these restrictions, the fuel mass (equation (2)) balance and energy balance become:

$$\dot{m}_f = V\dot{\omega}_f M_f \quad (5)$$

$$mc_v\dot{T} = -P\dot{V} - hS(T - T_w) - \dot{Q}_{comb} \quad (6)$$

Note that the third term of the right hand side of equation (4) has been written in global form as in equation (3). This source term modeling the heat release by combustion can be written as [15]:

$$\dot{Q}_{comb} = \dot{m}_f Q_{LHV} \quad (7)$$

where: $Q_{LHV}$ is the lower fuel heating value.

The conservation equations (5) and (6) defined the proposed model and are rewritten function of fuel mass fraction as:

$$\dot{Y}_f = \frac{V}{m}\dot{\omega}_f M_f \quad (8)$$

$$mc_v\dot{T} = -P\dot{V} - hS(T - T_w) - \dot{Y}_f m Q_{LHV} \tag{9}$$

## 3. Global Reaction Rate

The energy equation (9) implies that the fuel is consumed by a global step similar to a degradation reaction:

$$fuel \xrightarrow{\Omega} products \tag{10}$$

where: Ω represents the global coefficient reaction rate.

The fuel mass balance (8) can be written according to reaction (10) as:

$$\dot{Y}_f = -\Omega \cdot Y_f \tag{11}$$

Note that fuel mass fraction balance (8) is replaced by equation (11). With this formulation, the term Ω must be able to model the dynamic of the different stages of fuel consumption during HCCI process. Thus, this term cannot be modeled by an Arrhenius law; the idea here is to derive its behavior from a similar reduced model (0D) including a reaction mechanism. This step is necessary in order to find a global function able to reproduce the oxidation stages. Besides, the dynamical evolution of the reduced model (equations 9 and 11) must be kept in the specific coefficient reaction rate expression. Consequently, this coefficient has to be modeled function of one of the state variables (fuel mass fraction or temperature). As the temperature has a direct effect on the reactivity of the system, this state variable has been chosen to model the global specific coefficient rate Ω noted Ω(T).

The general shape of the function Ω(T) versus the in-cylinder temperature is plotted in Figure 1 at 2000 rpm for two different equivalence ratios (0.3 and 0.5). These plots have been obtained from a compression and expansion strokes one zone model simulations including a reduced reaction mechanism of n-heptane [19] developed previously. The engine geometry used here is given in Table 1 (single cylinder Renault diesel engine with small compression ratio). Note that the shape of the function Ω(T) has been observed for several engine operating conditions (variable conditions: inlet temperature, inlet pressure, equivalence ratio, EGR rate and engine speed).

Figure 1 shows that the global fuel reaction rate coefficient Ω(T) takes into account the two stages of combustion. From this figure, we can note that the profile of the reaction progress is governed by two characteristic temperatures: first stage temperature noted $T_{LT}$, and auto ignition or second stage temperature noted $T_{HT}$. These two temperatures need to be reached to start the combustion process and represent respectively the low temperature period and the high temperature period observed during the two stages oxidation process of most fuels under HCCI conditions. Most importantly, these two temperatures are function of engine inlet conditions.

**Table 1. Engine geometry**

| | |
|---|---|
| Bore | 68 mm |
| Stroke | 79 mm |
| Connecting rod length | 132.65 mm |
| Displaced volume | 287 $cm^3$ |
| Compression ratio | 11.68:1 |
| IVC | 166 CAD (BTDC) |
| EVO | 100 CAD (ATDC) |
| Injections timing | 70 CAD (BTDC) and 90 CAD (BTDC) |

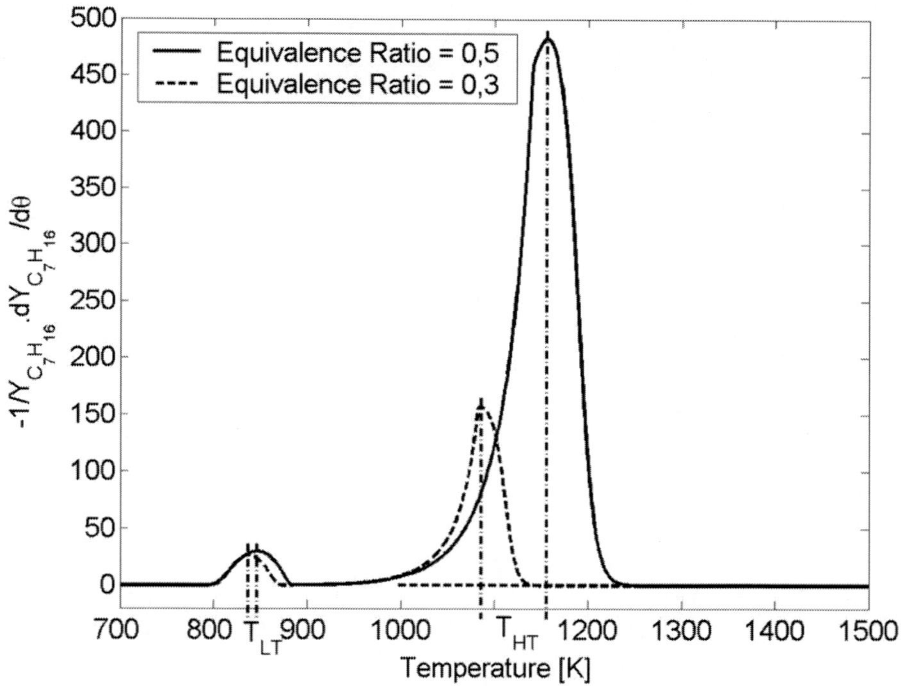


Figure 1. Effect of equivalence ratio on Ω(T) profile in a variable volume chamber versus in-cylinder temperature. Engine speed 2000 rpm, inlet temperature 400K, inlet pressure 1.5 bar, EGR rate 50%.

In order, to distinguish these two periods the global function Ω(T) is written as follows:

$$\Omega(T) = \Omega_{LT}(T) + \Omega_{HT}(T) \tag{12}$$

where:

$\Omega_{LT}(T)$ represents the low temperature regime function or low temperature reaction rate coefficient.

$\Omega_{HT}(T)$ represents the high temperature regime function or high temperature reaction rate coefficient.

Consequently, from the general profile of Ω(T) presented in Figure 1 the function Ω(T) is modeled in order to represent the different phases of the system evolutions:

- No reactivity.
- Slight increase and decrease of the reactivity (first stage).
- No reactivity.
- High increase of the reactivity (main ignition).

The two first phases are described by the low temperature regime function $\Omega_{LT}(T)$, and the two last phases are described by the high temperature regime function $\Omega_{HT}(T)$. The global shape of Ω(T) is then split up into two plots representing respectively the function $\Omega_{LT}(T)$ and $\Omega_{HT}(T)$ as shown in Figure 2. These functions can be written as following:

$$\Omega_{LT}(T) = A_{LT} \exp\left( - \frac{(T - T_{LT})^2}{C_{LT}^2} \right) \tag{13}$$

$$\Omega_{HT}(T) = A_{HT} \frac{\exp(C_{HT}(T - T_{HT}))}{1 + \exp(C_{HT}(T - T_{HT}))} \tag{14}$$

where: $A_{LT}$, $A_{HT}$, $C_{LT}$ and $C_{HT}$ are model parameters.

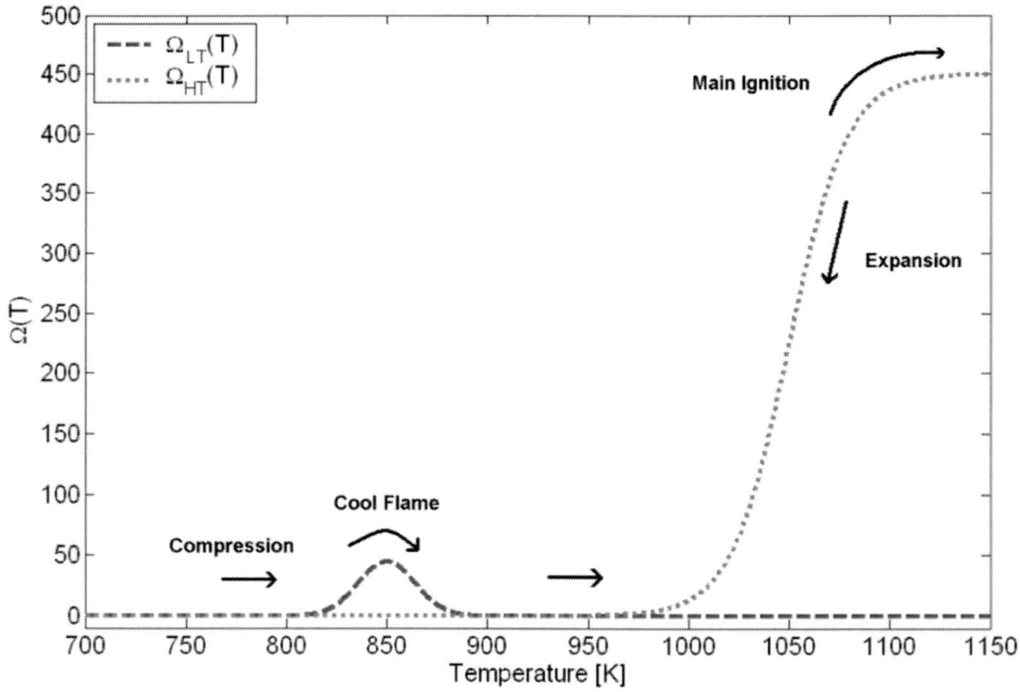


Figure 2. $\Omega_{LT}(T)$, $\Omega_{HT}(T)$ and Ω(T) profiles versus in-cylinder temperature.

Figure 2 shows an example of $\Omega(T)$ evolution vs. temperature T. Below 800 K, the reaction progress rate is equal to 0. During this period, the temperature increase is only caused by compression. Above this temperature, $\Omega_{LT}(T)$ first increases until $T=T_{LT}$ and then decreases, simulating the low temperature oxidation. After this phase, $\Omega(T)$ is again equal to 0, corresponding to the NTC regime with a low reactivity. Above 950 K, $\Omega_{HT}(T)$ increases towards a maximum value corresponding to $T_{HT}$ in order to simulate main ignition. $\Omega_{HT}(T)$ is only able to model the rapid increase in reactivity in terms of magnitude and ignition delay up to $T_{HT}$. In the proposed model, $\Omega_{HT}(T)$ is however coupled to the fuel mass fraction (equation (11)). Consequently, the product $\Omega(T).Y_f$ becomes equal to 0 when the fuel is completely oxidized, thus deleting the source term in equation (9). The system temperature is then governed by wall heat losses and displacement work.

In order to solve the model equations (9) and (11), the calibration of the model parameters introduced in the expressions of $\Omega_{LT}(T)$ and of $\Omega_{HT}(T)$ (equations 13 and 14) is required and developed in the next section.

## 4.1. Model Calibration

The objective of the calibration phase is to build simple relations of the parameters $A_{LT}$, $A_{HT}$, $C_{LT}$, $C_{HT}$, $T_{LT}$ and $T_{HT}$ in order to make the model autonomous. As mentioned above the two functions ($\Omega_{LT}(T)$ and $\Omega_{HT}(T)$) model respectively the dynamic of the first stage oxidation and the rapid transition to the main ignition.

Besides, these functions are principally driven by the two temperatures $T_{LT}$ and $T_{HT}$ corresponding respectively to the maximum level of the first stage and the maximum level of the second stage of the combustion process as shown in Figure 2. The numerous published investigations on HCCI mode have confirmed that this process is controlled by chemical kinetics, reduced in this study to a global reaction driven by $\Omega(T)$.

On another hand, the combustion progress due to chemical kinetics (i.e., $\Omega(T)$) is directly dependent of the engine operating conditions (equivalence ratio, EGR rate, inlet pressure...). Consequently, the best way to calibrate the model is to build relationships between the engine independent variables (i.e., inlet conditions) and these model parameters.

For the calibration step, numerical simulations have been used. These simulations take into account the oxidation process, using a reduced n-heptane mechanism developed previously with 25 species and 26 reactions (the reduced mechanism is given in appendix A) and validated with experimental data (see reference 19). These simulations have been done with SENKIN application of CHEMKIN code [23], applied to compression and expansion strokes from the IVC to the EVO with the engine geometry presented in Table 1. The cylinder charge is assumed homogeneous before injection (air and EGR) and after injection; the fuel is introduced in the combustion chamber at 70 CAD (BTDC). No heat transfer to the wall was taken into account during the calibration phase; this assumption will be discussed below. The choice of numerical calibration has been guided by the fact that to build a reliable calibration of the model, a wide range of engine conditions must be swept with single parameter variations, which is very difficult experimentally. We have done 2634 simulated points; for each point only one parameter varies. Table 2 indicates the different inlet conditions used at varying engine speed (1000 rpm, 2000 rpm and 3000 rpm).

In the proposed model the parameters $A_{LT}$ and $A_{HT}$ correspond to the magnitude of the combustion progress (i.e., heat release) of each stage. These two parameters have been calibrated from combustion heat release rate. The apparent heat release is easily calculated [16] from the first law of thermodynamics for a closed system which is quasi static (i.e., uniform in pressure and temperature) as follows:

$$\dot{Q}_{app} = \frac{1}{\gamma - 1}.V.\dot{p} + \frac{\gamma}{\gamma - 1}.P.\dot{V} \tag{15}$$

where: $Q_{app}$ is the apparent heat release which represents the difference between the chemical energy or heat release by combustion and the heat transfer to the walls. In equation (15), $\gamma$ is the ratio of specific heats (calculated here function of the mixture composition and of the temperature).

As mentioned above with the assumption of an adiabatic system the heat release rate by combustion is then equal to the apparent heat release and this quantity can be easily calculated from the SENKIN solutions using equation (15). From the equations (8) and (15) in adiabatic conditions, an expression of $\Omega(T)$ versus the rate of heat release by combustion can be written:

$$\Omega(T) = \frac{1}{m.Q_{LHV}.Y_{C_7H_{16}}}.\frac{dQ_{comb}}{d\theta} \tag{16}$$

In the relations (13) and (14), $A_{LT}$ and $A_{HT}$ represent respectively the maximum value of $\Omega_{LT}$ (T) at the temperature $T_{LT}$ and of $\Omega_{HT}$ (T) at the temperature $T_{HT}$, which allows writing:

$$A_{LT} = \frac{1}{m.Q_{LHV}.Y_{C_7H_{16}}(\theta_{cf})}.\frac{dQ_{comb}}{d\theta}\bigg|_{\theta=\theta cf} \tag{17}$$

$$A_{HT} = \frac{1}{m.Q_{LHV}.Y_{C_7H_{16}}(\theta_{mi})}.\frac{dQ_{comb}}{d\theta}\bigg|_{\theta=\theta_{mi}} \tag{18}$$

where: $\theta_{cf}$ and $\theta_{mi}$ correspond respectively to the timing of the maximum heat release rate by combustion of the cool flame and of the main ignition.

From equation (17) a map has been built for all the operating points summarized in Table 2 which gives a correlation of $A_{LT}$ function of inlet conditions:

$$A_{LT} = 54{,}71.\left(\frac{N}{N_{ref}}\right)^{-0{,}8019}\left(\frac{T_{in}}{T_a}\right)^{-5{,}995}.\left(\frac{P_{in}}{P_a}\right)^{1{,}574}.(X_{EGR})^{-0{,}904}.(X_{C_7H_{16}})^{0{,}3446} \tag{19}$$

During this step, we have found that the parameter for the high temperature regime is directly related to the parameter of the low temperature regime as following:

$$A_{HT} = a.(A_{LT})^b \tag{20}$$

where: a and b are only function of the equivalence ratio (fuel mass fraction or initial molar fraction). Usual extrapolation method has been used to obtain their expressions:

$$a = 9075 \cdot \frac{e^{(750\cdot(X_{C_7H_{16}}-0{,}0045))}}{1+e^{(750\cdot(X_{C_7H_{16}}-0{,}0045))}} \tag{21}$$

$$b = 0{,}6373 \cdot \frac{e^{(525\cdot(X_{C_7H_{16}}-0{,}0053))}}{1+e^{(525\cdot(X_{C_7H_{16}}-0{,}0053))}} \tag{22}$$

If the function Ω(T) is plotted versus CAD (Figure 3) instead of in-cylinder temperature as in Figure 1, we observe that each low temperature regime $T_{LT}$ is related to a timing of the cool flame in the cycle named $\tau_{cf}(T)$. The transition to the main ignition (i.e., $T_{HT}$) occurs after a delay called phasing in Figure 3.

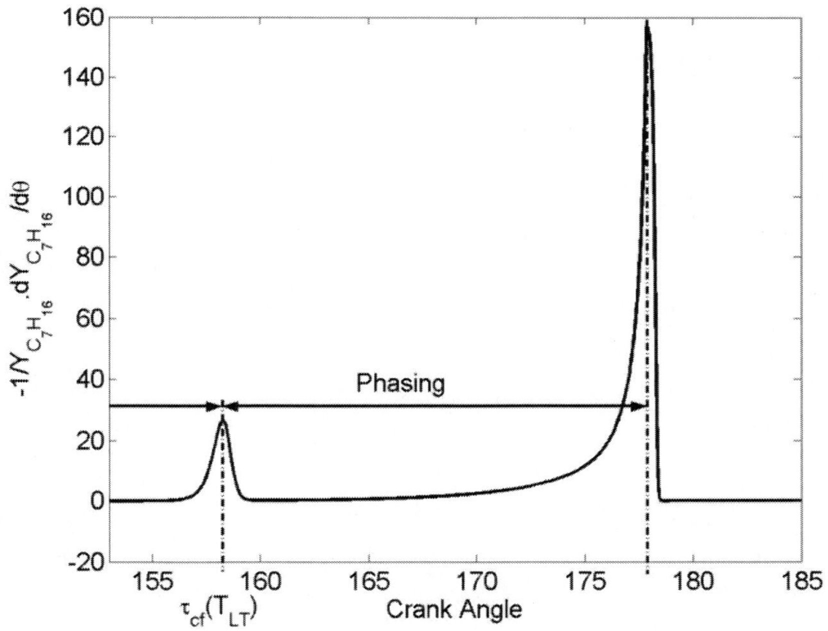


Figure 3. Ω(T) profile in a variable volume chamber versus CAD. Engine speed 2000 rpm, inlet temperature 400K, inlet pressure 1.5 bar, EGR rate 50%, equivalence ratio 0.3.

However, these two parameters vary with the inlet conditions and correspond to the timing of the maximum rate of heat release due to the combustion, respectively of the cool flame and of the main ignition. Consequently, if the relation expressing $\tau_{cf}(T)$ is known, then $T_{LT}$ can be easily calculated (using equation (13)). Similarly, if the phasing is known then the high temperature regime can be estimated (using equation (14)). Note that the model parameter $C_{HT}$ represents the slope of Ω(T) during the increase of the high temperature regime phase.

**Table 2. Simulated operating points**

| | Pressure [bar] | Temperature [K] | EGR rate [%] | Equivalence Ratio |
|---|---|---|---|---|
| Min | 1 | 350 | 20 | 0,2 |
| Step | 0,25 | 25 | 15 | 0,1 |
| Max | 2,5 | 450 | 65 | 0,9 |

The relation of the timing $\tau_{cf}(T(\theta))$ has been obtained from the numerous simulations, by creating a map of the timing of the maximum rate of heat release by combustion of the cool flame versus CAD with varying inlet parameters. The expression obtained for the cool flame delay (the time between the IVC and the maximum rate of heat release by cool flame) function of inlet conditions and the temperature is:

$$\tau_{cf}(T(\theta)) = 9{,}875.10^{-2}.\left(\frac{N}{N_{ref}}\right)^{-0{,}9193}.\left(\frac{T_{in}}{T_a}\right)^{-0{,}025}.\left(\frac{P_{in}}{P_a}\right)^{-0{,}05}.(X_{EGR})^{0{,}04}.(X_{C_7H_{16}})^{0{,}046}.\exp\left(-2{,}18.10^{-03}.T\right) \quad (23)$$

where: N is the engine speed and $N_{ref}$ is a reference engine speed (here fixed equal to 1000 rpm).

The different constants included in relation (23) have been obtained using a classical genetic algorithm.

In an engine, this timing is function of the inlet conditions and the temperature changes during the delay period due to the compression resulting from the piston motion. To account for the effect of changing conditions on the delay the empirical integral relation has been used in order to calculate $T_{LT}$ when the cool flame delay is reached:

$$\int_{IVC}^{\theta_{cf}} \left(\frac{1}{\tau_{cf}(T(\theta))}\right) \cdot d\theta = 1 \quad (24)$$

From their experimental study Ryan et al. [24] observe that the timing of the main ignition is dependent on the timing and the magnitude of the first stage (or cool flame). The authors propose a linear correlation to calculate the phasing function of the maximum rate of heat release during the cool flame stage.

This study agrees with the observation of Ryan et al., the map representing the shape of the phasing versus the maximum rate of heat release of the cool flame shows a correlation between those parameters. However, the correlation that we have obtained differs from Ryan et al. correlation and is written as:

$$P_h = 81{,}74.\left(\left.\frac{dQ_{comb}}{d\theta}\right|_{\theta=\theta_{cf}}\right)^{-0{,}07246} - 48{,}1 \quad (25)$$

The parameter $C_{LT}$ has very slight variation when the equivalence ratio varies; we have derived a simple relation to calculate its variation:

$$C_{LT} = 100 \cdot \frac{\phi}{2} + 5 \tag{26}$$

With these two above equations the model is completely calibrated. However, the parameter $C_{HT}$ has been kept constant and equal to 0.04; we have observed that its variation is very slight.

The model parameters are then function only of engine inlet conditions. Given the initial pressure and temperature, the EGR rate and the equivalence ratio, the model is able to predict the evolution of the mixture temperature and fuel mass fraction during the cycle. However, with the formulation proposed in this paper (equations (9) and (11)), the global reaction progress rate $\Omega(T)$ models both the dynamics of the fuel oxidation and the dynamics of the temperature evolution. We have observed that the model reproduces accurately either fuel oxidation dynamics or temperature dynamics; it is difficult to get a good prediction of both phenomena with a unique reaction progress rate.

These two dynamics are controlled mainly by the parameter $A_{LT}$. Numerical simulation results using the reduced mechanism [21] have shown that the ratio between $A_{LT}$ calculated from fuel mass fraction ($A_{LTf}$) and $A_{LT}$ calculated from temperature ($A_{LTT}$) is not equal to 1.

Therefore, two distinct progress rates have been introduced:

$$\Omega_f(T) = \Omega_{LTf}(T) + \Omega_{HT}(T) \tag{27}$$

$$\Omega_T(T) = \Omega_{LTT}(T) + \Omega_{HT}(T) \tag{28}$$

where the low temperature progress rates are written as:

$$\Omega_{LTf}(T) = A_{LTf} \exp\left( -\frac{(T - T_{LT})^2}{C_{LT}^2} \right) \tag{29}$$

$$\Omega_{LTT}(T) = A_{LTT} \exp\left( -\frac{(T - T_{LT})^2}{C_{LT}^2} \right) \tag{30}$$

where the parameters $A_{LTf}$ and $A_{LTT}$ are linked together, the Figure 4 represents the evolution of the ratio between these two parameters versus $A_{LTT}$. A simple expression of this ratio is then proposed:

$$\frac{A_{LTf}}{A_{LTT}} = 7{,}76.(A_{LTT})^{-0{,}3794} + 1{,}872 \tag{31}$$

Consequently, with these adaptations the fuel mass balance (11) and energy balance (9) can be rewritten and the modified model is:

$$\dot{Y}_f = -\Omega_f(T) \cdot Y_f \tag{32}$$

$$mc_v\dot{T} = -P\dot{V} - hS(T - T_w) + Y_f\Omega_T(T)mQ_{LHV} \tag{33}$$

## 5. Results and Discussion

The ODE system has been coded in C language using CVode software package [25]. The CVode is efficient for stiff problems, as is the case of the proposed model.

All the simulation results presented here have been done using the geometry of the engine presented in Table 1. The EGR composition has been fixed equal to: $CO_2$ (12%), $H_2O$ (2%), $N_2$ (83.5%), $O_2$ (2.5%). All the simulation results have been obtained with the assumption that after fuel injection the mixture is completely homogeneous (no computation of the injection process was used).

Computation times in all the simulation cases were less than one second (maximum computation time around two seconds).

The experiments have been done with a Renault single cylinder engine. The specifications of this engine are given in Table 1. The engine was running with premium diesel fuel in near homogeneous mode, the premium fuel has a cetane number of 55 very close to the cetane number of n-heptane (around 56) used to calibrate the developed model.

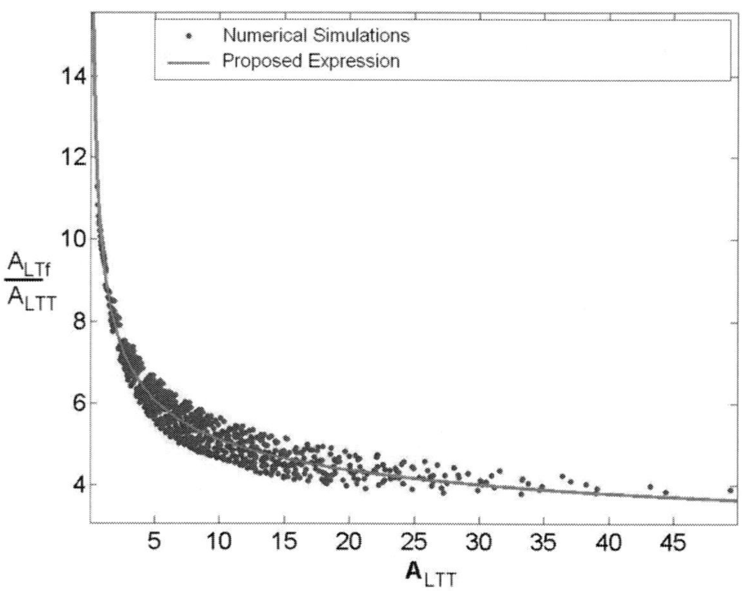


Figure 4. Evolution of the ratio between $A_{LTf}$ and $A_{LTT}$ versus $A_{LTT}$.

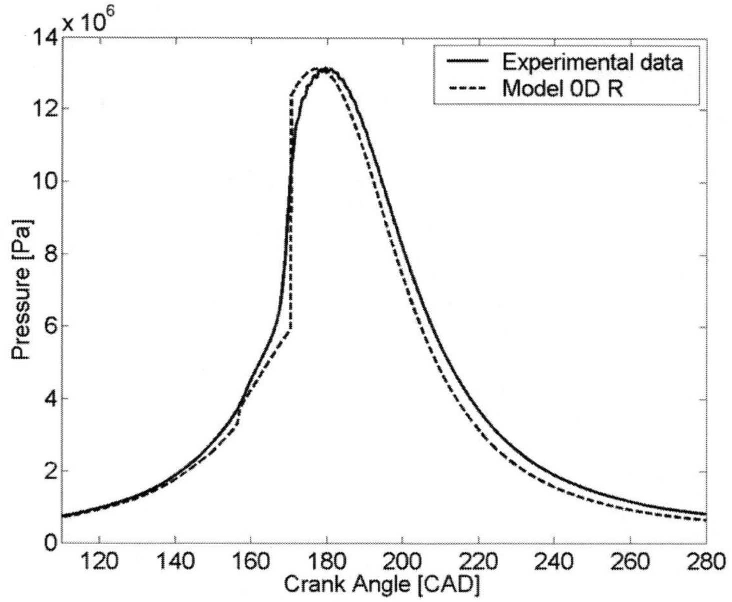


Figure 5. Comparison of the calculated cylinder pressure versus CAD with measured data. Inlet pressure 2.42 bar, EGR rate 55.3 %, inlet temperature 387.7 K and injected fuel mass 17.55 mg/cycle.

The amount of EGR was regulated by a valve in the pipe connecting the exhaust and inlet system and was evaluated by measuring the $CO_2$ concentration in the inlet and exhaust gases. The cylinder pressure was measured with a pressure transducer mounted in the cylinder head. The numerical simulations have been done taking into account the heat transfer to the walls by forced convection (see equation 34). The convective heat transfer coefficient is calculated from a modified Woschni's correlation [26], which is more suitable to HCCI process than the traditional Woschni correlation [27]. The wall temperature has been fixed equal to 500 K representing an average wall temperature estimated from the test bench engine during a warm-up operation.

Figure 5 presents the in-cylinder pressure profiles predicted by the model and measured on the engine when the engine speed is equal to 2000 rpm. The engine has been operated with an inlet pressure equal to 2.42 bar, an inlet temperature equal to 387.7 K, one single injection at 70 CAD (BTDC) with an injected fuel mass equal to 17.55 mg/cycle and an EGR rate equal to 55.3%. In this figure, the overall shape of the predicted pressure curve is close to the measured one. However, the first stage delay is advanced by one CAD and the main ignition is delayed by 2.5 CAD with the calculated profile. Consequently, the location of the maximum calculated pressure is positioned too early (around 2 CAD); the difference between the calculated maximum peak pressure and the measured maximum peak pressure is lower than 0.4 bar which is very small. The same trends as above are observed on Figure 6 with the same injection timing (70 CAD BTDC) and engine speed but different inlet conditions: inlet pressure equal to 2.454 bar, inlet temperature equal to 385.25 K, injected fuel mass equal to 14.97 mg/cycle and EGR rate equal to 62.9%. Here again, the difference between calculated and measured maximum peak pressure is very small (lower than 0.8 bar).

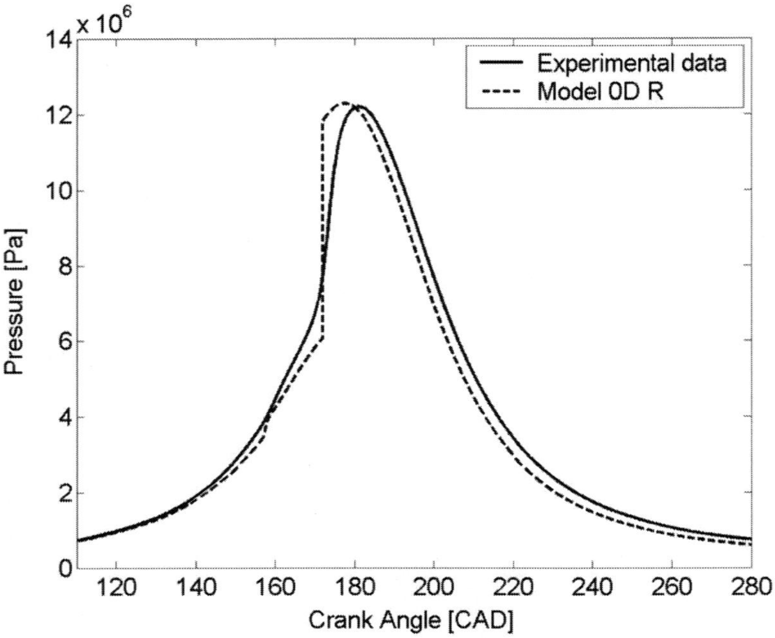


Figure 6. Comparison of the calculated cylinder pressure versus CAD with measured data. Inlet pressure 2.454 bar, EGR rate 62.9 %, inlet temperature 385.25 K and injected fuel mass 14.97 mg/cycle.

The comparison between the model and the measured in-cylinder pressure when the engine speed is fixed equal to 1500 rpm is plotted in Figure 7. The injection occurs as above at 70 CAD (BTDC) with an injected fuel mass equal to 11.8 mg/cycle and an EGR rate equal to 64.8%.

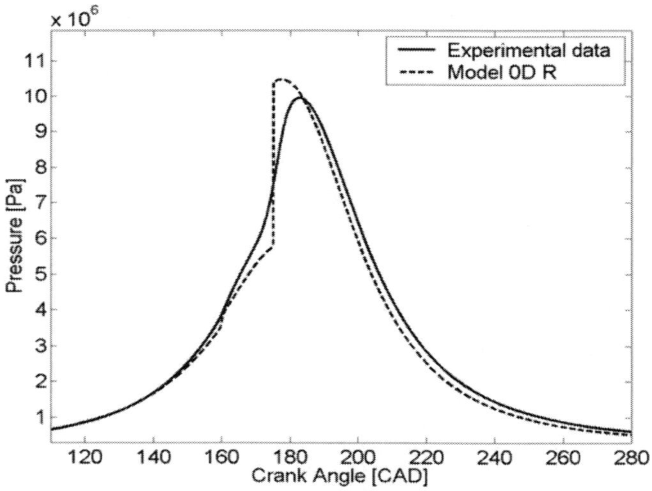


Figure 7. Comparison of the calculated cylinder pressure versus CAD with measured data. Engine speed equal to 1500 rpm, inlet pressure 2.179 bar, EGR rate 64.8 %, inlet temperature 361 K and injected fuel mass 11.8 mg/cycle.

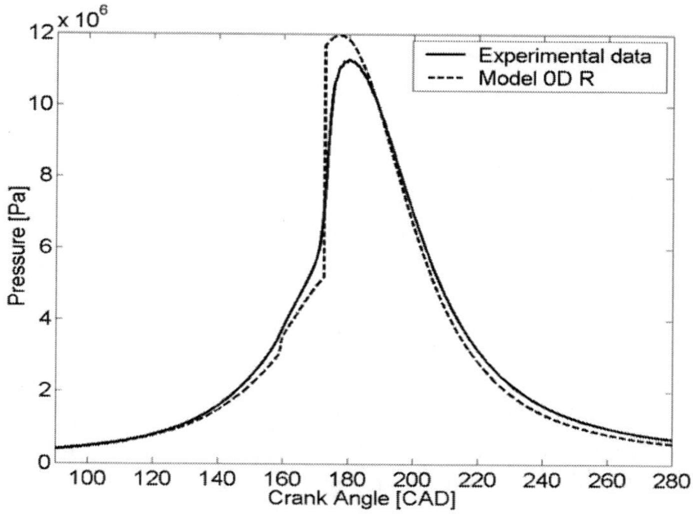


Figure 8. Comparison of the calculated cylinder pressure versus CAD with measured data. Engine speed equal to 2000 rpm, fuel injection at 90 CAD BTDC, inlet pressure 2.06 bar, EGR rate 54.9%, inlet temperature 387 K and injected fuel mass 16.59 mg/cycle.

The inlet pressure and the inlet temperature are respectively equal to 2.179 bar and 361 K. For a different engine speed, the predicted in-cylinder pressure is also very close to the measured one (difference in pressure lower than 1 bar mostly due to the stiffness of the calculated pressure increase). The injection timing is part of the model by itself. However the model calibration has been done with fuel injection at 70 CAD BTDC. The behaviour of the model when the fuel injection timing varies has been tested and compared to the engine experiments. Figure 8 shows the comparison between calculated and measured pressures with a fuel injection at 90 CAD BTDC at 2000 rpm, with an inlet pressure equal to 2.06 bar, an inlet temperature equal to 387 K, an injected fuel mass equal to 16.59 mg/cycle and an EGR rate equal to 54.9%. The model predicts the cool flame delay and main ignition delay with a negligible error. The calculated peak pressure is overestimated by 0.5 bar and in advance by 1.5 CAD compared to the experimental data. The differences observed between the measured and calculated pressure profiles presented in this section are due mostly to completely homogeneous mixture used during the simulations. This assumption is the main cause of the rapid pressure increase obtained with the model during the main ignition. However, model against experimental data shows overall satisfactory agreement. The differences between calculated and measured mean effective pressure (IMEP) reported in table 3 are very small (lower than 4%).

**Table 3. Comparison of the calculated mean effective pressure (IMEP) with measured data**

| Engine speed (rpm) | Injected fuel mass (mg) | Injection timing CAD BTDC | Experimental IMEP (bar) | Calculated IMEP (bar) |
|---|---|---|---|---|
| 2000 | 17.55 | 70 | 8.9 | 9 |
| 2000 | 14.97 | 70 | 7.7 | 7.8 |
| 2000 | 16.59 | 70 | 8.1 | 8.4 |
| 1500 | 11.8 | 90 | 5.9 | 6 |

## CONCLUSION

The first objective of this work is to show if it is possible to model the HCCI process with much reduced chemistry. The proposed model is reduced to two state variables: the temperature and the mass fraction of burning fuel in the combustion chamber. The chemistry is modeled by a global degradation reaction where the reaction rate coefficient, usually modeled through the Arrhenius law is driven in this approach by a global function Ω(T). A drawback of the chosen approach is the introduction of some new parameters, which need to be correlated to give reliable results; this has required an important and large parametric study. The proposed model is then autonomous, meaning that the model parameters $A_{LT}$, $A_{HT}$, $C_{LT}$, $C_{HT}$, $T_{LT}$ and $T_{HT}$ are function of the engine operating conditions only.

The first attempt of this modeling work has been to describe the two stages process by one single global expression; as this approach was not successful; the global function has been split up in two functions. In this second approach the global reaction progress rate Ω(T) models both the dynamics of the fuel oxidation and the dynamics of the temperature evolution. The results of this latter approach have shown that it is difficult to get a good prediction of both phenomena with a unique reaction progress rate. Thus, a modified model including two distinct progress rates to model the first stage oxidation has been proposed in this paper.

Generally the numerical results have proven the feasibility of the proposed methodology. Compared to the model with included chemistry, the delays for cool flame, phasing and main ignition are correctly reproduced. However, stiffness has been observed on the simulated in-cylinder pressure evolutions during main ignition due principally to the assumption that the mixture is completely homogeneous. The model has been compared to Renault engine experimental data. This comparison shows that the model gives reasonable accuracy with an engine running with premium diesel fuel in terms of ignition delay and mean effective pressure (IMEP).

Thanks to its very short computation time (around two seconds), the proposed model is appropriate for integration into model-based HCCI engine control systems, since it requires only the engine inlet conditions. This work is under progress to include initial heterogeneities of temperature and of fuel concentrations in the combustion chamber.

## ACKNOWLEDGMENTS

This research was funded by RENAULT AUTOMOBILES; the authors wish to thank this company.

## APPENDIX A

26 reactions and 25 species mechanism for n-Heptane ignition for HCCI process [Rate constants are written as A $T^b$ exp(-E/RT); Units are, cm – mole – s – K – cal]

| N° | REACTION | A | b | E(cal) |
|---|---|---|---|---|
| 1 | $C_7H_{16}+O_2=C_7H_{15\text{-}2}+HO_2$ | 2.800E+14 | 0.0 | 47180.0 |
| 2 | $C_7H_{16}+OH= C_7H_{15\text{-}2}+H_2O$ | 4.800E+09 | 1.3 | 690.5 |
| 3 | $C_7H_{15\text{-}2}+O_2=C_7H_{15}O_2$ | 2.000E+12 | 0.0 | 0.0 |
| 4 | $C_7H_{15}O_2=C_7H_{14}O_2H$ | 6.000E+11 | 0.0 | 20380.0 |
| 5 | $C_7H_{14}O_2H+O_2=C_7H_{14}O_2HO_2$ | 2.340E+11 | 0.0 | 0.0 |
| 6 | $C_7H_{14}O_2HO_2=C_7KET21+OH$ | 2.965E+13 | 0.0 | 26700.0 |
| 7 | $C_7KET21=C_5H_{11}CO+CH_2O+OH$ | 1.000E+16 | 0.0 | 42400.0 |
| 8 | $C_5H_{11}CO=C_2H_5+C_3H_6+CO$ | 1.000E+11 | 0.0 | 9600.0 |
| 9 | $C_7H_{15}=CH_3+2C_3H_6$ | 3.000E+13 | 0.0 | 29800.0 |
| 10 | $C_7H_{15}=C_2H_5+C_2H_4+C_3H_6$ | 1.200E+13 | 0.0 | 29600.0 |
| 11 | $C_3H_6+OH=CH_3CHO+CH_3$ | 3.500E+11 | 0.0 | 0.0 |
| 12 | $CH_3CHO+OH=CH_3+CO+H_2O$ | 1.000E+13 | 0.0 | 0.0 |
| 13 | $CH_3+HO_2=CH_2O+H+OH$ | 4.300E+13 | 0.0 | 0.0 |
| 14 | $CO+OH=CO_2+H$ | 3.510E+07 | 1.3 | -758.0 |
| 15 | $O+OH=O_2+H$ | 4.000E+14 | -0.5 | 0.0 |
| 16 | $H+O_2+M=HO_2+M$<br>Enhanced third body efficiencies:<br>$O_2$=0, $H_2O$=0, CO=0.75, $CO_2$=1.5, $C_2H_6$=1.5, $N_2$=0 | 2.800E+18 | -0.86 | 0.0 |
| 17 | $H+O_2+N_2=HO_2+N_2$ | 2.600E+19 | -1.24 | 0.0 |
| 18 | $HO_2+HO_2=H_2O_2+O_2$ | 2.000E+12 | 0.0 | 0.0 |
| 19 | $OH+OH\ (+M)=H_2O_2\ (+M)$<br>High<br>Low<br>Troe parameters: 0.7346, 94, 1756, 5182<br>Enhanced third body efficiencies:<br>$H_2O$=6, CO=1.5, $CO_2$=2, $C_2H_6$=3, $N_2$=0.7, $CH_4$=2 | <br>7.600E+13<br>4.3E+18 | <br>-0.37<br>-0.9 | <br>0.0<br>-1700 |
| 20 | $CH_2O+OH=HCO+H_2O$ | 2.430E+10 | 1.18 | -447.0 |
| 21 | $CH_2O+HO_2=HCO+H_2O_2$ | 3.000E+12 | 0.0 | 8000.0 |
| 22 | $HCO+O_2=HO_2+CO$ | 3.300E+13 | -0.4 | 0.0 |
| 23 | $CH_4+O=CH_3+OH$ | 1.020E+09 | 1.5 | 8604.0 |
| 24 | $CH_4+HO_2=CH_3+H_2O_2$ | 1.000E+13 | 0.0 | 18700.0 |
| 25 | $C_2H_4+OH=CH_2O+CH_3$ | 6.000E+13 | 0.0 | 960.0 |
| 26 | $C_2H_5+O_2=C_2H_4+HO_2$ | 2.000E+10 | 0.0 | -2200.0 |

## REFERENCES


[1] Onishi S., S. J. Hong, K. Shoda, P. Do Jo and S. Kato, "Active thermo – Atmosphere combustion (ATAC) – *A new combustion process for internal combustion engines*". 1979, SAE paper 790501.

[2] Najt P.M., D.E. Foster "*Compression ignition homogeneous charge combustion*". 1983, SAE paper 830264.

[3] Thring R.H, "*Homogeneous charge compression ignition (HCCI) engines*". 1989, SAE paper 892068.

[4] Ogume H., Ichikura T., Lida N., "*A study on adaptability of alternative fuels for lean burn two stroke ATAC engine*". 1997, SAE paper 972097.

[5] Rudolf H. Stanglmaier and Charles E. Roberts, "*Homogeneous charge compression ignition (HCCI): benefits, compromises, and future engine applications*". 1999, SAE paper 1999-01-3682.

[6] Thring R.H, "*Homogeneous charge compression ignition (HCCI) engines*". 1989, SAE paper 892068.

[7] Westbrook C.K. (Lawrence Livermore National Laboratory): "*Chemical Kinetics of Hydrocarbon Ignition in Practical Combustion Systems*". Proc.of the Comb.Inst, vol.28, 2000.

[8] Warnatz J. (University of Heidelberg, Germany): "*Rate coefficients in C/H/O system" in Combustion Chemistry* (W.C.Gardiner Jr ed.), Springer-Verlag, 1984.

[9] Van Blarigan P., Goldsborough S., "*A Numerical Study of a Free Piston Engine Operating on Homogeneous Charge Compression Ignition Combustion*". 1999, SAE paper 990619.

[10] Fiveland S.B., Assanis D.N., "*A Four-Stroke Homogeneous Charge Compression Ignition Engine Simulation for Combustion and Performance Studies*". 2000, SAE paper 2000-01-0332.

[11] Zheng J., Yang W., D.L. Miller, Cernansky S.K., "*A Skeletal Chemical Kinetic Model for the HCCI Combustion Process*". 2002, SAE paper 2002-01-0423.

[12] Aceves Salvador M., Flowers Daniel L., Westbrook Charles K., Smith J. Ray, Pitz William, Dibble Robert, Christensen Magnus, Johansson Bengt, "A Multi-Zones Model for Prediction of HCCI Combustion and Emissions". 2000, SAE paper 2000-01-0327.

[13] Xu H., Fu H., Williams H., Shilling I., "*Modelling Study of Combustion and Gas Exchange in a HCCI (CAI) Engine*". 2002, SAE paper 2002-01-0114.

[14] Yelvington P. E., Green W. H., "*Prediction of the Knock Limit and Viable Operating Range for a Homogeneous-Charge Compression-Ignition (HCCI) Engine*". 2003, SAE paper 2003-01-1092.

[15] Salvador M. Aceves, Joel Martinez-Frias, Daniel L. Flowers, Ray Smith, Robert W. Dibble, John F. Wright, Randy P. Hesse, "A *Decoupled Model of Detailed Fluid Mechanics Followed by Detailed Chemical Kinetics for Prediction of Iso-Octane HCCI Combustion"*. 2001, SAE paper 2001-01-3612.

[16] Scott B. Fiveland, Denis N. Assanis, "*Development of a Two-Zone HCCI Combustion Model Accounting for Boundary Layer Effect*". 2001, SAE paper 2001-01-1028.

[17] Komninos N.P., Hountalas D.T., Kouremenos D.A., "*Development of a New Multi-Zone Model for the Description of Physical Processes in HCCI Engines*". 2004, SAE paper 2004-01-0562.

[18] Song-Charng Kong, Craig D. Marriott, Rolf D. Reitz, Magnus Christensen, "*Modeling and Experiment of HCCI Engine Combustion Using Detailed Chemical Kinetics with Multidimensional CFD*". 2001, SAE 2001-01-1026.

[19] Noel. L, Maroteaux. F, Ahmed. A," *Numerical study of HCCI combustion in diesel engines using reduced chemical kinetics of n-heptane with multidimensional CFD code*". 2004, SAE paper 2004-01-1909.

[20] Song-Charng Kong and Rolf D Reitz, "*Numerical study of premixed HCCI engine combustion and its sensitivity to computational mesh and model uncertainties*". Combustion Theory Modelling 7 (2003) 417-433.

[21] F. Maroteaux, L. Noel, "*Development of a reduced n-heptane oxidation mechanism for HCCI combustion modelling*". Combustion and Flame 146 (2006), pp 246-267.

[22] John B. Heywood, "*Internal combustion engine fundamentals*". McGraw-Hill Book Company, 1988.

[23] Chemkin@reactiondesign.com

[24] Thomas W. Ryan III, Timothy J. Callahan and Darius Mehta, "*HCCI in a variable compression ratio engine-effects of engine variables*". 2004, SAE paper 2004-01-1971.

[25] Alan C. Hindmarsh, Radu Serban CVode v 2.3.0. Center of Applied Scientific Computing Lawrence Livermore National Laboratory. http://www.llnl.gov/CASC/sundials/description/description.html. April 2005.

[26] Junseok Chang, Orgun Güralp, Zoran Filipi, Dennis Assanis, Tang-Wei Kuo, Paul Najt and Rod Rask, "*New Heat Transfert Correlation for an HCCI Engine Derived from Measurements of Instantaneous Surface Heat Flux*", SAE 2004-01-2996, 2004.

In: Energy-Environment-Economics
Editor: Riad Benelmir

ISBN: 978-1-62948-862-2


*Chapter 16*

# Reference Calculations for Gas Turbine Installations

***V. Cenusa, Riad Benelmir* and A. Merabtine***
University of Lorraine, Vandoeuvre-Les-Nancy, France

## Abstract

This paper presents a mathematical model for the thermodynamic characterization for a gas turbine cycle validated by data issued from manufacturers catalogs. This comparison data is the exhaust gases composition at the gas turbine outlet. The relative error is less than 1%. The computer code is based on the "Delphi" utility software which is a Windows interface in order to facilitate exploitation by end users. Thermodynamics properties of moist air and of combustion gases are computed. An exergy analysis of the heat recovery generator is available in order to asses the feasibility of combined cycles.

**Keywords:** Gas turbine, combined cycle, exergy

## Nomenclature

| | | |
|---|---|---|
| A | $=\dot{v}_{air} / \dot{v}_{cb}$ | $[kmol_{air}/kmol_{cb}]$ |
| AB | silencer | |
| c1 | coefficient | |
| CC | combustion chamber | |
| CR | heat recovery steam generator (HRSG) | |
| Ex | exergy | [kW] |
| $Ex_{max\_rec}$ | maximum recuperated exergy from HRSG | [kW] |
| $Ex_{rec}$ | recuperated exergy from HRSG | [kW] |
| F | $= \dot{v}_{g.e.} / \dot{v}_{cb}$ | $[kmol_{g.e.}/kmol_{cb}]$ |

* E-mail: riad.benelmir@unniv-lorraine.fr.

| | | |
|---|---|---|
| FA | air filter | |
| GE | electric alternator | |
| h | enthalpy | [kJ/kg] |
| $h_{CR}$ | exhaust gases enthalpy from the HRSG | [kJ/kg] |
| $h_{min}$ | exhaust gases enthalpy from the HRSG at saturation temperature $t_{sat}$ (without the condensed water vapors within the evacuated gases). | [kJ/kg] |
| ISO | International Standard Organizations | |
| ITG | gas turbine installation | |
| K | compressor | |
| $k_{el}$ | correction coefficient for the electric alternator losses at the ITG | |
| $k_m$ | correction coefficient for the mechanical losses at the ITG | |
| w | specific work | [$kJ/kg_{air}$] |
| $\dot{m}$ | mass flow rate | [kg/s] |
| M | molar mass | [kg/kmol] |
| $n_{ITG}$ | number of ITG | [-] |
| p | pressure | [bar] |
| $P_{air}$ | thermal power of inlet air at air filter (AF) | [kW] |
| $P_b$ | electrical power output | [kW] |
| $P_{b\_CCGV}$ | electrical power of the combined cycle | [kW] |
| $P_{b\_ITV}$ | electrical power output for the steam turbine | [kW] |
| $P_C$ | thermal energy flow of the combustion | [kW] |
| PCI | lower heating value of the fuel (LHV) | [$kJ/Nm^3$] |
| $P_{g.e.}$ | thermal energy flow of the exhaust gases | [kW] |
| Pt | thermal energy flow | [kW] |
| $p_{tg}$ | power of the turbo-generator | [kW] |
| $pt_{max}$ | maximal available thermal energy flow at HRSG | [kW] |
| $Pt_{rec}$ | available thermal energy flow at HRSG | [kW] |
| $q_{sp}$ | heat rate | [$kJ/kWh_e$] |
| s | entropy | [kJ/kg/°C] |
| $s_{CR}$ | gas entropy at the exit of the HRSG | [kJ/kg/°C] |
| $s_{min}$ | exhaust gas entropy at $t_{sat}$ | [kJ/kg] |
| T | temperature | [K] |
| t | temperature | [°C] |
| $t_{CR}$ | evacuated gases temperature at the exit of the HRSG | [°C] |
| TG | gas turbine | |
| $t_{sat}$ | saturation temperature of the steam within the exhaust gases | [°C] |
| v | specific volume | [$m^3/kg$] |

| | | |
|---|---|---|
| $V_M$ | normal molar volume of methane | [$Nm^3$/kmol] |
| $y_{st}$ | inside coefficient | |

## Greek Symbols

| | | |
|---|---|---|
| $\dot{\nu}$ | molar flow | [kmol/s] |
| $\lambda$ | excess air | [-] |
| $\gamma$ | molar ratio | [%] |
| $\varphi$ | air relative humidity | [%] |
| $\Delta$ | losses | |
| $\eta$ | efficiency | |
| $\delta_K$ | compression pressure ratio | [-] |
| $\gamma_{CO2\_air}$ | molar fraction of air's $CO_2$ over air | [$kmol_{CO2}/kmol_{air}$] |
| $\gamma_{CO2\_cb(2)}$ | molar fraction of air's $CO_2$ over fuel | [$kmol_{CO2}/kmol_{cb}$] |
| $\gamma_{CO2\_cb(3)}$ | molar fraction of flue gas's $CO_2$ over fuel | [$kmol_{CO2}/kmol_{air}$] |
| $\gamma_{CO2\_g.e.}$ | molar fraction of gas's $CO_2$ over flue gas | [$kmol_{CO2}/kmol_{g.e.}$] |
| $\gamma_{H2O}$ | molar fraction of air's $H_2O$ over air | [$kmol_{CO2}/kmol_{air}$] |
| $\gamma_{H2O}$ | molar fraction of air's $H_2O$ over fuel | [$kmol_{CO2}/kmol_{air}$] |
| $\gamma_{H2O}$ | molar fraction of flue gas's $H_2O$ over fuel | [$kmol_{CO2}/kmol_{air}$] |
| $\gamma_{H2O}$ | molar fraction of gas's $H_2O$ over flue gas | [$kmol_{CO2}/kmol_{air}$] |
| $\gamma_{N2}$ | molar fraction of air's $N_2$ over air | [$kmol_{CO2}/kmol_{air}$] |
| $\gamma_{N2}$ | molar fraction of air's $N_2$ over fuel | [$kmol_{CO2}/kmol_{air}$] |
| $\gamma_{N2}$ | molar fraction of flue gas's $N_2$ over fuel | [$kmol_{CO2}/kmol_{air}$] |
| $\gamma_{N2}$ | molar fraction of gas's $N_2$ over flue gas | [$kmol_{CO2}/kmol_{air}$] |
| $\gamma_{O2}$ | molar fraction of air's $O_2$ over air | [$kmol_{CO2}/kmol_{air}$] |
| $\gamma_{O2}$ | molar fraction of air's $O_2$ over fuel | [$kmol_{CO2}/kmol_{air}$] |
| $\gamma_{O2}$ | molar fraction of flue gas's $O_2$ over fuel | [$kmol_{CO2}/kmol_{air}$] |
| $\gamma_{O2}$ | molar fraction of gas's $O_2$ over flue gas | [$kmol_{CO2}/kmol_{air}$] |
| $\eta_{el}$ | ITG alternator efficiency | |
| $\eta_{CC}$ | combustion chamber efficiency | |
| $\eta_{CCGV}$ | electrical global efficiency of the combined cycle | |
| $\eta_{cogeneration}$ | global efficiency of the ITG cogeneration installation | |
| $\eta_{CR}$ | HRSG exergy efficiency | |
| $\eta_{ITG}$ | ITG electrical efficiency | |
| $\eta_K$ | compressor efficency | |
| $\eta_m$ | mechanical efficiency of the ITG (including gear efficiency) | |
| $\eta_{alt\_TV}$ | mechanical efficiency multiplied by alternator efficiency of the steam turbine | |
| $\eta_{ITV}$ | steam cycle exergy efficiency | |
| $\eta_{TG}$ | gas turbine efficiency | |
| $\Delta p_{AB}$ | pressure drop at the AB | [bar] |
| $\Delta p_{CC}$ | pressure drop at the combustion chamber | [bar] |

| | | |
|---|---|---|
| $\Delta p_{FA}$ | pressure drop at the air filter | [bar] |
| $\Delta P_{alt}$ | sum of the mechanical losses ad ITG alternator losses | [kW] |
| $\Delta P_{alt+CC}$ | sum of the mechanical losses, the alternator losses, and the CC losses for the ITG | [kW] |
| $\Delta P_{el}$ | ITG alternator losses | [kW] |
| $\Delta P_{g.e.}$ | losses of the exhaust gases sensible heat | [kW] |
| $\Delta P_{CC}$ | combustion chamber losses at the ITG | [kW] |
| $\Delta P_{Cu}$ | copper losses at the ITG alternator | [kW] |
| $\Delta P_{Fe}$ | iron losses from the ITG alternator | [kW] |
| $\Delta P_m$ | mechanical losses | [kW] |

## Subscripts

| | |
|---|---|
| 0 | air inlet at the air filter |
| 1 | air exit from the air filter |
| 2 | air exit from the compressor |
| 3 | combustion gases from the combustion chamber |
| 4 | combustion gases from the gas turbine |
| 5 | combustion gases from the silencer |
| air | air |
| b | output terminal of the alternator |
| cb | fuel |
| CCGV | gas/steam combined cycle |
| $CH_4$ | methane |
| $CO_2$ | carbon dioxide |
| CU | copper |
| el | alternator |
| ex | exergy |
| Fe | iron |
| g.e. | combustion gases |
| $H_2O$ | water |
| i | index |
| ITG | gas turbine installation |
| ITV | steam turbine installation |
| TV | steam turbine |
| m | mechanic |
| max | maximum |
| max_rec | maximum available |
| min | minimum |
| $N_2$ | nitrogen |
| $O_2$ | oxygen |
| rec | available |
| ref | reference |

| | |
|---|---|
| sat | saturate |
| t | theoretic |
| tg | turbo-generator (turbo-alternator) |

## INTRODUCTION

In cogeneration of electrical and thermal energy and steam/gas combined cycle [1-6], the gas turbine is a key component, less significant in simple cycles [7]. Technology of Gas Turbine (GT) Plants is one of the most dynamic in the world. In this field, improvements where realized mostly in special materials and coatings resisting at high temperature conditions [8] as Directionally Solidified (DS) materials [9] and Thermal Barrier Coating (TBC), in burners with low $NO_X$ emissions as EnVironmental (EV) burners of Alstom ([10-13]), Dry low $NO_X$ (DLN) burners of General Electric [14, 15], Dry Low Emissions (DLE) burners of Westinghouse-Rolls Royce, Hybrid Burner (HB) of Siemens KWU [16, 17] and PreMix (PM) burners of Mitsubishi Heavy Industries [9], and also in cooling technologies of high temperatures parts.

GT plants are classified in two categories following the technology of construction [18]: (a) Aero-derivative machines fitted to ground operation (from few MW of tens of MW) which are derived from the aviation turbo-reactor with important adaptation due to high velocity of turbine exhaust gases and from the naval turbo-propeller with minor adaptation due to a lower velocity; (b) Heavy-Duty machines which are developed especially for power production (from tens to hundreds of MW).

Aerderivative GT in aviation engines have been designed for maximum thermal efficiency (minimum fuel consumption) since the aircraft had a limited quantity of fuel to cover large distances. Produced power is usually lower than 50 MW and the combustion gases have low thermal potential with respect to combined cycle applications but sufficient enough for cogeneration applications. The pressure ratio varies between 27 and 30.

Heavy-Duty GT and Aeroderivative GT with turbo-propulsion have been designed for maximum specific power production. In Heavy-Duty GT, the flue-gas, with a mass flow rate between 350 t/h and 560 t/h, have a temperature ranging from 550 to 590 °C. This thermal potential, greater than in Aeroderivative machines, is appropriate for combined cycles applications. The pressure ratio varies between 11 and 17. A particular case of Heavy-Duty machines is sequential combustion GT where the pressure ratio can reach calues comparable with the Aeroderivative machines. For example the GT24 and GT26 gas turbines of Alstom Power have a pressure ratio of 30. These turbines have also an important net efficiency (approximately 58 %) in combined cycle.

GT plants cover a large domain of electrical power production (3 to 265 MW) and reach up to 40 % efficiency. Most of these gas turbines are aeroderivative machines, as the LM6000 gas turbines of Nuovo Pignone and Motoren und Turbinen-Union, the LM6000 Sprint gas turbines of GE S & S Energy Products, the TRENT and WR21 gas turbines of Rolls-Royce and the Mercury50 gas turbine of Solar Turbines Inc.

Combines cycles with GT and Steam Turbines (ST) reach up to 57% efficiency. Gas turbines used in these combined cycles are "heavy-study" machines as the GT24 and GT26 gas turbine of Alstom Power, the MS9001FA gas turbine of Nuovo Pignone and General

Electric Power Systems, the V94.3A and W501G gas turbines of Siemens AG Power Generation Group KWU.

This paper presents a model for a gas turbine installation (ITG) validated by manufacturer's data, as well as an exergy analysis of a potential heat recovery steam generator (HRSG) and of the combined gas/vapor combined cycle.

## 1. ITG Cycle Design

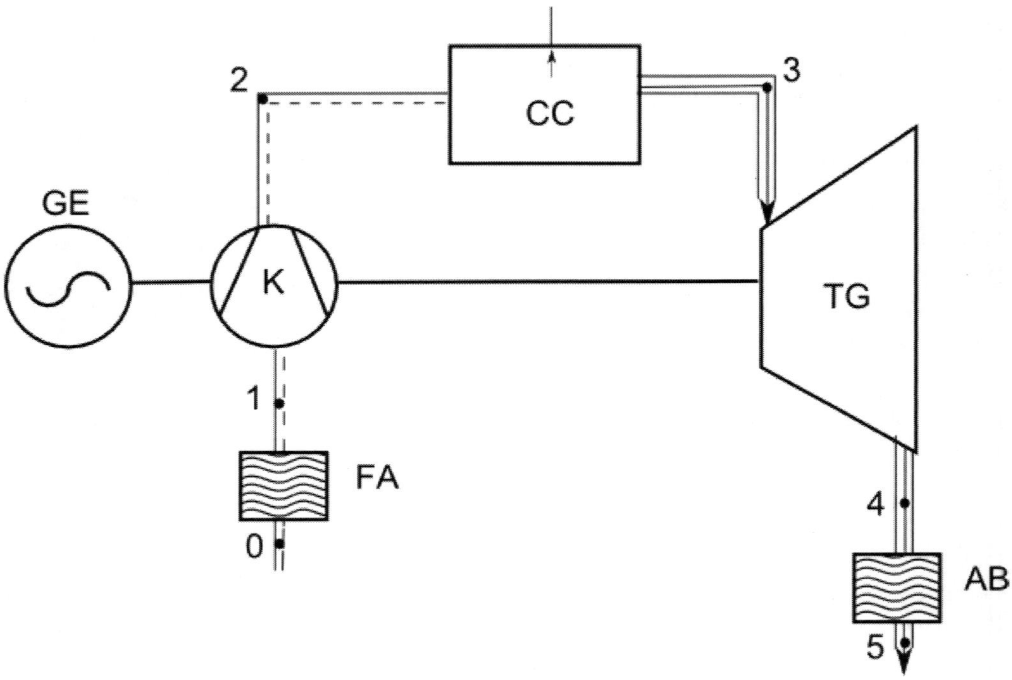


Figure 1. Schema of the ITG cycle.

### *1.1. Assumptions*

- The combustible is methane (other combustibles are not available in the computing program);
- Air and combustion gases thermodynamic properties are functions of temperature, pressure, and molar composition;
- Fixed pressure drop in the combustion chamber, in the air filter and in the silencer;
- Fixed efficiency of the combustion chamber;
- No consideration of air cooling of the gas turbine blades.

### *1.2. Initial Data*

*Manufacturer data*

- Electrical power output $P_b$ [kW]
- heat rate $Q_{sp}$ [kJ/kWh$_e$]
- Mass flow rate of exhaust gases $\dot{m}_{g.e.}$ [kg/s]

- Outlet temperature of gas turbine gases — $T_5$ [°C]
- Compression pressure ratio — $\delta_K$ [-]
- Lower heating value of fuel (LHV) — PCI [kJ/Nm$^3$]

*Operating conditions*

- ISO (International Standard Organizations)

$t_0$=15[°C]; $p_0$=1.013 [bar]; $\varphi_0$=60%; $\Delta p(FA,AB)$=0[bar]

- or other custom conditions:

$t_0$[°C]; $p_0$ [bar]; $\varphi_0$; $\Delta p(FA,AB)\neq 0$[bar]

*User data*

| | |
|---|---|
| Combustion chamber efficiency | $\eta_{CC}$ [%] |
| ITG alternator efficiency | $H_{el}$ [%] |
| Compressor efficiency | $\eta_K$ [%] |
| Pressure drop in the combustion chamber | $\Delta p_{CC}$ [%] |
| User data for HRSG and combined cycle exergy calculations (optional) | |
| Exit temperature of the CR | $t_{CR}$ [°C] |
| Reference temperature | $T_{ref}$ [°C] |
| Exergetic efficiency of CR | $\eta_{CR}$ [%] |
| Vapor cycle exergetic efficiency (turbine, pipes,…) | $\eta_{ITV}$ [%] |
| Steam turbine's alternator efficiency | $H_{alt\_TV}$ [%] |
| Number of ITG | $n_{ITG}$ [-] |

### *1.3. Variables*

*Combustion chamber*

- excess air
- Fuel mass flow rate
- Air mass flow
- Exhaust gas mass flow rate
- Molar fractions (air, combustion gas)
- Combustion power
- Combustion thermal losses
- Flue gas temperature

*Installation*

- Cycle thermodynamic state points
- Specific mechanical work
- Power of turbo-generator
- Global gross efficiency
- Thermal power at air filter inlet

### *1.4. Losses*

*Turbo-compressor group*

- Exhaust gas thermal power
- Isentropic efficiency
- Mechanical efficiency
- Alternator losses

## 2. The Mathematical Model

The Mathematical model of the ITG cycle is based on the available manufacturer data. The computer code is realized with "Delphi programming utility" which eases user interface due to its windows facilities. The data base of this program [a] contains gas turbine installations sorted by the manufacturer, the model and the following parameters: $P_b$, $q_{sp}$, $t_5$, $\delta_K$, and $\dot{m}_{g.e.}$. The user may add a specific configuration. The routine which computes inlet parameters of the air filter needs the atmospheric conditions: These parameters may be ISO conditions ($p_0$=1.013 [bar], $t_0$=15 [°C], $\varphi_0$=60 [%]) or specified by user. Air molar composition, molar mass of humid air and other air properties are calculated.

Global gas turbine installation efficiency ($\eta_{ITG}$):

$$\eta_{ITG} = \frac{3600}{q_{sp}} \tag{1}$$

Combustion thermal power:

$$P_c = \frac{P_b}{\eta_{ITG}} \tag{2}$$

Fuel molar and mass flow rate:

$$\dot{v}_{cb} = \frac{P_c}{PCI.V_M} \text{ [kmol/s]} \tag{3}$$

with $V_M$=22.414 [$Nm^3$/kmol]

$$\dot{m}_{cb} = \dot{v}_{cb}.M_{CH_4} \text{ [kg/s]} \tag{4}$$

with $M_{CH4}$=16 [kg/kmol]

The routine for the design of the combustion chamber computes the following parameters [3]:

Humid air / fuel:

$$A = \frac{2.\lambda}{\gamma_{O_2\_air}.\left(1-\gamma_{H_2O\_air}\right)} \text{ [kmol}_{air}\text{/kmol}_{cb}\text{]} \tag{5}$$

$$\gamma_{i\_cb(2)} = \gamma_{i\_air}.A \quad i \equiv \{N_2, O_2, H_2O \text{ and } CO_2\} \tag{6}$$

$$\gamma_{N_2\_cb(3)} = \gamma_{N_2\_cb(2)} \tag{7}$$

$$\gamma_{O_2\_cb(3)} = \gamma_{O_2\_cb(2)} - 2 \tag{8}$$

$$\gamma_{H_2O\_cb(3)} = \gamma_{H_2O\_cb(2)} + 2 \tag{9}$$

$$\gamma_{CO_2\_cb(3)} = \gamma_{CO_2\_cb(2)} + 1 \tag{10}$$

Humid combustion gases/fuel

$$F = \sum_i \gamma_{i\_cb(3)} \text{ [kmol}_{g.e.}\text{/kmol}_{cb}\text{]} \tag{11}$$

Molar composition of combustion gases:

$$\gamma_{i\_g.e.} = \frac{\gamma_{i\_cb(3)}}{F} \tag{12}$$

Molar mass of humid combustion gases:

$$M_{g.e.} = \sum_i \left(\gamma_{i\_g.e.}.M_i\right) \text{[kmol}_{cb.}\text{/kmol}_{cb}\text{]} \tag{13}$$

The excess air coefficient ($\lambda$) is unknown and requests an iterative calculation (figure 4): The convergence is reached when the computed mass flow rate equals the mass flow rate established by the user (the relative error is $10^{-8}$ %).

Air mass flow rate:

$$\dot{v}_{air} = A.\dot{v}_{cb} \tag{14}$$

$$\dot{m}_{air} = \dot{v}_{air}.M_{air} \tag{15}$$

Combustion gases mass flow rate:

$$\dot{m}_{g.e.} = F.\dot{v}_{cb}.M_{g.e.} = \dot{m}_{air} + \dot{m}_{cb} \tag{16}$$

Thermal energy flow existing the gas turbine:

$$P_{g.e.} = \dot{m}_{g.e.}.h_5 \tag{17}$$

Thermal energy flow of the inlet air:

$$P_{air} = \dot{m}_{air}.h_0 \tag{18}$$

Heat losses with the exhaust gases:

$$\Delta P_{g.e.} = P_{g.e.} - P_{air} \tag{19}$$

Global losses (shaft, alternator, combustion chamber):

$$\Delta P_{alt+CC} = P_c + P_{air} - P_{g.e.} - P_b = P_c - \Delta P_{g.e.} - P_b \tag{20}$$

The program verifies the data introduced by the user by means of the first principle of thermodynamics ( $\Delta P_{alt+CC} > 0$). In the case of negative value, the program returns a warning message and stops.

The combustion chamber efficiency ($\eta_{CC}$) introduced by the user and relation (2) are used to compute the losses within the combustion chamber ($\Delta P_{CC}$):

$$\Delta P_{CC} = (1 - \eta_{CC}).P_C \tag{21}$$

The sum of the power losses in the alternator are given by:

$$\Delta P_{alt} = \Delta P_{alt} + \Delta P_{CC} \tag{22}$$

If the losses expressed by (22) are negative, the program stops: in this case the user could introduce another realistic value of $\eta_{CC}$ or change all the initial values (figure 4). If the losses are positive, one can compute the ITG turbo-generator power:

$$P_{tg} = P_b + \Delta P_{alt} \tag{23}$$

Global (shaft and electrical) efficiency of the alternator group:

$$\eta_{alt} = \frac{P_b}{P_{tg}} \tag{24}$$

The mechanical (shaft) efficiency is computed using relationship (24) taking into consideration the alternator's electrical efficiency (given parameters). The program pursues an estimative computation for the mechanical and global efficiencies of the alternator using correlations (25) and (26) for the mechanical and electrical losses at the alternator [19]:

$$\Delta P_{el} = k_{el}.e^{\left(0.0365\ln^2(P_b)-0.0253\ln(P_b)+2.7366\right)} \tag{25}$$

$$\Delta P_m = k_m.e^{\left(-0.0111\ln^2(P_{int})+0.83.\ln(P_{int})-1.7318\right)} \tag{26}$$

Compressor calculations

$$\eta_K = \frac{h_{2t} - h_1}{h_2 - h_1} \tag{27}$$

$$h_2 = h_1 + \frac{h_{2t} - h_1}{\eta_K} \tag{28}$$

$$p_2 = p_1.\delta_K \tag{29}$$

$$p_1 = p_0 - \Delta p_{FA} \tag{30}$$

The gases temperature at the outlet of the combustion chamber is function of enthalpy, pressure and composition of the combustion gases. This temperature results from the energy balance of the combustion chamber. Usually, turbine manufacturers do not provide this temperature which may be defined differently [20].

$$p_3 = p_2 - \Delta p_{CC} \tag{31}$$

Turbine calculations:

$$\eta_{TG} = \frac{h_3 - h_4}{h_3 - h_{4t}} \tag{32}$$

$$p_4 = p_0 + \Delta p_{AB} \tag{33}$$

Specific mechanical work of the ITG:

$$w_{ITG} = w_{TG} - w_K = \frac{\dot{m}_{g.e.}}{\dot{m}_{air}}.(h_3 - h_4) - (h_2 - h_1) = \frac{F.M_{air}}{F.M_{g.e.}}.(h_3 - h_4) - (h_2 - h_1) \quad (34)$$

Mechanical and electrical losses at the alternator:

$$\Delta P_m = P_{tg}.(1 - \eta_m) \quad (35)$$

$$\Delta P_{el} = P_{tg}.\eta_m.(1 - \eta_g) \quad (36)$$

where

$$\Delta P_{el} = \Delta P_{Cu} + \Delta P_{Fe} \quad (37)$$

## 2.1. The Model Validation

*Manufacturer data* [21]

| | |
|---|---|
| Inter air pressure at air filter | $p_0$ = 1.013 [bar] |
| Dry thermometer temperature | $t_0$ = 5 [°C] |
| Air relative humidity | $\varphi_0$ = 60 [%] |
| Pressure lost in air filter | $\Delta p_{FA}$ = 0.01177 [bar] |
| Pressure lost in silencer | $\Delta p_{AB}$ = 0.02452[bar] |
| Outlet alternator power | $P_b$ = 40630 [kW] |
| Fuel specific consume | $q_{sp}$ = 11375 [kJ/kWh] |
| Exhaust gases mass flow rate | $D_{g.e.}$ = 145.6 [kg/s] |
| Turbine outlet temperature | $t_5$ = 533 [°C] |

Measured data from the manufacturer [21] and computed data by the program, table 1. From table 1, we can obviously highlight the validation of this model since the relative error is less than 1 %.

**Table 1. The comparison between the manufacturer data and the user data (Molar composition of the exhaust gases)**

| | $O_2$ (%) | $H_2O$ (%) | $CO_2$ (%) | $N_2$ (%) | AR (%) |
|---|---|---|---|---|---|
| Measured data | 13.91 | 6.73 | 3.18 | 75.82 | 0.9 |
| | | | | 76.18 | |
| Computed data | 13.96 | 6.77 | 3.16 | 76.11 | |
| Absolute data | - 0.05 | - 0.04 | 0.02 | 0.07 | |
| Relative error | - 0.36 | - 0.59 | 0.63 | 0.09 | |

### *Other Computed Data by the Program*

**Table 2. characteristic points of the installation**

| | p (bar) | t (°C) | h (kJ/kg) | s (kJ/kg/°C) | v ($m^3$/kg) |
|---|---|---|---|---|---|
| 0 | 1.013 | 5 | 5.129 | 0.017588 | 0.789769 |
| 1 | 1.00123 | 5 | 5.129 | 0.020962 | 0.799053 |
| 2 | 12.01476 | 320.13 | 327.546 | 0.082835 | 0.142029 |
| 3 | 11.41402 | 1042.13 | 1194.673 | 0.982734 | 0.336137 |
| 4 | 1.038 | 533 | 578.792 | 1.080318 | 2.266495 |
| 5 | 1.013 | 533 | 578.792 | 1.095319 | 2.321356 |

**Table 3. Mass flow rate**

| flow | value |
|---|---|
| $\dot{m}_{air}$ (kg/s) | 143.03 |
| $\dot{m}_{cb}$ (kg/s) | 2.57 |
| $\dot{m}_{g.e.}$ (kg/s) | 145.6 |

**Table 4. Pressure losses**

| Pressure losses | value |
|---|---|
| $\Delta p_{FA}$ (bar) | 0.01177 |
| $\Delta p_{AB}$ (bar) | 0.02452 |
| $\Delta p_{CC}$ (bar) | 0.60074 |

**Table 5. Power and losses**

| Power | Value |
|---|---|
| $P_c$ (kW) | 128380 |
| $P_{air}$ (kW) | 734 |
| $P_{g.e.}$ (kW) | 84272 |
| $P_{t.g.}$ (kW) | 43557.2 |
| $P_b$ (kW) | 40630 |
| $\Delta P_{alt+CC}$ (kW) | 4211 |
| $\Delta P_{CC}$ (kW) | 1283.8 |
| $\Delta P_{alt}$ (kW) | 2927.2 |
| $\Delta P_m$ (kW) | 2207.7 |
| $q_{sp}$ (kJ/kWh) | 113575 |

**Table 6. Efficiencies**

| Efficiencies | value |
|---|---|
| $\eta_{ITG}$ (%) | 31.65 |
| $\eta_K$ (%) | 89 |
| $\eta_{TG}$ (%) | 88.16 |
| $\eta_{CC}$ (%) | 99 |
| $\eta_m$ (%) | 94.93 |
| $\eta_{el}$ (%) | 98.26 |

**Table 7. Air data**

| Parameters | value |
|---|---|
| $\varphi_0$ (%) | 60 |
| $\lambda$ (-) | 3.225 |
| A ($kmol_{air}/kmol_{cb}$) | 30.878 |
| $M_{air}$ (kg/kmol) | 28.905 |
| $l_{ITG}$ ($kJ/kg_{air}$) | 304.53 |

**Table 8. Exhaust gases data**

| Parameters | Value |
|---|---|
| F ($kmol_{g.e.}/kmol_{cb}$) | 31.878 |
| $M_{g.e.}$ ($kmol_{cb}/kmol_{cb}$) | 28.502 |

**Table 9. Exhaust gases data**

| Parameters | Value |
|---|---|
| $\delta_K$ (-) | 31.878 |
| $\eta_K$ (%) | 28.502 |

**Table 10. Molar composition of the Air and of the exhaust gases**

| Molar composition | | | $O_2$ | $H_2O$ | $CO_2$ | $N_2$+Ar |
|---|---|---|---|---|---|---|
| Air | $Kmol/kmol_{air}*100$ | % | 20.89 | 0.52 | 0.02 | 78.57 |
| e.g. | $Kmol/kmol_{cb}*100$ | % | 13.96 | 6.77 | 3.16 | 76.11 |

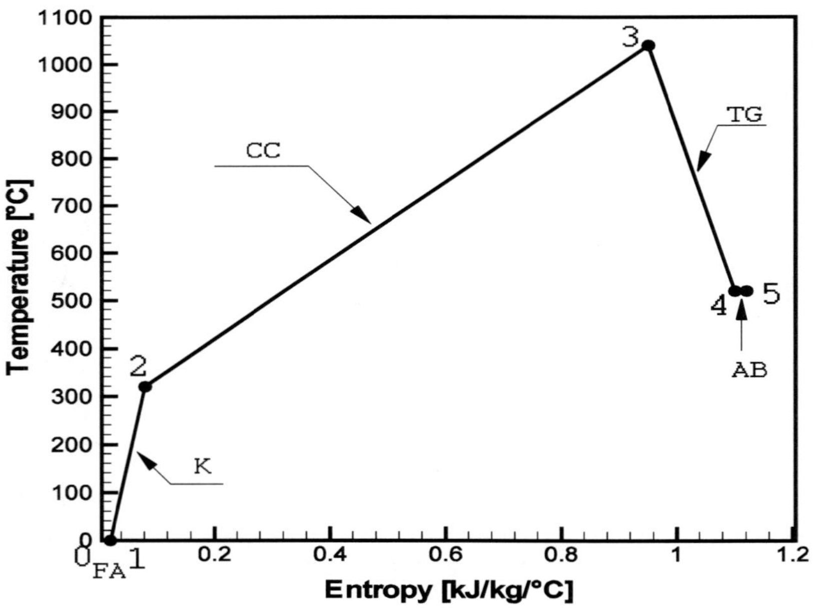


Figure 2. ITG: Brayton's cycle.

## 3. Preliminary Exergetic Analysis of the Combined Cycle

An ITG seldom works in a simple cycle due to low electrical efficiency. Recuperation of the available thermal energy at the outlet of the gas turbine is realized if the standard silencer is replaced by a heat recovery system generator at the exhaust of the ITG. The steam flow may be used to feed with heat an industrial consumer (cogeneration) or expended in a steam turbine, increasing the global (electric) efficiency of the ITG.

The maximum available thermal exergy ($Ex_{max\_rec}$) in the heat recovery steam generator is possible when the exhaust gases are cooled down to the saturation temperature of water vapour within the exhaust gases ($t_{sat}$). This temperature ($t_{sat}$) depends on the partial pressure of the water vapour ($p_{H2O}$):

$$p_{H_2O} = \gamma_{H_2O} \cdot p_0 \tag{38}$$

The available maximum thermal power and available exergy in the a heat recovery steam generator for $N_{ITG}$ gas turbines are expressed by:

$$Pt_{max} = N_{ITG} . \dot{m}_{g.e.} . (h_4 - h_{min}) \tag{39}$$

$$Ex_{mzx\_rec} = N_{ITG} . \dot{m}_{g.e.} . \left[ (h_4 - h_{min}) - T_{ref} (s_4 - s_{min}) \right] \tag{40}$$

where $T_{ref}$ [K] is the reference temperature which may be the saturation temperature in the condenser of the steam turbine cycle.

The actual temperature at the CR outlet ($t_{CR}$) depends on the secondary fluid (water/steam) parameters, as well as on the pressure levels of the steam production. This temperature is a parameter which as to be greater than $t_{sat}$ and $t_{ref}$. The available thermal power and exergy in the CR are:

$$Pt_{rec} = N_{ITG} . \dot{m}_{g.e.} . (h_4 - h_{CR}) \tag{41}$$

$$Ex_{rec} = N_{ITG} . \dot{m}_{g.e.} . \left[ (h_4 - h_{CR}) - T_{ref} (s_4 - s_{CR}) \right] \tag{42}$$

We calculate the alternator output power of the steam turbine ($P_{b\_ITV}$) considering the following parameters:

- The exergy efficiency of the CR ($\eta_{CR}$);
- The exergy efficiency of the steam cycle ($\eta_{ITV}$) (taking into account the internal efficiency of the steam turbine, the losses in the pipes and other exergy losses of the steam cycle);
- The alternator efficiency ($\eta_{alt\_TV}$) for the steam turbine.

$$P_{b\_ITV} = \eta_{CR} . \eta_{IRV} . \eta_{alt\_TV} . Ex_{rec} \tag{43}$$

The total electric power output of the combined cycle is:

$$P_{b\_CCGV} = N_{ITG}.P_b + P_{b\_ITV} \tag{44}$$

The global electric efficiency of the combined cycle ($\eta_{CCGV}$)

$$\eta_{CCGV} = \frac{P_{b\_CCGV}}{N_{ITG}.P_c} \tag{45}$$

If the main purpose of the installation is not the electric power, this available thermal power ($Pt_{rec}$) may be used for cogeneration. In this case, the efficiency of the installation is:

$$\eta_{cogen.} = \frac{N_{ITG}.P_b + c1.Pt_{rec}}{n_{ITG}.P_c} \tag{46}$$

The coefficient "c1" takes into account the heat losses through the CR insulation (c1=0.975 in our calculations). The "structure coefficient" for electricity and heat combined production is:

$$y_{st} = \frac{n_{ITG}.P_b}{c1.Pt_{rec}} \tag{47}$$

The performance indices are presented in table 11 for ideal conditions ($\eta_{CR}$=100 %, $\eta_{ITV}$=100 %, $\eta_{alt\_TV}$=100 %) for an existing ITG, highlighting the limits of the cycle ($t_{CR}$=60 °C and $N_{ITG}$ = 1).

**Table 11. Cycle limits for a given ITG**

| Gas turbine | V94.3A | GT8C | FT8 TwinPac | LM6000 Sprint |
|---|---|---|---|---|
| manufacturer | Siemens (KWU) | ALSTOM | Rolls-Royce | GE S&S |
| $P_b$ [kW] | 265000 | 52600 | 51380 | 43076 |
| $\eta_{ITG}$ [%] | 38.50 | 34.20 | 38.34 | 41.37 |
| $t_5$ [°C] | 584 | 517 | 455 | 450 |
| $\eta_{CCGV}$ [%] | 63.45 | 58.67 | 59.58 | 61.36 |
| $\eta_{cogen.}$ [%] | 92.48 | 90.74 | 91.22 | 91.45 |
| $y_{st}$ [-] | 0.7132 | 0.6049 | 0.7251 | 0.8263 |

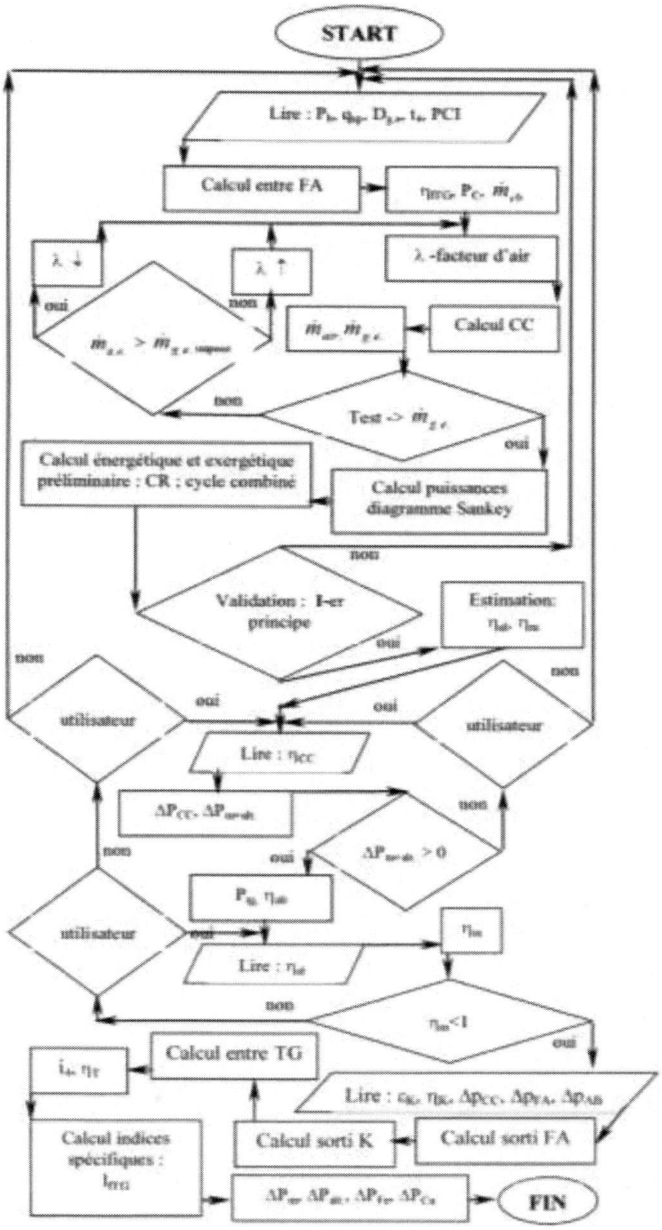


Figure 3. ITG: Organigram.

## CONCLUSION

This paper presents a reference calculation for gas turbine installations based on data issued from the manufacturer catalog. The product is a computer code using a Windows interface in a step by step process of data input. This code could carry either simulation or optimization of gas turbines for cogeneration or combined gas turbine/steam turbine cycles. An exergy analysis of heat recovery is available for combined cycle applications.

# APPENDIX

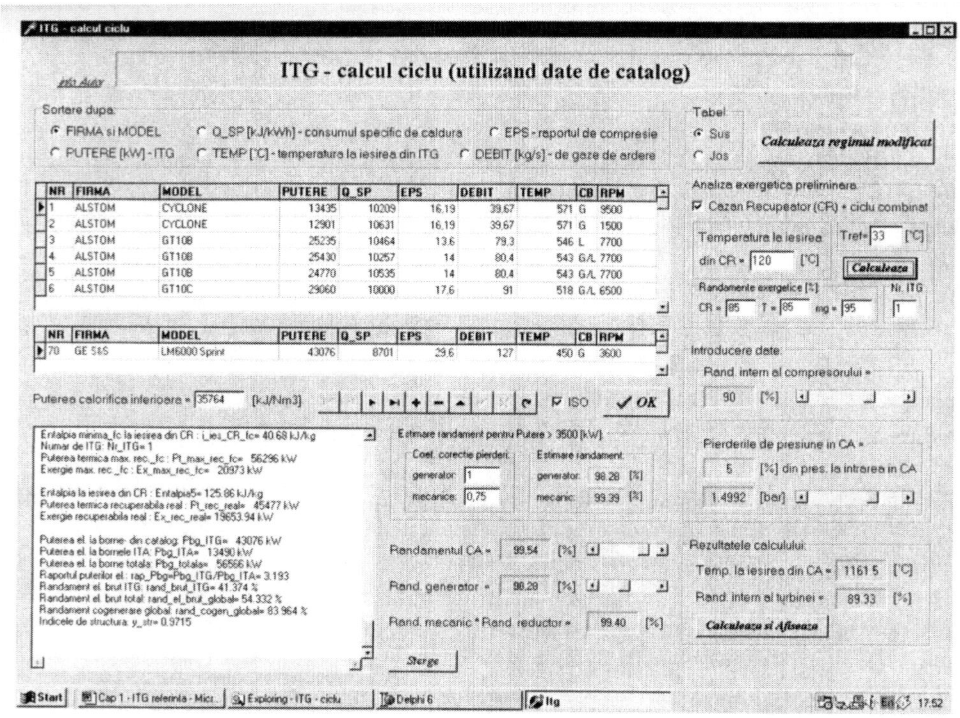


Figure 4. ITG: Program interface used for the calculus.

$p_0$=1.013 [bar], $t_0$=5 °C et $\varphi_0$=60 [%]

$\gamma_{N_2}$ = 0.78981435 [$kmol_{N2}/kmol_{air\ sèche}$]

$\gamma_{O_2}$ = 0.20995065 [$kmol_{O2}/kmol_{air\ sèche}$]

$\gamma_{CO_2}$ = 0.000235 [$kmol_{CO2}/kmol_{air\ sèche}$]

water vapor saturation pressure in atmospheric air

$p_{sat}(t_0)$=0.00872858 [bar]

water vapor partial pressure in atmospheric air

$$pp_{H_2O} = \frac{\varphi_0 \cdot p_{sat}(t_0)}{100} = 0.00523715 \text{ [bar]} \qquad (1')$$

$$\gamma_{H_2O\_air} = \frac{pp_{H_2O}}{p_0} = 0.00516994 \text{ [kmol}_{H2O}\text{/kmol}_{air}] \qquad (2')$$

$$\gamma_{N_2\_air} = \gamma_{N_2} \cdot (1 - \gamma_{H_2O\_air}) = 0.78573106 \text{ [kmol}_{N2}\text{/kmol}_{air}] \qquad (3')$$

$$\gamma_{O_2\_air} = \gamma_{O_2}.\left(1-\gamma_{H_2O\_air}\right) = 0.20886522 \text{ [kmol}_{O2}\text{/kmol}_{air}] \quad (4')$$

$$\gamma_{CO_2\_air} = \gamma_{CO_2}.\left(1-\gamma_{H_2O\_air}\right) = 0.00023379 \text{ [kmol}_{CO2}\text{/kmol}_{air}] \quad (5')$$

$$M_{air} = \sum_i \gamma_{i\_air}.M_i \text{ [kg/kmol]} \quad (6')$$

$$h_0 = \frac{\sum_i \gamma_{i\_air}.M_i.h_{i\_0}}{M_{air}} \text{ [kg/kmol]} \quad (7')$$

$$h_0 = \frac{\sum_i \gamma_{i\_air}.M_i.s_{i\_0}}{M_{air}} \text{ [kg/kmol]} \quad (8')$$

where "i" ≡ $N_2$, $O_2$, $H_2O$, $CO_2$

## REFERENCES


[1] V. cenusa, R. Benelmir, M. Feidt, A. Badea, *On gas turbine and combined cycles*, 56° Congresso Nazinale ATI, Napoli, 10-14 September 2001, 4 (2001), 237 – 247.

[2] J. Paren, C. Parietti, Centrales à cycles cobinées VEGA 109F 3 pressions resurchauffe, *Revue Technique GEC ALSTHOM*, 4 (1991), 15 – 26.

[3] M. Dupleac, *Perspective si optimizari pentru introducerea ciclului mixt cu amestec de gaze si abur*, Teza de Doctorat, Universitatea Politechnica din bucuresti, Romania, 1997.

[4] Fl. Alexe, V. Cenusa, C. Marculescu, M. *Ene, optimizarea centralelor electrice de termoficare de medie putere cu ciclu combinat gaze – obur, Producerea, transportul si distributia energiei electrice si termice,* Novembre (2000), 16 – 23.

[5] Fl. Alexe, V. Cenusa, C. Marculescu, M. Ene, optimizarea centralelor electrice de termoficare de medie putere cu ciclu combinat gaze – obur, Sesiunea stiintifica : 50 de ani de invatamant energetic in Romania, Bucuresti, Romania, 10-11 Novembre, 2 (2000), 171 – 178.

[6] G. Darie, M. Dupleac, M. Sava, V. Censua, *Cicluri combinate gaze – abur*, Editura AGIR, Bucuresti, Romania, 2001.

[7] F. Rouget, L. Michelat, G. Babaud, Analyse thermomécanique transitoire appliquée aux turbines à gaz « heavy-duty ». L'échappement de la MS 9001F de Gennevillers, Revue Technique GEC ALSTHOM, 16 (1995), 33 – 48.

[8] M. Toulios, D. McKenzie, L'utilisation performante des matériaux par l'analyse inélastique avancée. *Application aux turbines à gaz, Revue Technique* GEC ALSTHOM, 17 (1995).

[9] T. Murase, Y. Tsukuda, E. Akita, Y. Iwasaki, Current operating status of M501G gas turbine, *ECOS 2000*, Part 4, (2000), 1875 – 1868.

[10] F. Joos, P. Brunner, B.S. Werning, K. Syed, Développement du système de combustion séquentielle pour la famille des turbines à gaz GT24/GT26, Revue ABB, 4, (1998), 4 – 16.

[11] H. Neuhoff, K. Thorén, Les nouvelles turbines à gaz GT24/GT26 – rendements élevés grâce à la combustion séquentielle, *Revue ABB*, 2, (1994), 4 – 7.

[12] M. Vogel, La turbine à gaz – le composant principal de la technique moderne des centrales thermiques, *Revue ABB*, 5, (1994), 11 – 15.

[13] F. Joos, P. Brunner, M. Stalder, S. Tschirren, Expériences pratiques avec le système de combustion séquentielle de la famille des turbines à gaz GT24/GT26, *Revue ABB*, 5, (1998), 12 – 20.

[14] M.F. Cannon, M. Molière, S.M. Depieto, M. Perkavec, Les technologies « bas $NO_x$ » pour turbines à gaz. Expériences et références d'EGT, *Revue Technique GEC ALSTHOM*, 19, (1996), 1 – 16.

[15] J. Paren, C. Parietti, Centrales à cycles cobinées VEGA 109F 3 pressions resurchauffe, *Revue Technique GEC ALSTHOM*, 4 (1991), 15 – 26.

[16] W. Krebs, R. Walterskötter, P. Flohr, G. Walz, S. Hoffmann, Effect of burner design parameters on thermoacoustic stability of annular gas turbine combustors, *ECOS 2000*, Part 4, (2000), 2061 – 2077.

[17] J. Hermann, A. Orthmann, S. Hoffmann, P. Berenbrink, Active and passive control of combustion oscillations in a 260 MW Heavy-Duty gas turbine, *ECOS* 2000, Part 4, (2000), 2089 – 2102.

[18] Fl. Alexe, V. Cenusa, C. Marculescu, M. Ene, optimizarea centralelor electrice de termoficare de medie putere cu ciclu combinat gaze – obur, Producerea, transportul si distributia energiei electrice si termice, Novembre (2000), 16 – 23.

[19] F. Kostiuk, *Parovaie I gazovaie turbine*, Ed. GosEnergoIzdat, Moscova, 1986.

[20] G. Darie, instalatii energetice de turbine cu gaze, Editura Bren, Bucuresti, Romania, 1998.

[21] Manufacturer: General Electric.

In: Energy-Environment-Economics
Editor: Riad Benelmir
ISBN: 978-1-62948-862-2


*Chapter 17*

# BETA TYPE REFRIGERATING STIRLING MACHINE: NUMERICAL AND EXPERIMENTAL STUDY

***Nadia Martaj*[1]*, Lavinia Grosu*[1]*, Pierre Rochelle*[1]*, Stephan Savarese*[2] *and Rachid Bennacer*[3]**

[1]Laboratoire LEME, Université Paris Ouest Nanterre-La Défense, Ville d'Avray

[2]Comsol, France

[3]Dpt GC/ LMT-Ecole Normale Supérieure – Cachan, Cachan Cedex, France

## ABSTRACT

The problems involved in the destruction of the ozone layer, due to the use of organo-chlorinated compounds describing an inverse cycle, lead us to take into account the development of new alternatives. Thus, refrigerating Stirling or Ericsson cycles, using air or another inert gas, might become beneficial alternatives. This study relates to a Beta type refrigerating Stirling machine, comparing experimental measurements with results of two numerical simulations, first one in 0D and second one in 2D (CFD).

## NOMENCLATURE

$A$ heat transfer area, [m$^2$]

$c_p$ specific heat at constant pressure, [J/kg K]

$C$ stroke, [m]

$D$ diameter, [m]

$k$ heat exchange coefficient, [W/(m$^2$.K)]

$m$ mass, [kg]

$N$ engine speed , [rev/min]

$p$ pressure, [Pa]

$\dot{Q}$ heat flux, [W]

$Q$ heat [J]

$r$ gas constant, [J/(kg K)]
$T$ temperature, [K]
$t$ time, [s]
V volume, [$m^3$]
$W$ work, [J]

$\dot{W}$ mechanical power, [W]

## Greek Symbols

$\varepsilon$ porosity , -
$\phi$ phase lag, [°]

## Subscripts

c compression
d displacer
e expansion
g gas
h hot/high
l low
p working piston
r regenerator
t total
w wall

## 1. Introduction

The use of the refrigerating Stirling machine, using air or another inert gas as working fluid, represents an interesting alternative to mitigate the problem of the ozone layer destruction by organo-chlorinated refrigerants [1-4]. The theoretical cycle is composed of two isothermal processes within heat exchangers, and two isochoric processes associated to the working gas flow through the machine regenerator. This cycle has, theoretically, a performance coefficient equal to that of Carnot cycle, assuming a perfect regenerator (Figure 1).

This cycle could be studied using finite time, finite speed or more generally finite physical dimensions thermodynamic method [5-8]. The engine have three heat exchangers and the working gas is everywhere in the engine; temperature depends not only on time but also on the position in the engine.

Thus, it is useless to compare gas evolution in the Stirling engine with a theoretical thermodynamic cycle because of this space-time no homogeneity of the working gas. In 1871,

Schmidt [9] proposed an analysis, known under the name ideal analysis, which inspired models developed later [10-12].

In order to take into account the local phenomena, it is necessary to develop coupled analysis, more complete and more realistic than the preceding analyses (by simulating the thermo mechanical processes and aerodynamics). This analysis needs higher computing time [13-16].

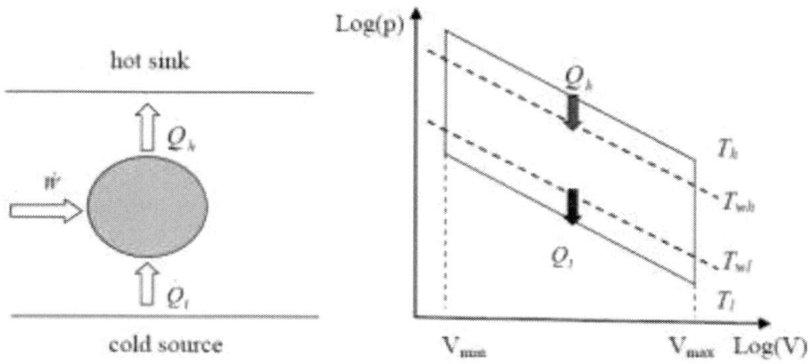


Figure 1. Inverse exo-irreversible Stirling cycle.

## 2. Studied Refrigerating Machine

The experimental device (figure 2) is composed of a cylinder of highly resilient glass inside which two pistons undergo a reciprocating alternative movement with a phase lag of 110°.

This cylinder is surrounded by a double-wall sleeve inside which a water flow constitutes the hot source. On the top part of the cylinder, a thermocouple measures the temperature of the cold space.

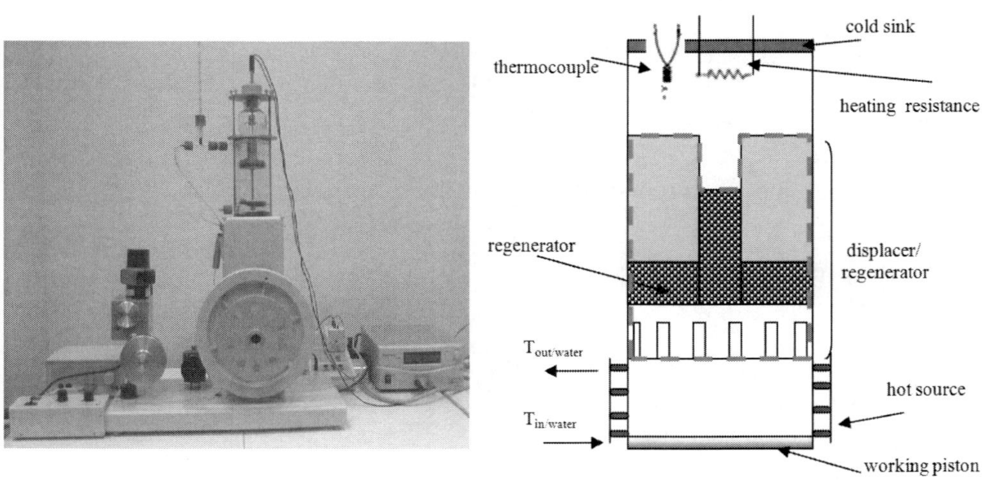


a) b)

Figure 2. Stirling Beta machine a) Image of the bench test b) Scheme of the transparent cylinder of the refrigerating.

The displacer forces the passage of the gas from the bottom part to the top part of the cylinder and conversely. It also holds a porous material which is used for the heat storage/release, thus playing the regenerator role.

## 3. MODELLING

Two models were carried out, one named "0-D" based on the thermodynamics laws, and a second, named "2-D", more elaborated, based on the numerical solution of the mass, momentum and energy conservation equations, applied to the working gas in the regenerator (porous media) and to compression and expansion spaces.

### 3.1. 0-D Modelling

This analysis relies on the division of the refrigerating machine in 3 spaces: expansion, compression and regeneration spaces (Figure 3).

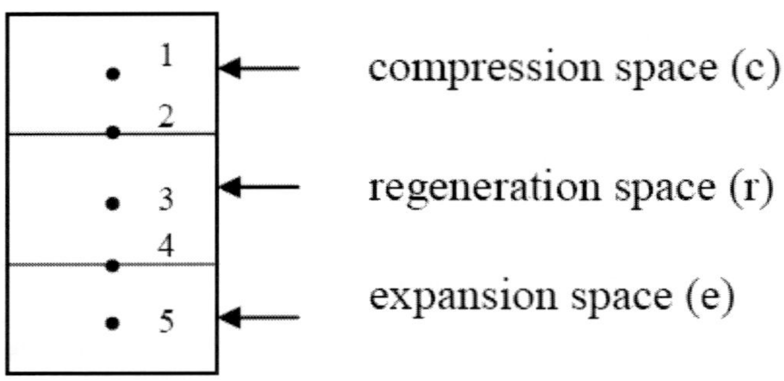


Figure 3. Representation of the machine 3 spaces with their boundaries.

The working gas is supposed perfect. We also assume constant temperatures inside the compression and expansion spaces ($V_c$ and $V_e$), in contact with the heat sources. The regenerator temperature, $T_r$ $=T_3$, is the logarithmic average of the of these two spaces temperatures.

The fluid mass in each volume is estimated using the perfect gas law:

$$m_c = \frac{p\,V_c}{rT_1} \; ; \; ; \; m_e = \frac{p\,V_e}{rT_5} \tag{1}$$

The total mass in the cylinder is .

Thus, the instantaneous pressure, assumed to be uniform in the whole engine, is given by the following relation:

$$p = m\ r / \left( \frac{V_c}{T_h} + \frac{V_r}{T_r} + \frac{V_e}{T_l} \right) \tag{2}$$

The elementary masses in each volume are calculated by using equations (1):

$$\begin{cases} dm_c = dm_1 = \dfrac{p\,dV_c + V_c dp}{r T_1} \\ dm_e = dm_5 = \dfrac{p\,dV_e + V_e dp}{r T_5} \\ dm_r = dm_3 = m_r \dfrac{dp}{p} \end{cases} \tag{3}$$

The elementary mass flows through the interfaces and the associated temperatures are given by the following equations:

$dm_2 = dm_c$ , if up flow if not $T_2 = T_1$

$dm_4 = -dm_e$ , if up flow if not

The pressure differential *dp* is given by using equation (2):

$$dp = -p \left( \frac{dV_c}{T_1} + \frac{dV_e}{T_5} \right) / \left( \frac{V_c}{T_1} + \frac{V_r}{T_3} + \frac{V_e}{T_5} \right) \tag{4}$$

The elementary heat transfers in these 3 spaces are obtained using the energy conservation equation applied to each volume:

$$\begin{aligned} \delta Q_c &= \left( \frac{c_v}{r} + 1 \right) p\,dV_c + \frac{c_v}{r} V_c\, dp + c_p T_2\, dm_2 \\ \delta Q_r &= V_r \frac{c_v}{r} dp + c_p\,(T_4\, dm_4 - T_2\, dm_2\,) \\ \delta Q_e &= \left( \frac{c_v}{r} + 1 \right) p\,dV_e + \frac{c_v}{r} V_e\, dp - c_p T_4\, dm_4 \end{aligned} \tag{5}$$

Elementary work in compression space, $\delta W_c = -\,p\ dV_c$ , and expansion space, $\delta W_e = -\,p\ dV_e$ , allows to determine, after integration, the work done during a cycle.

$$W = W_e + W_c \tag{6}$$

The temperatures of these two spaces are given by the Fourier-Newton law, using the heat transfer coefficients obtained by experimental tests.

$$\left|\dot{Q}_h\right| = k_h\, A_h\,(T_h - T_{wh}) \rightarrow T_h = T_{wh} + \frac{\left|\dot{Q}_h\right|}{k_h\, A_h} \tag{7}$$

$$\dot{Q}_l = k_l\, A_l\,(T_{wl} - T_l) \rightarrow T_l = T_{wl} - \frac{\dot{Q}_l}{k_l\, A_l} \tag{8}$$

The equations of this 0-D model are solved using the Simulink tool.

- Mask of the Stirling refrigerating machine

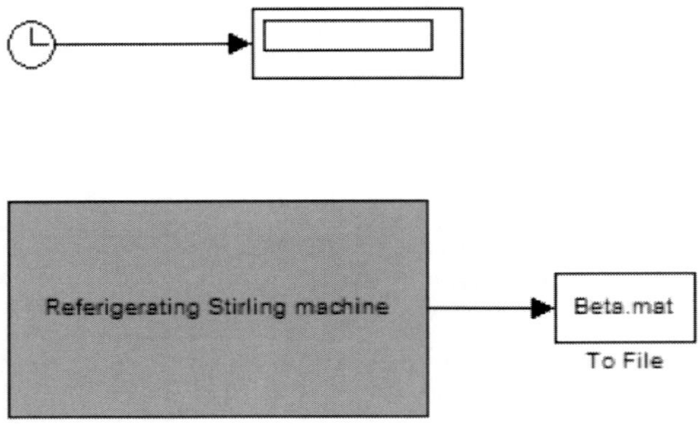


- The pressure and its differential

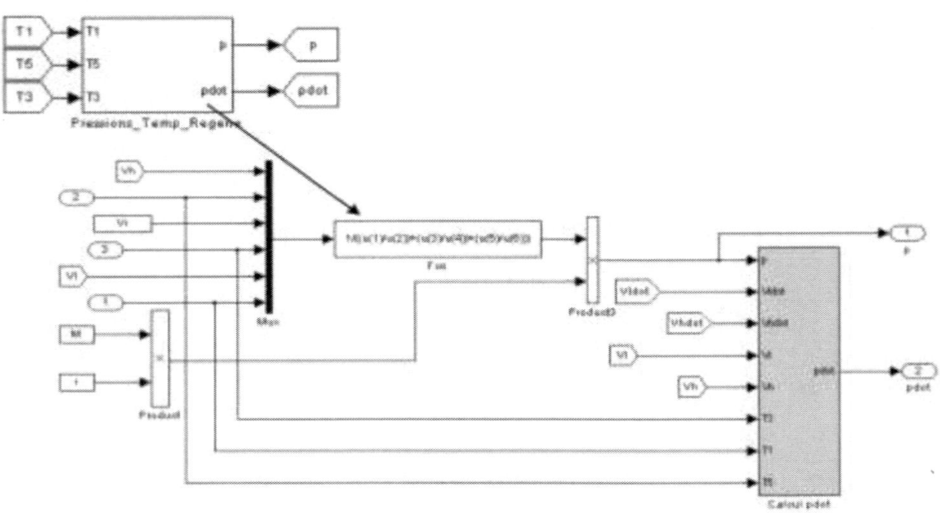


Figure 4. Block Simulink pressure.

### 3.2. 2-D Modelling

The assumptions considered for 2D modelling are as follows:

- the gas flow is turbulent,
- the thermo physical properties of gas are time dependant (function of the instantaneous local temperature),
- the physical properties of the porous material (regenerator) are uniform and constant,
- energy transport is mainly controlled by conduction and convection.

The Navier Stokes equations are numerically solved under the finite element approximation by using the simulation software COMSOL Multiphysics.

## 4. RESULTS AND DISCUSSIONS

The initial data used by the 2 modelling methods, are presented in the Table 1.

**Table 1. Relative data of the simulated points**

| $T_{wl}$ | $T_{wh}$ | N | $D_m$ | $D_d$ | $\phi$ | C |
|---|---|---|---|---|---|---|
| K | K | rev/s | m | m | ° | m |
| 278.85 | 297.30 | 2.65 | 0.065 | 0.065 | 110 | 0.04 |

The results obtained by the 0-D model are the instantaneous variables: temperature, pressure, volume and mass. For the 2D model, in addition to these variables, we also obtain the instantaneous gas velocity, the charts temperature and pressure fields, as illustrated in figure 5.

The cycles obtained with these two models are presented on figure 6. The indicated work is calculated by integration on a cycle. The results are presented in table 2.

**Table 2. Comparison of the results**

| | 0D | 2D | experiment |
|---|---|---|---|
| W by cycle [J] | 5.9 | 9.2 | 5.7 |

One remarks that despite the 0-D Model seems relatively simple it permits to estimate with good precision the delivered work providing the use of parameters obtained by experiments. In our case, we use experimental heat exchange coefficients estimate on our bench test, which can explain such agreement.

For the 2-D model, we have considered the regenerator as a porous matrix with a given porosity (0.6 in our case). A significant difference between experiment and numerical simulation can be observed and explained by the value of the porosity which has been approximated.

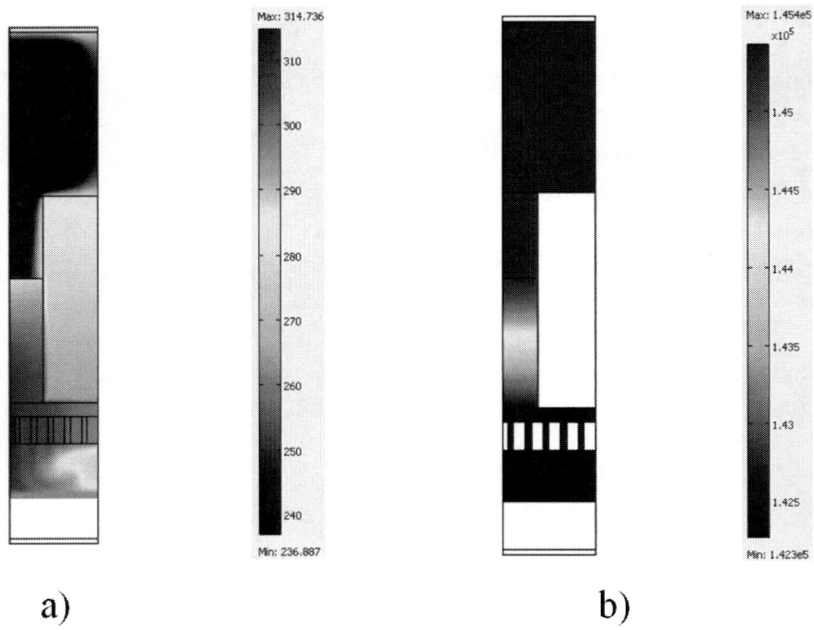


Figure 5. temperature (a) and pressure (b) fields for the displacer at the bottom dead center point and the working piston with phase lag of 110°: and for a porosity of 0.6.

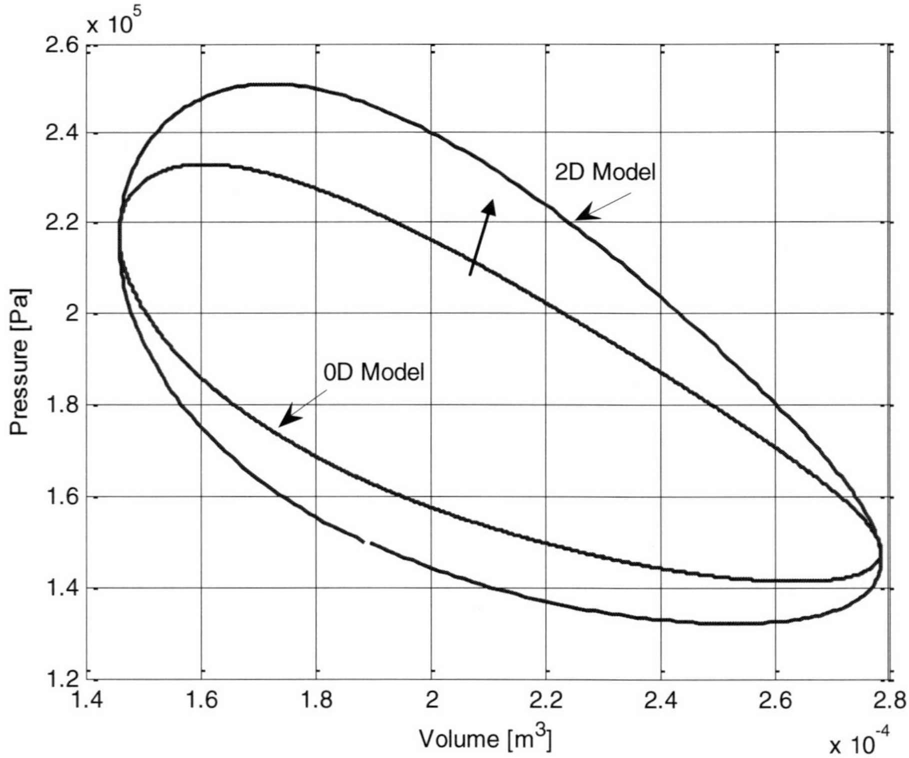


Figure 6. (p, V) diagram obtained by 2-D and 0-D numerical simulation.

## CONCLUSION

This work presents an experimental and numerical study of a beta type refrigerating Stirling machine. The confrontation of the results permits to validate 0-D isothermal model for known heat transfer coefficient. In a future work, we will be able to change the characteristics of the regenerator (porosity, permeability, etc.) and to carry out several experimental tests in order to find heat exchange coefficient variation laws, which are considered like a constant in our models. These considerations could improve and validate the 2-D model, approaching experimental and numerical simulation results.

## REFERENCES

[1] G. T. Reader, C. Hooper, "Les machines Stirling", Edit. E and F.N. SPON, N*ew Fetter Lane*, London, (1983).

[2] Ph. Nika, F. Lanzetta, "Développement d'une machine frigorifique Stirling de petite taille, adaptée à des niveaux thermiques modérés", *Journal de physique*, vol. 5, N°6, (1995), 835-861.

[3] Ercan Ataera, H. Karabulut, "Thermodynamic analysis of the V-type Stirling-cycle refrigerator", *International Journal of Refrigeration* 28, (2005), 183–189.

[4] N. S. Petrescu, L. Grosu, M. Costea, P. Rochelle, C. Dobre, C. Petre : "Analyse théorique et expérimentale d'une machine à froid de Stirling", Colloque francophone COFRET 2010, *IASI,* Roumanie, Vol. LVI, Fasc.3a, (4 – 7 mai 2010), 441 – 453.

[5] L.G. Chen, H.J. Feng and F.R. Sun, "Optimal piston speed ratio analyses for irreversible Carnot refrigerator and heat pump using finite time thermodynamics, finite speed thermodynamics and direct method", *Journal of Energy Institute* Vol.84, N°2, (2011), pp. 105-112.

[6] S. Petrescu et al., "Les cycles des machines à froid et des pompes à chaleur à vitesse finie", *Rev. Entropie,* 232, (2001), 48-54.

[7] M. Feidt, "Optimisation énergétique des systèmes et procédés", 2nd ed., *TechandDoc,* Paris (1996).

[8] L. Grosu, "Contribution à l'optimisation Thermodynamique et Économique des Machines à cycle inverse à deux et trois réservoirs de chaleur". PhD thesis, *Institut National Polytechnique de Lorraine*, (2000).

[9] G. Schmidt. "Theorie der Lehmann'schen calorischen Maschine. " Z. Des Ver. deutsher Ingenieue, XV(2), (1871) 98-112.

[10] S. Bonnet. "Moteurs Thermiques À Apport de Chaleur Externe : Étude d'un Moteur Stirling et d'un Moteur Ericsson", PhD thesis, Université de Pau et des Pays de l'Adour, (2005).

[11] N. Martaj, L. Grosu, P. Rochelle, "Thermodynamic study of a low temperature difference Stirling engine at steady state operation", *Int. J. of Thermodynamic*, Vol. 10, No.4, (2007), 165 – 176.

[12] L. Grosu, P. Rochelle, "Application de la méthode de Schmidt avec régénération imparfaite aux 3 types de moteur Stirling-nouvelles solutions analytiques", Congrès SFT, Vannes, Tome 2, (26-29 mai 2009), 895-900.

[13] N. Martaj, P. Rochelle, L. Grosu, R. Bennacer and S. Savarese, "Moteur Stirling à faible différence de températures (LTD) : confrontation simulations numériques et expérimentation", *Congrès SFT, Toulouse*, (3-6 juin 2008), 729-735.
[14] N. Martaj, P.Rochelle, L.Grosu, R.Bennacer, S.Savarese, "Multiphysics CFD simulation of an Stirling engine with a porous regenerator", Diffusion in Solids and Liquids DSL, Paris, (05-07 july 2010).
[15] H.Z. Hassan, A.A. Mohamad, R. Bennacer, "Simulation of an adsorption solar cooling system", *Original Research Article Energy*, Volume 36, Issue 1, (January 2011), 530-537.
[16] N. Martaj, R.Bennacer, P.Rochelle, L.Grosu, "Simulation dynamique d'un moteur: cas du Stirling de type gamma", *Mécanique and Industries* 11, (2010), 69–74.

In: Energy-Environment-Economics
Editor: Riad Benelmir
ISBN: 978-1-62948-862-2


*Chapter 18*

# EVALUATION OF A COLD MAP FOR A MOROCCAN CLIMATE BY A SOLAR ADSORPTION REFRIGERATION MACHINE WORKING WITH ACTIVATED CARBON- AMMONIA PAIR

***H. El Kalkha[1], A. Ezzarfi[1], A. Mimet[1], F. Lemmini[2], M. Pérez-Garcia[3] and M. El Ganaoui[4]***
[1]Energetic Laboratory, Sciences Faculty, Tetouan, Abdelmalek Essaidi University, Morocco, North Africa
[2]Laboratoire d'Energie Solaire, Faculté des Sciences, Université Mohammed V, Rabat, Maroc, North West Africa
[3]Dpto. De Fisica Aplicada- Universidad de Almeria, España, Spain
[4]Université de Lorraine, Fédaration Jacques Villermaux Méca. Ener. Procédés, France

## ABSTRACT

The aims of the present paper are to evaluate solar performances of a solar adsorption refrigerating machine and the amount of cold produced according to sunspots of different regions of Morocco.

We have developed a model that simulates the transient behavior of the machine. Indeed, the essentially variable solar radiation implies that the machine is at any moment in a transitional state. The model calculates, for initial conditions and a given solar flux for several sites in Morocco, the spatiotemporal distribution of the temperature, adsorbed mass and pressure in the adsorber, this allowed us to calculate with a good accuracy the cycled mass of refrigerant, the quantity of cold produced in the machine and the solar performance coefficient of solar refrigerating machine. The solar collector used for converting solar energy to heat is a solar collector flat plate.

**Keywords:** Solar energy, adsorption, refrigerating machine, Performance Coefficient

## NOMENCLATURE

$T$ - absolute temperature (°C)
$T_v$ - temperature of the glass (°C)
$T_p$ - temperature of the wall (°C)
$T_m$ - temperature of mixed adsorbent-adsorbate (°C)
$T_a$ - ambient temperature (°C)
$T_s$ - sky temperature (°C)
$T_{ev}$ - evaporation temperature (°C)
$T_{cond}$- condension temperature (°C)
$T_{g1}$ - temperature at start of regeneration(°C)
$T_{g2}$ - temperature at the end of desorption process (°C)
$T_{ads}$ - adsorption temperature (°C)
P - pressure in the reactor (bar)
$P_s$ - pressure of saturation (bar)
M - mass (kg)
$m_a$ - adsorbed mass in Kg of ammonia per Kg of activated carbon (kg/kg-CA)
x - relative pressure (= P/ PS )
G - globale irradiation ($W/m^2$)
L - length of the cylinder wall (m)
R - cylinder radius(m)
dr - layer width
$Q_f$ - quantity of cold ($J/ m^2$)
$Q_s$ - quantity of heat ($J/ m^2$)
$L(T_{ev})$ - latent heat of ammonia at evaporation temperature (J/kg)
$S_c$ - collecting area sensor ($m^2$)
C - characteristic gas (ammonia) constant
$C_s$ - specific heat of solid phase (J/kg K )
$C_g$ - specific heat of gaz phase (J/kg K )
$C_a$ - specific heat of adsorbed phase (J/kg K )
$C_l$ - liquid specific heat (J/kg K )
$C_v$ - thermal capacity of the glass ( $J/K.m^2$)
$C_p$ - thermal capacity of the wall ( $J/K.\ m^2$)
u - internal massic energy (J/kg)
$u_s$ - internal energy mass of the solid phase (J/kg)
$u_g$ - internal energy mass of the gaz phase (J/kg)
$u_a$ - internal energy mass of the adsorbed phase (J/kg) $h_a(T)$ adsorbed phase enthalpy at temperature T.
$h_g$ - enthalpy of gaseous ammonia (J/kg)
$h_s$ - emanating out products enthalpy
$h_e$ - entering products enthalpy
$P_m$ - mechanical power exchanged with the surrounding medium, which is null in our case.
$q_{ms}$ - massic flow of the fluids coming out of the layer.
$q_{me}$ - massic flow of the fluids entering in the layer.

$q_m$ - mass flow going through a layer (kg/s)
$q_v$ - Net flow energetic of the glass (W/ m2)
$q_p$ - Net flow energetic of the wall (W/ $m^2$)
$h_{va}$ - coefficient of heat exchange between the glass and the atmosphere ( W/K. $m^2$)
$h_{vs}$ - coefficient of heat exchange between the glass and the sky ( W/K. $m^2$)
$h_{pv}$ - coefficient of heat exchange between the wall and the galss ( W/K. $m^2$)
$h_{pa}$ - coefficient of heat exchange between the wall and the atmosphere ( W/K. $m^2$)
$h_{pm}$ - coefficient of heat exchange between the wall and the mixture ( W/K. $m^2$)
$a_p$ - net absorption coefficient of the wall
$a_v$ - net absorption coefficient of the glass
t - time (s)
$V_v$ - wind speed (m/s)
$e_p$ - insulation thickness(m)

## Greek Letters

$\delta u$ - layer internal energy
$\Phi$ - heat flux exchanged with surrounding medium.
$\Delta H_{ads}$ (T,m) - adsorption heat of ammonia on activated carbon (J/kg)
$\Delta H_{cond}$ - condensation heat of the adsorbate(J/kg)
$\Delta m$ - cycled mass of ammonia (kg/kg-CA)
$\varepsilon$ - porosity of activated carbon
$\rho$ - volumic mass
$\rho_{app}$ - apparent density of activated carbon (kg/$m^3$)
$\rho_s$ - density of the solid phase (kg/$m^3$)
$\rho_g$ - density of the gaz phase (kg/$m^3$)
$\rho_a$ - density of the adsorbed phase (kg/$m^3$)
$\alpha$ - volume fraction of the adsorbed phase
$(1 - \varepsilon)$ : - volume fraction of the solid phase
$(\varepsilon - \alpha)$: - volume fraction of the gaseous phase
$\lambda e$ - equivalent thermal conductivity
$\alpha_{vb}$ - adsorption coefficient of direct normal radiation by the glass
$\alpha_{vd}$- adsorption coefficient of diffuse radiation by the glass
$\alpha_{pb}$ - adsorption coefficient of direct radiation by the wall normal
$\tau_{rv}$ - energy transmission coefficient of direct radiation by the glass
$\tau_{dv}$- energy transmission coefficient of diffuse radiation by the glass
$\varepsilon_v$ - emittance of the glass
$\varepsilon_p$ - emittance of the wall
$\sigma$ - Boltzmann constant = 5.68 $10^{-8}$ (W/ $m^2$ .$K^{-4}$)

## Indices/Exhibitors/Abbreviations

n - number of layers
cmo - monolayer
ads - adsorption
ev - evaporation
cond - condensation
r - radiative
c - convective
$COP_{sol}$ - solar coefficient of performance
AC - activated carbon

# 1. Introduction

Adsorption refrigerating systems have been extensively studied [1-9] The main interest of these systems is their perfect coupling to thermal energy constituting, in fact, a very attractive solution for both industrial waste heat recovery and exploitation of renewable sources such as, specially, solar energy. The core element of this type of cooling systems is the adsorption unit, where pressure drops requested for the cooling cycle are provoked by a heat activated sequential adsorption/desorption process of the refrigerant (adsorbate) on a suitable solid substrate (adsorbent).

Different adsorbent/adsorbate pairs have been studied for their use in solar cooling machines [10]. For example, in LIMSI, the zeolite/water pair was chosen for the refrigeration [11] and the activated carbon/methanol pair for the production of ice [12]. The activated carbon/ammonia pair, has been also chosen by other authors [13-16]

The use of ammonia ($NH_3$/R717) as adsorbate has the advantages of a low molar mass, a high latent heat of vaporization and a high thermal conductivity and an operating pressure greater than atmospheric pressure[17, 18]. It has also an excellent heat and condensation transfer coefficients in addition to a low cost and high chemical stability, biodegradability and not ozone depletion risk [19, 20].

Activated carbon (AC) is a microporous adsorbent. It has a high specific surface (about $1000m^2/g$), it is widely available and it does not undergo structural modification below 180°C. The experiments showed that the pair activated carbon/ ammonia has very remarkable properties of adsorption cooling devices [21].

# 2. Materials and Methods

## 2.1. Solar Adsorption Refrigeration Machine

The solar adsorption refrigerating machine, object of this work (Figure.1), is constituted by:

- A reactor (adsorber) enclosed in a solar collector flat plate, containing the adsorbent-adsorbate mixture, where the phenomena of adsorption and desorption are produced;
- A condenser, where the refrigerant is liquefied ;
- An evaporator, in which the refrigerant vaporizes, producing the cold ;
- Non-return valves (V1) and (V2).
- An expansive valve (V3) ;
- A storage tank of the refrigerant liquid.

In this machine, solar energy is directly absorbed by the reactor located inside the flat solar collector. The suitability of this configuration has been already assessed by several authors [13], [22-24].

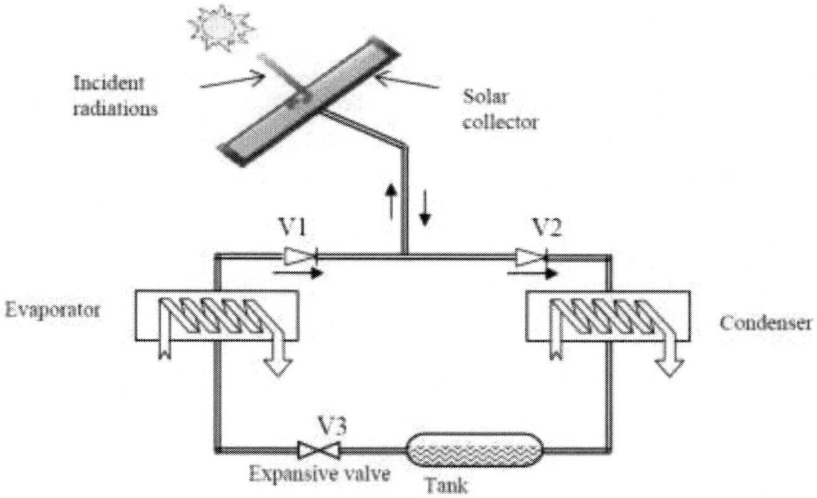


Figure 1. Solar adsorption refrigeration machine.

## 2.2. Adsorption Refrigeration Cycle Description

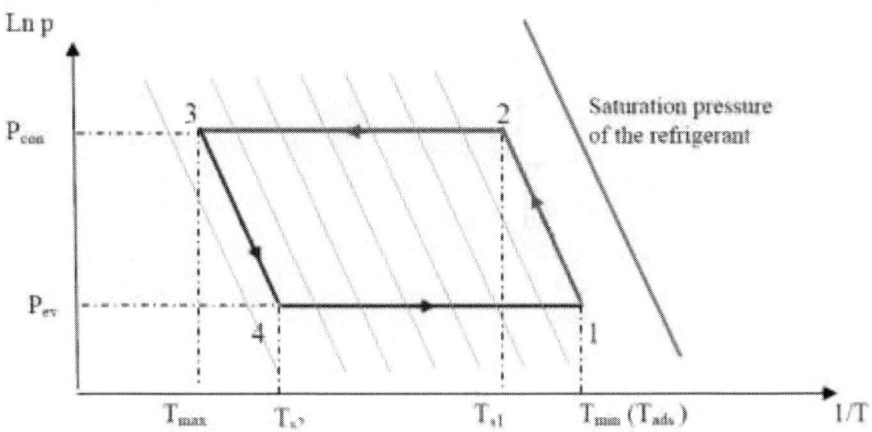


Figure 2. Thermodynamic cycle of adsorption refrigerating machine.

An ideal cycle (1-2-3-4) of adsorption refrigerating machine is represented on the diagram (Ln p,1/T) (Figure 2). The cycle represents the evolution of the mixture adsorbent-adsorbate contained in the adsorber, It consists of two isosteric phases (1-2 and 3-4), where the adsorbed mass remains constant, and two isobaric phases (2-3 and 4-1).

- Phase (1→2) : isosteric heating

At the beginning of the cycle (point 1), the mixture adsorbent-refrigerant is at its minimal temperature $T_{ads}$ and at the pressure of evaporation $P_{ev}$. When heating the reactor, the pressure and temperature of the mixture increase, to when the pressure reaches the condensation pressure prevailing in the condenser (point 2). The temperature reached is the desorption temperature threshold $T_{s1}$. The adsorbed mass remains constant throughout this phase1→2.

- Phase (2→3) : isobaric condensation-desorption

The mixture pressure reaches the saturation pressure of a pure refrigerant at the condenser's temperature. The desorption begins and the pressure remains constant until the temperature of the mixture increases at the maximum temperature where the refrigerant condenses, $T_{max}$.

- Phase (3→4) : isosteric cooling

In this phase, the reactor is cooled, and by evidence, the pressure decrease and follows an isoster. The pressure of the reactor decrease until it reaches the pressure the evaporator, $P_{ev}$, the temperature reached is called adsorption temperature threshold $T_{s2}$. The adsorbed mass remains constant.

- Phase (4→1) : isobaric evaporation-adsorption

When the mixture reaches the point 4, the evaporation of the refrigerant begins producing the cold in the evaporator, until the moment where the temperature reaches the minimal temperature, $T_{min}$. During this phase, the mixture follows an isobar imposed by the pressure of the evaporator, which corresponds to the saturation pressure in the evaporator, $T_{ev}$.

## 2.3. Solar Adsorption Cycle Model

The porous medium is a fixed bed of grains of activated carbon adsorption reacting with ammonia. The porous medium is enclosed within cylindrical tubes (adsorber) and heated by solar radiation as showed in figure.3". The adsorber is located in a solar collector flat plate, which consists of a cubic box, with a single glazed, a transparent cover and an absorbent surface.

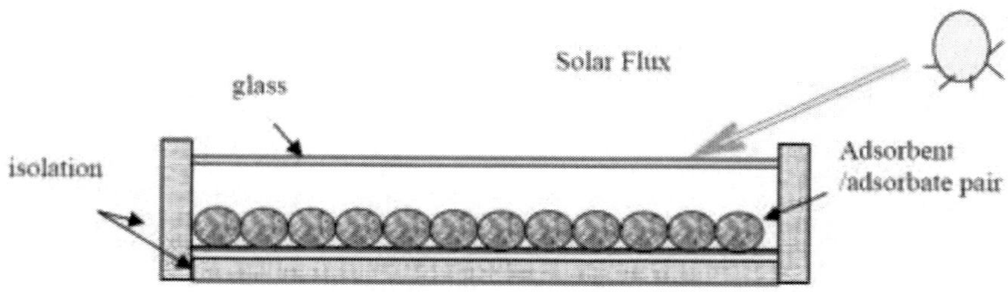


Figure 3. Description of a solar collector.

The solar collector flat plate converts solar radiations to heat energy, which is transfered to the porous medium during the day and loses the heat during the night

Modeling adsorption refrigerating machines requires the prior knowledge of the porous medium characteristics. In this study, we propose a model of heat and mass transfer in the porous medium of activated carbon reacting by adsorption with ammonia.

The phases existing in the porous medium are: Solid phase constituted by carbon grains, gaseous phase and adsorbed phase.

For this modelization, the following assumptions have been adopted:

- The porous medium properties have a cylindrical symmetry;
- All the phases are all the time in local thermal, mechanical and chemical equilibrium;
- The pressure is uniform;
- The heat transfer is radial and the convection heat transfer owing to the radial mass transfer is neglected;
- The conduction heat transfer in the medium can be characterized by an equivalent thermal conductivity coefficient $\lambda_e$.

### *2.3.1. Equations of Heat and Mass Transfer*

To obtain the equation of heat and mass transfer in the porous medium, we use the first thermodynamics principle applied for a layer of porous medium (activated carbon and ammonia) of width dr.

$$\frac{d(\delta u)}{dt} + \sum_s (q_{ms} h_s) - \sum_e (q_{me} h_e) = \phi + P_m \tag{1}$$

The medium porosity is define by:

$$\varepsilon = \frac{\text{Volume available for the fluid (gaseous and adsorbed phases)}}{\text{Total volume of the medium}}$$

The internal energy $\delta u$ appearing in equation (1) may be expressed as a function of the medium phases: solid, adsorbed and gaseous, Equation 2, where activated carbon [s], ammonia gas [g], adsorbed ammonia [a]

$$\delta u = 2\pi rdrL\ [(1-\varepsilon)\rho_s u_s + (\varepsilon-\alpha)\rho_g u_g + \alpha\rho_a u_a] \tag{2}$$

Equation (1) is expressed for a slice thickness dr as follows:

$$\frac{\partial}{\partial t}\left[2\pi rdrL\left[(1-\varepsilon)\rho_s u_s + (\varepsilon-\alpha)\rho_g u_g + \alpha\rho_a u_a\right]\right] - q_m(r+dr,t).h_g(T(r+dr),P) + q_m(r,t).h_g(T(r),P) = \lambda_e 2\pi rdrL\left[\left(\frac{\partial^2 T}{\partial r^2}\right)_r + \frac{1}{r}\left(\frac{\partial T}{\partial r}\right)\right] \tag{3}$$

with

$$\alpha = \frac{m\rho_{app}}{\rho_a} \tag{4}$$

The ammonia gas enthalpy is given by :

$$h_g(T) = h_a(T) + \Delta h_{ads}(T,m) \tag{5}$$

Heat transfer induce a mass transfer in the porous medium. In fact, the hottest layers will desorb the ammonia gas which will be adsorbed by the colder layers . This mass transfer will participate in the heat transfer process, since the gas that will be adsorbed on the cold layer is hotter, and also because of the exothermal character of adsorption. The difference between ammonia gas mass entering and going out of a layer is equal to the adsorbed ammonia gas remaining in the layer.

The ammonia mass conservation equation for a slice thickness dr is formulated as follows:

$$\frac{\partial}{\partial t}\left[2\pi rdrL\left[(\varepsilon-\alpha)\rho_g + \alpha\rho_a\right]\right] = -q_m(r,t) + q_m(r+dr,t) = \left(\frac{\partial q_m}{\partial r}\right)dr \tag{6}$$

Energy equation (2) combined with the mass conservation equation (6) yields:

$$\left[(1-\varepsilon)\rho_s C_s + (\varepsilon-\alpha)\rho_g C_g + \alpha\rho_a C_a\right]\frac{\partial T}{\partial t} - \frac{\partial((\varepsilon-\alpha)\rho_g)}{\partial t}.\frac{P}{\rho_g} - \rho_{app}\frac{P}{\rho_a}\frac{\partial m_a}{\partial t} - \Delta H_{ads}\rho_{app}\frac{\partial m_a}{\partial t} = \lambda_e \Delta T \tag{7}$$

We have adopted for calculating the mass of ammonia adsorbed as a function of temperature T and pressure P, the model of adsorption BRUNAUER, EMMET and TELLER (BET) [25] for its good description of the phenomenon adsorption for the couple ammonia / activated carbon [13].

$$\frac{m}{m_{cmo}} = \frac{C.x}{1-x} \cdot \frac{1-(n+1)x^n + n.x^{n+1}}{1+(C-1)x - Cx^{n+1}} \tag{8}$$

C is proportional to:

$$C \exp\left(\frac{\Delta H_{ads} - \Delta H_{cond}}{RT}\right) \tag{8.1}$$

Equation (4) written for all the layers contained in the cylinder gives a non linear system of partial derivatives equations, which is completed by initial and boundary conditions.

#### *2.3.2. Initial Condition*

A uniform initial temperature distribution in the porous medium is supposed:

$$T(r,0) = Ti \tag{9}$$

#### *2.3.3. Boundary Conditions*

for r = 0

$$\left(\frac{\partial T}{\partial r}\right)_{r=0} = 0$$

and equation (7) for r ⟶ 0 becomes:

$$\left[(1-\varepsilon)\rho_s C_s + (\varepsilon-\alpha)\rho_g C_g + \alpha\rho_a C_a\right] \_ \frac{\partial((\varepsilon-\alpha)\rho_g)}{\partial t} \cdot \frac{P}{\rho_g} - \rho_{app}\frac{P}{\rho_a}\frac{\partial m_a}{\partial t} - \Delta H_{ads}\rho_{app}\frac{\partial m_a}{\partial t} = 2\lambda_e \frac{\partial^2 T}{\partial r^2} \tag{10}$$

for r = R

We represent on the figure. 4, the nodal diagram of glass, metal wall and the porous medium (adsorbent/adsorbate).

The dynamic equations of heat transfer are written as:

$$\begin{cases} C_v \dfrac{dT_v}{dt} = q_v - h_{va}(T_v - T_a) - h_{vs}(T_v - T_s) + h_{pv}(T_p - T_v) & (11) \\ C_p \dfrac{dT_p}{dt} = q_p - h_{pv}(T_p - T_v) - h_{pa}(T_p - T_a) + h_{pm}(T_p - T_m) & (12) \end{cases}$$

where :

$$\begin{cases} q_v = a_v.G \text{ ; is the glass absorbed radiations} & (13) \\ q_p = a_p.G \text{ ; is the wall absorbed radiations} & (14) \end{cases}$$

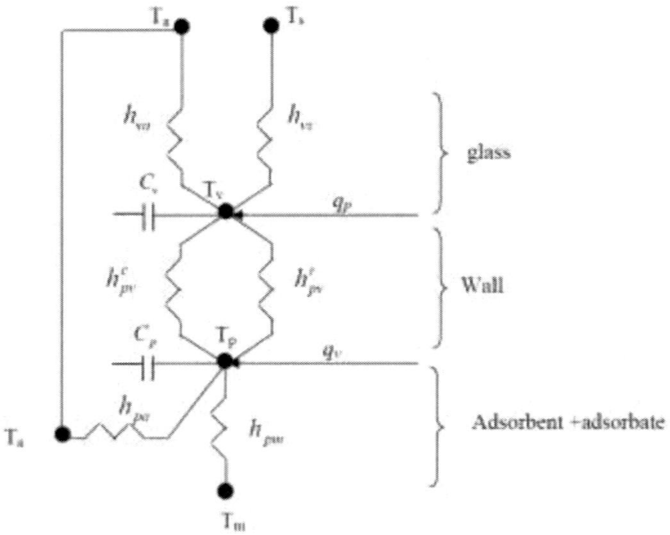


Figure 4. Nodal diagram of glass, metal wall and the porous medium (adsorbent/adsorbate).

The glass and the wall absorption coefficients of global radiation are given by the following expressions [26]:

$$\begin{cases} a_v = \alpha_{vb} + \left(\alpha_{vd}.(1-\alpha_{pb})\tau_{rv}\right)/\left(1-\tau_{dv}.(1-\alpha_{pb})\right) & (15) \\ a_p = \alpha_{pb}\tau_{rv} / \left(1-\tau_{dv}.(1-\alpha_{pb})\right) & (16) \end{cases}$$

The heat exchange coefficients are given by:

The convective exchange coefficient between glass and atmosphere is determined by the Correlation of Watmuff:

$$h_{va} = 2.8 + 3.0 * V_v \tag{17}$$

The radiation exchange between glass and sky is given by [27]:

$$h_{vs} = \varepsilon_v.\sigma\left(T_s^2 + T_v^2\right)\left(T_s + T_v\right) \tag{18}$$

Only radiation exchange between window and wall is taken into account [27].

$$h_{pv} = \sigma . \frac{\left(T_p^2 - T_v^2\right)\left(T_p - T_v\right)}{\frac{1}{\varepsilon_p} + \frac{1}{\varepsilon_v} - 1} \tag{19}$$

Only heat conduction exchange between wall and atmosphere is taken into account.

$$h_{pa} = \frac{\lambda}{e_p} \tag{20}$$

The heat coefficient exchange between wall and porous medium $h_{pm}$ was determined experimentally [13], the value found is equal to 33,65 $W/m^2.K$.

The equations system obtained are coupled and highly nonlinear. The discretization is done using the implicit finite difference method using the implicit scheme [28]. The resolution of algebraic equations resulting from the discretization is done using the Gauss-Seidel line by line [28], TDMA (Tridiagonal Matrix Algorithm) by linearisation. We have developed a numeric code (in Fortran) to solve our model for determining the temperature, pressure and mass adsorbed within the porous medium. . These equations are supplemented by initial and boundary conditions, giving the wall temperature $T_p$ of the metal wall of the glass and $T_v$.

### *2.3.4. Quantity of Cold Produced*

It is the useful effect produced by the evaporator, which is equal to the latent heat of vaporization of refrigerant unless the sensible heat required to cool the refrigerant condensing temperature to the temperature of evaporation, its expression is:

$$Q_f = \Delta m \left[ L(T_{ev}) - \int_{T_{ev}}^{T_{cond}} C_l dT \right] \tag{21}$$

where Δm is the cycled mass, given by:

$$\Delta m = m_a\left(T_{ads}, P_s\left(T_{ev}\right)\right) - m_a\left(T_g, P_s\left(T_{cond}\right)\right) \tag{22}$$

where $m_a(T, P)$ is the adsorbed mass of ammonia at temperature T and pressure P, calculated using the BET model.

### *2.3.5. Solar Performance Coefficient*

The performance evaluation of solar machine is determined from the amount of heat $Q_f$ and $Q_s$ where $Q_s$ is the amount of global irradiation received by the collector surface of the sensor, its expression is given by:

$$Q_s = S_c \int_{Sunrise}^{Sunset} G(t)dt \tag{23}$$

The solar coefficient of performance is given by:

$$COP_{sol} = \frac{Q_f}{Q_s} \tag{24}$$

# 3. Numerical Simulation and Results

We have realized a program written in FORTRAN 90 wich allows us to compute the spatial temporal distribution of temperature, pressure and adsorbed mass.

In our simulation, we have used the hourly solar data and climate (ambient temperature and global radiation on inclined surface) corresponding to a clear a typical day Tetouan (Morocco, 35°34' N, 5°22' W), from the climatological database [29] (Figure 5). The inclination sensor is equal to the approximate altitude because it can receive a maximum solar energy [30]. The software METEONORM uses the Perez model for calculating the radiation on an inclined surface from the values of the global horizontal radiation [29].

The condensation temperature is that corresponding to the ambient temperature recorded the same day, the evaporation rate is set at 0 °C

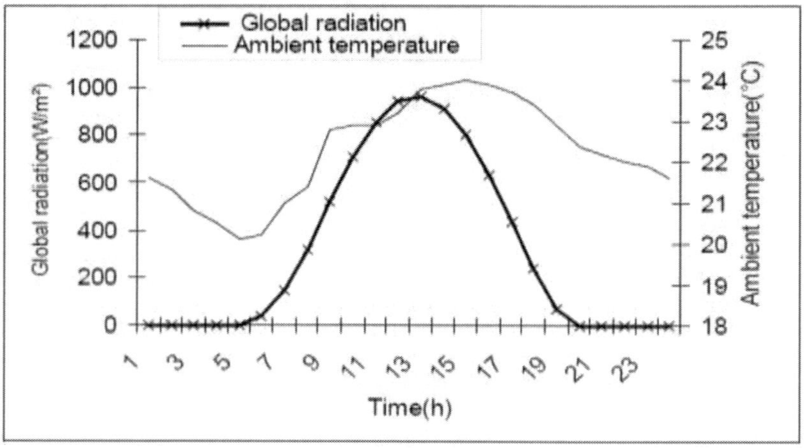


Figure 5. Global radiation and ambient temperature in Tetouan (18th July 1990 (clear typical day)).

## 3.1. Results

We present the figure.6, the variation of the wall temperature, glass temperature and adsorbate-adsorbent (porous medium) temperature. We note that the maximum temperature reached in the porous medium is equal to 91 °C.

In Figure 7, we present the evolution of the pressure within the adsorber. We note the existence of four phases, the pressure increases when the adsorber is heated and is closed. It becomes constant, from the temperature threshold, which corresponds to pressure of the beginning of the condensation phase.

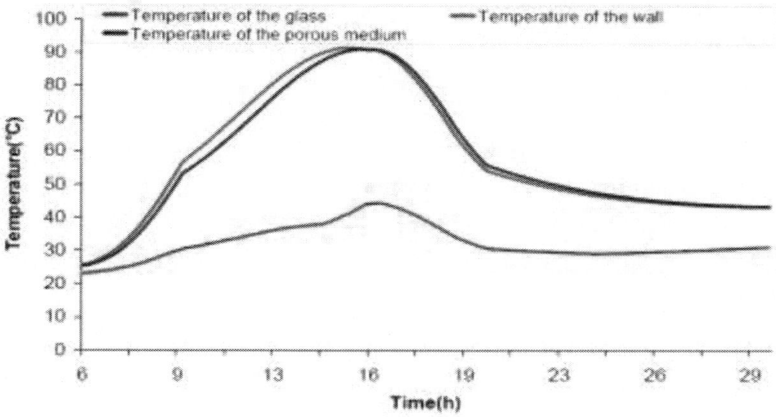


Figure 6. Temperature variation of the glass, wall and porous medium according to the time.

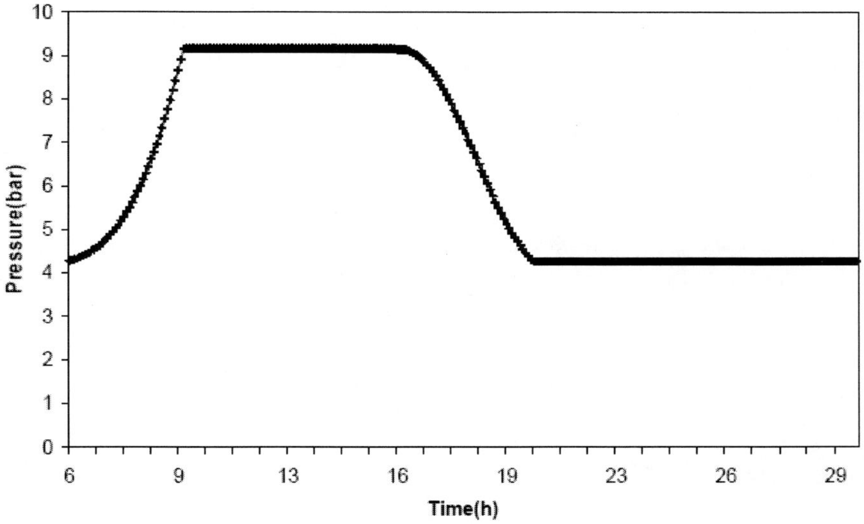


Figure 7. Pressure evolution according to the time.

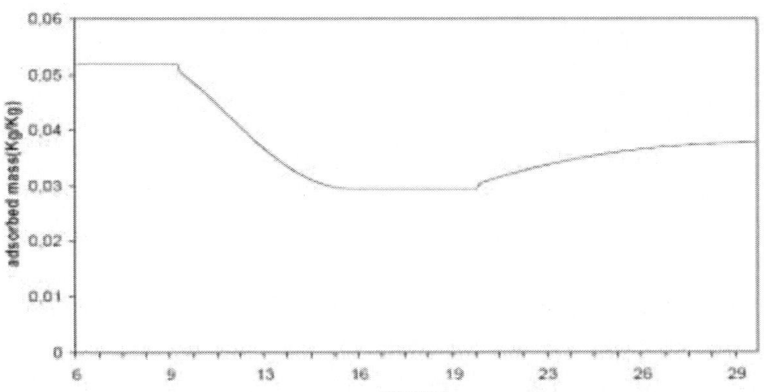


Figure 8. Adsorbed mass variation according to the time.

The pressure in the adsorber begins to decrease with the temperature during the cooling phase of the adsorber. We note that the pressure becomes constant when the temperature reaches the threshold temperature of the beginning of adsorption phase which corresponds to the production of cold in the evaporator. The Figure 8 represents the evolution of adsorbed mass in the adsorber according to the ime, it allows us to determine the cycled mass of ammonia in the refrigerating. In Figure 9, we present the machine cycle in the diagram (Lnp, 1/T), it consists of two isobars phases and two isosteric phases.

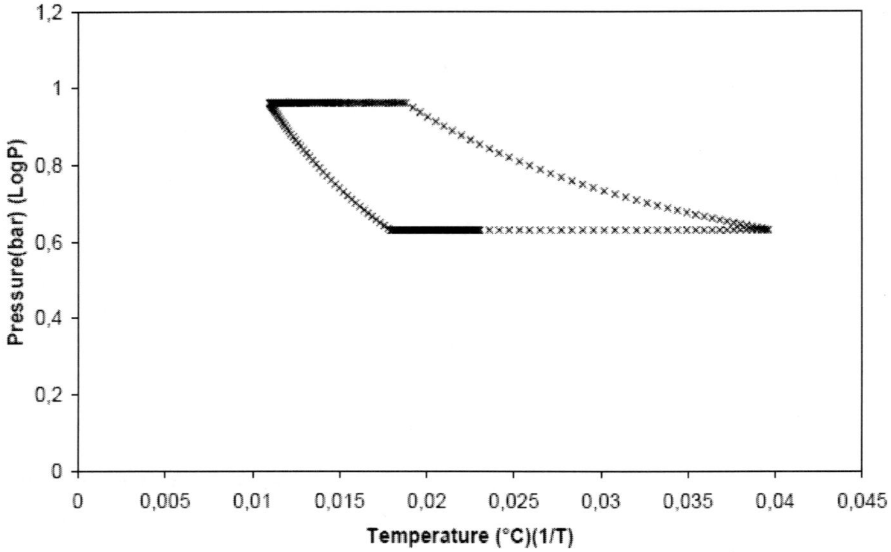


Figure 9. Computed cycle of adsorption refrigeration machine ($T_{ev}$=0°C ; $T_{cond}$= $T_a$ ).

### 3.2. Quantity of Cold Produced and Solar Performance Coefficient

We present in the table n°1 the data, which allows us to computer solar performance coefficient computed for a radiation typical day of June.

**Table 1. COP and the quantity of cold produced**

| $T_{con}$ (°C) | $T_{ev}$ (°C) | $T_{max}$ (°C) | $Q_s$ (Kj/m$^2$) | $Q_f$(Kj/m$^2$) | COP |
|---|---|---|---|---|---|
| 22 | 0 | 91 | 27009,00 | 3014,32 | 0,112 |

## 4. Presentation of Solar Maps and Cold Maps of Morocco

Our objective in this study is to determine the quantity of maps cold which present the cold produced for several regions in Morocco. The quantity of cold produced in the refrigeration machine will be used as a benchmark design. It's calculated for a solar adsorption refrigerating machine for varying climatic parameters and for different typical days.

## 4.1. Solar Maps of Morocco

The figure 10 present solar map used in this work, which describes the isoradiations on different regions of Moroccan territory in 1990.

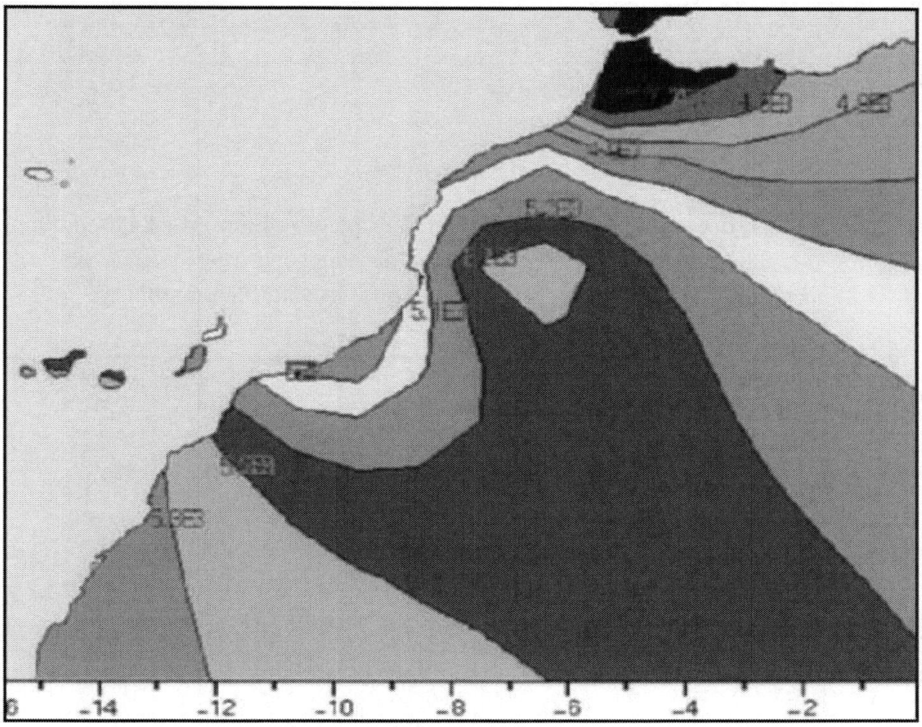


Figure 10. Horizontal global radiation annual average daily ( Wh/m$^2$).

**Table 2. Amount of daily global radiation and initial temperature of the morning for each region of the typical day in the month of May 1990**

| | The overall amount of radiation received (Kj/m$^2$) | Initial temperature of morning (°C) |
|---|---|---|
| Agadir (30°33’ N, 9°40’ W) | 27649,8 | 16 |
| Marrakech (31°49’ N, 8° W) | 17105,4 | 19,9 |
| Mekness (33°53’ N,5°37’ W) | 22527 | 17,9 |
| Oujda (34°41’ N, 1°50’ W) | 20973,6 | 15,1 |
| Rabat (34°02’ N, 6°51’ W) | 26843,4 | 15,7 |
| Tétouan (35°35’ N, 5°22’ W) | 25032,6 | 15,3 |

**Table 3. Amount of daily global radiation and initial temperature of the morning for each region of the typical day in the month of June 1990**

| | The overall amount of radiation received (Kj/m2) | Initial temperature of morning (°C) |
|---|---|---|
| Agadir (30°33' N, 9°40' W) | 27806,4 | 14,7 |
| Marrakech (31°49' N, 8° W) | 27954 | 17,6 |
| Mekness (33°53' N,5°37' W) | 20161,8 | 17,8 |
| Oujda (34°41' N, 1°50' W) | 21333,6 | 14,8 |
| Rabat (34°02' N, 6°51' W) | 11624,4 | 15,9 |
| Tétouan (35°35' N, 5°22' W) | 33130,8 | 16,4 |

## 4.2. Cold Maps of Morocco

We present in Figures (.11, .12, .13, .14) different maps, which given us the quantities of cold (Kj/m$^2$) calculated in different regions of Morocco according to the solar radiations maps) *for* a typical clear days of the months (May, June, July and September).

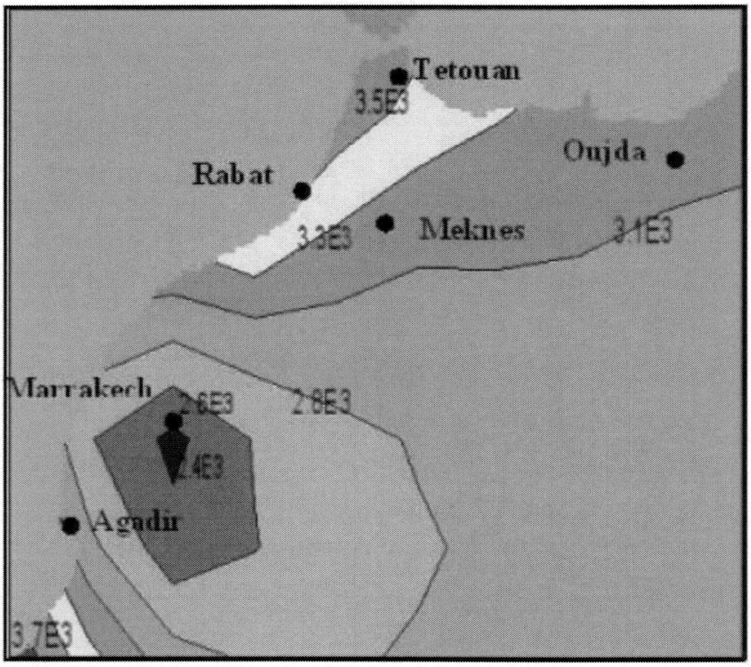


Figure 11. Graphical representation of the quantity of cold produced a typical day light in May 1990 (Kj/m$^2$).

The solar adsorption refrigerating machine is powered by solar energy, so it is very sensitive to climatic conditions. It is therefore evident that the performance solar coefficient of this machine is highly dependent on the amount of heat received. It Appears Clearly In The figures (.11, .12, .13, .14) than the amount of cold produced in the evaporator $Q_f$ and solar performance coefficient $COP_s$, are functions incident radiation. Indeed, when the energy increases, the mass desorbed ammonia $m_{min}$ at the end of the desorption phase-condensation decreases, due to the increase of the maximum temperature (regeneration), and therefore, the mass cycled $\Delta m$ increases. Therefore, the amount of cold produced at the evaporator and the solar performance coefficient must increase.

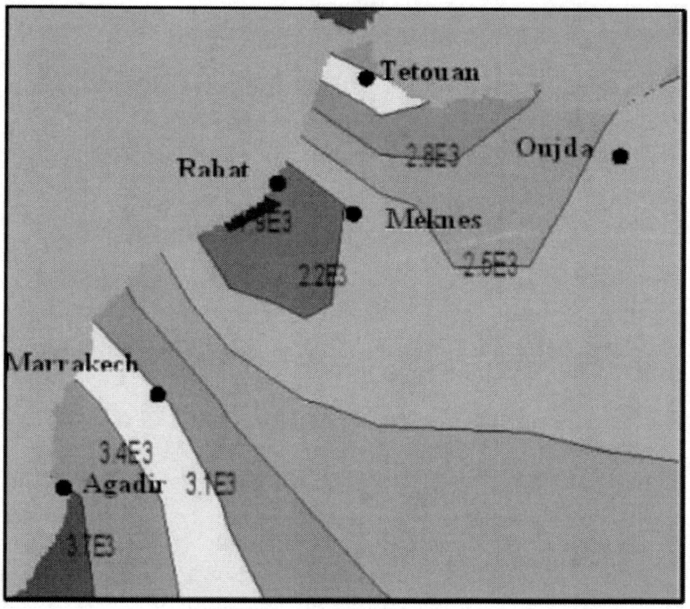


Figure 12. Graphical representation of the quantity of cold produced a typical day light in June 1990 ( Kj/m$^2$).

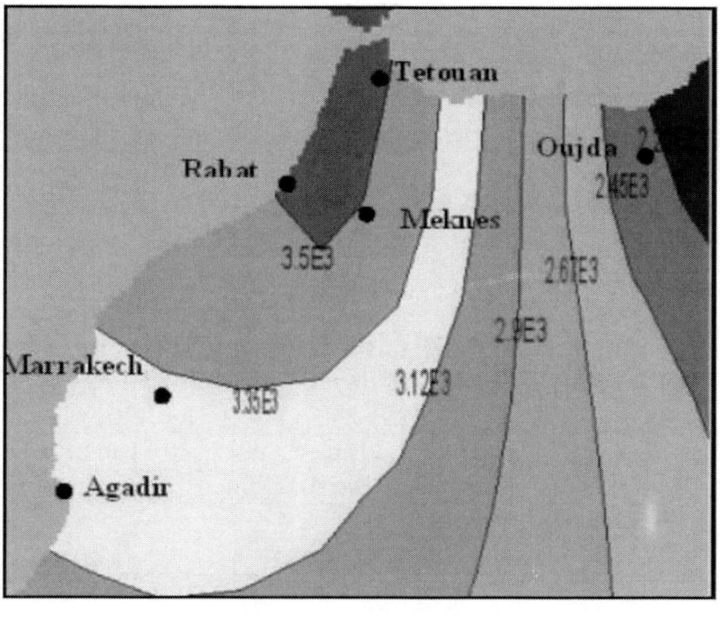


Figure 13. Graphical representation of the quantity of cold produced a typical day light in July 1990 (Kj/m$^2$).

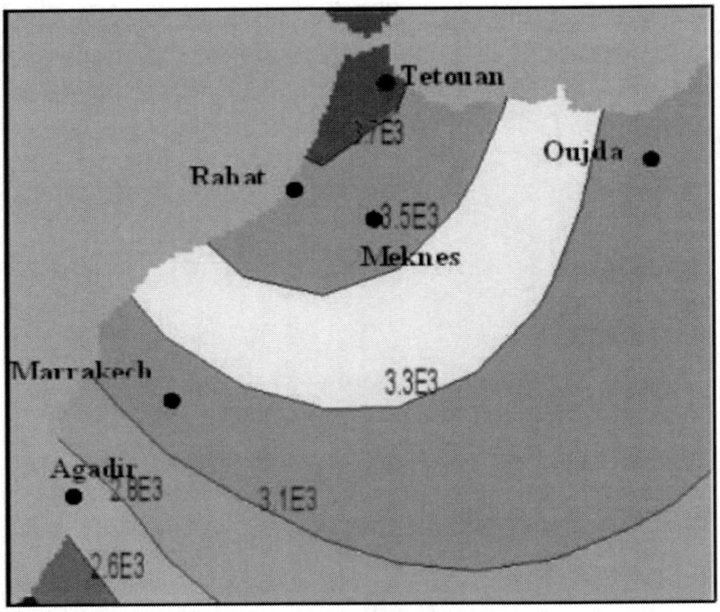


Figure 14. Graphical representation of the quantity of cold produced a typical day light in September 1990 (Kj/m$^2$).

The quantity of cold produced in the evaporator $Q_f$ and the solar performance coefficient $COP_s$ are a function of the adsorption temperature (initial or early morning). Indeed, the increase the adsorption temperature causes a decrease in the adsorbed mass at this temperature and the evaporation pressure $m_{max}$, and therefore the mass cycled $\Delta m$ decreases, which causes a decrease in $Q_f$ and $COP_s$.

The quantity of cold produced depends on the values of global radiation received in one day and on initial temperature of the morning.

The analyse of the results show that to increase the cycled mass of refrigerant through the machine (and consequently increasing the quantity of cold and solar coefficient performance) requires that the maximum temperature reached by the adsorber during the day mus be too large and the minimum temperature too low. For that, we must increase the solar flux or decrease the minimum temperature of adsorption, by several ways of improving solar cell of the machine.

**Table 4. Amount of daily global radiation and initial temperature of the morning for each region of the typical day in the month of July 1990**

| | The overall amount of radiation received (Kj/m2) | Initial temperature of morning (°C) |
|---|---|---|
| Agadir (30°33' N, 9°40' W) | 25162,2 | 19,8 |
| Marrakech (31°49' N, 8° W) | 27333 | 19,9 |
| Mekness (33°53' N,5°37' W) | 30088,8 | 19,2 |
| Oujda (34°41' N, 1°50' W) | 24316,2 | 23,8 |
| Rabat (34°02' N, 6°51' W) | 29367 | 19,1 |
| Tétouan (35°35' N, 5°22' W) | 26456,4 | 17,4 |

**Table 5. Amount of daily global radiation and initial temperature of the morning for each region of the typical day in the month of September 1990**

| | The overall amount of radiation received (Kj/m2) | Initial temperature of morning (°C) |
|---|---|---|
| Agadir (30°33' N, 9°40' W) | 17859,6 | 18,9 |
| Marrakech (31°49' N, 8° W) | 25650 | 20,1 |
| Mekness (33°53' N,5°37' W) | 25876,8 | 18,2 |
| Oujda (34°41' N, 1°50' W) | 27504 | 21,4 |
| Rabat (34°02' N, 6°51' W) | 28785,6 | 18,1 |
| Tétouan (35°35' N, 5°22' W) | 30214,8 | 18,6 |

## CONCLUSION

This work proposes a study of the transient behavior of an adsorber of solar adsorption refrigeration machine using the pair activated carbon ammonia, the model is based on the coupled equations of heat transfer and mass transfer in porous media constituted by the Activated carbon-ammonia mixture.

We have studied the evolution of temperature, pressure and mass adsorbed as a function of solar flux during a cycle of 24 hours. Conclusion, the adsorbed mass varies with solar flux and the minimum temperature reached in the adsorption adsorber.

We have demonstrated the feasibility of the machine in the region of Tetouan. The model has highlighted the importance of a study of the cycled mass, of the total cooling and solar coefficient of performance of the machine according to the site of sunshine and also depending on the nature of the solar.

We applied the model to determine the map of solar refrigeration using solar radiation data for Moroccan solar adsorption refrigerating machine using the pair activated carbon-ammonia.

Our study requires the need for a comparative study of solar panel with simple plan that solar panel plan selective glass, double glazing or solar panel plan his empty. Another way to explore is the study of cylindroparaboliques hubs.

## REFERENCES


[1] A.O. Dieng, R.Z. Wang, Literature review on solar adsorption technologies for ice-making and airconditioning purposes and recent developments in solar technology, *Renewable and Sustainable Energy Reviews* 5 (2001): 313–342.

[2] A. Boubakri, J.J Guilleminot and F. Meunier, Adsorptive Solar Powered Ice Maker: *Experiments and Model. Solar Energy* 69 (3) (2000): 249-263.

[3] A. Al Mers, A. Mimet and M. Boussouis, Numerical study of heat and mass transfer in a cubical porous medium heated by solar energy "Boubnov-Galerkin method", *Computational Methods for Flow and Transport in Porous Media*. J.M. Crolet (ed) (2000).

[4] A. El Fadar, A. Mimet, A. Azzabakh, M. Pérez-Garcia and J. Castaing, Study of a new solar adsorption refrigerator powered by a parabolical trough collector, *Applied Thermal Engineering* 29(5-6) (2008):1267-1270.

[5] A. El Fadar, A. Mimet, M. Pérez-Garcia, Modeling and performancesludy of a continuous adsorption refrigeration system driven by parabolical trough solar collector, *Solar Energy* 83(6) ( 2008): 850-861.

[6] M. Louajari , A. Mimet, A. Ouammi, Study of the effect of finned tube adsorber on the performance of solar driven adsorption cooling machine using activated carbon-ammonia pair, *Applied energy* 88 (3) (2010.): 690-698.

[7] M. Louajari , A. Mimet, A. Ouammi, Sustainable development of a solar adsorption cooling machine, *Management of environmental* 21 (2010) : 589-601.

[8] F. Lemmini, F. Meunier, Simulation of adsorptive solar refrigerator operating in Morocco, Solar Science, *Journal of Islamic Academy of Sciences* 3 :4 (1990) : 273-279.

[9] F. Lemmini, J. Buret-Bahraoui, M. Pons, F. Meunier, Simulation des performances d'un réfrigérateur solaire à adsorption: 2. Fonctionnement avec stockage de froid dans deux climats différents, *International Journal of Refrigeration* 15(3) (1992) : 168-176.

[10] L.W. Wang, R.Z. Wang, R.G. Oliveira A review on adsorption working pairs for refrigeration, *Renewable and Sustainable Energy Reviews* 13 (2009): 518–534.

[11] J. J. Guilleminot and F. Meunier, Etude expérimentale d'une glacière solaire utilisant le cycle zéolithe-eau. *Rev. Gen. Therm.* 239 (1981): p. 825.

[12] A. Boubakri, Ph. Grenier, and M. Pons, Utilisation of activated carbon and methanol pair to solar ice production, *Proceedings JITH Conference Rabat* 1, Morocco,1985, p. 69.

[13] A. Mimet, Etude théorique et expérimentale d'une machine frigorifique à adsorption d'ammoniac sur charbon actif, Thèse de Doctorat, FPMs, Mons, Belgique, 1991.

[14] R. E. Critoph and F. Gong, A rapid cycling ice-maker for use in developing countries. In *Proceeding of the 2nd World Renewable Energy Congress*, UK, 1992.

[15] Al Mers A, Azzabakh A, Mimet A, El Kalkha H. Optimal design study of cylindrical finned reactor for solar adsorption cooling machine working with activated carbon–ammonia pair, *Appl. Therm. Eng*. 26 (2006):1866–75.

[16] F. Lemmini, J. Buret-Bahraoui, M. Pons, F. Meunier, Simulation des performances d'un réfrigérateur solaire à adsorption : 1. Comparaison des performances pour deux types de charbon actif, *International Journal of Refrigeration* 15(3) (1992): 159-167.

[17] Institut Internationnal du foid, Tables et diagramme pour l'industrie du froid. Propriétés thermodynamiques du R717, Paris, 1981.

[18] S. Follin, Approche de Dubinin pour la caractérisation de charbons actifs et évaluation des performances des cycles frigorifiques à adsorption. Application à la production de froid par le couple charbon actif/dioxyde de carbone, université de Perpignan, thèse de Doctorat, 1996.

[19] F. Meunier, Le froid solaire par adsorption, Cahier.AFEDES, n°5, 1978.

[20] N. Douss, Etude expérimentale de cycles à cascades à adsorption solide, Thèse de doctorat, Université Paris VII, France, 1988.

[21] A. Mahamane, Etude de l'adsorption de vapeur sur solide microporeux, adaptation da la théorie de solution, Thèse de doctorat, *FPMS,* Mons, Belgique, 1988.

[22] M. Pons and J.J. Guilleminot, Design of an experimental solar-powered, solid-adsorption ice maker. Solar Energy Eng., *ASME Trans*. 108 (1986) :332-337.

[23] R.H.B. Exell and S. Kornsakoo, The development of a solar powered refrigerator for remote village. In *Solar World Forum*, proc I.S.E.S. Congress, Brighton, England, 1981, pp.1049-1053.

[24] M.Pons and Grenier. Experimental data on a solar-powered ice maker using activated carbon and methanol adsorption pair. *J. Solar Energy Eng*., ASME Trans. 109(1987): 303-310.

[25] S. Brunauer, *The adsorption of gases and vapors*. Ed. Oxf. University Press, 1943.

[26] J. Bougard and A. Pilatte, Programme National Belge R. D. Energie. Centre de recherche sur l'énergie solaire. Belgique.

[27] K.S. Ong, Thermal Performance of Solar Air Heaters: Mathematical Model and Solution Procedure', *Solar Energy* 55(2) (1995) : 93 - 109.

[28] S. V. Patanka, Numerical heat transfer and fluid flow, Hemisphere, New York, 1980.

[29] Meteotest. *Meteonorm version 6.1 – handbook*; www.meteonorm. Com, 2008.

[30] M. Aggour, Mesures et correlation du rayonnement solaire et des caractéristiques des panneaux photovoltaïques, thése de doctorat, Faculté des sciences, Rabat, 1987.

# INDEX

## C

## F

## G



## H



## I

## N



## O



## P

## Q



## R



## S

## T

## U



## V



## W

**X**



**Y**

RENEWABLE ENERGY: RESEARCH, DEVELOPMENT AND POLICIES

# DISTRIBUTED WIND ENERGY GENERATION

ANTONIO COLMENAR-SANTOS
ENRIQUE ROSALES-ASENSIO
AND
DAVID BORGE-DIEZ
EDITORS

nova
science publishers
New York

## Library of Congress Cataloging-in-Publication Data

ISBN: 978-1-53614-207-5

*Published by Nova Science Publishers, Inc. † New York*

# Contents

# Preface

This book constitutes the refereed proceedings of the 2018 International Conference on Distributed Wind Energy Generation, which was held on 29th May 2018. 2018 International Conference on Distributed Wind Energy Generation intends to provide an international forum for the discussion of the latest high-quality research results in all areas related to Distributed Wind Energy Generation. The editors believe that readers will find following proceedings interesting and useful for their own research work.

This book contains the Proceedings of the 2018 International Conference on Distributed Wind Energy Generation held online (https://enriquerosales.wixsite.com/virtualconferences), on 29th May, 2018. It covers significant recent developments in the field of Distributed Wind Energy Generation from an applicable perspective.

***ADVISORY BOARD:***

Organizing Committee Chair:
*Enrique Rosales Asensio, PhD*
Departamento de Física, Universidad de La Laguna, La Laguna, Spain
Email: erosalea@ull.edu.es

***PROGRAM COMMITTEE CHAIRS:***

*Enrique González Cabrera, PhD*
Departamento de Ingeniería Química y Tecnología Farmacéutica, Universidad de La Laguna, La Laguna, Spain
Email: eglezc@ull.edu.es

*Antonio Colmenar Santos, PhD*
Departamento de Ingeniería Eléctrica, Electrónica, Control, Telemática y Química Aplicada a la Ingeniería, Universidad Nacional de Educación a Distancia, Madrid, Spain
Email: acolmenar@ieec.uned.es

*David Borge Diez, PhD*
Departamento de Ingeniería Eléctrica y de Sistemas y Automática, Escuela Técnica Superior de Ingenieros de Minas de León,
León, Spain
Email: dbord@unileon.es

***SCIENTIFIC COMMITTEE:***

*Manuel Castro-Gil, PhD*, Universidad Nacional de Educación a Distancia, Madrid, Spain

*Clara M. Pérez-Molina, PhD*, Universidad Nacional de Educación a Distancia, Madrid, Spain

*Francisco Mur-Pérez, PhD*, Universidad Nacional de Educación a Distancia, Madrid, Spain

*José María Pecharromán Lázaro*, ENDESA, Palma de Mallorca, Spain

*Pedro Miguel Ortega Cabezas*, PSA, Madrid, Spain

*África López-Rey, PhD*, Universidad Nacional de Educación a Distancia, Madrid, Spain

*Jorge Blanes Peiró, PhD*, Universidad de León, León, Spain

May 2018

*Editors*

In: Distributed Wind Energy Generation ISBN: 978-1-53614-207-5
Editors: A. Colmenar-Santos et al. 

*Chapter 1*

# EU Plans for Renewable Energy

***Salvador Ruiz-Romero*[1,*], *Rosario Gil-Ortego*[1] *and Carlos de Palacio*[2]**

[1]Departamento de Ingeniería Eléctrica, Electrónica, Control, Telemática y Química Aplicada a la Ingeniería, Universidad Nacional de Educación a Distancia (UNED), Madrid, Spain

[2]ABB, Madrid, Spain

## Abstract

Recent success in the installation of renewable energy is largely due to public financial incentives. Data reported in this chapter show the incessant growth of renewable energy implementation and the key process for achieving the complete replacement of mixed electric generation in the European Union; specifically, this chapter focuses on the case of Spain. These data also emphasize the creation of skilled jobs is taking place and the source of economic growth that renewable energy will promote. The potential for solar and wind energy in Spain, if exploited efficiently [1], will result in energy independence and protection against negative environmental effects.

---

* Corresponding Author Email: s.ruiz.romero@gmail.com.

**Keywords:** renewable, wind energy, solar energy, renewable generation mix

## INTRODUCTION

The European Union has established changes in energy policies regarding the production of electric power using renewable energy [2].

On September 27$^{th}$, 2011 was issued the Directive 2001/77/CE of the European Parliament and Council, which focused on the promotion of electricity generated from renewable sources of energy within the electricity market in the EU.

The general community recognizes the importance of promoting renewable energy sources because they contribute to environmental protection and sustainable development, in addition to fulfilling the Kyoto objectives.

The White Paper promotes this type of energy on the basis of certainty and diversification of energy supplies, environmental protection, cost reduction and promotion of social cohesion.

Additionally, Directive 2001/77/CE indicates the essential need to establish national binding objectives as well as ambitious objectives for renewable energy, including setting national indicative objectives for the appropriate consumption of different types of energy and establishing compatibility with other commitments within the framework of climate change, complying with the Kyoto Protocol.

The need of public funds to support renewable energy sources is recognized in the communal directives of aid to member states for the environment, taking into account the need to internalize the external costs of electricity generation by the member states.

The increased market penetration of electricity generated from renewable sources will allow economies of scale, reducing costs.

The connection costs for new producers of electricity generated by renewable energy must be objective, clear and nondiscriminatory; they must also adequately reflect the benefits that the integrated producers contribute to the supply chain.

Priority access of the national electricity system must be allowed to use the electricity generated from sources of renewable energy by the operators of the network transport system.

Issued in Spain on May 25th, 2007, Royal Decree 661/2007 was intended to regulate the production of electric power in special schemes. It was established that in several sectors there was a need of decreasing energy reliance on outside sources, a need to better use available energy sources, and a need for environmentally friendly options, all of which can be met by using renewable energy and increasing efficiency in the electricity generation, to obtain a sustainable level of development, cost reduction, and in social and environmental aspects.

The relevant objectives for the promotion of renewable energy in Spain are included in the Renewable Energy Plan, for the periods of 2005-2010 and 2011-2020.

Spain has recently become one of the leading countries in the development of technology for the electricity generation from renewable energy sources. Currently, more than 25% of the electric power generated in Spain comes from renewable energy sources. This growth is due to the existence of a solid, stable and predictable economic policy and legal system, and to the contributions of stakeholders, regional government, and the technical and economic operators of the system and companies.

Of particular relevance is the growth of wind, solar thermoelectric and photovoltaic technologies, for which the objectives set for 2010 have been exceeded.

## RENEWABLE ENERGY IN EUROPE

The EU is not very rich in native natural resources. The process of extraction in this region is more difficult than in other places. Unfortunately, our resources continue to be decreased.

The exhaustion rate will depend on global prices and technological advances. However, the potential of renewable energy has not yet been fully exploited due to high production costs. If this problem is solved, renewable energies are the only sources that will have a strong future in Europe [3].

Increasing the use of renewable energy would allow the EU to reduce the consumption of fossil fuels by 200-300 million tons per year and carbon dioxide emissions by 600-900 million tons per year.

The greatest tool that the EU owns to fight against climate change is the Emission Allowance Trading Scheme created in 2005. The EU countries can buy and sell their rights to emission within bounds fixed for all of the European Union. This system—the first of its kind – allows countries to reduce their emissions in a profitable manner.

Around 80% of the energy consumed in the EU comes from fossil fuels (oil, gas and coal). A significant and increasing share of these resources is obtained from outside the EU. The EU's dependence on imported oil and gas, which is currently around 50%, could increase to 70% by 2030, which will be added to the vulnerability of the EU in view of supply cuts and rising prices caused by international crises. Furthermore, the EU must reduce its consumption of fossil fuels to reverse the process of global warming. The future depends on a combination of energy reductions obtained by means of a more efficient energy use, alternative energy sources (especially renewable energy sources within the EU) and an increase in international cooperation, according to CIEMAT [4].

The consumption of energy could be reduced by one-fifth by 2020 by encouraging changes in consumers' behavior and using efficient energy technology. This would correspond to savings of approximately 60 million euros yearly and would help to fulfill the international commitments to take part in climate change. If the EU does not act, its energy consumption may increase by 10% by 2020.

The cost of fuel for solar and wind energy is zero. Thus, all costs saved by a wind or solar farm could support jobs related to the building or installation of such farm.

Data showed by BP (www.bpsolar.com) indicate that the building of solar farms could generate 6 times more jobs than would arise if the same amount of money is invested in oil uses. Complete implementation of the plans detailed in the White Paper of the EU about renewable energies would result in a mass creation of jobs for its member countries.

Several European industry sectors have estimated in more than 1.6 million the number of jobs created in the renewable energy sector and related industries, following the policy implementation about renewable energies.

An independent study, TERES II (www.eurorex.com/teresii), has estimated that the fact of providing 12% of the energy through renewable sources would result in the creation of 500,000 jobs [5].

According to the EWEA, [6], [7] in the report Wind at Work (January 2009), the wind energy sector provided jobs for 154,000 people in 2007, 108,600 of which were direct jobs.

The report showed that the manufacturers of wind turbines are the main employers, representing 37% of the direct jobs, followed by the manufacturers of parts and the project developers. With respect to the member countries, 75% of the direct jobs in the wind energy sector are currently located in the three leading countries: Denmark, Germany and Spain; however, other countries, including France, the United Kingdom and Italy, are now beginning to catch up.

This has led to a rapid increase in the demand for renewable energy specialists who are able to design, install and maintain such farms [8].

Wind at Work revealed the vast potential of the wind energy sector as a source of employment that can bring significant long-term benefits to European economies.

The wind energy sector also generates indirect jobs, such as suppliers of raw materials.

According to the EWEA, more than 375,000 people may be directly employed in the sector by 2030: 160,000 onshore and 215,000 offshore.

According to the baseline of the EWEA (EWEA, 2008), there will be an expected 180 GW of wind energy operating in the EU in 2020 and 300 GW in 2030. Based on the EWEA baseline and the cost of wind energy with onshore and offshore capability provided by the EWEA for 2030, we discover that the number of jobs in the wind energy sector in the EU will be more than twice the amount for 2007, 154,000, increasing in almost 330,000 by 2020. Onshore energy production will continue to be the major employment source during most of the period, but by 2025, the jobs in the offshore energy sector will exceed the onshore jobs, according to the EWEA.

To represent the change performed by the introduction of renewable energy in recent years [9], Figure 1 shows the variation in the net power installed according to the different technologies in the EU during the years 2000-2009.

It can be appreciated that gas plants and wind power have had the greatest increase in power generation, increasing by 81,067 MW and 65,102 MW, respectively. The use of photovoltaic energy can also be stressed, exceeding 13,000 MW, in relation to other technologies that do not even reach 3,000 MW.

On the other hand, the production of nuclear, coal and oil technologies has been jointly decreased by 32,000 MW.

The distribution of the installed capacity of various technologies is shown in Figure 2.

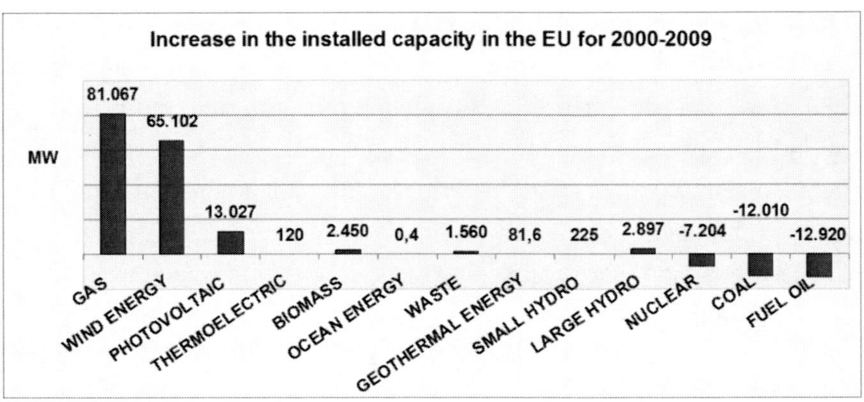


Source: EWEA.

Figure 1. Increase (MW) in the installed capacity in the EU during 2000-2009.

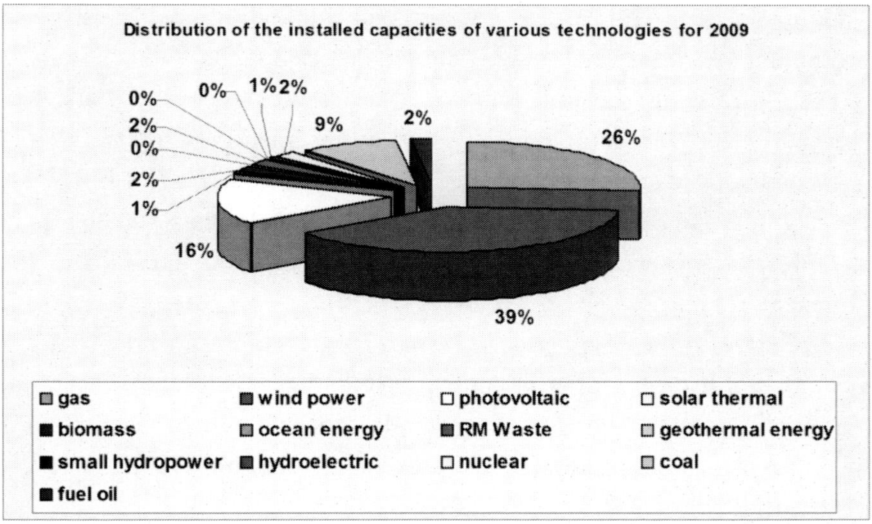


Source: EWEA.

Figure 2. Distribution of the installed capacities of various technologies for 2009.

Figure 2 shows that nearly 40% of the new energy installed in the EU comes from wind power; together with the photovoltaic energy, accounts for 55% of the new installed capacity.

The installation of steam power plants using gas exceeds the fourth part of the new generation, while nuclear and coal showed low rates of 2% and 9%, respectively.

The following extracted data from the Observ'ER [10] barometer on the status of renewable energy in Europe show the contribution of the different sources of renewable energy to the general energy pool in the EU.

The installed capacity of wind energy increased to 75,124.9 MW in 2009 with an electricity production of 131 TWh, representing an increase of 9.5% compared with the previous year.

Spain and Germany were the leading countries, with 24% and 19%, respectively, in energy installed during 2009; these two countries also led in accumulated installation and production.

The photovoltaic capacity installed in 2009 was 5,668.7 MWp on-grid and 16.4 MWp off-grid, and the total is 15,926.6 MWp on-grid and 144.6 MWp off-grid.

Germany led with 9,785.3 MWp on-grid and 45 MWp off-grid, representing 1% of the total power output consumed in 2009 in Germany.

The surface for installations of thermal solar energy collectors increased in 2009 by 4,166,056 $m^2$, with an installed capacity of 2,916.2 MWth and total accumulations of 32,551,532 $m^2$ and 22,786.1 MWTh. Germany led the EU countries with approximately 25% of the total installed capacity, according to the data extracted from the Observ'ER barometer.

In relation to the small hydroelectric energy produced (< 10 MW), it was increased by 259.2 MW in net capacity in 2009, with a production of 42.2 TWh. The total capacity used by the EU for this type of installation is 12,742.7 MW; Italy produced one-fourth of this amount, 10,382.4 GWh, with an installed capacity of 2,588 MW.

The geothermal energy exhibited an installed capacity of 897 MWe, with a net capacity of 744.9 MWe and a production of 5,596.6 GWh in 2009, where almost all of the total installations and production occurred in Italy, with an installed capacity of 843 MWe, a net capacity of 695.1 MWe and a production of 5,341.8 GWh in 2009.

Biogas was used as the primary energy source for the production of 8.3 Mtep in 2009; in Germany, the value was 4,213.4 ktep, of which 25,167.4 GWh were used for electric production in the EU and 12,562 GWh in Germany.

The EU plans for 2020 have an estimated electric production of 63.3 TWh from biogas, according to the data extracted from the Observ'ER barometer.

The consumption of biofuels for transportation in the EU countries in 2009 was 12,097,001 tep, registering a growth of 18.4% between 2008 and 2009, where Germany with 2,894,407 tep and France with 2,511,490 tep were the main consumers.

The production of primary energy coming from waste incineration was 7.7 million tep in 2009, with an increase of 2.9% from 2008.

The production of electricity supplied from waste incineration in 2009 increased to 15.4 TWh, where Germany led the countries with almost 25%, corresponding to 4,166 TWh.

The capacity of the 800 plants installed for solid biomass in Europe is 7.1 GW, with a primary energy production of 72.5 Mtep in 2009, representing an increase of 3.2% from the previous year and a total electric production of 62.2 TWh. The plans of action for the member states for 2020 should result in a production of 152.2 TWh, according to the data extracted from the Observ'ER barometer.

Regarding the installations of concentrated solar energy, Spain led the member countries with an installed energy of 532 MW in 2010, and future projects should result in a further increase of more than 2,000 MW.

The energy produced by the end of 2010 was 587.4 MW in the European Union, almost twice the value at the end of 2009, which was 282.4 MW.

The energy detailed in the NREP (National Renewable Energy Action Plan) for 2020 is 5,079 MW for Spain, 600 MW for Italy, 540 MW for France, 500 MW for Portugal, 250 MW for Greece, and 75 MW for Cyprus.

The energy capacity in the EU from ocean waves was 245.9 MW in 2009. The expected energy of the member states for 2020 is a total of 2.1 GW, with the United Kingdom contributing 1.3 GW, according to the data extracted from the Observ'ER barometer.

The amount of renewable energy included in the gross consumption of energy for 2009 in the Union is 9.4%.

The portion of renewable energy in the gross consumption of electricity for 2009 in the European Union is 18.2%.

The renewable energy portions in the consumption of primary energy for 2009 within the European Union are as follows: 66.6% biomass, 19.7% hydraulic, 7.2% wind, 4.8% geothermal and 1.7% solar.

The amounts of renewable energy in electric generation for 2009 within the European Union are as follows: 55.8% hydraulic, 22.4% wind, 18.3% biomass, 2.5% solar and 1% geothermal.

The goal in the gross consumption of energy by 2020 is 20% of the share to be renewable energy, according to the data extracted from the Observ'ER barometer.

## RENEWABLE ENERGY IN SPAIN

Currently, Spain is the leading country in Europe and the second-leading country in the world in solar thermoelectric installed power, the second-leading country in Europe and the fourth-leading country in the world in wind power, the second-leading country in photovoltaic energy in Europe and the rest of the world, and the third-leading country in Europe in small hydropower, according to project "Renovable Made in Spain" [11].

In order to be a leader in renewable energy, Spain has carried out great investments in I+D+i (investigation, development and innovation), approximately six times more than the average of the Spanish companies, and they particularly stand out for wind power, photovoltaic and solar thermoelectric technology.

More than 1,000 companies in this sector provide 89,000 direct jobs and 99,000 indirect jobs, according to the data from the *Instituto Sindical de Trabajo, Ambiente y Salud* (ISTAS) [12]. More than half million jobs could be created if Spain reaches its goal of 20% renewable energy consumption by 2020, as planned by the directives of renewable energy. This would allow Spain to reduce its energy dependence and reduce its $CO_2$ emissions to comply with the Kyoto Protocol. These facts provide enough arguments to prove that the renewable energy must play a key role in the new cost-reducing model to which Spain is aiming [13].

It is remarkable from a technical point of view that the evolution that has been created involves the management of significant renewable installed energy, particularly wind energy [14], with some 19,000 MW at the beginning of 2010. In 2004, it was considered of high risk to integrate more than 14% of wind energy, and some warned that exceeding this value would significantly increase the possibility of a substantial blackout. In November 2009, *wind energy generated more than 50% of the total electricity for five hours*, with a peak of 53%; the month's average was 22.7% [15].

This is a very relevant piece of information, because no other country as large as Spain with a similar consumption of electricity and some weak electric interconnections could have successfully exceeded more than 50%

contribution of renewable energy for such a long time. To make this possible, the REE, working with experts in the sector, started the Control Center of Renewable Energy (CECRE) in 2007, which allows for the integration of the maximum amount of renewable energy without compromising the electric system.

On Spain's electric network REE website, the wind power production is shown in real time.

Https://demanda.ree.es/eolica.html [16].

Of the renewable energy technologies, wind power stands out again, with a growth of 18.5% in generation, which has increased its participation in demand coverage to 16%.

This increase in generation of renewable energy combined with the decrease in the production of steam power plants has contributed in reducing the $CO_2$ emissions of the electric sector, which was approximately 58.7 million tons in 2010, approximately 20% less than the value for 2009, according to the 2010 REE draft report.

Figure 3 shows a graph of time-dependent growth in energy sales by producers under special schemes for electricity production [17].

As shown in Figure 3, a significant increase occurred almost every year in energy from special schemes.

Figure 4 shows the detailed installation capacity for each autonomous region according to the different kinds of energy.

The regions of Castile-La Mancha, Castile and León, and Andalusia have more than 5,000 MW each, representing more than 50% of the total installed capacity.

Regarding the energy sources, wind energy is the most remarkable for the majority of the communities; Castile and León, Castile-La Mancha and Galicia together have 50% of the wind energy installations.

The portion of energy sales by the different special scheme technologies for the demand for 2010 is shown in Figure 5.

Wind energy contributes 15% to the total of 24% renewable energy.

Cogeneration contributes 9% to the total of 33% special scheme.

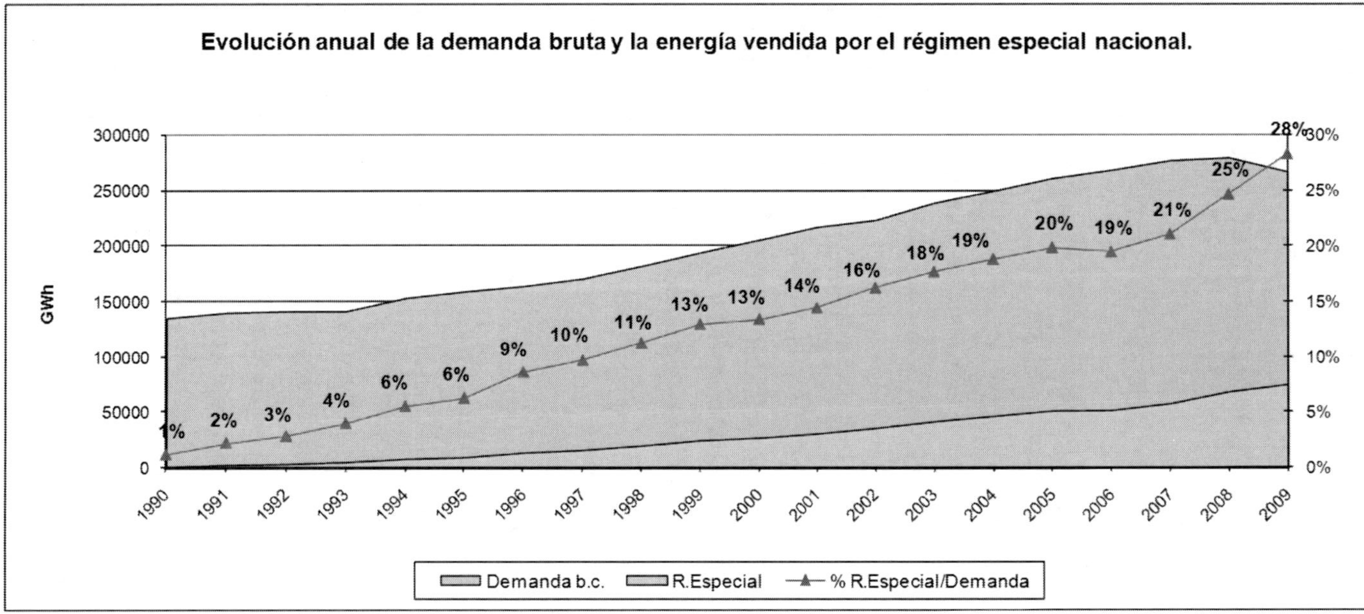


Source: CNE.

Figure 3. Annual progress of the gross demand and energy sold through special scheme in Spain.

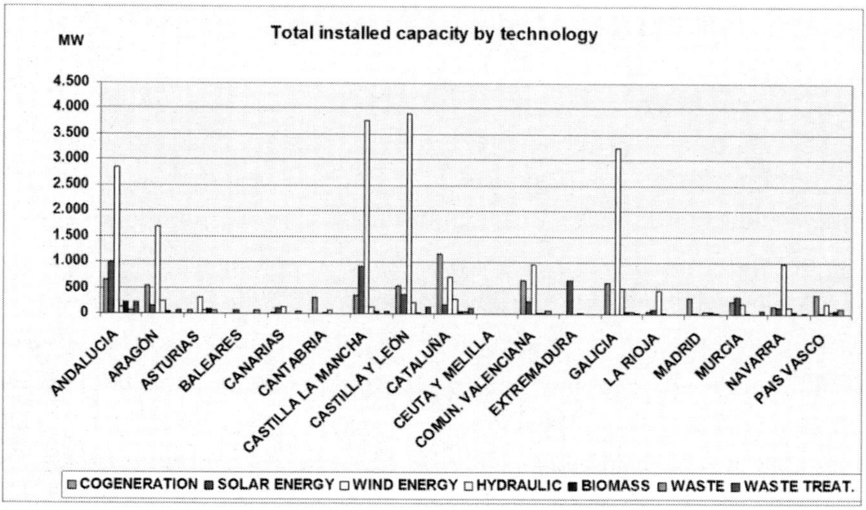


Source: CNE.

Figure 4. Total installed capacity by special scheme technology for each autonomous region.

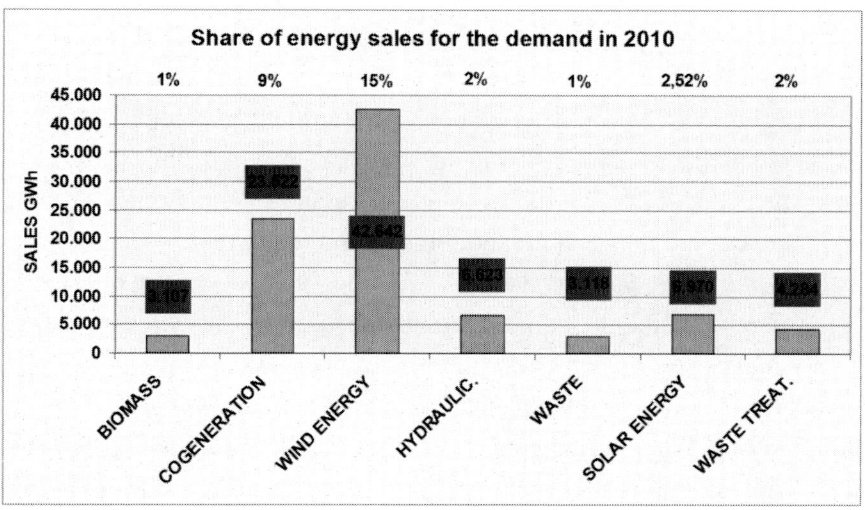


Source: CNE.

Figure 5. Share of energy sales for the demand in 2010.

## THE FOUR AIMS FOR ENERGY IN SPAIN

### Wind Energy

According to Waste magazine [18], the use of wind as an energy source is unsuitable in that it requires a series of conditions for the location, which significantly restricts the diffusion of this system, although technological innovations have produced a competitive model at international level for the trade generation of electricity. Almeria's coast and the Strait Of Gibraltar are the Spanish zones with the greatest potential; in those zones, there are wind speeds that reach 8.5 m/s, much higher than those considered suitable for good economic performance.

According to data from the project "Renovable Made in Spain", the Spanish wind industry has created a competitive business with a high international presence, led by the developers of wind farms, such as Iberdrola Renovables and Acciona Energía, among others, including the manufacturers of wind turbines. A number of industries are related to the supply chain of parts (in Spain, there are more than 75 industrial centers related to the wind sector, 18 of which are assembly lines for wind turbines manufacturing) and to the activities of operation and maintenance, which have been created and developed along with the growth of this sector. It is obvious that this sector has become an important job creator, with more than 40,000 current jobs that are directly and indirectly related to it.

There is one type of wind energy that has not yet been exploited in Spain: offshore wind. In some European countries, this energy source currently makes an important contribution to renewable energy.

In 2009, a study of the Spanish coast environmental strategy for the installation of offshore wind farms [19] was conducted, detailing suitable zones for development after examining relevant studies on corresponding environmental impacts.

Figure 6 displays the suitable zones, zones with contributing factors and zones excluded from the incorporation of offshore wind farms.

Half of the wind energy will be generated offshore by 2030 according to the European Wind Energy Association (EWEA).

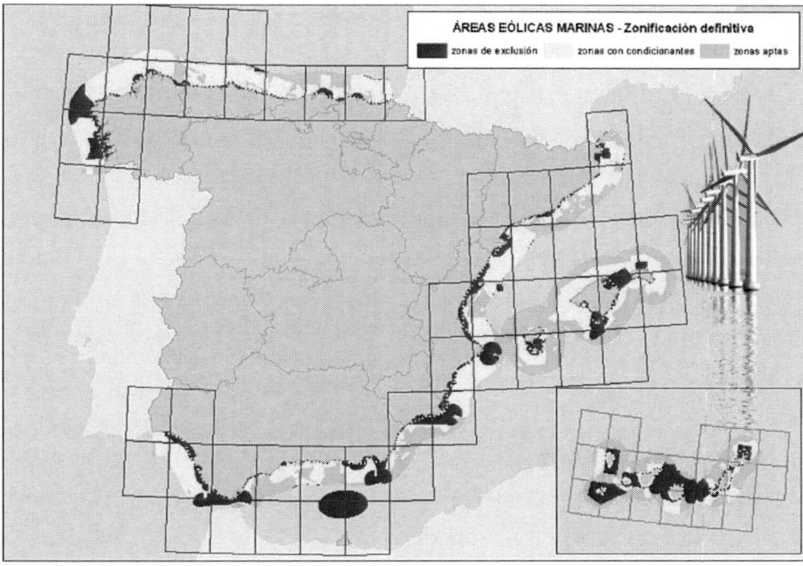


Source: AEE.

Figure 6. Zones for offshore wind farms.

The northern region of Europe has strongly benefited from offshore wind energy.

Nine countries contribute to the almost 3000 MW of wind power installed at sea.

Meanwhile, the first steps are being taken to harness offshore wind energy in Spain; initiatives include the launch of the Offshore Wind Atlas on the Canary Islands, the Institute of Hydraulic and Setting's Center of Ubiarco in Cantabria, and the Energy Research Institute Zefir project in Catalonia.

The government must change this trend with clear policies to encourage offshore wind energy. The Spanish coast offers great energy potential which should be taken advantage of [20].

Catalonia is committed to the technological development of offshore wind energy in deep waters.

The Zefir project [21] has made an investment of 143 million euros to build of an offshore wind farm with four wind turbines, anchored to the marine bottom, harnessing 10 MW to 20 MW of power at a 3 km distance from the coast between Ebro's delta and Salou. It is expected that this wind farm will begin operation in 2012 and be self-funded by means of energy sale. The second phase of this project includes the installation of 8 marine wind turbines located 20 km from the coast.

This project may be the starting point of the development of a new offshore wind energy sector in Catalonia, which will allow the creation of qualified jobs and new investments.

## Solar Thermal Electricity (STE)

In Spain are operating the four main technologies for solar thermal power [22]. The most commercially developed options are the parabolic trough; there are currently 17 working plants, and 14 are of this type, 5 of which are in Badajoz, with a total power of 700 MW. There are 2 plants with a central tower, both installed in Seville, with 31 MW of combined power output. Finally, there is 1 plant consisting of linear Fresnel collectors in Murcia, producing 1.40 MW.

Therefore, there is currently an installed capacity of 745.4 MW.

The initiatives of the Stirling dish, with advanced construction, which is normally used for the Stirling engine, will result in 1 MW when installed in Cuenca.

Also under construction is a parabolic trough, that will contributing to the majority, with 17 plants; Ciudad Real and Cordoba will have 4 each, contributing 850 MW in total. One central tower plant of 17 MW is under construction in Seville, along with a linear Fresnel collector with 30 MW of power in Murcia. Thus, in a short time, 898 MW of additional energy will be available, leading to a total of 1,643.4 MW.

Apart from the pre-assigned plants, there are another 23 in various stages of development: 16 parabolic troughs, 7 of which are in Badajoz, with a total power of 750 MW, and 7 dish Stirling plants in Ciudad Real with 70 MW of total power.

All of these plants are in advanced stages of creation, and those stated in the register of pre-allocation, which are in various stages of promotion or construction, would result in a total of 60 plants with 2,463.4 MW of combined power output, according to Protermosolar's data.

The situation of the thermoelectric plants in the different phases (operational, advanced construction and pre-assigned) is shown in Figure 7.

The South of Spain is leading in thermal energy worldwide.

The majority of power originates from the southern half of the peninsula, where the solar thermal energy production is growing. Andalusia has currently the most operational plants, eight, in addition to 11 plants in the advanced stages of construction and 3 pre-assigned. Meanwhile, Extremadura and Castile-La Mancha share most of the remaining operating and pre-assigned plants, and other regions, including Murcia, Valencia and Catalonia, have little involvement.

Source: Protermosolar.

Figure 7. Location of solar thermal power plants.

The solar thermoelectric plants have the same network interface as the steam power plants, with a set of steam turbines and alternators providing the network stability that other renewable technologies cannot provide.

Because of their thermal storage system and their capability to be hybridized with other sources such as biomass or natural gas, the solar thermoelectric plants may follow the curve of demand even in the absence of solar radiation.

Thermal solar power plants generate more jobs in the manufacturing stage than in the energy production stage. Throughout its various phases (design, manufacturing of components and installation), each plant of 50 MW employs an average of direct 5,000 job-equivalent-years and a similar amount for indirect jobs. Additionally, the 50-MW plants that are being built in Spain employ around 500 people per plant on-site during their two years of construction. Once in operation, the plants require a final staff of 50 people per plant [23].

## Photovoltaic Energy (PV)

The general introduction of this technology in Spain started in 2008, when 2,708 MW were installed, with an increase of 500% compared to the value for 2007. The constantly decreasing costs of photovoltaic panels, which reflect daily effects in the economies of scale and improvements in the technological learning curves, have added to their financial appeal, to diversification in the activities of many companies, and to normative changes to push this technology into levels that were hard to imagine only a few years ago. This is not only limited to Spain; Europe doubled its power in 2008, with the addition of 4.6 GW, with the photovoltaic accumulation exceeding 9.5 GW, according to data from the project "Renovables Made in Spain".

In Figure 8, the increase in installed capacity and the number of installations of PV energy are shown.

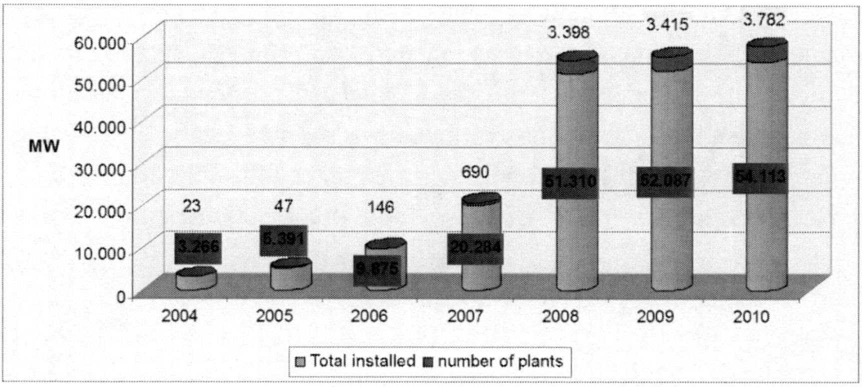


Source: CNE.

Figure 8. Total installed capacity and total number of installations of PV energy.

While there was considerable growth in 2008, the trend has stagnated as a result of regulatory changes and difficulty in financing the projects.

Nevertheless, the level of development attained seems enough to encourage researchers to find new ways of harnessing this solar energy. The costs of photovoltaic technology are decreasing fast, which will allow it to play an important role in the framework of the Spanish electric sector, along with the development of a system of distributed generation or continuity [24].

Spain has always been in the vanguard of photovoltaic investigation. The Institute of Solar Energy (IES) of the Polytechnic University of Madrid, which is known worldwide, has already completed 30 years of research in this field. More recently, the Instituto de Sistemas Fotovoltaicos de Concentración (ISFOC) of Puertollano (Ciudad Real) [25] has become the greatest producer in pilot plants for this technology, using high-efficiency cells, with performances that exceed 40%. These investigators are requested to participate in numerous projects worldwide.

The Spanish photovoltaic sector includes companies that exploit all the chain value of this industry, from manufacturing of cells to promotion of farms, inside and outside our borders. Their strength gives an idea that several companies dedicated exclusively to this activity have entered into the stock market.

A new system of electric generation was developed with a high concentration module and a tracking and controlling system with high precision. The module uses III-V cells with a triple junction, as used in space applications, with an efficiency of more than 39%; when supplemented with advanced optics designs, this allows for better results in comparison to other systems in zones with high temperatures [26].

## Biomass Energy

In Spain, biomass is obtained from a wide variety of sources (forest splinters, olive pits, nutshells of dried fruit, etc.), which ensures a continuous and abundant supply in any part of the country. Due to its use in both thermal applications (air conditioning, hot water or industrial processes) and power stations, it has undeniable advantages in addition of being able to replace fossil energy.

The use of forest biomass remains, especially firewood for thermal applications, a characteristic element in Spain, the European leader in so called fuel wood forests, timber production forests, whose principal use is energy production.

Among the thermal applications currently being produced in Spain, there has been a recent important growth of a new economic activity in the field of fuel densification (mainly pellets). Numerous industries from several economic sectors have invested in this field, and the production capacity is estimated to have been multiplied by 10 in the recent years, from 60,000 t/year in 2004 to a production capacity of 600,000 t/year in 2009.

The Energy Service Company (ESCO) [27] is the key in continuing to advance biomass thermal applications and is ensuring the good design and maintenance of the installations, which would guarantee the supply of biomass to the user in a competitive way with other options from a cost-reduction point of view.

With regard to electricity applications, by-products from the production of paper and pasta are used as bio-fuels in Spain, along with the transformation of wood and the production of oil like appropriate biomass from energetic cultivations, agricultural waste (straw, pruning of olive tree) or from the mountains.

On the other hand, various Spanish companies are making significant efforts in the development of gasification technology on a small scale, which have reached their first trade achievements.

Figure 9 shows the growth in yearly sales of biomass.

It should be noted that the proportional contribution has been consistent during the period of 2000-2010, at approximately 1% of the total energy sold.

The Renewable Energy Technology Center (CTAER) [28] is working on one of its main technology goals for the future, the hybridization of solar energy with biomass energy. The CTAER aims to establish the basic characteristics and the engineering of a solar-biomass hybrid thermoelectric power plant of experimental nature, such that in the second stage, it is only necessary to build it, measure its output and evaluate it, so that companies can make it into a trade product.

The CTAER is especially interested in solar-biomass hybrid systems, mostly due to their application in the generation of electricity, especially for distributed generation and cogeneration for the use of heat and cold.

## FINANCIAL SUPPORT SYSTEMS IN SPAIN

In addition to priority access to the networks defined in the European Directive, which have been launched in recent years in Europe and the rest of the world, different support systems for renewable energy have been created. Such systems can be classified, among other ways, according to these two main criteria [29]:

(a) If the regulatory intervention acts on the price or profit received or regarding the amount of power to be installed or energy generated.
(b) If such intervention acts on the initial investment stage or on a later stage of electricity generation.

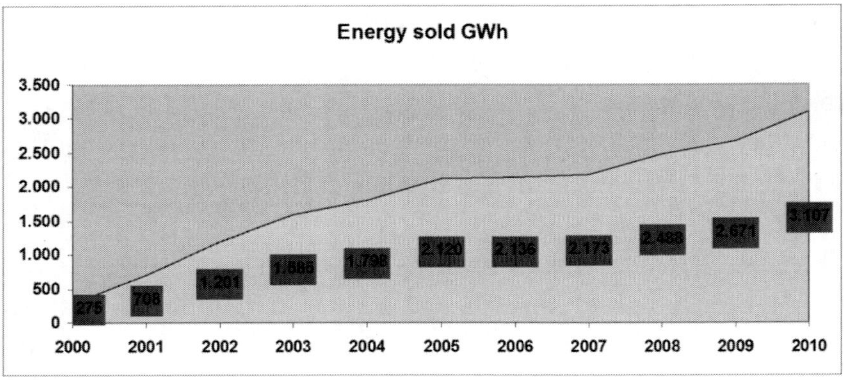


Source: CNE.

Figure 9. Annual energy produced from biomass.

Of all the support systems, the one currently with the highest prevalence in the European Union is that of rates or regulated bonuses, a system with different variants generally gathered under the name of Feed-in Tariffs (FIT), according to energy and the society.

Spain has chosen to support the sale price of renewable electricity, either by charging a flat rate (different for each technology) or by charging a bonus which is added to the market price, for facilities that choose this form of sale.

**Table 1. Remuneration and total bonuses received by producers for technology in 2010**

| Year | Technology | Total Compensation (Thousands €) | Average prices Total Compensation (cent€/kWh) | Premium equivalent (Thousands €) |
|---|---|---|---|---|
| 2010 | COGENERATION | 2.133.665 | 9,019 | 1.334.223 |
| | SOLAR PV | 2.879.366 | 45,076 | 2.649.343 |
| | SOLAR TE | 209.749 | 30,332 | 184.848 |
| | WIND ENERGY | 3.154.714 | 7,324 | 1.962.631 |
| | HYDRAULIC | 503.254 | 7,492 | 295.949 |
| | BIOMASS | 352.499 | 11,217 | 243.983 |
| | WASTE | 197.904 | 6,352 | 92.806 |
| | WASTE TREAT | 495.453 | 11,564 | 350,513 |
| Total 2010 | | 9.926.605 | 10,9 | 7.114.296 |

Source: CNE.

**Table 2. Retribution for put option for the producers of the special scheme in the year 2010**

| Energy Sale Options | Technology | Distributor Average Billing Price (cent€/kWh) | Average Market Settlement Price (cent€/kWh) | Total Compensation (Thousands €) | Average Price Total Compensation Total (cent€/kWh) |
|---|---|---|---|---|---|
| Participation Market Offers | COGENERATION | 3,229 | 3,383 | 238.306 | 6,612 |
| | SOLAR TE | 26,743 | 3,601 | 208.976 | 30,344 |
| | WIND ENERGY | 4,537 | 2,776 | 2.854.133 | 7,313 |
| | HYDRAULIC | 4,133 | 3,095 | 246.316 | 7,228 |
| | BIOMASS | 7,073 | 3,453 | 104.995 | 10,526 |
| | WASTE | 2,9 | 3,383 | 157.212 | 6,283 |
| Sales at tariffs through representative | COGENERATION | 6,071 | 3,383 | 1.893.809 | 9.454 |
| | SOLAR PV | 41,475 | 3,601 | 2.879.366 | 45,076 |
| | SOLAR TE | 23,769 | 3,601 | 773 | 27,37 |
| | WIND ENERGY | 4,661 | 2,776 | 231.925 | 7,437 |
| | HYDRAULIC | 4,951 | 3,095 | 199.535 | 8,046 |
| | BIOMASS | 8,492 | 3,453 | 219.320 | 11,945 |
| | WASTE | 0 | 3,383 | 0 | 0 |
| | WASTE TREAT. | 8,181 | 3,383 | 495.453 | 11,564 |
| Participation in other markets | COGENERATION | 3,824 | 3,383 | 1.550 | 7,207 |
| | WIND ENERGY | 4,632 | 2,776 | 68.657 | 7,408 |
| | HYDRAULIC | 3,825 | 3,095 | 57.403 | 6,92 |
| | BIOMASS | 5,663 | 3,453 | 28.185 | 9,116 |
| | WASTE | 3,25 | 3,383 | 40.692 | 6,633 |

Source: CNE.

The Spanish regulatory system allows renewable generators the option every twelve months to follow one or another variant, that is, the regulated rate or total bonus or incentive compensation options on top of the price of the electric market, if they choose to participate in it, according to energy and society.

The current rates and bonuses are established in the following provisions:

(a) Cogeneration: ITC/1723/2009 Order of June 26
(b) Renewable and Waste: ITC/3801/2008 Order of December 26

The remuneration and bonuses received by the producers of the different technologies are arranged according to group and royal decree.

In Table 1 which is shown below, it can be seen the retribution and bonuses earned by producers of the different technologies of the special scheme during 2010.

In Table 2 it can be seen the retribution for put options which have the producers of the special scheme in 2010.

The regional grants are directed primarily to individuals and small businesses to promote energy saving and renewable energy consumption, such as solar panels for water heating, biomass boilers for heating and other equipment that generates savings. In the case of installations connected to the distribution network, each time there are less autonomous communities which grant aid as they tend to disappear [30].

## CONCLUSION

The most important energy need for the future is efficiency, with responsible consumption based on the power of social awareness aiming toward a primary good.

This requires the diversification of technologies, better use of available resources, and the possible implementation of hybrid systems with renewable energy [31].

To attain 100% electricity demand satisfied by renewable energy, Spain has an important role at the European level.

Wind energy is expected to be the main engine of self-sufficiency, supported by the development and subsequent maturity of solar thermal plants and enhanced by a variety of other renewable sources such as biomass plants.

The diversity of power generation throughout the year allows the demand to be satisfied in different percentages of participation from each of the energy sources.

The following data extracted from the 2010 REE draft report prove this diversity.

Wind energy is a dominant resource in autumn and winter months; in the year 2010, its share was 20% of the total demand in February and 10% in September, with a total of 2,250 hours as yearly average of equivalent use.

Solar energy is the dominant resource in spring and summer; in 2010, it had a share of 4% of the total demand for the months of July and August and a participation of 1% in December and January, with a total of 1,755 hours as yearly average of equivalent use.

The remaining renewable energy sources exhibit equivalent participation during the year, which is why such diversification should be feasible.

Spain has a total installed capacity of 97,447 MW, with a yearly demand of 260,000 GWh, a maximum demand of half hourly power of 44,122 MW and a maximum yearly demand of daily energy of 895 GWh, according to data from the 2010 REE preliminary report.

These facts are reflected in Figure 10, which shows a possible mix of energy generation completely provided by renewable energy, which can be set as an objective for 2030.

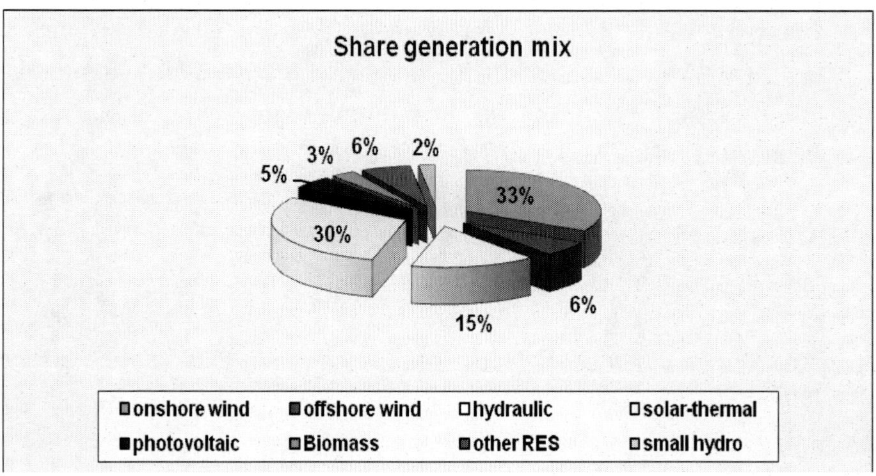


Figure 10. Generation mix of renewable energy.

As can be seen, onshore wind growth continues to be 1,000 MW yearly, followed by solar thermal electric power because of its storage power and the emergence of offshore wind technology.

In addition to improving process efficiency, it is important to make use of the excess wind energy in peak hours for reversible hydraulic pumping, to integrate thermoelectric systems such as solar-biomass or wind-photovoltaic systems, and to initiate energy distribution on intelligent networks.

## REFERENCES

[1] Jose, Ramón San Cristobal. A multi criteria data envelopment analysis model to evaluate the efficiency of the Renewable Energy Technologies, *Renewable Energy*, 36, (2011), 2742-2746.

[2] *Ministerio de Industria, Turismo y Comercio* [*Ministry of industry, Tourism and Commerce*] http://www.mityc.es/ENERGIA/ELECTRICIDAD/REGIMENESPECIAL/Paginas/Index.aspx [Consulted 04/27/2011].

[3] Dominik, Heide Lueder von Brenem; Martin, Greiner; Clemens, Hoffman; Markus, Speckman; Stefan, Bofinger. Seasonal optimal mix of wind and solar power in a future, highly renewable Europe. *Renewable Energy*, 35, (2010), 2483-2489.

[4] *Centro de investigaciones Energéticas, mediambientales y tecnicas* (CIEMAT) [*Energy Research Center, environmental and technical.*] http://www.energiasrenovables.ciemat.es/ [Consulted 04/27/2011].

[5] ECO logic Institute for International and European Environmental Policy. *Job Creation Potencial of Clen Tecnologies*. [*Job Creation Potential of Clean Technologies*]. http:// ecologic.eu/download/projekte/1800-1849/1848/ 1848_study. pdf.

[6] The European Wind Energy Association (EWEA). *Report Wind and Work 2009*. http://www.ewea.org/ fileadmin/ewea_documents/documents/publications/Wind_at_work_FINAL.pdf [Consulted 04/19/2011].

[7] The European Wind Energy Association (EWEA). http://www.ewea.org/ [Consulted 04/27/2011].

[8] Philip, Jennings. New directions in renewables energy education. *Renewable Energy*, 34, (2009), 435-439.

[9] Asociación Empresarial Eólica (AEE). *Spanish Wind Power Association. Annual Report 2010.* http://www.aeeolica.org/uploads/documents/aee_publica/AEE_Anuario_2010.pdf?phpMyAdmin=nkH26XnGN7Ws3Rn1f-QjR33eVc7 [Consulted 04/19/2011].

[10] Eurobserv'ER - the state of renewable energies in europe - 2010 edition. *Report Barometer prepared by Observer'ER in the frame of the Eurobserv'ER.* http://www.eurobserv-er.org/pdf/barobilan10.pdf [Consulted 04/19/2011].

[11] Project "Renovables Made in Spain", Instituto para la diversificacion y ahorro de energía (IDAE). *Ministerio de Industria, Turismo y Comercio.* [Institute for Energy Diversification and Efficiency. *Ministry of Industry, Tourist and Trade*]. http://www.renovablesmadeinspain.es/ficheros/documentos/pdf/RENOVABLESmadeinspainEs_En.PDF [Consulted 04/01/2011] http://www.renovablesmadeinspain.es/ [Consulted 04/01/2011].

[12] *Instituto Sindical de Trabajo Ambiente y Salud (ISTAS).* [*Union Institute by work environment and Health*]. http://www.istas.net/web/index.asp [Consulted 05/02/2011].

[13] Antonio, Gómez; Javier, Zubizarreta; Norberto, Fueyo. Spanish energy roadmap to 2020: Socioeconomic implications of renewable targets. *Energy*, 36, (2011), 1973-1985

[14] Ghassn, Zubi; Jose, L. Bernal; Agustín, Ana B. Fandos Marín. Wind energy (30%) in the Spanish power mix-technically feasible and economically reasonable. *Energy Policy*, 37, (2009), 3221-3226.

[15] Red Eléctrica Española (REE). *Spanish Electric Grid. Preliminary report 2010.* http://www.ree.es/sistema_electrico/pdf/ infosis/Avance_REE_2010.pdf [Consulted 04/19/2011].

[16] (REE) *Seguimiento tiempo real producción eólica.* [*Monitoring real time wind power production*]. https://demanda.ree.es/eolica.html [Consulted 04/19/2010] Update 01/25/2011.

[17] Comisión Nacional de Energía, (CNE). [National Energy Commission (CNE)]. *Documents*. http://www.cne.es/cne/contenido.jsp?id_nodo=23&&&keyword=&auditoria=F [Consulted 04/01/2011].

[18] *Waste magazine*. Specialized Publications http://waste.ideal.es/primeraenergias.htm [Consulted 04/27/2011].

[19] Asociación Empresarial Eólica (AEE). [Spanish Wind Power Association]. http://www.aeeolica.es/ [Consulted 04/02/2011].

[20] Snyder, B; Kaiser, MJ. Ecological and economic cost-benefit analysis of offshore wind energy. *Renewable Energy*, 34, (2009), 1567-1578.

[21] Institut de Recerca de Energia de Catalunya (IREC). [Institute Energy Research of Catalonia]. *Project Zefir*. http://www.irec.cat/index.php/es/los-proyectos/20-projectes-energia-eolica-marina [Consulted 05/ 02/2011].

[22] Asociación Española de la Industria Solar Termoeléctrica (Protermosolar). [Spanish Solar thermoelectric Industry Association]. http://www.protermosolar.com/ [Consulted 04/19/2011].

[23] European Solar Thermal Electricity Association (ESTELA). http://www.estelasolar.eu/ [Consulted 04/19/2011].

[24] Directive 2009/28/EC; Analysis (EPIA) European Photovoltaic Industry Association; (ASIF) Spanish Photovoltaic Industry Association; SET Plan; Analysis A.T. Kearney. *Energía Fotovoltaica: Una fuente de energía clave para Europa en 2020*. [*Photovoltaic Energy: A key energy source by Europe in 2020*]. Madrid, 2009 November 19. http://www.appa.es/descargas/ES_Exec%20Briefing_ATK.pdf [Consulted 05/02/2011].

[25] Instituto de Sistemas Fotovoltaicos de Concentración. [*Institute of Concentration Photovoltaic System*]. http://www.isfoc.es/ [Consulted 05/02/2011]

[26] *Abengoa solar Company, news 2011* http://www.abengoasolar.com/corp/web/es/acerca_de/general/noticias/historico/2011/solar_20110415.html [Consulted 05/02/2011] http://www.abengoasolar.com/corp/web/en/index.html [Consulted 05/02/2011].

[27] ESCO Sociedades de servicios energéticos. [*Energy Service Company*]. http://emsega.galeon.com/ [Consulted 05/02/2011].

[28] *Centro tecnológico Avanzado de Energías Renovables (CTAER).* [*Advanced Technologic Center of Renewable Energy*]. http://www.ctaer.com/ [Consulted 04/19/2011].

[29] Energy and society. *Report.* http://www.energiaysociedad.es/documentos/9_3_mecanismos_de_apoyo_a_las_energias_renovables.doc [Consulted 09/11/2011].

[30] Ayudas, energías. [*Energies aids*]. http://www.ayudasenergia.com/subvenciones-energia/ [Consulted 09/11/2011].

[31] Giuseppe, Marco Tina; Salvina, Gagliano. Probabilistic modeling of hybrid solar/wind power system with solar tracking system. *Renewable Energy*, 36, (2011), 1719-1727.

In: Distributed Wind Energy Generation
Editors: A. Colmenar-Santos et al.
ISBN: 978-1-53614-207-5


*Chapter 2*

# DISTRIBUTED GENERATION PERSPECTIVES

***Salvador Ruiz-Romero*[*], *Rosario Gil-Ortego* and *Antonio Molina-Bonilla***
Departamento de Ingeniería Eléctrica, Electrónica, Control, Telemática y Química Aplicada a la Ingeniería, Universidad Nacional de Educación a Distancia (UNED), Madrid, Spain

## ABSTRACT

European efforts to fight climate change, to improve the security of energy supply, and to drive innovation and competitiveness in the next decade will make distributed generation (DG) develop and grow considerably. DG growth is an integral component of a new vision of an effective and highly responsive "European Smart Grid" in which the actions of all stakeholders are fully integrated. According to IMPROGRES (Improvement of the Social Optimal Outcome of Market Integration of Distributed Generation (DG) and renewable energy resources (RES) in European Electricity Markets), DG/RES electricity generation in the EU-27 will increase from 490 TWh/year in 2005 to approximately 1280 TWh/year in 2030. The relative proportion to the total electricity generated will also grow from 15% to approximately 26%

[*] Corresponding Author Email: s.ruiz.romero@gmail.com.

during the same time period. However, there are a number of existing barriers that prevent the spread of DG in the European market. These barriers range from simple commercial issues, such as the fact that the energy produced is currently not cost-effective in comparison to the electricity generated on a large scale, to complex regulatory reforms. In Spain, these barriers are being overcome, grid parity is almost certain, and regulatory reform is on track to become a reality. In this article, several comparisons are made regarding different solutions already adopted in European countries for low power distributed generation, thereby providing possible solutions to the budding energy situation in Spain, as well as an overview of trends and growth forecasts for distributed generation and smart grid (SG) projects.

**Keywords:** distributed generation, renewable energy, self-supply, virtual power plant, smart grid, active grid management

## ABBREVIATIONS

| | |
|---|---|
| WBA | Wind Business Association, (Asociación Empresarial Eólica) |
| OBS | Official Bulletin of the State, (Boletín Oficial del Estado) |
| ECE | Estimated Cost of Energy |
| VPP | Virtual Power Plant |
| DER | Distributed Energy Resources |
| EPIA | European Photovoltaic Industry Association |
| EV | Electric Vehicles |
| FIT | Feed in Tariff |
| AGM | Active grid management |
| CIGRE | International Council on Large Electric Systems |
| DG | Distributed Generation |
| IEA | International Energy Agency |
| IDEC | Institute for Diversity and Energy Conservation, (Instituto para la Diversidad y Ahorro de la Energía) |
| PV | Photovoltaic |
| SME | Small and Medium Enterprises, (Pequeñas Y Medianas Empresas) |

| | |
|---|---|
| FMM | Fixed Marketing Margin |
| MPRS | Micro Combined Heat and Power, Micro CHP Micro-Production Registration System, |
| NAPRE | National Action Plan for Renewable Energy, (Plan de Acción Nacional de Energías Renovables) |
| RES | Renewable Energy Sources |
| WEEVR | REVE, Wind Energy and Electric Vehicle Regulation |
| SDGG | Smart Distributed Generation Grid |
| SG | Smart Grid |
| LRT | Last Resort Tariff |
| PLRT | Power Term for the LRT |
| ELRT | Energy Term for the LRT |
| PAT | TPA, Power Term for the Access Tariff |
| EAT | Energy Term for Access Tariff |
| ICT | Information and Communication Technology |
| V2G | Vehicle to Grid |
| WWA | World Wind Energy Association |

## INTRODUCTION

In recent decades, significant progress made in the field of small-scale power generation and electric storage has involved changes in the manner by which the generation and distribution of electric power have been carried out, reinvigorating the idea of distributed generation (DG).

Meeting the future energy demand while substantially reducing the associated greenhouse gas emissions requires an energy revolution.

The energy sector now represents approximately 65% of $CO_2$ emissions, which contribute directly to climate change according to data from the International Energy Agency (IEA) [1].

The most important benefits of DG include the improved security of supply, reductions in greenhouse gas emissions, reduced capacity, reduced transport energy losses, and increased flexibility in investments. However, there are technical, economic, legal, and regulatory barriers that still prevent the optimal development of DG in electrical systems according to the magazine CIER 57 [2].

With the onset of DG, there have been alterations to the traditional hierarchical structures of networks in which energy flows from the conventional and concentrated production centres to the final consumers. The impacts of DG are mainly detected through changes in power flows, taking into account both their magnitude and direction, according to the magazine CIER 57.

One type of renewable energy that can be expanded as modular DG is the use of grid-connected photovoltaic solar power (PV). PV has a distributed nature and is easily installed because it is produced in modular systems, making its consumption by the end-user quite practical.

Moreover, whether independently or by means of a hybrid system constructed with PV, micro-wind energy presents advantages for electricity use in remote locations in addition to its micro-distributed generation capacity. Therefore, micro-wind generation systems can be adapted to renewable resources and to the specific energy needs of a given location.

## Distribution Networks

The classic electricity supply model begins with large power generation units and ends in the distribution networks, decreasing in capacity in each section; all of this electricity distribution is carried out through the use of reliable transport networks, which now must be complemented with new distributed generation [3] technology to meet the changing demands of modern society. The demand for renewable electricity production in combination with the demand for increased energy efficiency is redefining the classic mechanisms of supply according to the ABB Review (2008) [4].

Although power generation was originally distributed, it later became centralised, and at present, most of the current systems for electricity production and transmission typically lose more than 60% of the power through heat dissipation before the useful energy is finally delivered to the end user. A promising way to reduce these losses is for the DG electricity to be closer in proximity to the end user. This approach has led to a huge increase in solution-based demands, such as micro-generation at households and industries that connect households directly with heating and cooling (Micro Combined Heat and Power, [Micro CHP]) [5]. In these instances, the usable energy is increased to 85% according to ABB (2008).

Consequently, distribution networks are changing from traditional passive networks developed for defined peak loads to alternative networks that dynamically adapt to absorb the increasing demands placed on them [6].

Many small generating units can be managed as single sources, which are called virtual power plants (VPP).

The use of energy storage solutions to ease the problems associated with limited capacity will continue to be incorporated into what is now seen as the future, i.e., smart grid technology, which is based on active grid management (AGM) [7] in conjunction with automated systems.

This type of automation system must be intelligent enough to accommodate generation profiles that change with the weather and time of day, as determined by the associated wind and photovoltaic production capacities. The result of this approach is a continuously variable distribution in terms of the direction and flow of energy in contrast to the classic unidirectional, relatively stable transport of the current distribution grid.

Building the next generation of active power networks requires a combination of new technologies, the optimal exploitation of existing infrastructure and changes in the operating practices of energy companies. In the context of complex research and development, in which many converging interests can be advanced, progress can be made only by working as a team.

In this context, therefore, various projects for the advance of European smart grids have been developed in which multiple companies and institutions of the Member States of the European Union are involved:

## The Aura-NMS Project

(Autonomous Regional Active Network Management Systems) [8]. This endeavour is a research and development project sponsored by the Engineering and Physical Sciences Research Council in the UK (EPSRC) that attempts to demonstrate new network operation concepts in the United Kingdom. The 3-year project is aimed to demonstrate the benefits of network management integration and energy storage technologies used in networks. Among the new features are

- Improved efficiency and supply reliability
- Optimal assets management
- Effective management of DG limitations
- Reduction of customer minutes lost (CML)
- The control decentralisation of network control centres (NCC)

## The Micro-Grids Project

This project [9], which is supported by the EU in Spain, was aimed at the implementation of local low-voltage distribution systems based on distributed energy sources, storage devices, and charge control, which can be connected to the main network or can be isolated. Thus, either a neighbourhood, a small town, or an industrial park could be partially or totally self-sufficient in energy, benefiting from improvements in energy efficiency and reductions in energy consumption and environmental impacts.

A micro-grid [10] is a self-sufficient, not-strictly-defined DG interconnection for use in industrial and residential low-voltage power supply in networks that lack continuous connections to larger and more powerful grids. Furthermore, the creation of *ad hoc* micro-grids from isolated pockets within the larger network offers the possibility of circumventing cascading blackouts. It is also able to maintain critical charges in line.

## The ADDRESS Project

(Active Distribution network with full integration of Demand and distributed energy RESourceS) [11]. This project is an integrated project belonging to the 7th Framework Programme for Research in the European Union.

The project's aim is to develop a commercial and technical framework that allows all of the benefits of active networks with distributed resources. The project was started in June 2008 and has a duration of four years with a total budget of 16 million Euros, of which 9 million Euros were provided by the European Commission.

The project develops new architectures for active networks based on the provision of distributed intelligence throughout the grid with emphasis on the end users. This architecture will allow users the opportunity to exercise real-time reaction (with approximately 20 min. intervals) based on price signals.

The competitive advantage of the project is focused on using the flexibility provided by the architecture of ADDRESS for the expansion of new possibilities associated with balancing the charge and generation of power. The results of this project will validate the idea that active distribution networks have the capability to accelerate the substantial implementation of clean distributed generation and to significantly optimise the network charge.

The project contributes more specifically to

- Conveying the energy prices in real-time to all users.
- Improving the security of supply.
- Increasing the efficiency, safety, reliability, and quality of power supply at the European level in the context of a common energy market.

## The Denise Project

This project includes the participation of 12 companies and 7 research organisations in Spain [12]. With a duration of four years and a budget of 30 million Euros, the project addresses research related to the control of distribution networks with a primary focus on the efficient management of supply and demand, as well as network security. This project hopes to lay the foundation for the smart, safe, and efficient electricity grid of the future.

The main objective of the project is to answer two problems currently facing the supplying company:

- The difficulty in supporting the increasing energy demand of the distribution network.
- Linking the ICT (Information and Communication Technologies) services with traditional electricity supply services.

During project implementation, an automated system to monitor and control energy consumption in homes was developed. Specifically, the implementation of a residential gateway was demonstrated that, when in contact with the energy supplier, performs the control and monitoring functions of the house, as well as of the overall energy efficiency and active management of demand in the house. The developed gateway has a touch screen control where the configuration tasks of the home network are performed. Additionally, the X10 protocol on PLC is used for the monitoring and control of the various elements of the house, thereby avoiding the installation of a new infrastructure for automated control.

## The Fenix Project

Flexible Electricity Networks to Integrate the expected "energy evolution" (FENIX) is an integrated project, belonging to the 6th Framework Programme for Research in the European Union [13], the ETP-

SG project, which seeks to encourage distributed energy resources (DER) and renewable energy sources (RES), maximising their contribution to the electrical system through their inclusion in the so-called Large Scale Virtual Power Plants (LSVPP) and their decentralised management. This inclusion allows the DER and RES to gain access and visibility in the electricity market. This project will allow the DER and RES to become more flexible and controllable than if they were isolated from each other.

The FENIX goal is to launch distributed generation from passive to active generation, supporting the operation of power grids and thus assuming a role similar and complementary to centralised generation:

- Supporting the grid by adding reactive power and voltage control.
- Contributing to the active power reserve when it is needed in the system.
- Helping to solve problems related to network surcharge.

This operation will result in higher penetration of renewable and distributed generation into society with associated benefits reaching society, as well.

FENIX represents a step toward future networks, the management of which will be significantly more sophisticated than the networks utilised today.

## The DER Project

IREC. This project is funded by the Government of Catalonia (2009-2011) and with the support of the EU.

DER-IREC22@ MICROGRID is an industrial research project aimed at developing new products and services around the DER and electric vehicle concept [14]. The experimental platform integrates an interconnected, smart micro-grid with a set of existing DER, including micro-generation conventional technologies (natural gas or diesel), new technologies (micro- turbines and energy storage devices), and renewable technologies (wind turbines, solar).

The project should result in

- Algorithms for the optimisation of micro-grid management according to different criteria, which consider the optimal management of energy.
- Algorithms for the optimisation of interaction between the combination of renewables and EV (electric vehicle) charging.
- New research for EV recharging points to incorporate smart charging technology.
- The evaluation of the impact of EV mobile network charging.
- Algorithms for energy exchange between the micro-network and the distribution network based on simulated price signals from the charge demand of the electric system.
- Methodologies for advanced management of the aggregate demand of a micro-grid charge or even a possible network of micro-grids.
- Communication systems between the micro-grid and the grid.
- Communication systems of the micro-grid virtualisation devices and improvements in existing protocols.
- Technologies that make the DER model scalable.
- Technologies that are applicable to large micro-grids.
- New standards to enable the interoperability of equipment from different manufacturers in the micro-grid.
- Legislation and regulations specific to micro-grids to be considered for planning, design, implementation, and operation.
- Characterisation of the impact of the environmental conditions.

This project was conceived as the beginning of a series of R&D efforts intended to create a platform for companies, research centres, and technology centres in Catalonia constructed around an experimental platform to increase prior knowledge in the field of smart grid electricity.

Objective:

- Study of Micro-networks (Micro-grids)
- Control Algorithms
- Multi-agent systems
- V2G (Vehicle to Grid)
- DER
- Renewables
- Device Virtualisation

## The GEBE Project

The GEBE (Gestor de Balances Energéticos con GD inteligente - Energy Balance Manager with smart DG) project in Spain [15] aims to design, build, and test a smart system for energy network management with distributed generation, the function of which is to optimise energy flows in accordance with economic parameters, thereby ensuring electrical security and stability.

For the project, the following partial objectives were developed (2010-2012):

- To design and build configurations of powered electronics for generation and storage systems that are capable of interacting with the smart distributed generation Grid (SDGG) manager, the production of which is at the service of the latter. These settings should be of the plug and play type.
- The design and adaptation of powered electronic systems for power generation and networking of these systems with demonstrators.
- To design and build energy measuring systems at very low cost and easy installation, both active and reactive, for instant knowledge of flows in every generation, storage and consumption system.
- To develop an SDGG System Manager with the ability to make real-time decisions regarding the generation and storage systems

based on different parameters and in consideration of both the technical and economic factors. The aim of the System Manager is to achieve maximum energy self-sufficiency from the SDGG resources alone.

- To define a communication system among the sensors, actuators, and (SDGG) Manager that would allow for real-time performance reporting of each SDGG component.
- Developing a Supervisor System for SDGG Managers with the ability to make real-time decisions on energy flows between self-governed SDGG; these decisions should be primarily based on economic parameters. To adapt the various RGDI in the project to achieve the maximum possible instantaneous energy independence.
- To demonstrate the economic and technical feasibility of the Supervisor System for SDGG Managers concept in the context of a liberalised electricity market.
- To carry out energetic and economic evaluations of a domestic application with renewable generation and energy storage in the form of hydrogen.
- To carry out the control system design for electricity generation and its networking with the demonstrators.
- To provide an energetic and economic analysis for the use of renewable energy and hydrogen as an energy vector for real domestic applications.
- To carry out the operational study of home systems with energy storage in the form of hydrogen.

## The Redox 2015 Project

This project is implemented by a collaboration of five electric companies and five research institutions in Spain [16]. One of the major problems that power supply systems face is the difficulty (and cost) associated with storing energy during low demand periods for use later during peak demand. This problem is particularly relevant from the point of view of renewable energy, if it is to constitute a realistic alternative to fossil fuel energy. The aim of the Redox 2015 Project is the research and development of a reliable and economically competitive redox flow battery for industrial and commercial use; the battery should be applicable to medium and low voltage power grids, allowing the high capacity storage of electricity while providing solutions to the problems described.

## The Redes 2025 Project

This project is partially funded by the Ministry of Science and Innovation in Spain [17] with a budget of 40 million Euros. It proposes technological developments in software and hardware and the application of new materials, having identified the main leading areas for development as information and communication systems, forecasting and optimisation, power electronics, materials, and sensors for active distribution. This project was started in 2009 and has a duration of four years. The implementation of this ambitious project is divided into 5 sub-projects. Power electronic applications include network control, optimal integration of distributed resources, electric power storage, new network applications based on superconducting materials, and information management on smart grids.

The project objectives are to

- Design and implement a simulation environment to investigate several scenarios of electricity networks in various time horizons theoretically.
- Develop tools and technical solutions for the operation, control, maintenance, and measurement of electrical networks, considering the evolution of today's networks into the networks of the future in such settings as the integration of renewables and distributed generation, and energy storage.

- Identify requirements to be met by both the teams and the abovementioned tools from the point of view of physical integration into electricity grids and their controllability, as well as from the point of view of the management systems for their correct functioning at different operating conditions.
- Propose and develop equipment, systems, and solutions based on required and identified needs.
- Implement and evaluate emerging technologies for application in power systems, such as storage systems, ICT-based solutions, power electronics, and superconducting applications, as alternatives to the proven, traditional technologies.
- Analyse market conditions and regulatory frameworks that allow the implementation of such solutions.

The development of several such smart grids is part of a larger programme at the European level, which is called the European Strategic Energy Technology Plan (SET-Plan) [18] and generally aims to boost low carbon technologies and other areas, including the capture and storage of $CO_2$, solar, wind, and nuclear fusion initiatives according to the ECOOO project's SmartGrids report of July 2010 [19].

The application of micro-grid solutions, such as networks or virtual power plants, along with the implementation of innovative technologies, such as bidirectional measuring devices with tele-management and tele-measurement, will allow greater presence in the electricity market for small electricity producers and consumers. Thus, distributed generation will achieve greater levels of energy efficiency because production and consumption will become increasingly more balanced, and large transport losses will be avoided, allowing for the reduction of investment costs, as well as a reduction in environmental impact. This approach to distribution will also improve the quality and security of supply.

Additionally, it should be noted that the various combinations of management systems [20] can be combined for different energy use needs, such as the use of hybrid systems [21] that involve renewable energy [22, 23, 24].

## The Regulatory Framework

According to data from the CIGRE (International Council on Large Electric Systems) [25], in various countries around the world, the percentage of installed DG power capacity has increased in relation to the total installed capacity. Thus, DG has become a reality in many countries, such as the USA, Canada, Colombia, Chile, UK, Germany, Sweden, and Portugal; additionally, all of these countries currently possess specific regulations. An example is the situation in Portugal, which has established specific legislation for micro-production.

The 2004/8/EC Directive of the European Parliament and of the Council, given on 11 February 2004, concerns the promotion of cogeneration based on a useful heat demand in the internal energy market. Furthermore, the Directive 92/42/EEC amendment aims to increase energy efficiency and improve the security of supply. These goals are currently being carried out by the creation of a framework for the promotion and development of high-efficiency cogeneration of heat and power based on useful heat demand and primary energy savings, while taking into account specific national circumstances. One of the stated objectives listed in this directive is the development of small generation plants [26].

Similarly, the 2009/28/EC Directive of the European Parliament and of the Council, given on 23 April 2009, concerns the promotion of renewable energy amending and repealing Directives 2001/77/EC and 2003/30/EC. This directive establishes the obligation to streamline and accelerate administrative procedures for authorisation and connection to the electricity distribution and transmission networks, urging simplified authorisation procedures. The directive also regulates the general procedures that govern access to networks and their operation in relation to renewable energy, bearing in mind its future development [27].

The new National Renewable Energy Action Plan (NREAPs) 2011-2020 [28] notified the European Commission on 30 June 2010, under the provisions of Directive 2009/28/EC of the European Parliament and the Council, given on 23 April, 2009, that it has established the objectives for the next decade in addition to the proposed regulatory changes required to achieve them.

According to the Energy Efficiency Plan 2011-2020 of the Institute for Diversity and Energy Saving, IDES (Instituto para la Diversidad y Ahorro de la Energía) [29], the energy services sector will triple by 2020 with the potential to generate direct and indirect employment at approximately 1 million jobs. The revenues from these services will increase from approximately 22,000 million Euros to more than 70,000 million Euros, and distributed generation and self-supply may be able to lead this development.

On 8 December, 2011, the Official Bulletin of the State (OBS) (Boletín Oficial del Estado) published the Royal Decree 1699/2011, given on 18 November, regulating the network connection of small power production facilities [30]. The purpose of this Decree is to carry out the development of Electricity Sector Law 54/1997, of 27 November in accordance with the current Directive 2004/8/EC, of 11 February 2004 and Directive 2009/28/CE, of April 23, 2009 by establishing the basic administrative and technical conditions for connection to the grid at low and high voltages (up to 36 kV) for the renewable energy and small power cogeneration facilities, further taking into account their special features and to establish a specific regulation that allows the development of these activities. This Royal Decree repeals Royal Decree 1663/2000 of 29 September on photovoltaic connections to low voltage networks, expanding the scope of the new regulation and maintaining the basic structure of its content. A new feature is intended to simplify the requirements for small power plants seeking to connect at points where a supply already exists. Similarly, production facilities with nominal power under 100 kW will be excluded from the administrative authorisation system, and the future and forthcoming regulation of the electricity supply generated within the network by a consumer for their own consumption will be announced. This regulation will encourage self-supply.

These measures are intended to develop DG, which presents a variety of system benefits, such as the reduction of network losses, reduced investment needs in new networks and, ultimately, a minimisation of the environmental impact of electrical installations. The standard will facilitate the development of renewable energy in homes and SMEs (Small and Medium Enterprises). It will be complemented by the rules of net balance, which, though under preparation, will regulate self-supply. The gradual entry of such small plants will modify the current centralised model of large electrical installations by promoting a new generation system that is increasingly more distributed with significant benefits for the system and consumers. The standard will allow homes and small businesses to access small-scale power generation and to consume the energy that it produces once the regulation of net balance is ready.

The future of net balance will depend largely on how access tariffs are regulated. This regulation is not yet defined and currently represents a notably important portion of the total price of energy (approximately 40%).

The current draft of the Net Balance regulation proposes two payments for deferred electricity consumption, including an access tariff and a net balance service cost. According to the same draft, a cost ceiling will be set for net balance service. The access tariff charged to customers today includes various items, such as network losses, generation capacity, transport, the system operator, the nuclear moratorium, the regulator, the deficit rate, distribution costs, and trade margin.

With the new generation scheme, losses in energy transport will be avoided, aiming to gradually increase the service from 5 MW in 2012 at a rate of 50 MW/year until 2017 and the years following until a total of 300 MW is reached according Enair [31].

## SMALL-SCALE GENERATION

### Photovoltaic (PV)

The self-supply of electricity and distributed generation are the two key elements of the immediate future of solar PV in Spain because of the evolution of the regulatory framework, as well as strong levels of solar radiation. This ensures profit levels without short or middle term aid, called grid parity.

Grid parity means that the cost of generating power without premiums is the same as the purchase price for grid power. This point is the moment at which users become indifferent with respect to buying or generating the energy they consume.

According to the EPIA (European Photovoltaic Industry Association) [32], solar PV will compete in price with conventional sources in 2020.

Figure 1 shows the stages of grid parity in the world between 2010 and 2014. Prices of the Feed in Tariff (FIT) and residential market prices, tariff of 2.0, are compared. This shows that Spain will not meet grid parity in 2012, although it is close to doing so as long as there is fair regulation and the kWh can be assessed.

Figure 2 compares commercial and residential scenarios. This figure shows that for 2016, not only Spain but other countries, such as Portugal, Turkey, Germany, and France, will have grid parity in the residential sector. This possibility means they will have to act accordingly not only by the increase in solar panels and the increase in supply through small-scale wind turbines but also by preparing the entire electrical system with smart grids.

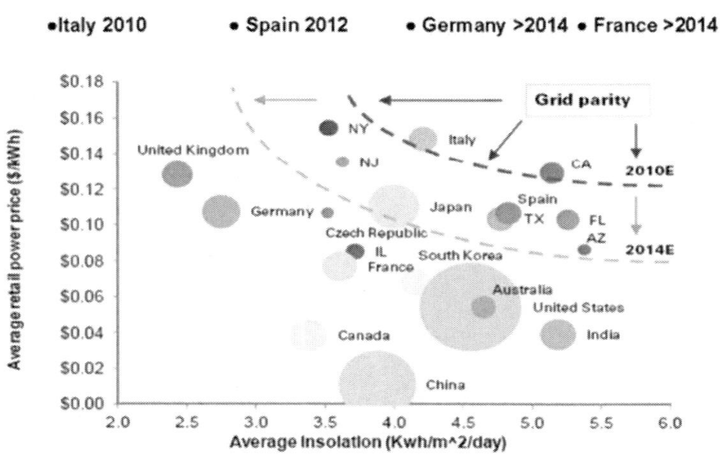


Source: Goldman Sachs. Jan. 2011.

Figure 1. Worldwide grid parity scenarios.

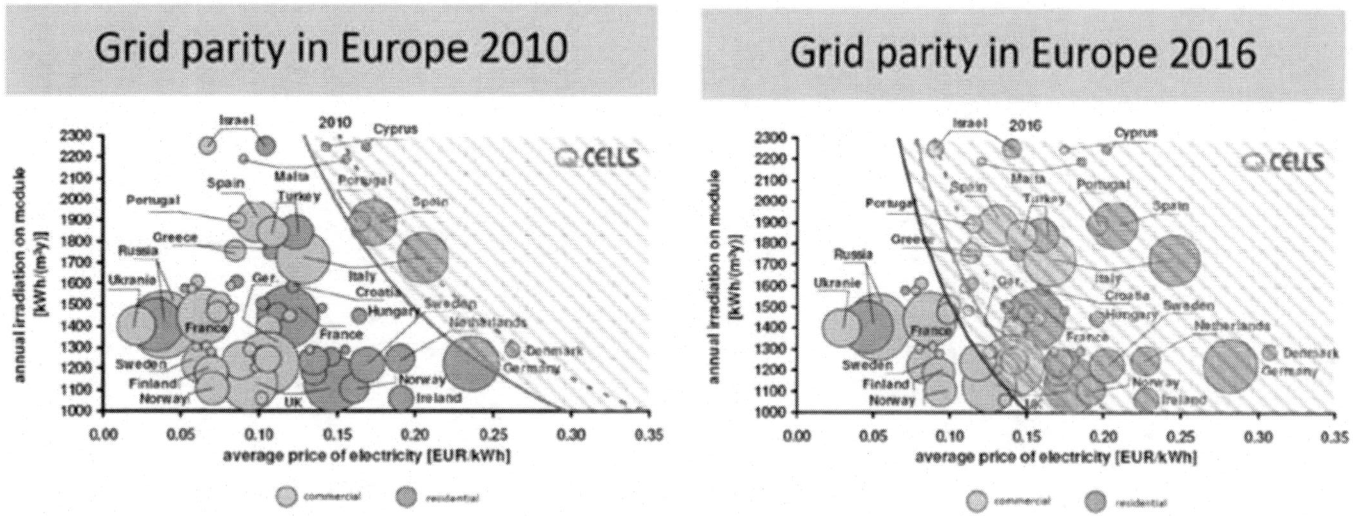


Source: Schleicher-Tappeser. Jan. 2011.

Figure 2. Grid parity in Europe between 2010 and 2016.

Net metering (net balance) allows for the consumption of PV electricity and the selling of excess electricity when there is no self-supply demand.

Net metering allows excess PV electricity to be injected into the grid.

In return, the producer receives financial compensation or a consumer credit from their power company.

In Spain, however, this system does not yet exist. A regulatory change must introduce [33] [34] this system if grid parity is to have an impact on the PV market.

Markets with net balance already exist. The following are several examples in Europe, as assessed by SUNPOWER Corporation [35] in the report "The road to self-sufficiency by Net Balance in Europe."

In the ***UK***, there are three types of fees:

1. The Generation Tariff, equivalent to the Feed in Tariff (FIT) in the Spanish model. This tariff is applied according to current legislation for all kWh produced.
2. The Export Tariff, also an FIT, is a low tariff rate costing only 0.038c€ for excess kWh generated.
3. The Import Tariff is the price per kWh that the end customer pays at the point of consumption.

Because the UK is still far from the grid parity scenario, it adds a complement called the export rate in addition to the rate for those who export. This leads to a greater savings bill and to interesting results that initiate a discussion regarding the self-supply model. At this point, the incentive comes by paying a fee for the total production, as in the Spanish model. The important point here is that the systems pay for the electricity they generate. In Spain, on the contrary, in the proposed net balance Royal Decree, after a year, the meter is zeroed, and no excess generation is paid.

In ***Germany***, there are two rates:

1. The FIT tariff is equal to the model used in Spain, which applies either to the total kWh production or only for kWh exported to the grid.
   In Spain, the tariff applies to 100% of the kWh generated, but there is an optional incentive called the differentiating tariff for anything self-supplied.
2. The Feed in Premium (FIP) tariff for self-supply, which is lower than the FIT tariff, generates savings in the system because the rate for consumption is lower than the tariff for 100% of production. This tariff is also intended to encourage engagement when the self-consumption exceeds 30%. This increases the model's appeal because the self-supply tariff is higher, and when the consumption is below 30%, the tariff is lower. This is intended to guide demand so that instant self-supply reaches at least 30%.

In ***Italy***, there is a self-supply model called Scambio sul Posto, which offers mixed incentives, i.e., there is a FIT tariff for 100% of what is produced, and as in the Spanish model, pre-registration is required.

Both the energy and the service are valued, which is part of the variable cost of the tariff, and taken together with the toll is called Scambio sul Posto. As a function of production time, this system allows credit generation in both economic and kWh measures. Therefore, the total incentive is truly significant because there is a fixed tariff portion, plus a lower bill, plus a credit in Euros or in kWh. This model allows for a steady rise in investment in the Italian market. This is reflected by the fact that in only three years, the Italian market in this area went from 0 to 13 GW installed. In Italy, this model is limited to systems with 200 kW of maximum power.

Table 1 shows a comparison of the different rates of the countries described.

**Table 1. Comparison chart of the 3 countries described**

| UK | | | Italy | |
|---|---|---|---|---|
| **The Generation Tariff** | **The Export Tariff** | **The Import Tariff** | **FIT** | **Net Balance** |
| FIT for all kWh produced | 0.038c€ for excess kWh generated | which is the price per kWh that the end customer is playing at the point of consumption | for all kWh of PV produced | function of credit compensation in both economic and kWh measures called Scambio sul Posto |
| **Germany** | | | | |
| **FIT** | **PRIT (Premium Feed in Tariff) for self-supply, which is lower than the FIT Tariff** | | | |
| for all kWh produced or only the kWh exported to the grid | Self-consumption | < 30 kW | 30 - 100 kW | 100 - 150 kW |
| | **Support** | | | |
| | > 30% of production | 0.1830 | 0.1690 | 0.1543 |
| | ≤ 30% of production | 0.1449 | 0.1308 | 0.1161 |
| | **Equivalent Economic Advantage** | | | |
| | > 30% of production | fit + 0.080 | fit + 0.080 | fit + 0.080 |
| | ≤ 30% of production | fit + 0.036 | fit + 0.036 | fit + 0.036 |

Source: SUNPOWER Corporation.

**Table 2. Electricity tariffs in Spain**

| LRT (last resort tariff) | | | |
|---|---|---|---|
| **PLRT (power term for the LRT)** | | **ELRT (energy term for the LRT)** | |
| **PAT** | **FMM** | **EAT** | **CE** |
| (power term for the access tariff) | (Fixed marketing margin) | (energy term for the access tariff) sale | (energy estimated cost) |

Source: Official Bulletin of the State (OBS).

Table 2 shows the tariff structure in Spain.

In ***Spain***, the tariff has a binomial structure, beginning with the ELRT (energy term of the LRT, Last Resort Tariff), which is energy variable and consists of supplements reflecting the energy cost. Access fees or tolls are the second component of the tariff. The ECE energy cost (estimated cost of energy) incorporates several costs, including the estimated cost of the ELRTs sale, service costs, system adjustment, derivative payment mismatch, payment for generation capacity, and a loss coefficient. That is, there are a number of concepts that are attributed to this energy term, which is both variable and priced in kWh. The access rates are composed of an energy access tariff term (ELRT), a fixed term called the power access tariff (PAT) and, finally, the marketing margin (FMM).

With the approximate 2011 prices of the various components according to the BOE [36] we have:

PLRT 16.63 € /kW year
FMM 4 €/kW year
ELRT 0.15 €/kWh
PAT 0.063 €/kWh

The energy term is between approximately 55%-60%, while tolls are 40%. Thus, if the application of new solar energy is to be feasible, the impact that the decision will have on net balance regarding tolls is predictable.

The net balance consumer would pay a fraction of the co-payment of tolls with an annual increase until it reaches the point at which the net balance consumer pays 50% of the tolls.

Moreover, the net balance must allow for the generation of credit, either in kWh or Euros, for the kWh produced to be valued. At least in the beginning, the value will be a ratio of 1 to 1 at the point of consumption. Furthermore, the compensation at least within two years should be allowed and not in a year, as proposed by the Net Balance Royal Decree according to SUNPOWER Corporation.

## Small Wind Turbines

Wind energy production significantly contributes to the reduction of fossil fuel imports.

According to the WEA (Wind Energy Association) [37] in their Macroeconomic Study of the Impact of Wind Energy in Spain report, during 2010, wind generation in Spain produced 43.692 GWh. This production prevented the emission of 22.8 million tons of $CO_2$ and was worth of €329.8 million in 2010. By using wind energy generation, since 2005, the emission of 110.3 million tonnes of $CO_2$ has been prevented, which meant not importing 8.9 million fossil fuel toe (tonne of oil equivalent), which was worth €1,616,100,000. The primary energy consumption in 2010 relied on oil as the energy source and stood at 63.684 Ktoes, which accounted for 48.8% of the total.

According to the Energy Trends to 2030 report [38] of the European Commission, the use of wind energy will further grow in this period. Specifically, 136,000 MW of wind power in the European Union is expected to be installed between 2011 and 2020, which would mean that 41% of the total power will be installed in the near future.

The European Commission estimates that wind power will account for 14% of electricity consumption in the EU by 2020, further implying that it will supply electricity to approximately 120 million households according to calculations by EWEA (the European Wind Energy Association) [39].

The total installed capacity of power in the EU will increase by 333,000 MW in the next decade according to the European Commission. From that power, 64% will be renewable, 17% will come from gas, 12% from coal, 4% from nuclear, and 3% from fuel.

Another important report, the 2050 Roadmap [40], gives a practical guide to a prosperous, low-carbon Europe and was produced by the European Climate Foundation, concluding that the electricity sector must reduce its emissions by 95% - 100% to reach 80% reduction in total European $CO_2$ emissions by 2050.

An increase in the competitiveness of the wind industry also requires effective storage of the wind energy produced. In addition to the hydraulic pump option, another alternative would be to use wind power production generated during hours when electricity demand is low to recharge electric vehicle batteries. Thus, the batteries function as distributed storage devices.

In Spain, research is being conducted on various possibilities for the use of electric cars, such as the WPREV project (Wind Power Regulation with Electric Vehicles) [41], funded by the Ministry of Industry, Tourism and Trade and the European Regional Development. This project has estimated that a car equipped with a battery with an average consumption of 14 kWh/100 km travelling 15,000 km annually would consume 2,100 kWh annually. If the electric car penetration is high, the positive impact of the replacement of fossil fuels by clean energy would be significant.

The small-scale wind energy involves the harnessing of wind resources using power turbines below 100 kW. According to international standards, technology mills must have a swept area not exceeding 200 $m^2$.

There are many virtues of the small-scale wind turbine energy. The turbine's installation requires notably little work, it produces energy at low wind speeds, it supports wind speeds up to 200 km/h, and with only 1 generator of 75k W, it can produce between 50 - 70% of the average electric consumption of a home for the 15 or 20 years of device life. Furthermore, the small turbines manufactured for urban use are increasingly more aesthetically pleasing, reducing visual impact.

From a technological point of view, there currently has been work on new high-efficiency electronic converters, designed specifically for small turbines, which apply the latest industry developments and technological improvements for an effective grid connection.

The sector is also developing major advances in wind turbines. Modern horizontal wind turbines must be oriented in the direction of the wind, either manually or by means of a control mechanism. The vertical axis turbines do not require the use of such a control system; independent of which side the wind blows, the rotor position is always correct.

Currently, new horizontal axis wind turbines are in development, especially within the power scales for which there is no current supply in Spain (from 15 kW - 100 kW). These turbines are more suitable for the countryside areas or industrial estates. The other turbines have vertical axes for integrated applications, mainly for buildings, according to data from the Association of Renewable Energy Producers APPA [42].

However, because there has been no specific regulation for grid-connected systems (except small experimental facilities), most facilities in Spain are isolated.

In the field of research, the Research Centre for Energy, Environment and Technology (CIEMAT) and CEDER are involved in small turbine research, leading a Strategic Sector Project through the guidance of a working group to develop new rules for quality certification specific to small wind turbines under the framework of the International Electrotechnical Commission and that of the International Energy Agency [43].

At the "Small Wind World Report 2012" [44], the World Wind Energy Association (WWEA) indicated for the small turbine that an installed power of 443 MW would be achieved worldwide by 31 December 2010 and gave a forecast of 31 December 2020 for more than 3400 MW.

The report considers small wind turbines to be all of the machines producing between 6 kW and 300 kW of power.

China leads the world market in the number of units accumulated, while the United States leads in accumulated power.

Figure 3 shows the cumulative power in small wind turbines through 2010 by country.

***China*** had 450,000 cumulative small wind turbine units in late 2010, which was three times the number registered in the United States. However, the U.S. devices have three times more power than the Asian ones.

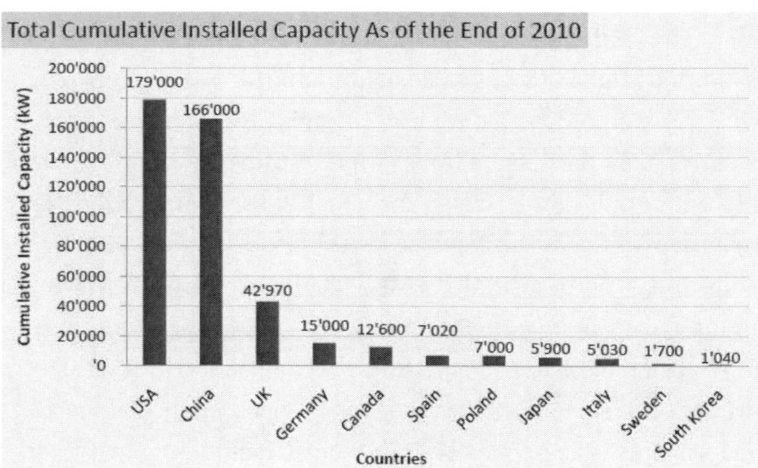


Source: WWEA.

Figure 3. Cumulative power of small wind turbines in late 2010 by country.

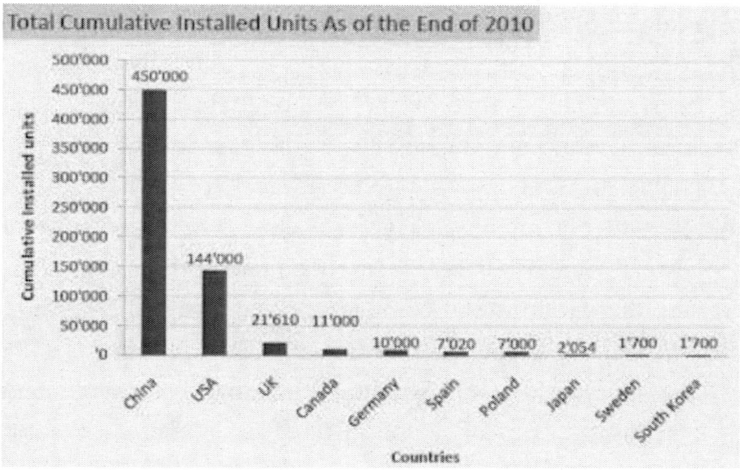


Source: WWEA.

Figure 4. Small wind turbine cumulative units to the end of 2010 by country.

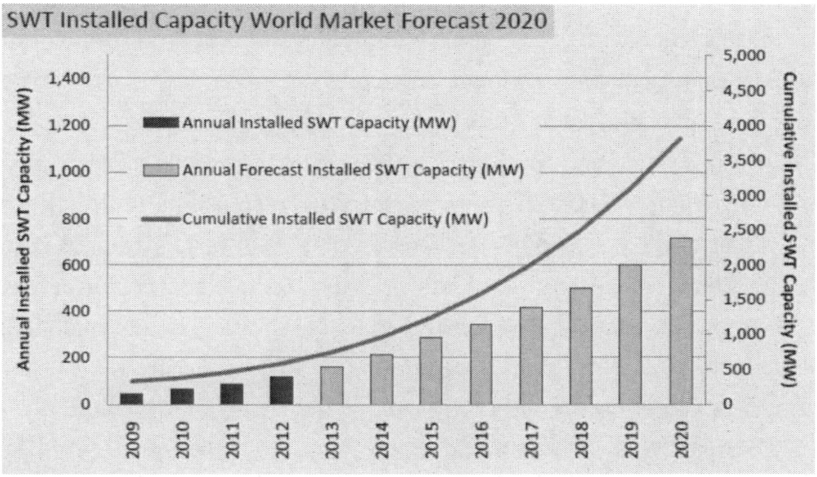


Source: WWEA.

Figure 5. The growth forecast for the worldwide installed capacity of small wind turbines.

Figure 4 shows the number of small wind turbine cumulative units up to the end of 2010 by country.

In late 2010, there were 656,000 small wind turbine units worldwide compared with 521,000 registered by the end of 2009 or the computed

460,000 in late 2008. The Small Wind World Report 2012 forecasts the production of 3,800 MW in 2020, estimating that small wind turbines will follow a path similar to that of the photovoltaic industry, which has experienced an average growth of 39% in the decade from 2001 – 2010. The small wind turbine industry trend in recent years has shown an aggressive growth rate of 35%. Figure 5 shows the growth forecast for 2020 of the worldwide installed capacity of small wind turbine.

According the wind turbine company Bornay [45], in other countries, such as the Netherlands, UK, or U.S., small-scale wind power is a reality because these countries have a special regulatory framework and have aimed to generate between 30% and 40% of the country's electricity by 2050. This use of small-scale wind power is accomplished through the use of distributed micro-generation in buildings, mainly fuelled by small wind turbines and solar photovoltaics through the "Low Carbon Buildings" programme.

In 2010, the ***UK*** had approximately 100,000 micro-generation facilities. Renewable Obligation Certificates are awarded to the owners, introducing tax incentives for those who produce green energy.

Small domestic wind turbines can provide enough clean electricity to power more than 800,000 homes in the UK, according to the Energy Saving Trust (EST). The report estimates that EST installation in houses located in dense urban environments does not reduce the performance of these devices. In total, it is estimated that the small turbines installed in domestic environments could provide up to 3.1% of total energy consumption in UK households [46].

In ***Portugal***, on 2 November, 2007, the Law Decree 363/2007 on Micro-generation [47] came into effect, which provides for the implementation of a simplified licensing regime for local grid connection to low voltage devices for residential or small producers using renewable energy sources (up to 1.5 kW in photovoltaics and 2.5 kW in small wind turbines). This piece of legislation is part of the "National Strategy for Energy," which aims to boost the micro-production of electricity significantly.

This Decree creates the Micro-Production Registration System (MRS), consisting of an electronic platform for interaction with producers, in which it will be possible to carry out all of the relationships necessary for the implementation of micro-producer activity by means of management. The micro-producer receives or pays, through a single transaction, the net value of the relative input of electricity produced, and the payments are relative to the amount of electricity consumed. According to this regulation, there are two regimes for the sale of electricity produced by micro-generation units, including the general system (which considers the production of power up to 5.75 kW) and the special system (which considers production up to 3.68 kW).

Production to the grid in Portugal is paid at 0.45 €/kWh.

Any consumer of energy can be an energy producer. However, consumers can only produce and sell energy to the grid up to half the power rating of their home. This regulation only allows access to the special system for entities with facilities that have at least 2 $m^2$ of solar thermal collectors.

In ***Italy***, Law 244 of 24/12/07 [48] provides a rate of 0.30 €/kW h for energy production during 15 years for wind turbines of less than 200 kW.

In ***Ireland***, regulation has meant that the state power company, Electricity Supply Board (ESB), can purchase electricity from homeowners who have a small wind turbine. A tariff of 0.19 €/kWh was established, which will reduce the payback period of the micro-generation facility. The board also has promoted a policy that reduces connection time and facilitates the connection process; this policy also plans for permit exemptions.

In ***Holland***, there are deductions for energy investments for for-profit organisations. It is possible to obtain a maximum of € 5,000 per turbine (for turbines below 25 kW). Within the Environmental Quality of Electricity Generation programme, there are benefits per kWh produced, the amount of which is fixed by the government and established in the subsidies ministerial programme.

After years of blockade and administrative obstacles, the Government of Spain has given its approval to the development of this energy source, which allows users to generate their own electricity at home without producing any $CO_2$. The NREAP, which was prepared by the Ministry of Industry to meet a request from the EU on energy planning over the next decade, paves the way: through 2020, 100,000 small wind turbines can be installed reaching up to 370 MW.

In conjunction with the provision of electricity in remote areas for businesses and homes, the APPA hopes that the new regulation will promote the connection of small wind turbines to the power grid. At current rates, the payback period is between 25-30 years.

The connection will require a retributive regulation, allowing a return on investment purchase of the micro-generators. The estimated total cost of a small 1.7 kW turbine is approximately 6,000 Euros.

There is a proposal for a premium that may allow repayment over 10 years, which is 2/3 of the expected life of the device; the range of the premium is between 0.4 €/kWh and 0.6 €/kWh with respect to the price determined each year. This repayment scheme could mean a return of 5.75 times the base price.

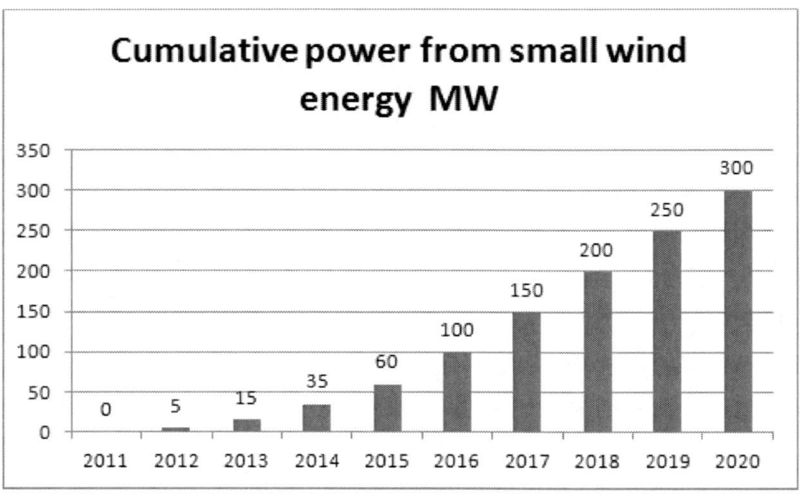


Source: IDEC.

Figure 6. The projected cumulative progression of small wind power 2011-2020.

Therefore, small wind energy will join the path of clean and efficient micro-generation initiated by solar photovoltaics.

Figure 6 shows the forecast for cumulative power from small wind energy by the IDEC for 2011-2020 under the Renewable Energy Plan PER (Plan de Energías Renovables).

## CONCLUSION

According to Spain's REE (Red Eléctrica de España), in advance of the 2011 report of the Spanish electricity system [49], the annual electricity demand in Spain at the end of 2011 was 255,179 GWh, representing a decrease of 1.2% over the previous year.

The installed capacity increased by 1,879 MW in 2011, bringing the total generation capacity at the end of 2011 to 100,576 MW, which is 1.9% higher than the previous year. The vast majority of this increase in power (93%) comes from new infrastructure established as renewable sources, mainly wind power (997 MW) and photoelectric and thermal (674 MW).

The hydraulic production was 22,954 GWh, which is 18% lower than the historical average and 37% lower than that recorded in 2010 (a year with a remarkably high rainfall). Hydroelectric reserves contained in the set of reservoirs ended the year with a filling level of 54% of their total capacity, up from 66% the previous year.

Overall, renewables have covered 33% of the electricity demand in Spain. Wind power remains at 16%, which is the same value as in the previous year.

On April 16, 2012, at 3:48 hours, 60.46% of mainland demand (21,098 MW) was covered by wind power (12,757 MW), surpassing the previous high of 59.63% registered on 6 November, 2011, at 2:00 hours. These numbers provide an idea of the management and the large percentage of wind energy demand that can be provided without compromising the security of supply. In a near-future electric scenario, DG in Spain will comprise an important part of the electrical generation mix [50], which, in parallel, will make it possible for the percentage of renewable energy generation to exceed the established objectives [51].

Figure 7 shows the evolution of power generation policies, which initially involved nuclear generation and later were supported by solar energy, and finally, this figure gives the 2011-2020 forecast on low power distributed generation.

The users will become vital power agents, not only because they may participate in preservation and efficiency measures but also because they will be part of an increasingly more distributed power generation network. Hence, a new player will participate in the energy scenario as a Pro-consumer.

Several solutions are proposed for the final decisions regarding DG regulation, involving:

- A simplification of administrative procedures for authorisation, network connection, and small power plant legalisation.
- A reduction in access fee costs for the net balance consumer, understanding that there is no justification for several current cost items, and for such costs to increase annually up to a maximum of 50% to reach grid parity.
- Rights generated by deferred consumption on the transfer of generated power will have at least two years for compensation, and in a 1 to 1 ratio until grid parity is achieved.
- Increased bonus when the consumption exceeds 30% at a FIP consumption tariff.
- For small wind power, a premium of 0.30 €/kWh should be created to increase the rate of return on investment.
- Inclusion of intelligent distribution grids having the objective of achieving an active network architecture in which conventional grids communicate with micro-grids and a decentralised management for a higher flexibility.
- Inclusion of electric vehicle charging and storage during periods of low demand.
- Implementation of bidirectional remote management meters.

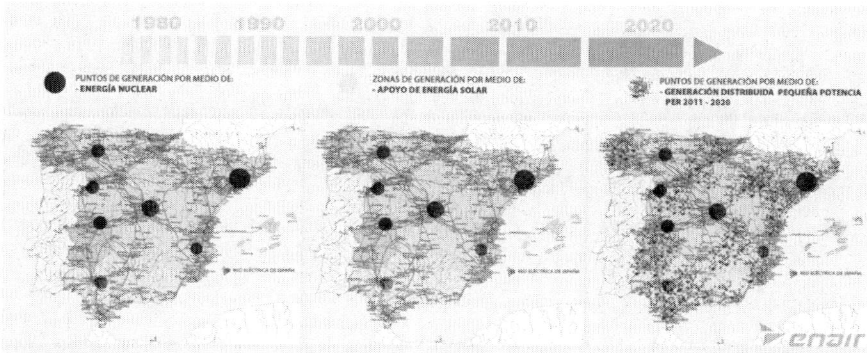


Source: ENAIR.

Figure 7. The evolution of power generation policies, DG forecast for 2020.

In brief, DG will participate in the contribution that renewable energy can deliver to the world energy system in high percentages. This participation is crucial for sustainability and compliance with environmental protocols.

## REFERENCES


[1] International Energy Agency (IEA). *Energy Technology Perspectives, 2010.*

[2] CIER Revista nº 57 (2010). *Valoración técnica y económica del impacto de penetración de Generación Distribuida a través de energía solar fotovoltaica* [*Technical and economic evaluation of the impact of Distributed Generation via photovoltaic solar energy*]. Marcos Facchini, Federico Morán/Universidad Nacional de San Juan.

[3] Lopes Ferreira, H; Costescu, A; L'Abbate, A; Minnebo, P; Fulli, G. Distributed generation and distribution market diversity in Europe. *Energy Policy*, 2011, 39, 5561-5571.

[4] ABB Revista 1/2008 Cherry Yuen, Duncan Botting, Andrew D.B. Paice, John Finney, Otto Preiss (*Cuando las redes se vuelven inteligentes* página 44 - *When grids get smart* page 44).

[5] Moreno-Munoz, A; de-la-Rosa, JJG; Lopez-Rodriguez, MA; Flores-Arias, JM; Bellido-Outerino, FJ; Ruiz-de-Adana, M. Improvement of power quality using distributed generation. *Electrical Power and Energy Systems*, 2010, 32, 1069-1076.

[6] Murat Fahrioglu, FL; Alvaradob, RH; Lasseter, T. Yong. Supplementing demand management programs with distributed generation options. *Electric Power Systems Research*, 2012, 84, 195-200.

[7] Angel, A. Bayod-Rújula. Future development of the electricity systems with distributed generation. *Energy*, 2009, 34, 377-383.

[8] Proyecto *AuRA-NMS* http://gow.epsrc.ac.uk/NGBOViewGrant.aspx?GrantRef=EP/E003583/1.

[9] CENER. *Proyecto Micro-redes* [*Project Micro-grids*] http://www.optimagrid.es/optimagrid/noticias/CENER_sanguesa_mayo_2011_Raquel_Garde.pdf.

[10] Alvaro, Llaria; Octavian, Curea; Jaime, Jiménez; Haritza, Camblong. Survey on microgrids: Unplanned islanding and related inverter control techniques. *Renewable Energy*, 2011, 36, 2052-2061.

[11] ADDRESS. *Proyecto ADDRESS* [*ADDRESS Project*], http://www.addressfp7.org/

[12] DENISE. *Proyecto Denise* [*Denise Project*], http://www.cedint.upm.es/es/proyecto/denise.

[13] FENIX. *Proyecto Fenix* [*Fenix Project*], http://www.fenix-project.org/.

[14] DER-IREC. *Proyecto DER-IREC* [*DER-IREC Project*], http://www.irec.cat/index.php/es/los-proyectos/19-projectes-electronica-de-potencia.

[15] GEBE. *Proyecto GEBE* [*GEBE Project*], http://www.gebe.sinter.es/.

[16] REDOX. *Proyecto Redox 2015* [*Redox Project 2015*]. http://www.futured.es/?p=1249.

[17] REDES 2025. *Proyecto Redes2025* [*Grid2025 Project*], http://www.redes2025.es/.

[18] SET-Plan. *The European Strategic Energy Technology Plan*, http://ec.europa.eu/energy/technology/set_plan/set_plan_en.htm.

[19] ECOOO. ECOOO proyectos sobre energía fotovoltaica [ECOOO projects on photovoltaic energy] (*PV projects*). http://www.ecooo.es/.

[20] Zeng, Jun; Liu, Junfeng; Wu, Jie; Ngan, HW. A multi-agent solution to energy management in hybrid renewable energy generation system. *Renewable Energy*, 2011, 36, 1352-1363.

[21] Józef, Paska; Piotr, Biczel; Mariusz, K1os. Hybrid power systems-An effective way of utilising primary energy sources. *Renewable Energy*, 2009, 34, 2414-2421.

[22] Pérez-Navarro, A; Alfonso, D; Álvarez, C; Ibáñez, F; Sánchez, C; Segura, I. Hybrid biomass-wind power plant for reliable energy generation. *Renewable Energy*, 2010, 35, 1436-1443.

[23] Maria, Stefania Carmeli; Francesco, Castelli-Dezza; Marco, Mauri; Gabriele, Marchegiani; Daniele, Rosati. Control strategies and configurations of hybrid distributed generation systems. *Renewable Energy*, 2012, 41, 294-305.

[24] Jorge, Martínez; Aurelio, Medina. A state space model for the dynamic operation representation of small-scale wind-photovoltaic hybrid systems. *Renewable Energy*, 2010, 35, 1159-1168.

[25] CIGRE. International Council on Large Electric Systems, http://www.cigre.org/.

[26] European Commission. *Directive 2004/109/EC of the European Parliament and of the Council of 15 December 2004 on the harmonisation of transparency requirements in relation to information about issuers whose securities are admitted to trading on a regulated market and amending Directive 2001/34/EC*, http://eurlex.europa.eu/LexUriServ/LexUriServ.do?uri=OJ:L:2004:052:0050:0050:ES:PDF.

[27] European Commission. *Directive 2009/28/EC of the European Parliament and of the Council of 23 April 2009 on the promotion of the use of energy from renewable sources and amending and subsequently repealing Directives 2001/77/EC and 2003/30/EC* http://eurlex.europa.eu/LexUriServ/LexUriServ.do?uri=OJ:L:2009:140:0016:0062:es:PDF.

[28] Ministerio de Industria de España. *Plan de Acción Nacional de Energías Renovables* (PANER) (*National Renewable Energy Action Plan* (NREAP)) 2011-2020. http://www.minetur.gob.es/energia/desarrollo/EnergiaRenovable/Documents/20100630_PANER_Espana version_final.pdf.

[29] IDEA. Instituto para la diversidad y ahorro de la energía [Institute for Diversity and energy saving], http://www.idae.es/index.php/mod.documentos/mem.descarga?file=/documentos_11227_PER_2011-2020_def_93c624ab.pdf.

[30] BOE. *Real Decreto 1699/2011* [*Royal Decree 1699/2011*], http://www.boe.es/boe/dias/2011/12/08/pdfs/BOE-A-2011-19242.pdf.

[31] Enair, 2011. http://www.enair.es/actualidad/noticia/real-decreto-1699-2011-que-regula-la-eolica-de-pequena-potencia-conectada-a-red.

[32] EPIA (European Photovoltaic Industry Association – Asociación de la Industria Fotovoltaica Europea). http://www.epia.org/.

[33] Rafael, Cossent; Tomás, Gómez; Pablo, Frías. Towards a future with large penetration of distributed generation: Is the current regulation of electricity distribution ready? Regulatory recommendations under a European perspective. *Energy Policy*, 2009, 37, 1145-1155.

[34] Stephanie, Ropenus; Henrik, Klinge Jacobsen; Sascha, Thorsten Schröder. Network regulation and support schemes e How policy interactions affect the integration of distributed generation. *Renewable Energy*, 2011, 36, 1949-1956.

[35] SUNPOWER Corporation (empresa fotovoltaica) [photovoltaic company]. http://www. sunpowercorp.es/.

[36] BOE-*Boletín Oficial del Estado*. [*State official newsletter*]. http://www.boe.es/.

[37] AEE (Asociación Empresarial Eólica) informe Estudio *Macroeconómico del Impacto del sector Eólico en España* (*Macroeconomic Study of the Impact of the wind sector in Spain*), (2010).

[38] *Energy Trends to 2030* http://ec.europa.eu/energy/observatory/trends_2030/doc/trends_to_2030_update_2009.pdf.

[39] EWEA The European Wind Energy Association. http://www.ewea.org/.

[40] Roadmap 2050. *Report Power Perspectives 2030* http://www.roadmap2050.eu/.

[41] Proyecto REVE - *Regulación Eólica con Vehículos Eléctricos.* [*Regulation of Wind Power with Electric Vehicles*]. http://www.evwind.es/.

[42] Asociación de productores de Energías Renovables APPA. [Association of Renewable Energy Producers]. http://www.appa.es/12 minieolica/12tecnologia.php.

[43] IEA *Wind task* 27 http://www.ieawind.org/task_27_home_page.html.

[44] *Small Wind World Report* 2012, World Wind Energy Association (WWEA). http://www.wwindea.org/webimages/WWEA%_20Small%20Wind%20World%20Report%20Summary%202012.pdf.

[45] Bornay (empresa de Aerogeneradores) [Wind Turbine company]. http://www.bornay.com/.

[46] http://www.guardian.co.uk/environment/2009/jul/08/domestic-wind-turbines.

[47] Decreto Ley [Decree Law] 363/2007 http://www.luxmagna.pt/legislacao/ DL363 2007.pdf.

[48] Ley 244 24/12/07 http://www.parlamento.it/parlam/leggi/07244 1.pdf.

[49] *Red Eléctrica de España REE, en el avance del informe 2011 del sistema eléctrico español* [*Red Eléctrica de España REE, in the progress of the 2011 report on the Spanish electricity system*], http://www.ree.es/sistema_electrico/ informeSEE-avance2011.asp.

[50] Rafael, Cossent; Tomás, Gómez; Luis, Olmos. Large-scale integration of renewable and distributed generation of electricity in Spain: Current situation and future needs. *Energy Policy*, 2011, 39, 8078-8087.

[51] Salvador, Ruiz Romero; Antonio, Colmenar Santos; Manuel, Alonso Castro Gil. EU plans for renewable energy. An application to the Spanish case. *Renewable Energy*, 2012, 43, 322-330.

In: Distributed Wind Energy Generation ISBN: 978-1-53614-207-5
Editors: A. Colmenar-Santos et al. 

*Chapter 3*

# URBAN WIND SYSTEMS

***Francisco Toja-Silva* and Elio San Cristóbal-Ruiz***
Departamento de Ingeniería Eléctrica, Electrónica, Control,
Telemática y Química Aplicada a la Ingeniería,
Universidad Nacional de Educación a Distancia (UNED),
Madrid, Spain

## ABSTRACT

The growth of the world energy demand, the limited fossil fuel reserves and the increasing greenhouse gas emissions require improvements in energy-generation technologies. Specifically, urban wind energy is a source with great potential that is currently being wasted. The characteristics of urban wind and perspectives and proposals to exploit it have been researched and analysed in the literature. The results show that urban winds have a strong multidirectional component that requires analysing the wind turbine behaviour. To explain the influence of the multidirectional wind on the turbine, a simulation of the air flow around a building section was performed, the sections of various wind turbines were superimposed on the velocity fields, and their aerodynamic behaviour was qualitatively studied. The results show that horizontal-axis wind turbines have better performance in flat-terrain

* Corresponding Author Email: f.toja.silva@gmail.com.

applications, whereas in high-density building environments, the superiority of vertical-axis wind turbines is demonstrated. The main benefits of urban wind power development are: distributed power generation, the use of a renewable energy source, and the technological and economic exploitation of building roofs.

**Keywords:** wind energy, urban wind power, wind turbines, wind turbulence, VAWT

## 1. Introduction

According to International Energy Agency (IEA) estimates [1], the world demand for primary energy is expected to double over the period 2000-2035 (Figure 1).

Faced with the exhaustion of traditional energy sources (fossil fuels and nuclear fission), to satisfy the predicted demand, energy sources must be exploited in new ways, especially renewable energy sources. An example is wind-based energy generation, which has grown exponentially since 1995 (Figure 2), from 4,778 MW in 1995 to 160,084 MW in 2009 [3].

Most of the power shown in Figure 2 comes from flat-terrain installations. However, the urban environment also has great potential for wind power that has not been harnessed [4]. In urban areas, there is a multiplication factor of the wind speed because of the presence of buildings, but the turbulence intensity and the multidirectionality also severely increase, which is an aspect that requires special attention [5, 6, 7, 8]. Additionally, these installations increase the profitability of the external surfaces, i.e., the roof and the walls, which currently serve only to enclose the building.

Another advantage of exploiting wind energy in urban environments is its proximity to the consumption points (distributed electric power generation). Distributed generation offers significant benefits in terms of high energy efficiency, lower emissions (of pollutants), reduced energy dependence and stimulation of the economy [9]. The optimisation of distributed generation requires voltage profile improvements, the reduction of electric energy flow in power lines (with the associated reduction of energy losses in power lines and electric devices), and the increase of the energy source availability [10]. The reduction of greenhouse gas emissions is another significant factor [11].

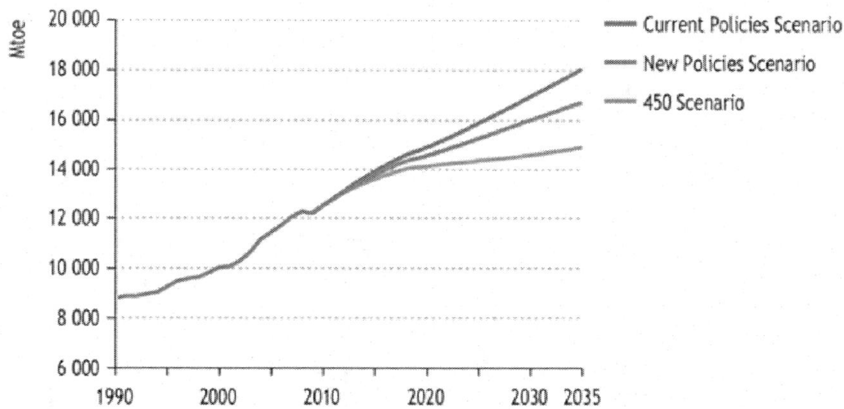


Figure 1. Estimate of the world demand for primary energy for the period 1990-2035, considering three base scenarios [2].

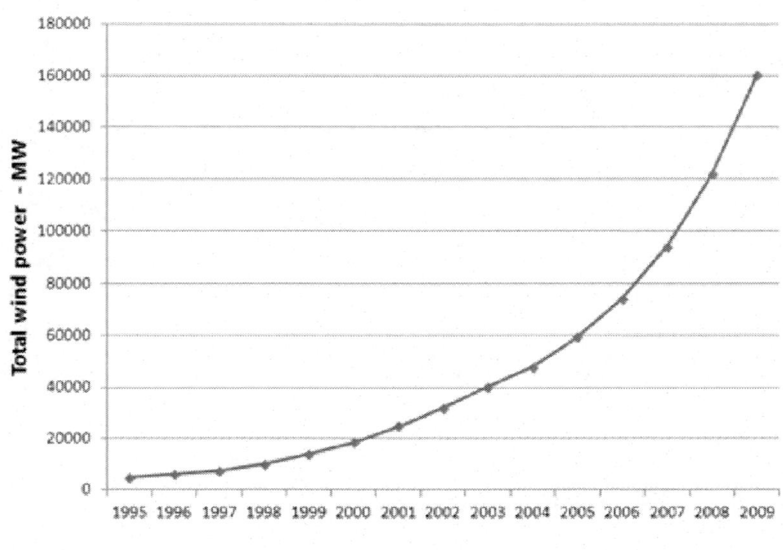


Figure 2. Evolution of global installed wind power [3].

**Table 1. Water consumption of various energy technologies [12]**

| Technology | Litters/kWh |
|---|---|
| Nuclear | 2.30 |
| Coal | 1.90 |
| Oil | 1.60 |
| Combined cycle gas | 0.95 |
| Wind | 0.004 |
| Solar | 0.110 |

Water consumption, a limited resource in some regions, is reduced with the exploitation of the wind power. Table 1 shows a comparison of the water consumption associated with various energy technologies.

Despite the great positive impact of wind power, this energy source has disadvantages.

One of the shortcomings is the visual impact. However, urban buildings and their auxiliary facilities (e.g., chimneys and aerials) share the visual impact with the wind turbines, minimising it. Additionally, the wind generators can be architecturally integrated.

Noise emissions, both audible and infrasound, are a significant environmental factor to consider. Most of the noise pollution comes from conversion and generation machinery, although the blades of horizontal-axis wind turbines (HAWT) also cause noise when they interact with the tower structure (especially with leeward working conditions) [5]. At high wind velocities, the noise due to the forced circulation of the wind around the building and its associated facilities is higher than the noise generated by the wind turbine (Figure 3) [13].

Wind-powered generators also generate infrasound (with frequencies above 16 Hz) and low frequency vibrations that can be transmitted to the building structure [5]. These vibrations can be tolerated by industrial buildings, but they can cause problems in residential buildings [5]. This aspect highly varies depending on both the generator and the building characteristics, and it must be analysed case by case.

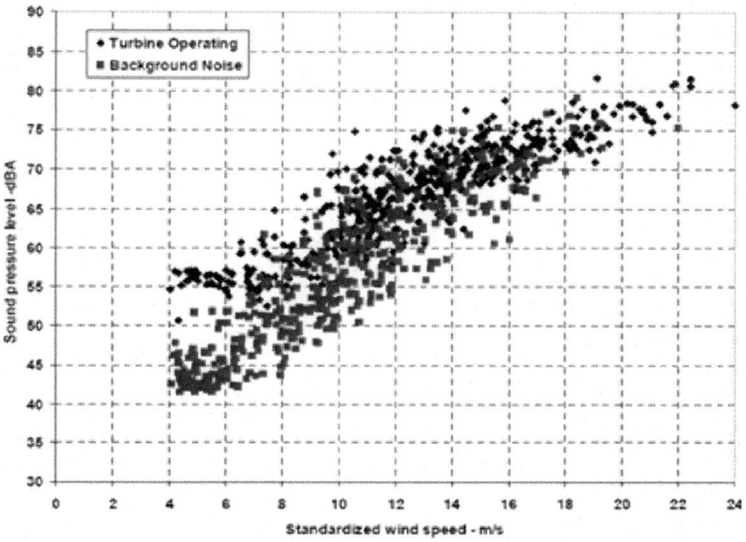


Figure 3. Example of sound levels vs. wind speed for small wind turbines [13].

The impact of wind turbines on birds is very important in flat-terrain wind facilities [12, 14], but its repercussions are smaller in urban environments because of other anthropogenic factors that have a greater impact.

Wind power can also affect TV and radio reception [12, 14, 15]. This is due to the periodic modulation of the electromagnetic fields by means of reflection, absorption and dispersion by the blades [12]. In urban environments, this impact is lower because of the building volumes, which are greater in size than the wind turbines. Any mobile or stationary structure generates interference with the electromagnetic signals [14]. However, both the low power and size of urban wind turbines lessen this impact because the intensity of the interference has a direct relationship with the obstacle size [15].

Mechanical safety is also a fundamental aspect to consider. For each wind turbine, an analysis of the resistance to fatigue of both the structural (including the building structure) and mobile components (especially the blades) must be conducted [5]. The detachment of a blade (or a part of it) can cause a very serious accident because of the substantial momentum [5].

However, according to Grauthoff [5], the probability of a blade breaking, striking a person and causing injuries (independent of the distance it covers) is extremely low, and the author did not advise establishing a safety perimeter around the wind turbine outside of the facility. Grauthoff [5] advises establishing a safety perimeter only in the case of hazardous industries.

Faced with the potential of urban wind energy exploitation, this work presents bibliographic research on the perspectives of various authors regarding this topic and their proposed solutions. The proposed solutions to urban wind energy exploitation have been analysed, and their advantages and disadvantages are discussed.

In Section 2, urban wind characteristics and the effects of buildings are discussed, and a computational fluid dynamics (CFD) simulation of the air flow around a vertical section of a flat-roof building is presented. In Section 3, proposals for wind energy exploitation on buildings and large structures are described, and turbine sections are superimposed over the velocity field obtained from the simulation to analyse their behaviour. Section 4 presents a discussion on the significance of the multidirectional character of the wind for the various types of turbines. Finally, in Section 5 the conclusions of the research are summarized.

## 2. Characteristics of the Wind in the Urban Environment

The terrain is rougher in urban environments, which modifies the incident velocity field. This is shown in Figure 4, where the urban boundary layer is also represented.

The average velocity of the wind is lower in urban environments than over flat terrain [16]. Additionally, in the urban environment the turbulence intensity is substantially higher, which transmits additional loads to the wind turbines [6].

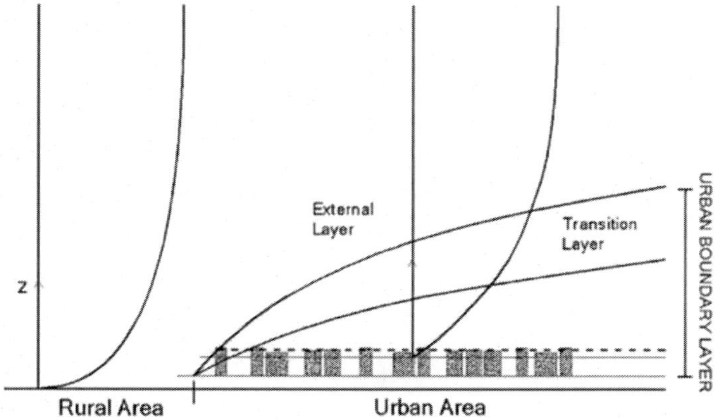


Figure 4. Urban boundary layer and velocity profile in urban environments (right) and flat terrain (left) [16].

Moreover, around the buildings, regions with significant wind velocity intensities are created [6, 7, 16, 17], which entail higher local average velocities [6, 7, 17].

Several authors such as Ledo et al. [6] and Lu and Ip [7] describe the influence of the buildings on both the wind velocity and turbulence intensity from results obtained in CFD simulations. These kinds of studies are essential for determining both the optimal location and the wind turbine model. In these studies, four types of roofs were analysed: flat, sloped, pitched and pyramidal roofs. The results show that, considering both velocity distributions and turbulence intensity, flat roofs are more attractive for installing wind turbines [6]. Sloped roofs are also interesting, and a wind turbine could be installed on the top edge because both high velocities and low turbulence intensities are present [7]. Abohela et al. [18] demonstrated the interest of both vaulted and domed roofs because of the lower turbulence and the intensification factor, although these shapes are used infrequently.

Ledo et al. [6] present an estimate of the power density available in flat, pitched and pyramidal roofs (Figure 5). The power density is clearly higher in the flat roof case.

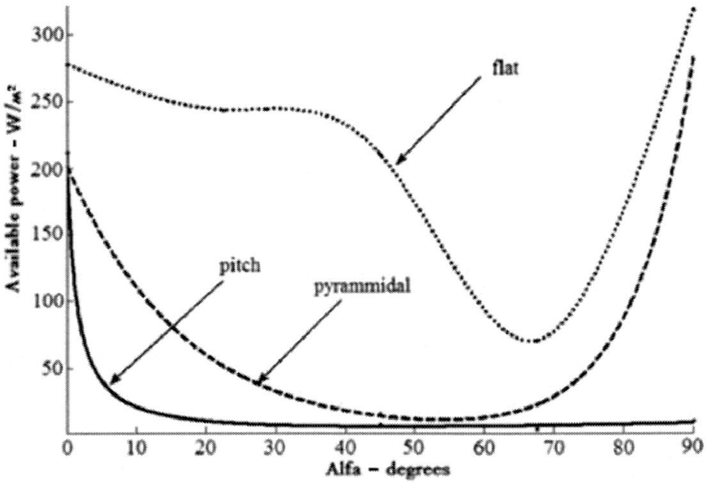


Figure 5. Power density dependence on the incident wind angle for flat, pitched and pyramidal roofs. In all three cases, the wind turbine position is at a corner of the roof [6].

The concentration factor of the wind caused by the building shape can increase the local average velocity of the wind from 1.5 to 2 times and the power density 3-8 times in certain zones of the building [7].

Applying the same criterion in the urban environment as in flat terrain, because of the velocity profile of the wind, the tower of the wind turbine must be between 20 (in low-density building zones) and 40 meters (in high-density building zones) higher and highly reinforced to resist the high intensity of the turbulence [5]. The height requirement can be satisfied by installing the wind turbine on a high-rise building [7].

Turbulence is transmitted to the wind turbine tower and from the tower to the building structure, so for high-power facilities, the structure may have to be reinforced. As a solution to this problem, there have been proposals to integrate the wind power system into the building design and use a higher number of low-power wind turbines to distribute the loads transmitted to the structure [5].

Because in the urban environment the wind velocity is highly variable and its behaviour is extremely turbulent, an urban wind turbine must be designed appropriately to operate under these conditions.

To analyse the influence of the multidirectional urban wind on the different types of turbines, a 2D simulation of the wind circulation around a vertical section of a long-short building was conducted using the computational fluid dynamics software Ansys Fluent-Workbench. Figure 6-left shows a diagram of the simulation case. The geometry was tested for incident wind velocities (*U*) of 1, 2, 4 and 10 m/s. Standard values were used for both the air density ($\rho$) and the viscosity ($\mu$). To validate the results, the central length of a symmetric building with a length (*L*) much larger than its height (*H*), or $L>>>H$, must be considered, as shown in Figure 6-right.

In the urban environment, the intensity of the turbulence and the multidirectional character of the wind are more important than the incident velocity. Significant variations in both the direction and the velocity of the wind appear close to the building surfaces for small variations of the incident wind velocity. Figure 7 shows the velocity maps obtained for incident wind velocities of 1, 2, 4 and 10 m/s.

The simulation results show that the air flow is highly unstable and multidirectional. Small variations in the incident wind velocity cause large variations in the local velocity distribution. A local intensification factor is clearly observed close to the building surfaces that can multiply the incident wind velocity by 6 in certain zones.

A vortex is formed on the building roof that is very sensitive to the incident velocity variation and is clearly defined for lower incident velocities. For higher incident velocities, both dragging and dispersion of this vortex are observed. Above approximately 2 m/s, the vortices on the roof are projected upward. Likewise, a vortex appears in front of the upstream wall of the building.

Results obtained in other studies such as those from Ledo et al. [6], Lu and Ip [7], Abohela et al. [18] or Watson [19] validate the results of this simulation. A stable atmosphere was assumed in these simulations [6, 7, 18, 19], although additional strong disturbances appear in actual cases. In the highly variable conditions of the actual urban wind, with sudden changes in both the direction and the velocity in a very short time, the presence of turbulence is high. This is the reason that the qualitative analysis of the wind turbine behaviour under urban wind conditions is of great interest.

To analyse the multidirectional wind conditions that the turbines are subjected to on a building roof, in Section 3.2 the turbine sections are superimposed on the velocity field obtained from the simulation. To obtain a clearer graphic representation, the map of the velocity distribution for an incident wind velocity of 1 m/s is used because the vortex on the building roof is clearly defined. Analogous conclusions are obtained with other conditions (velocity and direction) of the incident wind.

Figure 6. Diagram of the simulation case (left) and a symmetric building with a large length with respect to its height (right). The simulation comprises the central length of the building.

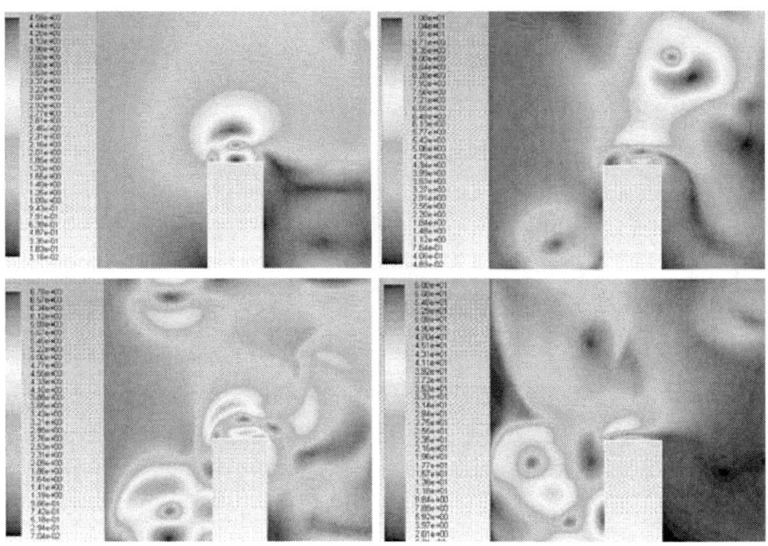

Figure 7. Instantaneous velocity (m/s) maps obtained in the simulation for incident wind velocities of 1, 2, 4 and 10 m/s (from left to right and from the top to the bottom).

# 3. Wind Energy Exploitation Systems in Anthropogenic Environments

This section is focused on the proposed solutions for wind power exploitation in anthropogenic environments. The urban environment is the core of this section (Section 3.2), but large structures are also considered (Section 3.1) as an intermediate case between the urban environment and flat terrain.

## 3.1. Structure-Associated Wind Power Systems

Large structures (such as bridges and oil platforms) can be considered as intermediate applications between flat terrain and the urban environment. That is, the structure disturbs the air flow although the external environment can be flat terrain.

Oppenheim [20] describes a representative example of wind turbine integration into a structure, specifically the Bolte Bridge in Melbourne (Australia). This study [20] shows the results of the analysis of three options:

1. The first option is a vertical-axis wind turbine (VAWT) with a capacity of 0.6 MW installed between two columns of the bridge structure. According to Oppenheim, the main advantage of the VAWT is aesthetic. VAWTs within this power range present considerable technical difficulties, and they are not produced on a commercial scale. Oppenheim [20] estimates that this turbine can generate 2 GWh/year and reduce $CO_2$ emissions by 3000 tonnes/year.
2. The second option is a diffuser-augmented wind turbine (DAWT), also installed between two columns of the structure. According to Oppenheim [20], this option is also aesthetically attractive, but it presents technical difficulties because of the impossibility of

changing the direction. The power generated by this turbine is unacceptably low because of the fixed direction of the rotor [20].

3. The third option is a horizontal-axis wind turbine (HAWT) with a capacity of 2 MW installed above the bridge structure (Figure 8-left). This option is the most technically viable, and the power generated is the highest of the three options, estimated to be 8 GWh/year, the equivalent to 10,500 tonnes/year of $CO_2$ emissions saved or 210,000 tonnes in the 20 years of operation of the wind turbine. This proposal is also economically feasible, with a return period of 8-9 years and a profit in the 20 years of operation of 10 million dollars [20].

The case of the wind energy exploitation with large structures is similar to flat-terrain applications regarding wind characteristics. Hence, the HAWT installation is advantageous because of the higher power coefficient (performance) under unidirectional wind conditions.

Figure 8. HAWT at the Bolte Bridge [20] (left) and an example of a BAWT (right).

### 3.2. Building-Associated Wind Power Systems

As explained in Section 2, the wind has a significant irregularity (undergoing a concentration effect in certain zones) and high-intensity turbulence in urban environments. With these considerations, the previous studies that assumed flat terrain do not apply. This presents an opportunity for innovation and development because innovative alternatives are required for wind energy exploitation in high building density zones. Several authors such as Dayan [21] and Mertens [22] mention the suitability of the VAWT in this environment because of its higher efficiency under turbulent wind conditions [20, 21, 22, 23, 24, 25].

The main alternatives for wind energy exploitation in urban settings are analysed in the following.

### 3.3. Building-Augmented Wind Turbines (BAWT)

Mertens [22] describes the advantages of the architectural integration of wind turbines into buildings, showing how the building can be designed to generate a multiplication factor of the wind to enhance wind energy exploitation. These are referred to as building-augmented wind turbines (BAWT). Although the strict meaning of this expression covers several methods of building wind exploitation, the most representative sense is that in which the wind turbines are integrated into the building morphology. Figure 8-right shows an example of a BAWT. The main advantages are the aesthetic factor and the wind power concentration. In contrast, the single direction of both the wind concentration configuration and the wind turbine (direction consistent with town-planning criteria) causes substantial waste when the wind direction differs from that of the design.

## 3.4. Horizontal-Axis Wind Turbines (HAWT)

HAWTs have lower performance under high-turbulence conditions [20, 26], and they are mainly recommended for very open areas or isolated buildings [27]. The high variability of the wind direction in the urban environment is an additional disadvantage because with lower wind velocities, the start-up time is greater [28]. The wind turbine must be installed at a greater height as its swept area increases to reduce the exposure to turbulence [27]. Consequently, this kind of wind turbine is more appropriate for installations on large structures (as mentioned in Section 3.1) or in zones with low building densities.

Using wind diffusers, HAWTs can operate with lower wind velocities, and they can better resist turbulence [29, 30]. This type of wind turbine is called a diffuser-augmented wind turbine (DAWT).

Wang et al. [29, 30] comment that by using a diffuser-concentrator, the wind velocity can be multiplied by 1.5 and the power by 2.2. These wind turbines (Figure 9) can be installed in a higher building density zone than a HAWT without a diffuser, but the orientation requirement is a disadvantage compared with the VAWT. Stable and unidirectional wind conditions were assumed in the studies of Wang et al. [29, 30]. These assumptions do not reflect the actual behaviour of the wind in the urban environment, where the turbulence is intense and the wind is very variable. These aspects are more favourable for VAWTs.

Figure 9. DAWT.

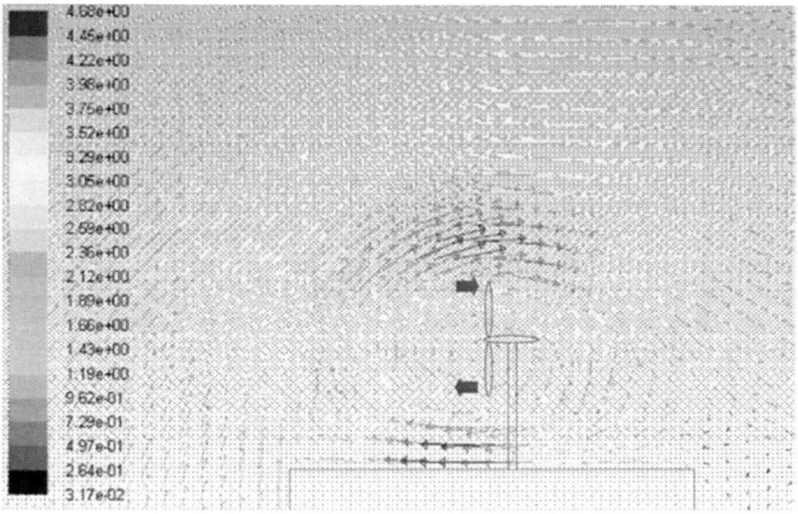


Figure 10. Superposition of a HAWT over the air flow for an incident velocity of 1 m/s.

Figure 10 shows the superposition of a HAWT over the air flow for an incident velocity of 1 m/s. The represented conditions show an incompatibility with wind turbine operation, and they could even cause damage to the turbine, with the consequent risk of an accident.

## 3.5. Vertical-Axis Wind Turbines (VAWT)

Riegler [31] comments that for applications of less than 10 kW, VAWTs have significant advantages over HAWTs. These advantages increase in high-turbulence zones such the urban environment. Additionally, VAWTs generate lower noise emissions, and they are less expensive in terms of both construction and maintenance. Riegler [31] additionally mentions the importance of the hybrid VAWT (Figure 11), which combines the principles of traditional VAWTs (such as Darrieus and Savonius wind turbines) to improve its weaknesses.

Eriksson et al. [26] present a comparative analysis of two VAWTs (a Darrieus and an H-rotor, or Giromill) and a HAWT. As shown in Figure 12, the HAWT has a higher power coefficient (although the three values are similar), but there are several factors that make the VAWT technology more advantageous [26]. In Table 2, a summary of the main differences between the three systems analysed by Eriksson et al. [26] is shown. Among the main technical advantages of the VAWT are the better behaviour under highly turbulent wind conditions, the absence of a direction control mechanism (omnidirectional turbines), the lower noise emissions (because of the lower rotational speed, as shown in Figure 12, and the position of the electric machinery at the base), the lower vibrations transmitted to the structure, the lower cost of both construction and maintenance, and the greater simplicity of the structures [26].

Howell et al. [33] present a study of a Giromill, or H-rotor, wind turbine (Figure 13), a variant of the Darrieus rotor. The results (Figure 14) show that the power coefficient depends on the material roughness. A study conducted by El-Samanoudy et al. [34] highlights the great influence of the blade type on the power coefficient (Figure 15).

Figure 11. Hybrid VAWT generator (Darrieus and Savonius) [32].

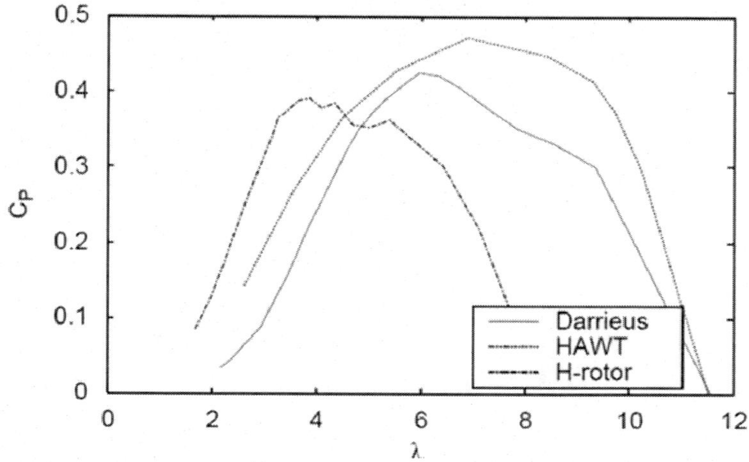


Figure 12. Power coefficients $C_p$ vs. the specific velocity $\lambda$ for three turbines: HAWT, Darrieus and H-rotor [26].

**Table 2. Summary of the main differences between three turbines: HAWT, Darrieus and H-rotor [26]**

| | H-rotor | Darrieus | HAWT |
|---|---|---|---|
| Blade profile | Simple | Complicated | Complicated |
| Yaw mechanism needed | No | No | Yes |
| Pitch mechanism possible | Yes | No | Yes |
| Tower | Yes | No | Yes |
| Guy wires | Optional | Yes | No |
| Noise | Low | Moderate | High |
| Blade area | Moderate | Large | Small |
| Generator position | On ground | On ground | On top of tower |
| Blade load | Moderate | Low | High |
| Self-starting | No | No | Yes |
| Tower interference | Small | Small | Large |
| Foundation | Moderate | Simple | Extensive |
| Overall structure | Simple | Simple | Complicated |

Figure 13. Giromill wind turbine with five blades.

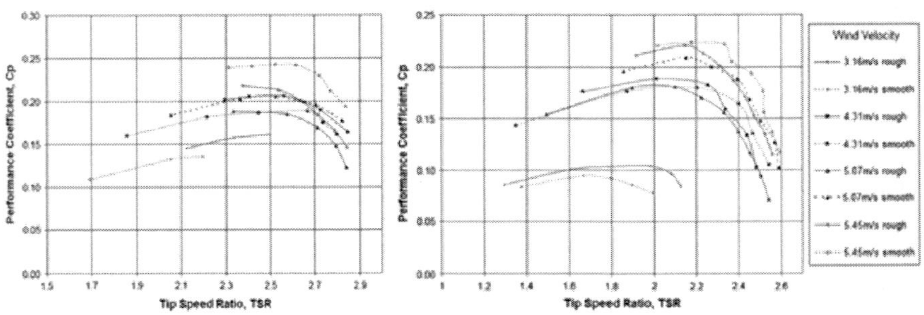


Figure 14. Power coefficient of a Giromill wind turbine with two (left) and three (right) blades [33].

Figure 16 shows the superposition of a generic VAWT (valid for Darrieus, Giromill or Savonius) over the air flow for an incident velocity of 1 m/s. The represented conditions are compatible with wind turbine operation. However, although the multidirectionality of the wind is a normal situation in urban environments, quantitative tests of wind turbine behaviour under these conditions are not found in the literature.

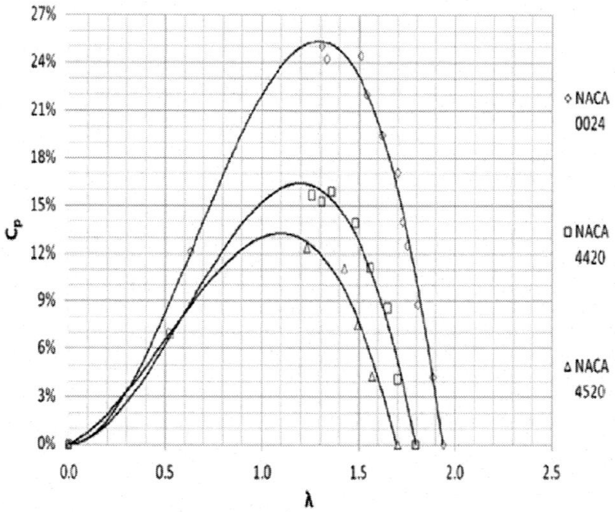


Figure 15. Power coefficients $C_p$ vs. the specific velocity $\lambda$ for three types of blades, Giromill wind turbine with 4 blades [34].

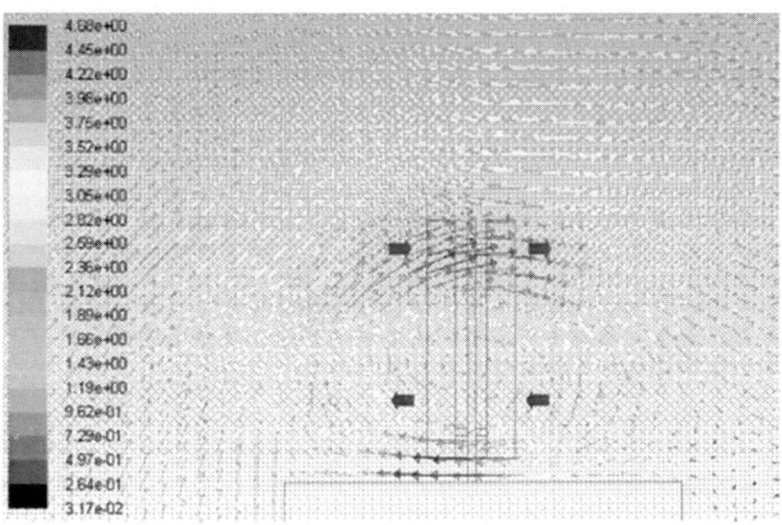

Figure 16. Superposition of a VAWT over the air flow for an incident velocity of 1 m/s.

Figure 17. Helical Darrieus wind turbine.

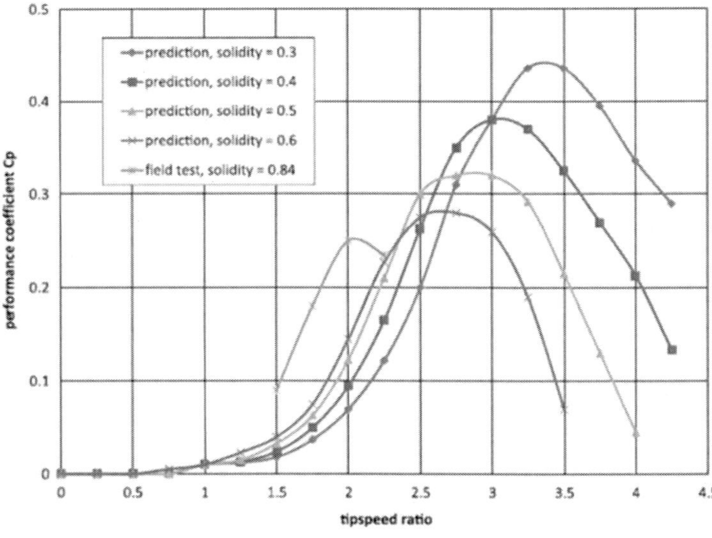


Figure 18. Power coefficient for various solidities, helical Darrieus wind turbine [35].

Kirke and Lazauskas [35] present a study of the helical Darrieus wind turbine (Figure 17). The helical shape results in a variable pitch that generates a high starting torque (the starting torque of the fixed-pitch Darrieus turbine is insufficient for self-starting), a higher efficiency and reduced vibrations, although the active velocity control systems are more complicated and expensive [35]. This study concludes that the lower solidity of the blades increases both peak efficiency and tip-speed ratio, as shown in Figure 18.

Sharpe and Proven [36] propose a Darrieus wind turbine with flexible blades. This wind turbine can be installed either vertically or horizontally, as shown in Figure 19. The streamlined support structure performs the function of wind concentration [19, 36]. The main advantages of the flexible blades are the efficiency increase and the reduction of the vibrations transmitted to the building structure [36].

Figure 19. Flexible Darrieus wind turbine in both horizontal (left) and vertical (right) positions [36].

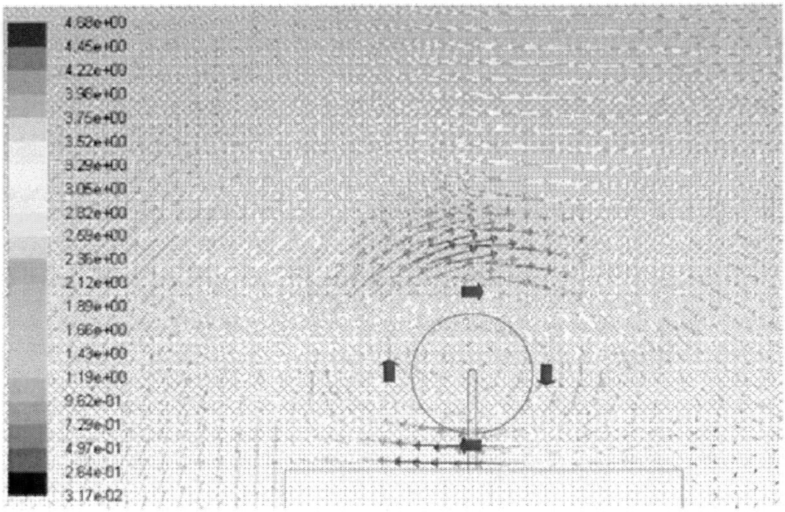


Figure 20. Superposition of a horizontal Darrieus wind turbine for an incident velocity of 1 m/s.

Figure 20 shows the superposition of a horizontal Darrieus wind turbine for an incident velocity of 1 m/s. The represented conditions are appropriate for turbine operation, and furthermore, its behaviour improves under omnidirectional wind conditions.

Because of the low solidity of the blades of the fixed-pitch flexible Darrieus, the wind turbine must work with a variable tip speed to take advantage of the highest power coefficient peaks (Figure 21) to achieve a substantial efficiency increase, which implies a higher level of sophistication in the active velocity control system.

D'Alessandro et al. [37] conducted a study of the Savonius wind turbine. Compared with other wind turbines, it has a lower power coefficient (Figure 22), although it has advantages in that it is self-starting and does not have to change direction (omnidirectional) [37]. Another great advantage is that it is extremely simple to construct, and hence it can be extremely low-cost (recycled containers, pipes or barrels can be used in its construction). This feature would be very attractive in zones with extremely low economic resources (Figure 23-left). The possibility of using recycled materials confers to this technology a highly sustainable character.

The modifications of Savonius wind turbines that increase the power coefficient have an important role, especially the variation of the number of blades, the interposition of obstacles and the helical shape [37, 38].

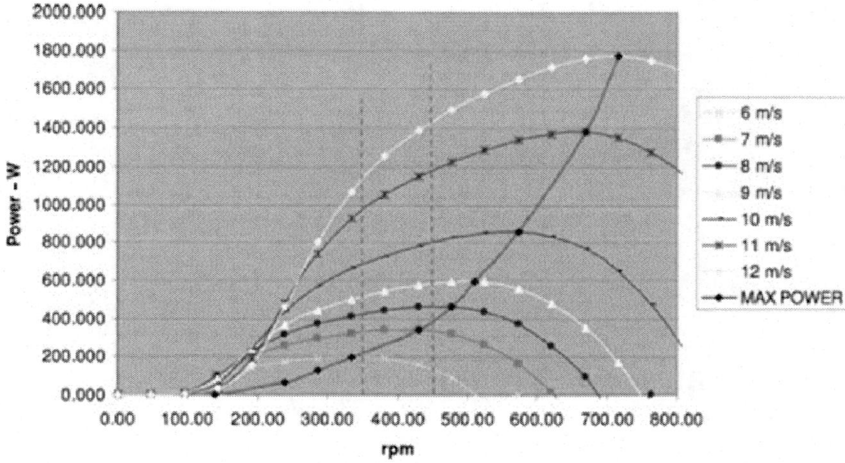


Figure 21. Power vs. rotation speed for various wind velocities, flexible Darrieus wind turbine [36].

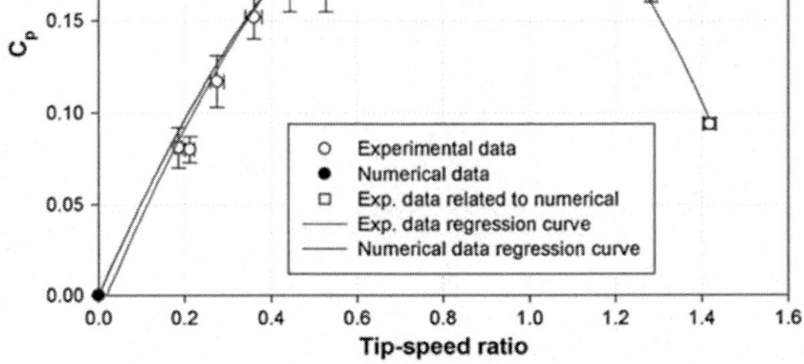


Figure 22. Power coefficient $C_p$ of the Savonius wind turbine [37].

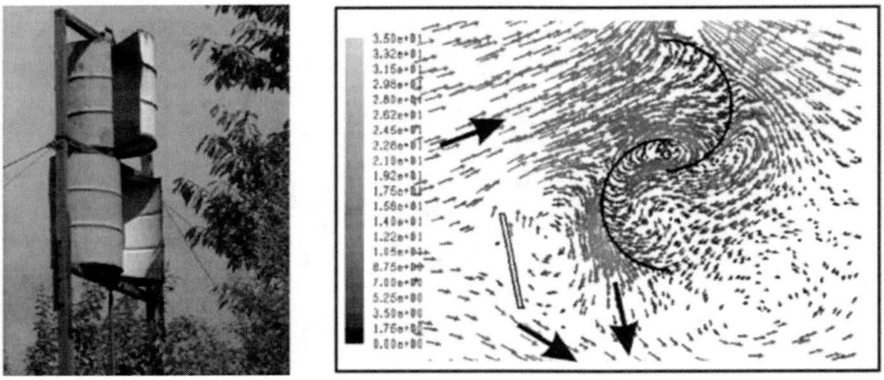

Figure 23. Extremely low-cost Savonius wind turbine (left), and a diagram of a Savonius wind turbine with two blades and a deflector sheet (right) [38].

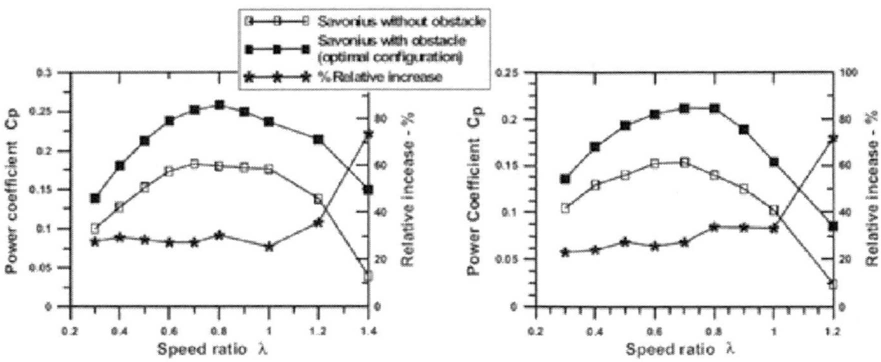


Figure 24. Power coefficient of a Savonius wind turbine with two (left) and three (right) blades with and without a deflector sheet [38].

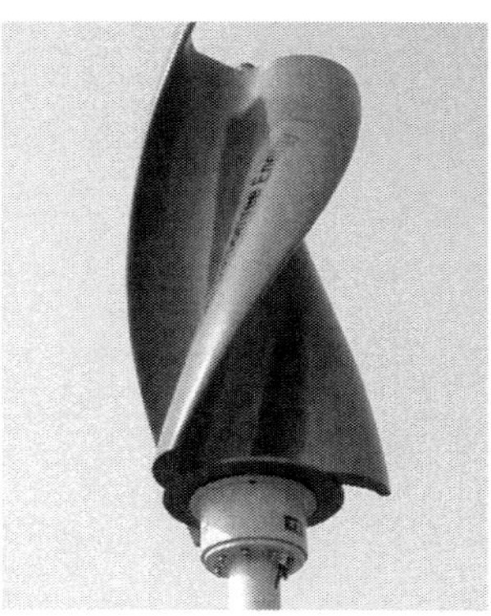

Figure 25. Helical Savonius wind turbine with three blades.

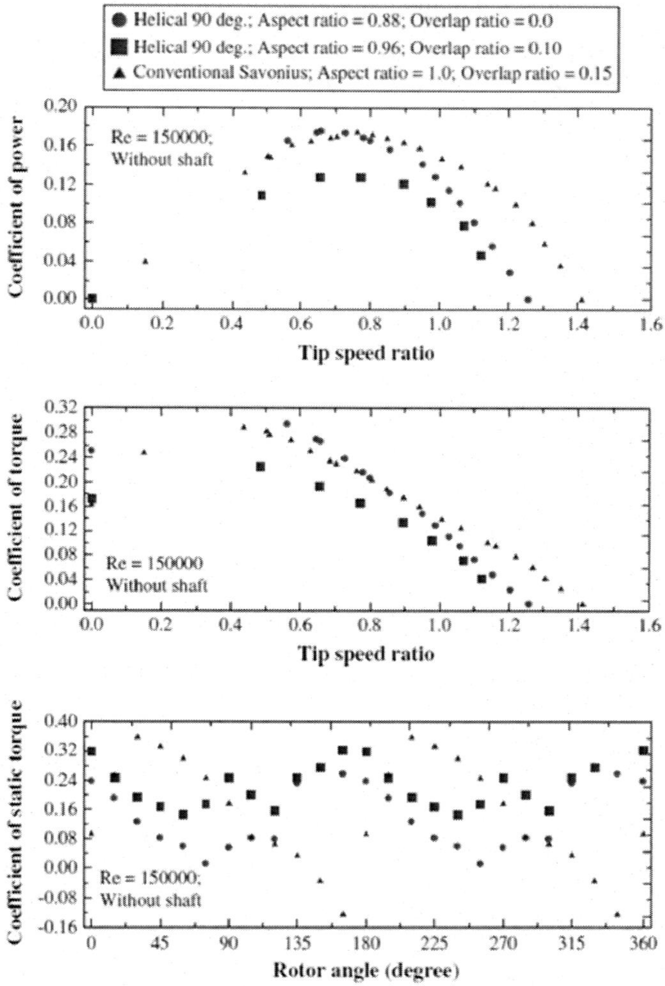


Figure 26. Coefficients of power, torque and static torque for a two-blade Savonius wind turbine with both helical and conventional blades [39].

Mohamed et al. [38] present an analysis of Savonius wind turbines with two and three blades and with a deflector sheet (Figure 23-right). The results show that the deflector sheet installation increases the wind turbine efficiency considerably (Figure 24) and that the power coefficient is higher in the case of the rotor with two blades.

Kamoji et al. [39] conducted a study of the helical Savonius wind turbine (Figure 25). Figure 26 shows that both the power and torque coefficients are slightly higher in the helical design than in the traditional Savonius under low wind velocity conditions. The great advantage of the helical blades is that the coefficient of static torque is always positive for all incident angles (Figure 26). The main disadvantage is the higher complexity of the construction, which implies a cost increase [38].

As mentioned previously, the low coefficient of power of the Savonius rotor is the most important disadvantage; hence, the improvement of this factor is important to the development of this technology [37, 38]. In Table 3, a summary of the main improvements to increase the performance of the Savonius rotor is shown.

Müller et al. [40] propose a vertical-axis, resistance-type wind turbine (Figure 27). The main advantages are the simplicity of the design, which implies a lower cost of construction, and the high efficiency (Figure 28), approximately 48%-61%. This system can be part of a wind-solar hybrid [41, 42]. A disadvantage is that because the wind turbine is partially covered, the sweep area is lower and it has a fixed direction. This problem can be solved by adding a moving power-augmentation guide vane (PAGV) that surrounds the turbine and orients the inlet air flow, which is capable of increasing the rotor rotational speed by a factor of 1.75, the torque by a factor of 2.88 and the power output by a factor of 5.8 [43].

**Table 3. Improvements to increase the performance of Savonius wind turbines [37]**

| Design modification | Gain | Comments |
|---|---|---|
| Helical rotors | Improvement of static torque | Complex design, high cost |
| Deflector plate | 20% | No further details since 1992 |
| Twisted-blade | 27% relative | Complex design, high cost |
| Guide-box tunnel | 50% (3 blades) | Complex design |
| Modified Savonius | 60% in static torque | Expected vibration problem |
| Guide vanes | Depends on wind speed | Problems for large tip-speed |
| Obstacle plate | 15% on peak value | Small parameter space used |

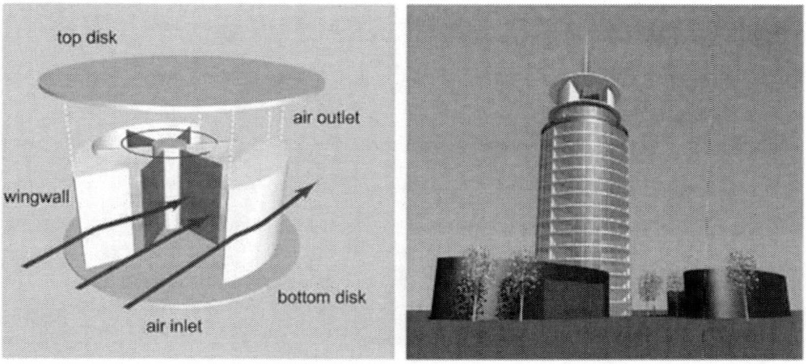


Figure 27. Vertical-axis resistance-type wind turbine [40].

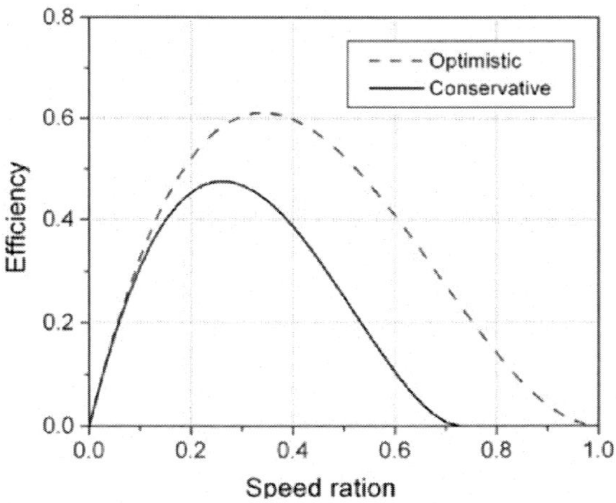


Figure 28. Efficiency of vertical-axis resistance-type wind turbines [40].

## 3.6. Ducted Wind Turbine

Grant et al. [44] present a ducted wind turbine, installed on the edge of the building roof (Figure 29). The power coefficient can be higher than the Betz limit, even exceeding unity (Figure 30). However, both the high cost/power ratio and the fixed direction make it difficult for this technology to compete against other alternatives [44]. Furthermore, this kind of wind turbine can only operate with a perpendicular wind direction.

Figure 29. Ducted wind turbine on the edge of the building roof [44].

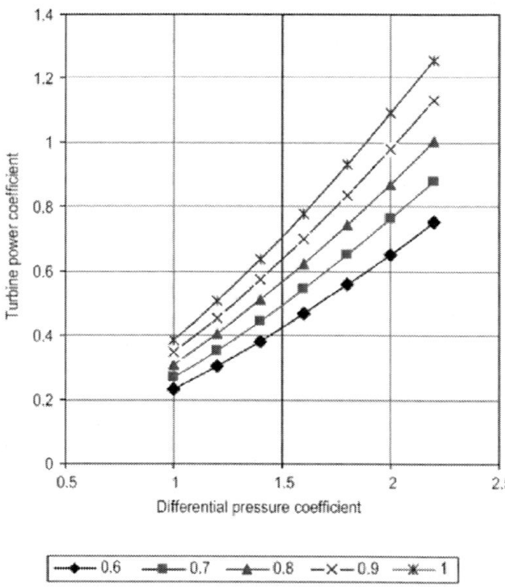


Figure 30. Power coefficient vs. the differential pressure coefficient of a ducted wind turbine for various values of the speed coefficient [44].

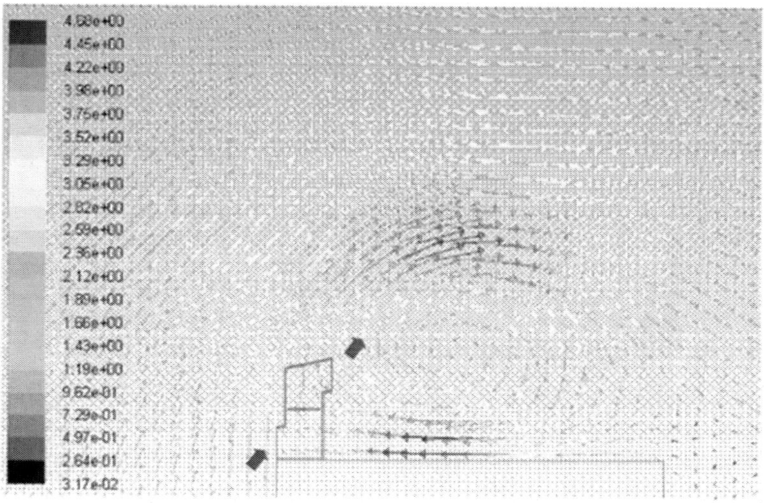


Figure 31. Superposition of a ducted wind turbine on the edge of the building roof for an incident velocity of 1 m/s.

Figure 31 shows the superposition of a ducted wind turbine on the edge of the building roof for an incident velocity of 1 m/s. The represented conditions are appropriate for turbine operation, although this kind of wind turbine can only operate with a perpendicular wind direction, as in the simulation.

## 4. The Significance of the Multidirectional Character of the Urban Wind for Turbines

The results of the qualitative analysis show that the use of a HAWT on the central area of a flat building roof presents problems from the aerodynamic point of view because the multidirectional wind is incompatible with normal turbine operation. Moreover, the HAWT is subjected to loads beyond its design specifications.

In contrast, because of their omnidirectional character, VAWTs (including horizontal Darrieus and ducted wind turbines) have aerodynamic advantages in urban environments. Both horizontal Darrieus and ducted wind turbines have additional aerodynamic advantages in building roof applications, where the multidirectional condition of the urban wind is more favourable than the unidirectional wind conditions.

In urban power applications, the multidirectional character of the wind plays a decisive role, even more than the incident velocity. For this reason, the VAWT technology is of great interest.

The results of this analysis are applicable to the case of buildings with a lower relative length, including skyscrapers, because of a vorticity phenomenon that occurs on the roof [7]. The vertical edges are also of interest for wind power exploitation in high-rise buildings because of the concentration factor of the velocity around them [36], as shown in Figure 19-right.

The results can be extrapolated to other roof shapes. Both pyramidal and pitched roof shapes present a multidirectional character of the wind even greater than that of the flat roof [6].

## CONCLUSION

In this work, a literature review of the main perspectives and proposals for wind power exploitation in urban environments has been given. With the obtained information, the characteristics of the urban wind and the proposed solutions for its exploitation have been analysed. This analysis provides an understanding of the particular characteristics of the urban wind, and it gives a wide overview of the various technologies that have been proposed for wind resource exploitation in this complex environment, with many challenges and opportunities for development.

In the urban environment, the multidirectional character of the wind plays a more important role than the incident velocity. The velocity fields have a highly multidirectional component, which requires both qualitative and quantitative analyses of the wind turbine behaviour under these conditions. The normal test conditions for these systems are a unidirectional incident wind and a stable atmosphere. Because of this, to analyse the effect of the multidirectional conditions of the urban wind on the turbines, a CFD simulation of the air flow around a building has been conducted. The results of the simulation show velocity fields in agreement with the highly multidirectional component.

The sections of various kinds of wind turbines have been superimposed over the velocity field obtained from the simulation, and their aerodynamic behaviour has been analysed in a qualitative manner.

The obtained results show that HAWTs have higher performance in flat terrain applications or similar conditions such as on large structures. The main advantage of HAWTs is their high efficiency, but their great disadvantage is that they can only operate under unidirectional wind conditions, so they are less appropriate for urban environments.

Because of their omnidirectional character, VAWTs are more suitable in areas with a higher building density – they can accommodate multidirectional and turbulent conditions. Both horizontal Darrieus and ducted wind turbines have better behaviour on a building roof than in flat terrain conditions, although they have technical and economic disadvantages.

Both pyramidal and pitched roof shapes present a higher multidirectional character of the wind than flat roofs. These shapes are less desirable for wind exploitation than the flat roof because the wind velocity on the roof decreases for a wide range of incidence angles, causing a decrease in wind potential.

The vertical edges are also a potential site for wind power exploitation in high-rise buildings because of the concentration factor of the velocity around them.

The wind energy in the urban environment is a resource with great potential that is currently wasted. The main advantages of its exploitation are:

- It creates an electricity supply in isolated areas far from the power grid, adapting itself to renewable sources and to the supply requirements. Either HAWTs or VAWTs can be chosen according to the topography and the wind characteristics.

- The energy is generated in a distributed manner (distributed energy micro-generation), avoiding both transport and distribution electrical losses.
- It allows the energy self-consumption to be either isolated or connected to the power grid. In the case of Spain, for example, self-consumption is regulated (RD 1699/2011), and future regulations to allow the injection of the excess electricity into the power grid are expected. Then, the electricity generated but not used at that moment will be injected to the power grid and measured by means of a bidirectional electricity meter. Therefore, the use of batteries is not necessary, decreasing both installation and maintenance costs. A balance between consumption and generation will be determined at the end of each month, yielding the net electricity consumption or generation.
- It can be combined with photovoltaic energy into hybrid facilities.

Micro wind generation in the urban environment is viable. The greatest impediment to its development is the lack of adequate regulation. The development of intelligent power grids will facilitate, from the technical point of view, distributed generation, and it will enhance the requirements of both small wind and photovoltaic facilities.

The conclusions of this work show the necessity of further work in the urban wind energy field. Specifically, the development of both qualitative and quantitative tests of different wind turbines are of great interest, mainly VAWTs (which present some advantages in urban environments), under multidirectional wind conditions because these are the actual conditions that the turbines are subject to in urban environments. Future work will enhance the proposed solutions for the urban wind exploitation, a field with many opportunities for development in a future with many uncertainties about the energy supply.

## REFERENCES


[1] World energy outlook 2010. *International Energy Agency*. http://www.worldenergyoutlook.org/ (02/14/2013).

[2] United States Energy Information Administration. http://www.eia.doe.gov/ (02/14/2013).

[3] *BP statistical review of world energy*. June 2010. http://www.bp.com/statisticalreview (02/14/2013).

[4] Walker SL. Building mounted wind turbines and their suitability for the urban scale - A review of methods of estimating urban wind resource. *Energy Build* 2011;43:1852-62.

[5] Grauthoff M. Utilization of wind energy in urban areas. Chance or utopian dream? *Energy Build* 1990/91;15-16:517-23.

[6] Ledo L, Kosasih PB, Cooper P. Roof mounting site analysis for micro-wind turbines. *Renew Energy* 2011;36:1379-91.

[7] Lu L, Ip KY. Investigation on the feasibility and enhancement methods of wind power utilization in high-rise buildings of Hong Kong. *Renew Sustain Energy Rev* 2009;13:450-61.

[8] Bahaj AS, Myers L, James PAB. Urban energy generation: Influence of micro-wind turbine output on electricity consumption in buildings. *Energy Build* 2007;39:154-65.

[9] Chicco G, Mancarella P. Distributed multi-generation: a comprehensive view. *Renew Sustain Energy Rev* 2009;13:535-51.

[10] El-Ela AAA, Allam SM, Shatla MM. Maximal optimal benefits of distributed generation using genetic algorithms. *Electr Power Syst Res* 2010;80:869-77.

[11] Ackermann T, Andersson G, Söder L. Distributed generation: a definition. *Elect Power Syst Res* 2001;57:195-204.

[12] Saidur R, Rahim NA, Islam MR, Solangi KH. Environmental impact of wind energy. *Renew Sustain Energy Rev* 2011;15:2423-30.

[13] Pantazopoulou G. Noise issues and standards for domestic wind turbines. Microwind: The Complete Picture - 13th January 2009. *Warwick Microwind Trial project* (Encraft).

http://www.warwickwindtrials.org.uk/resources/Giota+Pantazopoulou+-+BRE.pdf (02/14/2013).

[14] Bansal RC, Bhatti TS, Kothari DP. On some of the design aspects of wind energy conversion systems. *Energy Convers Manag* 2002;43:2175-87.

[15] Dabis HS, Chignell RJ. Wind turbine electromagnetic scatter modeling using physical optics techniques. *Renew Energy* 1999;16:882-7.

[16] Ricciardelli F, Polimeno S. Some characteristics of the wind flow in the lower Urban Boundary Layer. *J Wind Eng Ind Aerodyn* 2006;94:815-32.

[17] Ng E, Yuan C, Chen L, Ren C, Fung JCH. Improving the wind environment in high-density cities by understanding urban morphology and surface roughness: A study in Hong Kong. *Landsc Urban Plan* 2011;101:59-74.

[18] Abohela I, Hamza N, Dudek S. Effect of roof shape, wind direction, building height and urban configuration on the energy yield and positioning of roof mounted wind turbines. *Renew Energy* 2013;50:1106-18.

[19] Watson SJ. Predicting the yield of small wind turbines in the roof-top urban environment - an update. Microwind: The Complete Picture - 13th January 2009. *Warwick Microwind Trial project* (Encraft). http://www.warwickwindtrials.org.uk/resources/Simon+Watson+-+Loughborough+University.pdf (02/14/2013).

[20] Oppenheim D. Outside the square. Integrating wind into urban environments. *Refocus* 2004;May/June:32-5.

[21] Dayan E. Wind energy in buildings. Power generation from wind in the urban environment - where it is needed most. *Refocus* 2006; March/April:33-8.

[22] Mertens S. Wind energy in urban areas. Concentrator effects for wind turbines close to buildings. *Refocus* 2002; March/April:22-4.

[23] Holdsworth B. Options for microwind generation: part I. *Renew Energy Focus* 2009; March/April:60-3.

[24] Holdsworth B. Options for microwind generation: part II. *Renew Energy Focus* 2009; May/June:42-5.

[25] Holdsworth B. Options for microwind generation: part III. *Renew Energy Focus* 2009; September/October:36-40.

[26] Eriksson S, Bernhoff H, Leijon M. Evaluation of different turbine concepts for wind power. *Renew Sustain Energy Rev* 2008;12:1419-34.

[27] Syngellakis K, Carroll S, Robinson P. Small wind power. *Refocus* 2006; March/April:40-5.

[28] Wright AK, Wood DH. The starting and low wind speed behaviour of a small horizontal axis wind turbine. *J Wind Eng Ind Aerodyn* 2004;92:1265-79.

[29] Wang F, Bai L, Fletcher J, Whiteford J, Cullen D. The methodology for aerodynamic study on a small domestic wind turbine with scoop. *J Wind Eng Ind Aerodyn* 2008;96:1-24.

[30] Wang F, Bai L, Fletcher J, Whiteford J, Cullen D. Development of small domestic wind turbine with scoop and prediction of its annual power output. *Renew Energy* 2008;33:1637-51.

[31] Riegler H. HAWT versus VAWT. Small VAWTs find a clear niche. *Refocus* 2003;July/August:44-6. http://wapedia.mobi/en/Savonius_wind_turbine (02/14/2013).

[32] Howell R, Qin N, Edwards J, Durrani N. Wind tunnel and numerical study of a small vertical axis wind turbine. *Renew Energy* 2010;35:412-22.

[33] El-Samanoudy M, Ghorab AAE, Youssef SZ. Effect of some design parameters on the performance of a Giromill vertical axis wind turbine. *Ain Shams Eng J* 2010;1(2010-09-01):85-95.

[34] Kirke BK, Lazauskas L. Limitations of fixed pitch Darrieus hydrokinetic turbines and the challenge of variable pitch. *Renew Energy* 2011;36:893-7.

[35] Sharpe T, Proven G. Crossflex: Concept and early development of a true building integrated wind turbine. *Energy Build* 2010;42:2365-75.

[36] D'Alessandro V, Montelpare S, Ricci R, Secchiaroli A. Unsteady aerodynamics of a Savonius wind rotor: a new computational

approach for the simulation of energy performance. *Energy* 2010;35:3349-63.

[37] Mohamed MH, Janiga G, Pap E, Thévenin D. Optimization of Savonius turbines using an obstacle shielding the returning blade. *Renew Energy* 2010;35:2618-26.

[38] Kamoji MA, Kedare SB, Prabhu SV. Performance tests on helical Savonius rotors. *Renew Energy* 2009;34:521-9.

[39] Müller G, Jentsch MF, Stoddart E. Vertical axis resistance type wind turbines for use in buildings. *Renew Energy* 2009;34:1407-12.

[40] Chong WT, Naghavi MS, Poh SC, Mahlia TMI, Pan KC. Techno-economic analysis of a wind–solar hybrid renewable energy system with rainwater collection feature for urban high-rise application. *Appl Energy* 2011;88:4067-77.

[41] Chong WT, Fazlizan A, Poh SC, Pan KC, Ping HW. Early development of an innovative building integrated wind, solar and rain water harvester for urban high rise application. *Energy Build* 2012;47:201-7.

[42] Chong WT, Pan KC, Poh SC, Fazlizan A, Oon CS, Badarudin A, Nik-Ghazali N. Performance investigation of a power augmented vertical axis wind turbine for urban high-rise application. *Renew Energy* 2013;51:388-97.

[43] Grant A, Johnstone C, Kelly N. Urban wind energy conversion: The potential of ducted turbines. *Renew Energy* 2008;33:1157-63.

In: Distributed Wind Energy Generation ISBN: 978-1-53614-207-5
Editors: A. Colmenar-Santos et al. 

*Chapter 4*

# Wind Control Centres

***Jose-Maria Gallardo-Calles*, Javier Ontañón-Ruiz and Elio San Cristóbal-Ruiz***

Departamento de Ingeniería Eléctrica, Electrónica, Control, Telemática y Química Aplicada a la Ingeniería, Universidad Nacional de Educación a Distancia (UNED), Madrid, Spain

## Abstract

In this chapter, the problems that have motivated the development of wind control centres are presented, highlighting the Spanish case. This chapter describes typical telecontrol architecture from the local wind farm level to inter-control centre connectivity, with special emphasis on differences compared with other control centres and the functionalities that can be implemented in this architecture. A study on the use of virtual machines in wind control centres is also included for the purpose of shedding light on this modern trend, despite the strong inertia in the use of conventional servers.

**Keywords**: control centre, SCADA, virtual machines, wind energy; communications

* Corresponding Author Email: elio@ieec.uned.es.

## INTRODUCTION

Wind energy has experienced fast growth in the last thirty years. Before the renewable revolution, this method of generating energy was only exploited by small and medium-sized consumers, who were always connected to isolated networks. In the early 1980s, the construction of wind farms began; these farms were composed of tens of wind turbines connected to distribution and transmission networks. Recently, the worldwide wind power exceeds 150 GW and is mainly distributed in the USA, Europe and China [3, 14].

There are around 80 manufacturers of wind turbines in the world and almost 500 different models [1]. In recent years, the technology has evolved quickly, and there are many differences between the wind turbines currently being installed and those installed initially, such as the shape of blades, the type of electrical generator and the control system.

One of the main differences between wind energy and other traditional energy resources is its distributed nature, i.e., each plant or wind farm is composed of many generation units, or wind turbines, completely independent and connected in series. This distributed nature implies a challenge not only during the construction phase of the wind farm, but also during operation. The operation and maintenance effort per unit of power is much greater than in traditional generation plants and, therefore, the design of monitoring and control systems adapted to this technology is key to ensuring the profitability of these installations [11].

## LEGISLATIVE FRAMEWORK

The regulation of wind power is completely different in each country, but in all countries its participation is magnified over non-renewable resources using different market mechanisms. However, the unpredictable nature of wind power implies an important problem for grid operators because they must match the energy demanded by consumers with the production estimated by producers. To reach a balance between promoting renewable energy and protecting the quality of the electric system, governments have approved specific laws for renewable energy. Spain, one of the pioneering countries in integrating non manageable renewable energy in the grid, approved the "special scheme for renewable energy" in 1997. This law establishes advantageous conditions for connecting plants to the grid and special remuneration for renewable facilities whose rated power is less than 50 MW [2]. This scheme is still in force, but some modifications have been approved because renewable generation, driven mainly by the force of wind energy, has increased its share in the Spanish electric system, becoming the third largest energy source in Spain in 2009 [7]. The purpose of most modifications is to minimize the impact of non-manageable renewable energy on the quality of the electricity supply. New bonuses and penalties related to active and reactive power control have been introduced. The imbalance between forecast and real time power output also incurs a huge penalty.

In 2007, the Spanish government established that "all facilities of renewable energy whose rated power exceeds 10 MW must be assigned to a generation control centre, which acts as an interlocutor to the grid operator, transmitting information in real time and enforcing to execute the instructions received in order to ensure the reliability of the electric system. To be assigned to a control centre is mandatory for receiving the fee or established bonus" (Real Decreto 261/2007 25th May). Currently, the Transmission System Operator (TSO) of the Spanish electricity system, Red Eléctrica de España (REE), manages all renewable energy facilities over 10 MW from the Control Centre of Renewable Energy (CECRE). REE receives real-time information about the grid connection status of these facilities and their main production measurements and also sends commands to curtail power when it is necessary to guarantee grid stability.

The European Union launched a project called Twenties this year to connect more renewable energy to the grid safely. Twenties is an initiative to integrate new technologies that can connect more renewable energy to the grid [13], with the final purpose of cutting 20% of $CO_2$ emissions in 2020 [2]. This project is led by REE and Iberdrola Renovables, among other organizations and private companies, who will try to prove that it is possible to connect an additional 500 MW of wind power to the grid using a new voltage and power-frequency control system. This project will be coordinated between the Renewable Control Centre of Iberdrola Renovables (CORE) and CECRE; both groups are pioneers in the wind industry.

To fulfil the current and future laws, it is necessary to have wind farm control systems and remote control centres capable of managing and consequently acting upon the large amounts of information involved.

## Wind Farms Local Control Systems

To monitor and control wind turbines, meteorological masts and substations that continuously control devices must be connected to the wind farm local communication network, which supports the traffic between PLCs and SCADA (Supervisory Control and Data Acquisition). The SCADA system is responsible for collecting and managing all information received. Figure 1 below represents the typical control architecture of a wind farm.

Most wind turbine manufacturers include their own SCADA system as part of the wind farm supply. This system communicates with wind turbines not only to receive information but also to send, start and stop commands and active and reactive power set points. Wind turbine manufacturers also implement necessary regulation algorithms corresponding to the local legislation of each country. A strong knowledge of wind turbine technology is essential in developing these algorithms, so wind turbine manufacturers are not willing to share the key for communicating with wind turbine controllers directly [16].

The wind farm SCADA must be able to implement all communication protocols used by different devices to receive information [20]. This information is used by algorithms in real time and represented in an HMI (Human Machine Interface). Personnel responsible for O&M (operation and maintenance) can also send commands to turbines and substation switchgear through the HMI [5]. A subgroup of data received is saved in historical databases and can be requested later via external reports, following the requirements of the developers.

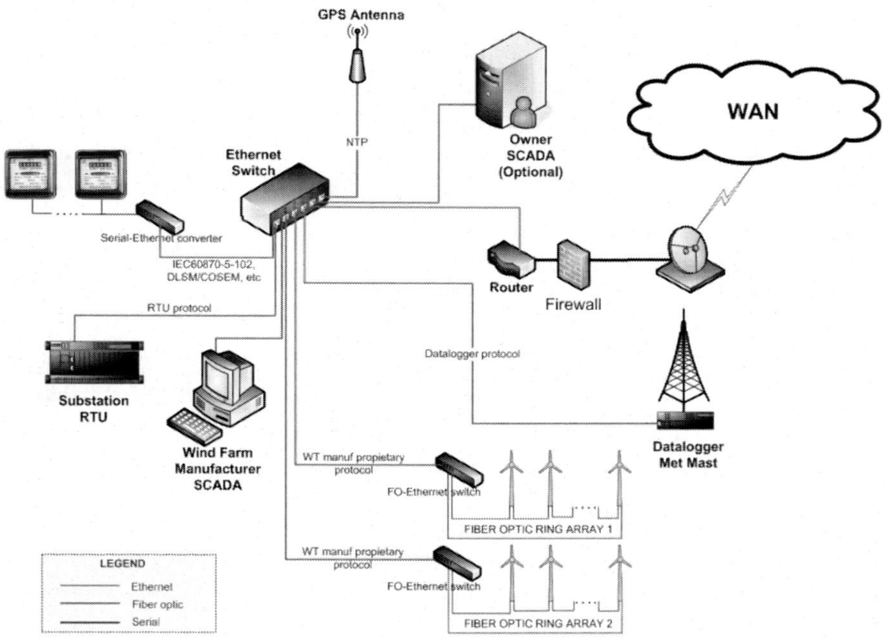


Figure 1. Typical control architecture of a wind farm.

Communication among devices represented in the diagram below is established via protocols that are frequently not standard [9]. Sometimes, wind turbine manufacturers define their own proprietary protocols for wind turbine PLCs to, among other reasons, prevent third parties from establishing communication with wind turbine controllers directly, overriding their wind farm SCADA [23].

Meteorological data have provided essential information due to penalties related to deviations between energy offers presented to the energy market and real time power output. The energy offer is estimated from the forecast and is calculated using historical data of meteorological masts. For this reason, developers are continuously increasing investments in control systems for meteorological masts [4].

The monitoring and operation of substations is also performed with the wind farm SCADA provided by the wind turbine manufacturers or developers. Communication protocols commonly implemented in substations are IEC 6870-5-101, DNP 3.0, Modbus RTU and TCP [21]. Some PLC manufacturers define their own proprietary protocols that are non-compliant with any standard [8].

Energy data registered by meters are also a variable required by the developers to verify that the energy paid by the distribution company corresponds to the energy produced. Energy meters are usually connected to a serial bus. A serial-ethernet adapter can be used to link the bus terminal to the wind farm SCADA, which can download energy trends from the meter via IEC 60870-5-102 over Ethernet or via other standard protocols, depending on the country.

The IEC (International Electrotechnical Commission) developed two new standards, IEC 61850-7 and IEC 61400-25, with the aim of establishing a common communication architecture for monitoring and controlling substations and wind farms, respectively. The final purpose was to make any SCADA system capable of communicating with any device in a standard way [10], highlighting three essential areas: data models, information interchanged and communication between systems. The main reason behind standardizing information is to make the operation of a large number of wind farms from one control centre easier, which is especially interesting for large developers.

## WIND CONTROL CENTERS

Most important wind developers have wind farms distributed geographically in one or more countries, and their individual rated power does not reach the threshold to make profitable 24x7 operations in situ, but the sum of all of them is very high. The production of a wind farm and, consequently, its profitability depends on how quickly incidents are detected and resolved. Therefore, developers install control centres to control and monitor their wind farms, from which they perform remote commands and warn qualified staff when an incident requires action.

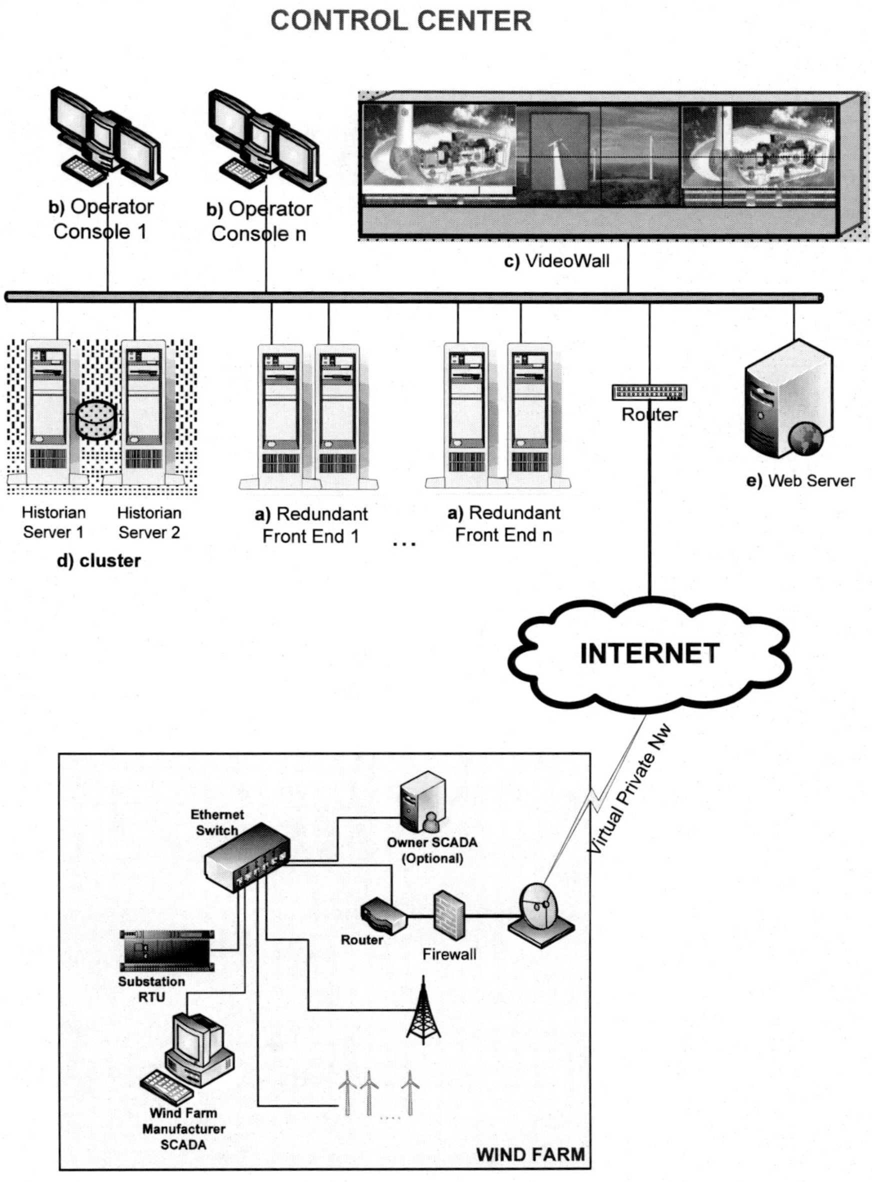


Figure 2. Typical architecture of a wind control centre.

Wind control centres, like any other type of control centre, are designed according to the amount of information managed, the criticality of data and the need to utilize that data in the future. Figure 2 shows a diagram of the typical architecture of a control centre capable of controlling an unlimited number of wind farms by adding identical systems in parallel while maintaining the basic original structure.

The main features of this system are the high availability configuration and access to information in real time and the historical mode very quickly, efficiently and in a standard way. In this design, various technologies participate; these technologies are discussed independently below.

## COMMUNICATIONS

For optimum availability of the whole system, it is necessary to ensure that communication links are stable and satisfy the requirements of the bandwidth and latency needed for operation. To achieve the stability objective, it is desirable to make devices and links redundant whenever possible and profitable. The computers at the control centre are usually placed in the same building; therefore, it is easy and inexpensive to establish an Ethernet LAN (local area network) of high capacity and even make it redundant. Many communication hardware providers on the market offer solutions for this architecture. Most also offer the capability to monitor the status of their equipment via standard communication protocols such as SMTP [20].

The issue related to communication links between the control centre and wind farms is more complex because minimum requirements and redundant links are not always achievable. When it is possible, sometimes it is not economically viable. Wind farms are often built far from urban centres where there is no Internet access; consequently, it is necessary to invest in radio links to access points or satellite links with non-guaranteed bandwidth at a high cost. Some developers use fibre optic lines next to the high voltage lines of the distribution company.

Whatever the solution, it is important to protect the highly sensitive information transmitted between wind farms and the control centre, especially if it is transmitted through the Internet. One possible solution is to use VPNs (virtual private networks), configured by routers at both ends, which establish a virtual local network over a public network that is not controlled, therefore maintaining the authentication, integrity and confidentiality of messages.

## COMPUTER SYSTEMS

Computer systems are the core of this architecture. They must be sized to receive, process and store the necessary information. In the diagram above, the following devices are represented:

a. n units of front end associations, composed of a couple of servers configured in hot-standby mode, which are responsible for receiving information from wind farms. This configuration allows the growth of the control centre by connecting additional couples of front ends in parallel as the number of wind farms increases. In the case of wind energy, the amount of information to be stored per unit of power is very high because each wind farm consists of many wind turbines and each turbine is composed of a multitude of sensors, controls, etc. Considering the example of a medium-size wind farm in Spain with 30 turbines of 1.5 MW each and assuming that each turbine has around 200 variables, the total number of variables to be controlled is 6000 variables for only 45 MW.
b. n units of operator consoles to represent the information acquired by the front ends using a simple and intuitive graphical interface. The number of monitors connected to the each operator console is commonly three or four, depending mainly on the method of operation.

c. A video wall composed of n rear projection cubes in matrix configuration, with a standard size of 100x100 cm per cube. Most common configurations are 3x2 cubes that make a total size of 3x2 meters. This system also requires a computer to reproduce the image and a projector to project the graphical output from the computer to the cubes. The application of operator consoles also runs on this computer, and functionalities such as video security systems can be implemented as well. New technologies allow the use of LCD monitors, with sizes from 40" to 70" or even greater, that can also be used in matricial configurations.
d. A cluster for storing historical data composed of two independent servers that increase the availability of the system by sharing a hard disk accessed by both servers. This access can be simultaneous (hot-hot standby) or alternative (hot-cold standby). The shared disk array must be redundant to prevent a crash of one hard disk from leading to a system failure. There are different types of redundant associations of disks, such as RAID (redundant array of independent disks), which offers different levels to improve data integrity against faults, performance or capacity.
e. A web server that shows real-time and historical information through remote connections using web clients, such as Microsoft Internet Explorer, concurrently. Real-time information is provided by the SCADA through graphic screens programmed with http language, and the historical information is exploited by reports that establish connections to historical servers.

Historical servers can store the information received from front ends or wind farm SCADAs directly. The first option is the best from the point of view of communication performance because it avoids sending the information to the control centre twice, first for monitoring and real-time applications and second for the historical servers. The second option only makes sense when there are applications to buffer the information in local SCADAs to fill information gaps that can arise due to temporal communication disruptions between the wind farm and the control centre.

In this case, regarding bandwidth usage, the cost of uploading the same information twice must be evaluated against the criticalness of historical data.

The decision regarding which type of historical database to install in the control centre depends mainly on the amount of information to store, the desired depth, the scan rate and the speed required in the data access using reports. Most of the data stored in a wind farm comprise analogue measurements that are continuously changing; therefore, large quantities of records are generated. Software based on numerical series, such as OSIsoft PI, are suitable for this purpose because they incorporate algorithms for data compression based on interpolation, increasing the speed of the method to extract information. In the case of alarms and events, it is not possible to compress the information, and relational databases such as Oracle or SQL Server provide satisfactory results at a lower cost than the former.

The architecture presented can be implemented in Windows and Linux operating systems.

Whatever the final architecture, the investment in hardware and software is very high in a control centre of a certain size, not only for the cost of equipment but also for necessary infrastructure. Once the control centre begins the operational phase, the main costs are determined by maintenance, operation and consumption of energy. As in other domains of information technology, virtualization is a strong trend to reduce both the initial investment in hardware and the energy consumption of data centres. The use of virtualization of multiple machines on a single high performance server is evaluated in this chapter.

A virtual computer is the abstraction of computer resources by creating software replicas of one or more computing devices, i.e., multiple operating systems are executed in a single physical server, called the host, simultaneously [17][18]. Each virtual machine uses the resources of the host machine without interfering in the operation of other systems. The limits of resources (RAM, hard disk, network interfaces, processing time, etc.) are established when each virtual machine is created.

Many suppliers provide software with these features, such as VMware with its family of products for different solutions, VirtualBox, Mac-on-Linux, Oracle VM and Microsoft Virtual PC 2007 [6].

The functionalities offered by virtual machines are increasing as technology and research advance, highlighting simplification to create virtual machines, backup and restore tools, hot back-ups, simple connection and disconnection of physical devices from the host machine, hot reconfiguration of resources of each virtual machine, etc. [6].

The main advantage of using virtualized systems is the cost savings. A comparison study between two systems, a conventional system and a proposed virtualized system, for a control centre is presented below using a large wind control centre of 5,000 MW as an example.

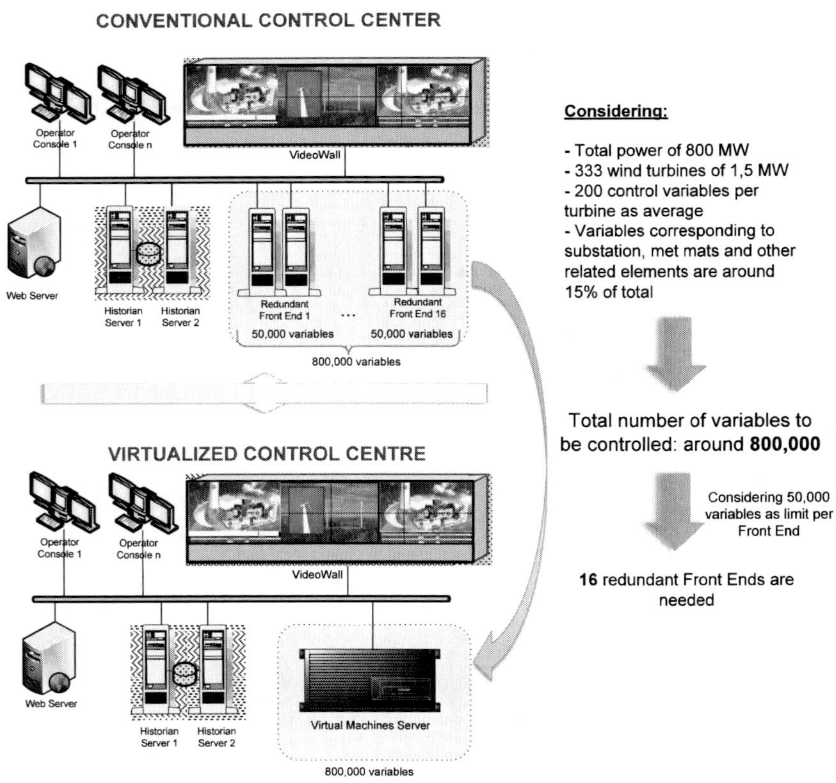


Figure 3. Conventional control centre vs. virtualized control centre.

A maximum limit of variables per front end exists because the information is acquired, processed, stored and sent to the operator consoles the front end in real time. Thus, including more variables than this limit in each front end could compromise the stability of the system and make the machine more critical. A limit of 50,000 variables is considered as an example, based on experience with different SCADA applications, but this limit depends on the particular software.

Front ends of a control centre are commonly industrial servers of medium-high performance (2 GHz dual processor, two 120 GB at 7500 rpm SATA hard drives, two 10/100 Mbps network cards, 2 GB of RAM, etc.). On the other hand, a high performance (e.g., two quad-core processors, twenty 1 TB at 15,000 rpm SAS hard drives in RAID 1 + 0 configuration, six network cards at 1 Gbps, 64 GB of RAM, four power supplies, etc.) and redundant server are required for a server to act as a host for virtual machines of the control centre proposed. However, the amount of equipment needed for a virtualized control centre is reduced significantly when virtual machines are used, which also implies other savings:

- Decreased number of support elements: PDUs (power distribution units), UPSs (uninterrupted power systems), cabinets, ventilation systems, etc.
- Reduced maintenance costs and replacements.
- Simplification of wiring, which implies lower costs and reduced communication incidences.
- Reduced energy consumption.
- Reduced space required for the data centre.

The operating system of industrial servers must be Microsoft Windows Server Standard Edition 32 bit or similar, but the virtual machine server requires Microsoft Windows Server Enterprise Edition 64 bit (or the equivalent Linux operating system) and the operating systems used by the virtual machines.

The hardware savings is high in this case because 32 servers are replaced by only one physical machine. Obviously, the savings is not 32:1 because the cost of the virtual machine server is much greater than that of each front end. The software does not produce any savings; in fact, it is necessary to invest in an additional operating system of higher features for the host machine. It is necessary to keep in mind that the license usage agreements do not distinguish between a virtual machine and a physical one, so the same number of licenses is needed.

However, it is important to mention that the use of virtual machines increases the criticality of the hardware and means that hardware maintenance will cause downtime or unavailability.

## DATA MODELS

The aim of a control centre is to control a group of wind farms with different technologies, configurations and sizes from a single point. An operator must be able to control wind farms under his charge without a strong knowledge of the specific characteristics of each one. Thus, it is essential to standardize information received from different sources and present it in a common way, independent of the wind turbine model, manufacturer, communication protocol implemented, etc. This standardization effort can be made at the wind farm level or at the front end of a control centre. The main advantage of performing standardization at the wind farm level is the simplification of configuring and maintaining the control centre. For this purpose, new communication gateways, independent of those from the manufacturer, are installed to establish communication with all elements of the wind farm through corresponding protocols, adapt the information to the standard model and send it to the control centre via a common protocol, which follows the IEC 61400 or IEC 61850 standards mentioned above.

## FUNCTIONALITIES IMPLEMENTED IN A CONTROL CENTRE

The architecture proposed above can not only operate wind farms but also implement additional functionalities that make it possible to comply with current legislation and provide additional O&M value.

In the particular Spanish case, REE Control Centre for Renewable (CECRE) establishes connections with the wind control centres. These centres have been installed and officially homologated by REE since 2007. To be included in the list of control centres certified to interact with CECRE, a control centre must fulfil a list of technical and human resource requirements such as 24x7 operations, redundant and dedicated communication links to primary and secondary CECRE, redundant servers for real time communication and capability of curtailing active power exported by wind farms. REE, as shown on the official web site [http://www.ree.es] [Accessed: 11th August 2010]," *(...) guarantees the continuity and reliability of electrical supply and the appropriate coordination between production and transport systems, executing its functions according to the principles of openness, objectivity and independency.*" To maintain the balance in the electrical supply, the grid operator manages the adjusting services, which adapt the production programs generated by diary and intra-diary markets to the quality and security requirements of the electric system. However, unexpected disruptions occasionally occur. Disruptions are commonly produced by wind energy exported to the grid due to its unpredictable nature, its limited capability to generate and consume reactive power and its vulnerability to voltage dips. CECRE, which is connected to the control centres of REE, CECOEL and CECORE, continuously monitors the status of the high voltage network and pays special attention to the effects related to wind power. Taking into account the information analysed, CECRE decides if it is strictly necessary to reduce wind power and consequently sends corresponding set points to control centres that must be attained within 15 minutes. When the danger is past, CECRE cancels the constraint and transmits the new situation to the control centres. When wind farms connected to the same control centre and the same high voltage node have an equivalent technology, that is, the capability to regulate power and performance against voltage dips, the overall power constraint can be shared among them in the most convenient way. The control centre can therefore optimize the production of its wind farms; for instance, if one wind farm cannot reach the set point sent by CECRE due to low wind speed or availability of wind turbines, for example, others can generate more power without exceeding the node constraint. This kind of algorithm can only be implemented at the control centre level because it is necessary to receive information from all wind farms to decide which ones must reduce or increase power and then simultaneously send corresponding set points to wind farm power regulators. The capability of controlling the power exported to a high voltage node can also be useful for maintenance purposes, which implies disconnecting one transformer of the substation and consequently reducing the total power that can be evacuated.

Beyond real time operation, the historical data analysis is essential to calculate the efficiency of a wind farm, the wind profile of the geographical area and the behaviour of wind turbines, as well as to record all actions executed by controllers or received from CECRE. These analyses are useful to evaluate the wind potential of an area and consequently decide if installing more wind farms is profitable and which wind turbine model is more appropriate for a specific environment. Such analyses will also help in making predictions as accurate as possible meteorological predictions for presentation to the energy markets. To improve predictions, promoters use powerful systems with complex algorithms that can "learn" from data saved in the past months and years and then estimate future wind curves. The error is reduced as more data are analysed [15].

## INTERCONNECTION BETWEEN CONTROL CENTRES

The control of a wind farm is not always limited to the control centre actions. For example, CECRE can reduce the power of a wind farm in an indirect way. To make this interaction possible, it is necessary to establish communication links between control centres and agree on a common communication protocol than can be understood by both centres [19]. The most common protocol for real time communication between electrical control centres is IEC 60870-6-TASE.2 or ICCP (Inter Centre Control Protocol). This protocol is used by REE, not only by CECOEL and CECORE but also by CECRE. The protocol is also implemented in USA control centres to coordinate the energy imported and exported among generation and distribution companies and county grid operators.

Another example of the interconnection between control centres is the redundant configuration of some centres. Some promoters build a secondary control centre to guarantee that wind farms will be properly controlled if the main control centre is affected by a catastrophe. To ensure that the secondary control centre goes to active mode immediately after the main centre goes down, communications with wind farms must be established very quickly, and historical databases must be synchronized [22].

Finally, the internationalization of some companies has resulted in the need for connectivity between control centres located in different countries. This connection is intended to support the operation of one control centre from another if the first centre is destroyed and to collect information for international companies for use in analysing the efficiency of affiliate companies, to promote technical studies among different countries, etc.

## Conclusion and Further Developments

Control centres arose from the promoters need to operate a large number of wind farms by optimizing technical and human resources to reduce costs and increase the availability of wind turbines.

However, recent legislation approved by governments worldwide gives grid operators more capabilities to control renewable energy, using generation companies as a gateway with the purpose of connecting more wind power to the grid but avoiding any risk to the stability of the electric system. Consequently, control centres become more important every day in the renewable energies environment.

The density of information per wind power unit is very high, so the architectural design of a wind control centre is challenging with regard to communications and systems. Improved communication infrastructure and the use of light protocols are alternatives for the future [22]. The progress in computer technology provides new systems solutions every day.

This chapter provides a comparison between traditional control centres and virtualized centres. The main conclusion is that the cost savings are substantial. In the case of the example studied, the savings are around a factor of thirty. The profitability of using virtual machines depends on the size of the control centre because more front end servers are replaced by only one server, which means more cost savings. Several factors, such as the density of information, power distribution per generation unit, and capability of managing information of front ends, determine when virtual machines are recommended instead of physical ones.

It is important to clarify that economical savings are not the only factor that motivates the use of this technology. The reliability of the system and the independence against the hardware platform are also relevant [24]. The latter is especially important due to the short computer life cycle, which implies that servers are considered obsolete within a few years.

Another research line is offshore wind power. Some countries have run out of high wind potential areas and have started to develop offshore wind farms to increase their renewable energy capacity. Northern European countries are the pioneers of this new technology, which experts consider the second renewable energy revolution. Although the performance of offshore wind turbines is almost the same as that of onshore turbines, there are many environmental differences, such as sustenance, highly corrosive environment, access to turbines for maintenance purposes, and submarine electric lines. All of these reasons imply a great challenge for the control systems, which must be able to include new variables in algorithms such as tower vibrations produced by non-rigid anchoring. Moreover, telecontrol and communication systems must be more reliable and powerful for several reasons: the large number of critical variables related to electric infrastructures and security systems, the increased size of wind farms and the inability to frequently perform maintenance tasks in situ [12].

## References

[1] EWEA (European Wind Energy Agency), August 2011 [Online]. Available: http://www.ewea.org/.

[2] *REE annual report 2010.* August 2011 [Online]. Available: http://www.ree.es.

[3] *The Wind Power.* August 2011 [Online]. Available: <http://www.thewindpower.net.

[4] Foleya Aoife M., Leahya Paul G., Marvugliac Antonino, McKeogha Eamon J. Current methods and advances in forecasting of wind power generation. *Renewable Energy* 2012; 37:1-8.

[5] Thomas MS. *Remote Control. IEEE power & energy magazine.* July/August 2010.

[6] Nieh J and Leonard OC. Examining VMware. *Dr. Dobb's Journal* August 2000.

[7] Ruiz-Romero S, Colmenar-Santos A, Castro Gil MA. EU plans for renewable energy. An application to the Spanish case. *Renewable Energy* 2012; 43:322e33.

[8] (REE) Seguimiento tiempo real producción eólica. *Monitoring real time wind power production.* https://demanda.ree.es/eolica.html [accessed 19.04.10] Update 25.01.11.

[9] García Márquez FP, Tobias AM, Pinar Péreza JM, Papaelias M. Condition monitoring of wind turbines: Techniques and methods. *Renewable Energy* April 2012.

[10] Tian Z, Jin T, Wu B, Ding F. Condition based maintenance optimization for wind power generation systems under continuous monitoring. *Renewable Energy* 2011; 36:1502–1509.

[11] Carta González JA, Calero-Pérez R, Colmenar Santos A, Castro-Gil MA. *Generación Eléctrica con Energías Renovables.* [*Electric*

*Generation with Renewable Energies*]. Pearson-Prentice Hall & UNED, 2009.

[12] Scheu Matti, Matha Denis, Hofmann Matthias, Muskulus Michael. Maintenance strategies for large offshore wind farms. *Energy Procedia 24*(2012) 281 – 288.

[13] National Grid plc. *Operating the electricity transmission networks in 2020.* Initial consultation. June 2009 [Online]. Available: http://www.nationalgrid.com/NR/rdonlyres/32879A26-D6F2-4D82-9441-40FB2B0E2E0C/39517/Operatingin2020Consulation1.pdf.

[14] Kaldellis John K., Zafirakis D. The wind energy (r)evolution: A short review of a long history. *Renewable Energy* 36 (2011) 1887-1901.

[15] Foleya Aoife M., Leahya Paul G., Marvuglia Antonino, McKeogha Eamon J. Current methods and advances in forecasting of wind power generation. *Renewable Energy.* 37 (January 2012) 1–8.

[16] Márquez Fausto Pedro García, Tobias Andrew Mark, Péreza Jesús María Pinar, Papaelias Mayorkinos. Condition monitoring of wind turbines: Techniques and methods. *Renewable Energy.* 46 (October 2012) 169–178.

[17] Bailey David. 3 SCADA systems. Software and protocols. Newnes. *Practical SCADA for Industry* 2003. 64–99.

[18] Bailey David. 2 SCADA systems. Hardware and firmware. Newnes. *Practical SCADA for Industry* 2003. 11–63.

[19] Clarke Gordon. 2 Fundamentals of SCADA communications. Newnes. *Practical Modern SCADA Protocols.* 2003. 12–62.

[20] Clarke Gordon. 15 Applications of DNP3 and SCADA protocols. Newnes. *Practical Modern SCADA Protocols*. 2003. 374–390.

[21] Clarke Gordon. 3 Open SCADA protocols DNP3 and IEC 60870. Newnes. *Practical Modern SCADA Protocols.* 2003. 63–65.

[22] Clarke Gordon. 13 Fieldbus and SCADA communications systems. Newnes. *Practical Modern SCADA Protocols.* 2003. 349–361.

[23] Sangeetha A. Lakshmi, NaveenkumarB., GaneshA. Balaji, Bharathi N. Experimental validation of PID based cascade control system

through SCADA–PLC–OPC and internet architectures. Elsevier. *Measurement* 45. 4. (May 2012). 643–649.

[24] Brown E.P.M. Supervisory control and data acquisition (SCADA) reliability and availability improvement when state estimation and automatic testing are used. Elsevier. *Electric Power Systems Research*. 12. 1. (February 1987). 71–82.

In: Distributed Wind Energy Generation ISBN: 978-1-53614-207-5
Editors: A. Colmenar-Santos et al. 

*Chapter 5*

# Wind Repowering

***Severo Campíñez-Romer[1,*]***
***and Carlos-Ignacio Cuviella-Suárez[2]***
[1]Departamento de Ingeniería Eléctrica, Electrónica, Control,
Telemática y Química Aplicada a la Ingeniería,
Universidad Nacional de Educación a Distancia (UNED),
Madrid, Spain
[2]ROCA, Barcelona, Spain

## Abstract

At the end of January 2012, the Spanish government suspended the economic incentives for electricity generation facilities using renewable energy sources, including wind energy plants. Spain maintains a high level of energy dependence that can only be reduced by applying measures to increase energy efficiency and using massive amounts of renewable sources. In addition, the target assumed by Spain, i.e., to have at least 20% of the primary energy to be supplied by renewable sources by 2020, has not yet been reached. In Spain, wind farms, a number of which have been in commercial operation for over 15 years, offer a broad market appropriate for repowering. The use of more efficient wind turbines by means of repowering provides benefits to the electricity sector

[*] Corresponding Author Email: s.campinez.romero@gmail.com.

as a whole by optimizing the use of natural resources and facilitating the grid integration of the energy generated. This chapter analyses existing wind farms to quantify and characterise the market suitable for repowering. We discuss whether repowering is a valid alternative from the point of view of feasibility to enable the continuation of the integration of wind energy in the Spanish energy mix and whether this feasibility is sufficient when the energy generated is charged at the electricity market price in terms of grid parity. The results support that repowering is a profitable alternative and is often even better than the construction of new wind farms under certain conditions.

**Keywords:** wind energy repowering, profitability, grid parity

## NOMENCLATURE

### General

| | |
|---|---|
| $C_P$ | Coefficient of Performance |
| FIT | Feed-in-tariff |
| N-WF | New Wind Farm |
| O&M | Operation and Maintenance |
| P&C&G Grid | Power, Communications and Grounding Grid in the WF |
| Rep-WF | Repowered Wind Farm |
| RES | Renewable Energy Source |
| SCADA | Supervisory Control and Data Acquisition |
| Spot price | Daily Electricity Market Price |
| WTG | Wind Turbine Generator or Generators |
| WF | Wind Farm/Wind Farms |

### Generation and Losses

| | |
|---|---|
| CF | Capacity Factor (%) |
| $E_L$ | Electrical losses (%) |
| $F_{AV}$ | Mechanical Availability Factor (%) |
| $W_L$ | Wake effect losses (%) |

**Financial**

| | |
|---|---|
| IRR | Internal Rate of Return |
| NPV | Net Present Value |
| K | Discount Rate for NPV Calculation |
| TDEP | Depreciation Period of the WF (years) |
| RPI | Retail Price Index (%) |
| $CF_z$ | Cash Flow for Year z |
| $CFA_z$ | Cumulative Cash Flow for Year z |
| NPV | Net Present Value (€) |
| z | Ordinal indicating the number of years the WF has been in commercial operation, which is used in the IRR calculation |
| FL | Financial Leverage |
| VAT | Value Added Tax (%) |
| TAX | Incomes Taxes (%) |

## 1. INTRODUCTION

Law 54/1997 of the Electricity Sector [1] established a new regulatory framework, with the aim of guaranteeing the electricity supply with the highest quality standards and the lowest costs. This new framework was designed based on free competition with only the intervention of the administration to create the specific regulation.

From a retributive perspective, for the power plants under the special regime set up in the Law 54/1997[1], the regulation established a system

[1] The article 27 of the Law 54/1997 defines the production of electricity in a special regime, such as that implemented in facilities up to 50 MW of installed power with the following characteristics.

- The facility uses cogeneration or other electricity production methods associated to non-electric activities, and they suppose a high degree of energetic efficiency.
- The facility uses as primary energy one of the renewable energies sources (RESs), biomass, or other type of biofuel, and the owner does not develop the production of electricity under the ordinary regime.

  The facility uses non-renewable waste as primary energy.

supported by a Feed-in-Tariff (FIT hereinafter) without time limits, which essentially consists of charging a bonus for renewable electricity fed into the grid over the price matched in the daily electricity market. As a second option, it was possible to choose a fixed tariff [2-4].

Under this stable regulatory framework and supported by an adequate legislation to facilitate the administrative authorisation of the plants [5], at the end of March 2012, wind energy reached a degree of supply higher than 25% of the electricity demand in the Spanish market central bus-bars[2] [6].

Considering the significant level of primary energy dependence of Spain (75.6% in 2011), which is well above the average for the 27 countries in the European Union, the reduction to a value of approximately 55% at the end of March 2012 [7, 8] indicates the success achieved in the implementation of the energetic mix of Renewable Energy Sources (RES hereinafter), especially for wind energy [9-11], which exceeded a participation rate of 18% in the demand supply in 2012 [12].

In January 2012, the Spanish Government suspended the economic incentives for new-generation facilities based on RES, including wind energy [13]. Accounting for the fact that the targets established by the European Union (EU hereinafter) and the Spanish Government to cover 20.8% of the energy demand with renewable energy sources by 2020 have not yet been achieved [8, 14], new strategies and regulatory policies are required to continue with the integration of RESs [15-17]. In this work, we examine wind energy as an important step toward attaining such a global objective; because of its degree of maturity, wind energy constitutes an actual possibility in a new setting without a FIT, thereby enabling scenarios that imply the lowest costs for the entire electricity sector [18].

The first Wind Turbine Generators (WTG hereinafter) have been in commercial operation for more than 15 years and can be considered that they are entering the last stage of their nominal lifetime [19].

[2] Energy fed into the grid from generators and international exchanges deducting the consumption required for generation and pumped storage.

The repowering of Wind Farms (WF hereinafter) provides benefits to the electricity system as a whole [20], can improve the social and environmental impact [21, 22] and may represent a reasonable option for the Spanish wind industrial sector to address the current critical situation [23].

This work is aimed at determining whether and under what conditions the repowering of a WF is a profitable alternative. We considered the feasibility of two possible alternatives to achieve a reasonable profitability for investors: maintaining the retributive system based on a FIT or charging the price determined in the daily electricity market (spot price hereinafter) for the electricity generated. The volume and the characteristics of the WFs that are suitable to be repowered are also determined.

Below, in Section 2, we analyse existing WFs to define, quantify and characterise the market formed by those suitable for repowering. Afterward, in Section 3, we estimate the expected production of a Repowered Wind Farm (Rep-WF hereinafter) and the changes with respect to an old WF.

In Section 4, we define and estimate the costs of the facilities that form a WF and how these facilities could be reused in a Rep-WF to reduce the construction costs. In addition, the works for dismantling, waste treatment and valorisation of the old WTG are also analysed and considered.

In Section 5, we perform an analysis of the expected profitability of a Rep-WF, introducing a retributive proposal for the electricity generated based on the spot price and comparing the results with those obtained in case of retribution under the present scheme supported by FIT. Sensitivity analyses in relation to the most important parameters are included to consider the effect of their variations in the results.

Finally, Section 6 presents the conclusions of the study.

# 2. Determination of the Market Volume and Its Characteristics

## 2.1. Lifetime and Financing

According to the standard IEC 61400 [24], a WTG should be designed for a lifetime of at least 20 years. During this period, with proper Operation and Maintenance (O&M hereinafter), the WTG will offer a level of mechanical availability[3] of nearly 100% during the first five to ten years – presently, the typical values guaranteed by the manufacturers are approximately 97% – and over 90% for the remainder of its lifetime.

For most WFs, Project Finance was the method chosen by investors. With this method, the recourse to investor from the lenders is limited or even eliminated. The project itself, the WF in this case, will exclusively respond for repaying the debt. The incomes for selling the energy are the main guarantee, so these incomes must be predictable and sufficient [25, 26].

For WFs, incomes can be estimated with a high level of confidence. Such an estimate is based on an adequate measurement campaign of the wind resources at the selected site, the retributive stability given by the regulatory framework during the entire commercial operation period and, finally, selecting an efficient and mature technology to ensure good functionality in the long term.

In the Project Finance method, the debt repayment period will depend on the cash flows generated by the WF. Therefore, to estimate this period, it is necessary to consider the special features of the project itself (its wind resource, the WTG, the costs, etc.). In general, periods in the range of 13 to 15 years are obtained [25].

From a financial approach, a suitable WF for repowering should have achieved a stage in which the investors had been able to repay the loan taken to finance the construction. However, from a technical point of view, the WTGs (of that generation) arriving to the end of their lifetimes will

---

[3] Mechanical availability is defined typically in a yearly period, as the percentage of time (year) in which the WTG is ready to produce electricity.

exhibit high failure rates, will require more maintenance and likely will suffer from a lack of relevant spare parts (generator, gearbox, blades, etc.).

Due to the evolution in the capital cost, the WFs first implemented in commercial operation were slightly more expensive to build [27]. This high capital cost considered separately could imply a negative impact on the repayment period; however, other factors also come into play. As will be demonstrated below, these WFs are located in regions of higher wind resources than the ones built later; in addition, the legislative evolution has allowed them to change to new frameworks, which are more advantageous in terms of retribution. Both of these facts imply a positive impact regarding incomes in the long term. For this reason, it is realistic to expect a lower repayment period that can be estimated in approximately 10 years[4], which is shorter than the previously mentioned period. Thus, to characterise the repowering market, the considered WF must have an operation period equal to or longer than 13 years because during this period, it can be hoped that the incomes have allowed the investor to repay the debt incurred to finance the construction and have produced a positive cash flow for 3 years. In addition, the WTGs will not have arrived at the end of their lifetime but it is expected that a probable increase in the failure rates and the costs for O&M will occur to maintain a high level of mechanical availability [28].

## 2.2. Power of WFs

At the end of 2012, the WFs in Spain that were in commercial operation represented an installed power capacity of 22.6 GW [12]. The installation of the first units started in 1992, but the most significant

---

[4] In order to minimise the financial costs it is important to reduce the repayment period in which the investor must pay an interest rate for the borrowed money. This period is very sensitive with the electricity generated because more energy implies bigger incomes and the possibility for the investor to choose a smaller repayment period or reduce it during the commercial operation. By means of simulations carried out according with the cash – flows calculation method of the Table 9, an increase of around 10% in the yearly electricity generation can reduce the payback period between three and five years depending on other variables (i.e., the capital costs).

growth has occurred since 1998 due to the stability provided by Law 54/97 [1] and subsequent regulations. This growth was also possible due to the evolution of WTG technology and the technical measures established to maximize their integration. For example, consider the forecasting of wind power generation [29, 30] or the affiliation at control centres [31].

Figure 1 shows the evolution of installed power in the period of 1998-2012 [12], which includes the years in operation. The first units, commissioned before 1999, were found to have a history of over 15 years in commercial operation.

The market volume of installed power for an operation period equal to or greater than 13 years (i.e., WFs installed up to the year 2000), would currently be approximately 2.3 GW. In addition, as shown in Figure 1, the evolution of the yearly installed power will make this volume increase at a rate of approximately 1 GW per year.

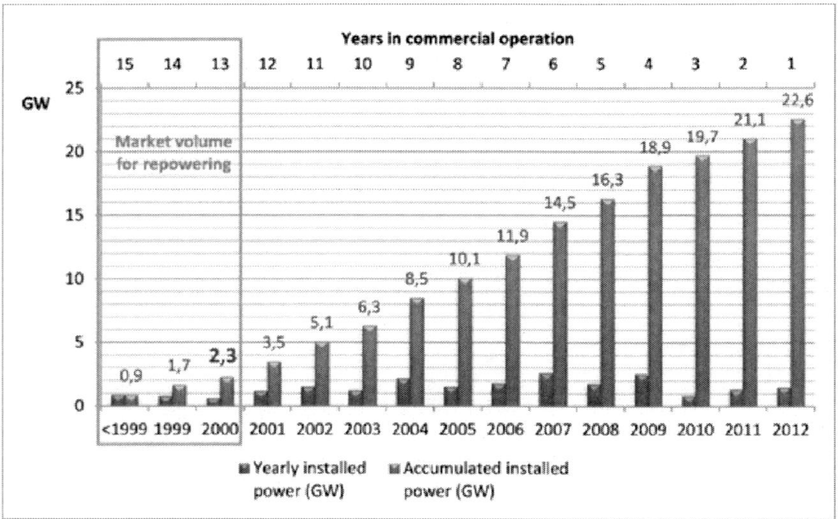


Source: National Energy Commission of Spain (CNE) [12].

Figure 1. Yearly and accumulated installed power in WFs and the number of years in commercial operation.

## 2.3. Characterisation of the Power of WFs and WTGs

During the considered period of 1998-2012, approximately 1,300 WFs were installed in the Spanish territory [12].

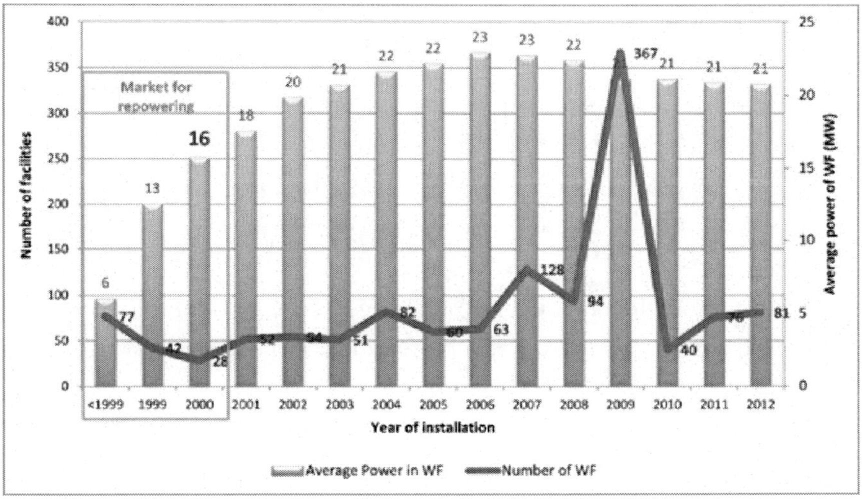


Source: National Energy Commission of Spain (CNE) [12].

Figure 2. Average power installed in WFs per year of installation.

Figure 2 shows the yearly evolution of the number of WFs installed and their average installed power. Note that the growth was regular and sustained, achieving a maximum in 2009 when 2.5 GW were installed in 367 WFs [12].

According to Law 54/97 [1], the capacity of the facilities under the special regime, including WF, cannot exceed 50 MW. Nevertheless, this maximum has yet to be approached; as Figure 2 shows, the average installed power of WFs increased in the first years to become stabilised at approximately 20 MW.

Since the beginning of the deployment of wind energy, WTG technology has progressed to units that are more efficient and have a higher unitary rated power. This progress has encouraged the possibility to take more advantage of wind energy by both selecting the best places in the site – with a higher mean wind speed – and reductions in the construction and O&M costs, which is strongly influenced by the number of WTGs in the WF and by the use of these new WTGs [32].

Thus, WTGs have passed from the first units with a capacity of several hundred kilowatts to the current ones offered by the manufacturers with a rated power above 3 MW. Figure 3 shows the evolution of the average rated power of a WTG installed up to 2012.

According to Figure 2, for those WFs installed up to the year 2000, the value for the average installed power would be 16 MW. Nevertheless, for the purpose of characterizing a WF for repowering, an output of 18 MW was selected because it is the closest value to 16 MW which is a multiple of 2 and 3 MW and therefore facilitates the simulation with WTG of 2 and 3 MW.

Regarding the unitary rated power of a WTG, as shown in Figure 3, for a WF that is 13 or more years in commercial operation, the typical value for the unitary rated power can be set between 559 and 651 kW.

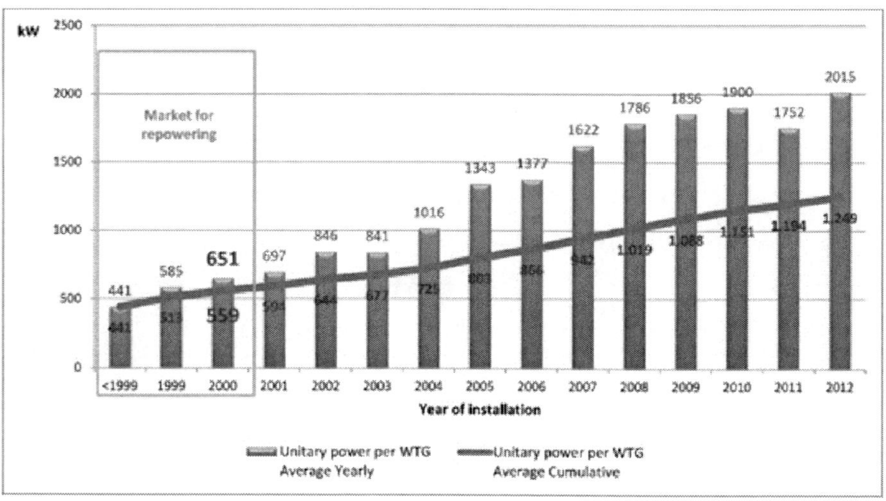


Source: Spanish Wind Energy Association (AEE) [23, 33, 34].

Figure 3. Unitary rated power per WTG in a WF.

## 2.4. Generation of Electricity

Figure 4 shows the annual evolution of the net WF capacity factor (CF hereinafter)[5] in the period of 1999-2012 differentiated by the year the WF was placed in commercial operation. Note that the WFs installed during the initial years continually offer the highest CF during the entire period.

In addition, Figure 5 shows the temporary evolution of the yearly average net CF and its accumulated value for a WF in Spain, differentiated by the year they were placed in operation. The WTGs used in the first WFs were less efficient and less technologically advanced, besides considering that it is expected a relevant decline in their performance with age [28], one may conclude that the wind energy potential of the sites where the first WFs were built is substantially better than those of the later WFs.

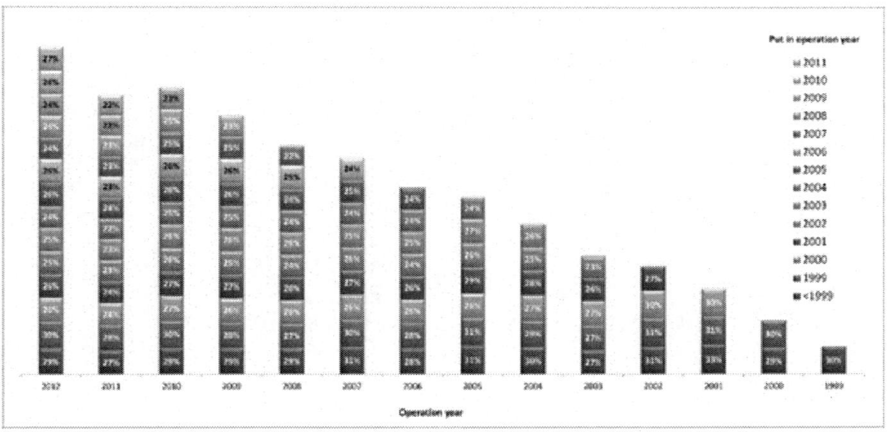


Source: CNE [12].

Figure 4. Yearly capacity factor of WFs deployed during the year they were placed in operation.

---

[5] The CF is defined, in a yearly period, as the percentage of the year during which the WF should have been working at nominal power to generate the entire production obtained in the year. The gross value (Gross CF) does not include the wake losses ($W_L$), the electric losses ($E_L$) up to the point where the electricity fed into the grid is measured, and the production losses due to the mechanical availability of the WTG ($F_{AV}$). The net value (Net CF) includes the effect of all the aforementioned losses (expressed as a percentage); therefore the relation between the Net CF and the Gross CF is the following:

$$Net\ CF = Gross\ CF \times W_L \times E_L \times (1 - F_{AV})$$

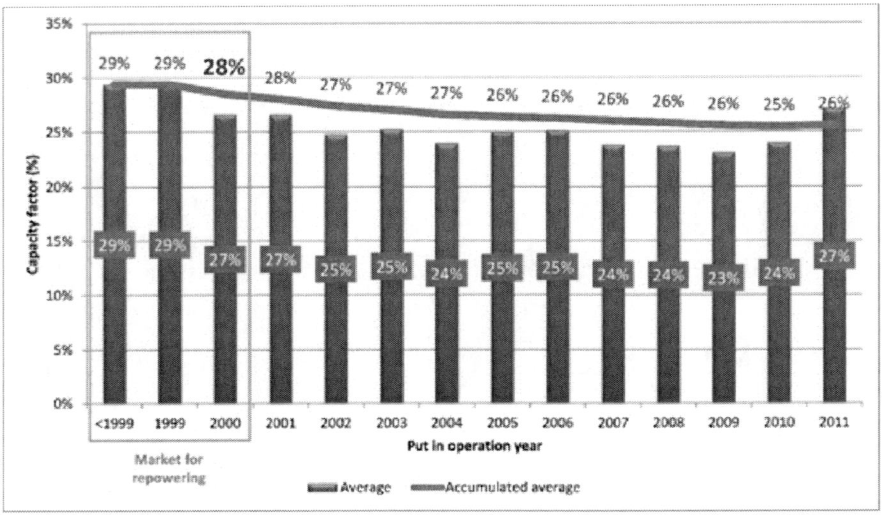


Source: CNE [12].

Figure 5. Net average CF of WF's in the period 1998–2012 deployed for the year of entry in operation.

By characterizing the WFs suitable for repowering, the average net CF considered for a WF with a minimum commercial operation period of 13 years is calculated and shown to be approximately 28% in Figure 5.

## 2.5. Characterisation of WFs Suitable for Repowering

Table 1 summarises the results obtained from the characterisation of WFs suitable for repowering that were described in the previous sections.

**Table 1. Market volume and characteristics**

| | |
|---|---|
| Year placed in operation | <2000 |
| Total installed power | 2.3 GW |
| WTG average size | 559-651 kW |
| WF average power size | 18 MW |
| Average net CF | 28% |

## 3. Analysis of the Electricity Production of Repowered WFs

The mechanical power, $P_{wind}$, represents the kinetic energy of the wind going through the rotor of a WTG, which depends on the non-disturbed wind speed (v), the air density (ρ) and the area swept by the rotor (S) according to the following expression:

$$P_{wind} = \frac{1}{2}\,\rho\, S\, v^3 \qquad (1)$$

A WTG is able to extract only a portion of this kinetic energy to transform it into electricity. The limit of this transformation is $\frac{16}{27}\% \approx 59.25\%,$ which is calculated according to Betz's law [36]

The formula for the mechanical power extractable from the wind stream by a WTG is:

$$P_{util} = C_p \times P_{wind} = C_p \frac{1}{2}\,\rho\, S\, v^3 \qquad (2)$$

where $C_p$ is the coefficient of performance and determines how efficiently the WTG takes advantage of the wind stream going through the rotor. The value of $C_p$ varies according to the wind speed and depends on the characteristics of the blades.

In addition, the increase of the wind speed with the height must be taken into consideration [37]. This increase depends on the particularities of the site. With the appropriate wind measurement campaign, the choice of the most convenient hub height can be determined.

Therefore, on the basis of (2), the possibilities for an improved use of wind resources in a specified location are the following:

1. Increase the diameter of the rotor by increasing the length of the blades.

2. Increase the wind speed through the rotor by installing the WTG hub higher – with a higher tower.
3. Increase $C_p$ by installing the WTG with more efficient rotors.

In fact, the repowering of a WF takes advantage of these three possibilities when the older WTGs are replaced by those that are more technologically advanced and efficient, with larger blades and higher towers. In addition, new WTGs will have a larger rated power, making it possible to install more power in the best positions, i.e., those with more wind speed.

To quantify the effect of the replacement of a WTG, we performed a simulation of the expected production from wind measurements obtained over the last three years from an actual site located in the province of Lugo (Spain). As shown in Figure 6, there are many other sites in the Spanish territory with similar wind conditions in terms of the average speed, so that the analysis performed in this section is valid for all of these sites.

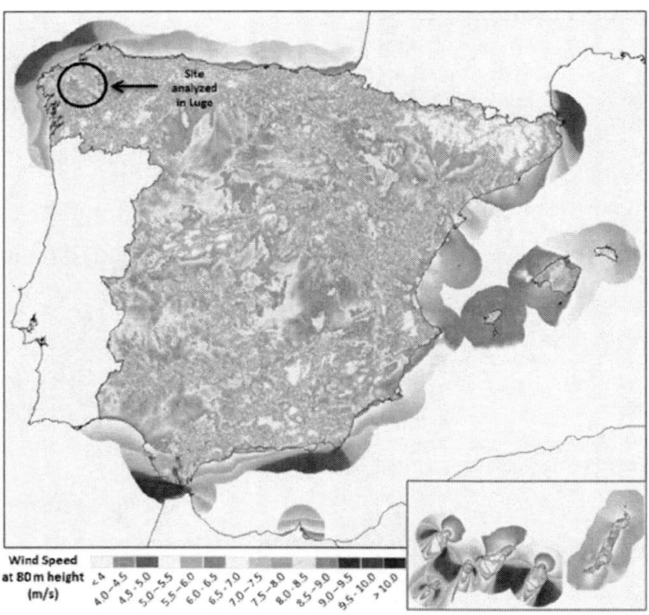


Source: IDAE [38].

Figure 6. Spanish wind map. Yearly average wind speed at 80-m height.

The utilisation of WTGs with higher unit power may enable an increase in the power of the WF after repowering. For an example, we have considered a distance between WTGs, measured perpendicular to the predominant wind direction, of three times the diameter of the rotor (a typical value for the design to reduce wake losses [44]). With this assumption, for WTGs with 46 m of rotor diameter and 660 kW of unit power, it is possible to install up to 2.4 MW/km, while for WTGs with 90 m of rotor diameter and 3 MW of unit power, it is possible to install up to 5.5 MW/km.

However, in the repowering process, it is very important to make an effort to maintain the validity and applicability of the permits, licenses and authorisations already obtained for the old WF. This will avoid additional work (and their associated costs) that could extend the development process.

This issue has a special relevance in relation to installed power because its variation can substantially modify the affected area of the WF and the permits and contracts subscribed for access to the electric grid. The little good in terms of RES integration is the installed power. In the process of sectorial regulation, as well as to fix domestic targets established in the development plans [8], the installed power was largely used to delimit the integration limits. In addition, an increase in WF power will involve important modifications in evacuation facilities (substations and line) with the associated costs.

In contrast, unless there were physical constraints in the site that could limit the installation of the required number of new WTGs to achieve the old WF capacity, it is convenient not to reduce it, thus maximizing the use of the affected area and the potential of the site.

Therefore, for the aim of this chapter, the installed power in a Rep-WF was considered to not vary with respect to the authorised power for the old WF.

## 4.1. Capital Costs

The capital costs vary widely depending on multiple factors. Among others, the following factors can be pointed as the most important: the market status (supply and demand), the site (country and local characteristics), the characteristics of the connection point to the electric grid, the existence of nearby WTG factories and the local presence of experienced subcontractors.

With the aim of counting on a solid base for the calculation of the capital costs and considering the possible variations we did the following:

- A survey among several international companies with a large experience in the Spanish market, including engineering, WTG manufacturers, civil works and electric infrastructure subcontractors. The detailed definition of the scope of works included in the survey and the prices obtained can be found in the Appendix.
- Include sensitivity analysis for the results in relation with the main capital cost corresponding with the WTG.

Table 3 shows a comparative for the main costs between several sources and the result of the survey we carried out for this chapter applied to the characteristics for both the Rep-WF and the N-WF. As shown, the results are consistent with the other sources but with a focus in the market.

## 4.2. Development

Development is performed prior to the construction of the WF and consist mainly of the study of the wind resource, the technical definition of the project and the management of the permits, licenses and authorisations necessary for the construction and the commercial operation.

**Table 3. Comparison of capital and O&M costs between several sources and the estimate considered in the simulations**

| Source | Range | WTG Cost [a] | WF cost | WTG O&M Cost |
|---|---|---|---|---|
| [45] (Valid for 2010) | min | | 1,501 €/kW | 0.011 €/kWh |
| | typ | 1,176 €/kW | | |
| | max | | 1,704 €/kW | 0.020 €/kWh |
| [46] (Valid for 2012) | min | 752 €/kW | | |
| | typ | | 1,592 €/kW | |
| | max | 1,069 €/kW | | |
| [47] (Valid for 2014) | min | 560 €/kW | 800 €/kW | |
| | typ | | | |
| | max | 770 €/kW | 1,100 €/kW | |
| [7, 8] (Valid for 2010) | min | 744 €/kW | 1,000 €/kW | 0.090 €/kWh |
| | typ | | | |
| | max | 992 €/kW | 1,300 €/kW | 0.120 €/kWh |
| Calculation based in the survey carried out for this chapter. See Appendix | typ for a Rep-WF | 850 €/kW | 983 €/kW | 0.004 €/kWh |
| | typ for a N-WF | | 1,108 €/kW | 0.006 €/kWh |

[a] When the data source was in USD, data were converted to € using the following USD/€ exchange rates: 1.233 in 2010 (value in 2010/06/15), 1.263 in 2012 (value in 2012/06/15) and 1.349 in 2014) (value in 2014/02/10).

**Table 4. Summary of the development costs**

| Item | New WF | | | Rep- WF | | |
|---|---|---|---|---|---|---|
| | €/kW | € | % | €/kW | € | % |
| *Engineering* | *2.2* | *40,000* | *12%* | *2.2* | *40,000* | *15%* |
| *Wind resource assessment* | *4.8* | *87,000* | *25%* | *0.4* | *7,000* | *3%* |
| Meteo tower rent | 4.4 | 80,000 | | 0.0 | 0 | |
| Final report | 0.4 | 7,000 | | 0.4 | 7,000 | |
| *Personnel* | *12.2* | *220,000* | *63%* | *12.2* | *220,000* | *82%* |
| Hourly average cost | 6.7 | 120,000 | | 6.7 | 120,000 | |
| Other personnel costs | 5.6 | 100,000 | | 5.6 | 100,000 | |
| Total development costs | 19.3 | 347,000 | 100% | 14.8 | 267,000 | 100% |

Source: Self-elaboration based on Table A.1.

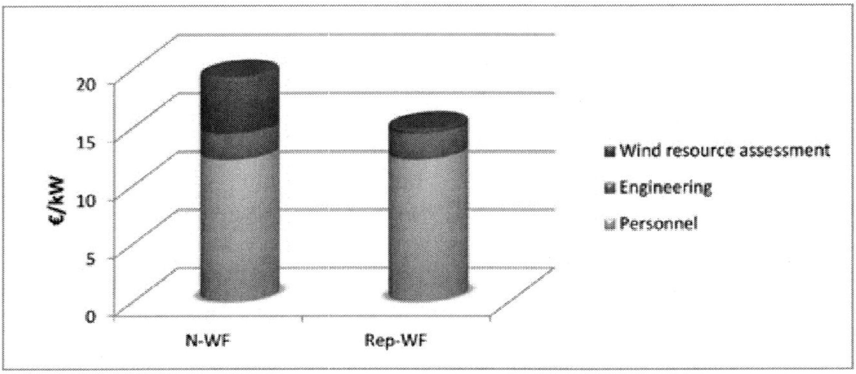


Figure 10. Deployment of the development costs per kW.

Table 4 and Figure 10 present a summary of the development costs to take into consideration in an N-WF or in the repowering of an existing one. The only difference is the wind resource assessment required to make an appropriate choice for the new WTG as well as a good production estimate. In the case of a Rep-WF, it will not be necessary to perform a new measurement campaign of wind resources because the data were obtained during the commercial operational period of the old WF.

## 4.3. Construction

For a given WTG, the facilities of a WF will depend on its geographical location (distance to the public road network, soil characteristics, rainfall, etc.) and the way to connect it to the electric grid (voltage level, evacuation line, specifications for the connection point, etc.). However, several basic elements allow us to estimate the costs of construction with an appropriate accuracy for the objectives of this chapter.

A basic scheme of the facilities, which is considered for the definition of the scope of the construction and the estimation of their associated costs, is shown in Figure 11. These facilities are as follows:

- The foundations of the WTG and the assembly areas to support the cranes and heavy transport vehicles during the assembly and maintenance of the WF.
- The roads to allow access to all the facilities from the public transport network. These roads should be suitable for the transit of large-dimension transport vehicles and cranes.
- The power, communications and grounding grid (P&C&G Grid hereinafter). The P&C&G is composed of high-voltage cables and accessories to take the power generated in each WTG to the substation, communications cables (usually optical fibre) to provide communications between the WTGs, a meteorological tower, a Supervisory Control And Data Acquisition system (SCADA hereinafter) and ground network cables (usually bare copper wire). The associated civil works are also included.
- The electric substation, in which the electricity generated by the WTG is collected, typically at a voltage level between 15 and 33 kV, and transformed to the voltage level of the electric grid where the WF connects.
- The evacuation line to carry the electricity from the substation up to the connection point with the electric grid.

The extent of the roads and the P&C&G grid is directly conditioned by the distance between the WTGs. Obviously, the best positions, in terms of wind resource, should be selected. With the objective to maximise generation with the minimum cost, several methods have been suggested to optimise WF layouts [48], however, very often, especially in sites where wind resources are homogenous over all the area available and for the purpose of minimizing losses caused by the wake effect [43] the WTGs should be located with a distance in the range of two or three times the length of the WTG rotor diameter in the line perpendicular to the prevailing wind direction. At the same time, the typical distance between two WTG lines – both perpendicular to the prevailing wind direction – is approximately 10 times the length of the WTG rotor diameter [44].

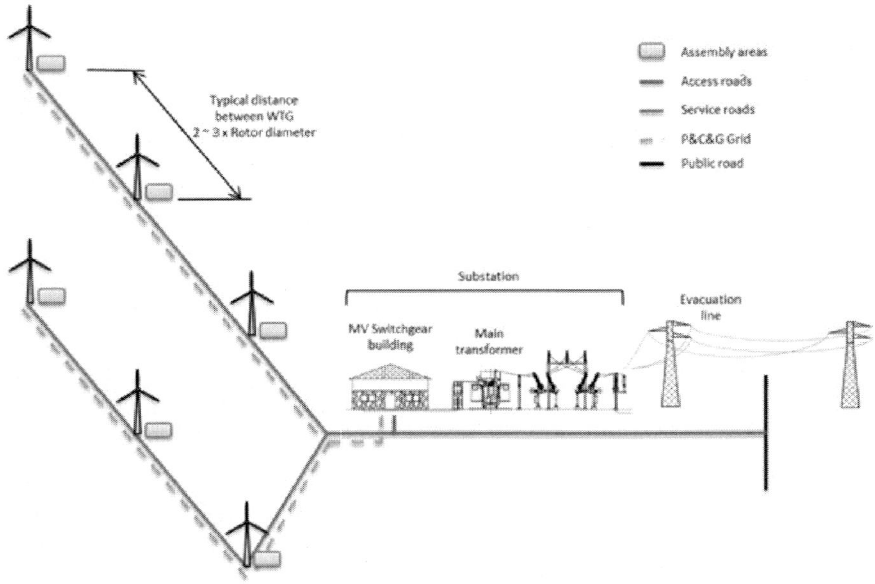


Figure 11. Facilities in a WF.

Table 5 and Figure 12 detail the construction costs considered for an N-WF and for a Rep-WF. As one might expect, the construction of a Rep-WF is less expensive for the following reasons:

1. The roads must be constructed for an N-WF, while they only require renovation and adaption to the requirements of the new WTGs (dimensions and weights) for a Rep-WF.
2. According to the assumption previously mentioned in this section, the installed power will not change; therefore, there is no work required in the substation and the evacuation line for a Rep-WF.
3. The cost of the permits, which is calculated over the construction costs, will be higher for an N-WF.
4. For a Rep-WF, the only land permits required are for new roads and P&C&G grid facilities.

**Table 5. Summary of the construction costs**

| Item | New WF | | | Rep WF | | |
|---|---|---|---|---|---|---|
| | €/kW | € | % | €/kW | € | % |
| *Main equipment* | *853.9* | *15,370,000* | *78%* | *853.9* | *15,370,000* | *88%* |
| WTG | 850.0 | 15,300,000 | | 850.0 | 15,300,000 | |
| Meteorological mast | 3.9 | 70,000 | | 3.9 | 70,000 | |
| *Foundations and assembly areas* | *56.1* | *1,010,000* | *5%* | *56.1* | *1,010,000* | *6%* |
| Foundations | 53.3 | 960,000 | | 53.3 | 960,000 | |
| Assembly areas | 2.8 | 50,000 | | 2.8 | 50,000 | |
| *Roads* | *26.0* | *468,000* | *2%* | *17.3* | *312,000* | *2%* |
| Access road length | 12.5 | 225,000 | | 8.3 | 150,000 | |
| Services road length | 13.5 | 243,000 | | 9.0 | 162,000 | |
| *P&C&G grid* | *15.3* | *275,400* | *1%* | *15.3* | *275,400* | *2%* |
| Civil works | 11.3 | 202,500 | | 11.3 | 202,500 | |
| Supply and installation | 4.1 | 72,900 | | 4.1 | 72,900 | |
| *Substation* | *55.6* | *1,000,000* | *5%* | *0.0* | *0* | *0%* |
| High voltage level | 30.6 | 550,000 | | 0.0 | 0 | |
| Medium voltage level | 8.3 | 150,000 | | 0.0 | 0 | |
| Main transformer power | 16.7 | 300,000 | | 0.0 | 0 | |
| *Evacuation line* | *34.7* | *625,000* | *3%* | *0.0* | *0* | *0%* |
| *Construction Permits* | *47.2* | *849,957* | *4%* | *25.9* | *466,825* | *3%* |
| County Construction license | 2.2 | 40,110 | | 1.7 | 30,490 | |
| Urban license | 26.0 | 468,710 | | 23.6 | 424,185 | |
| Land owners permits | 19.0 | 341,137 | | 0.7 | 12,150 | |
| Total construction costs | 1,088.8 | 19,598,357 | 100% | 968.6 | 17,434,225 | 100% |

Source: Self – elaboration based on Table A.2.

### *4.3.1. Dismantling Works and Valorisation of the Valuable Waste*

For a market that is able to install more than 22 GW, the dismantling of an old WTG, which are much lighter and smaller than the new ones, is not a barrier for repowering. The services offered by specialised companies and the technical resources for the elevation and transport of heavy loads exceeds the necessities to perform these tasks.

The dismantling works and valorisation of valuable waste will include:

- The dismantling of the WTGs into manageable parts.
- Ensuring the adequacy of the transportation.
- The classification and transport of non-valuable waste and the delivery to authorised agents for their treatment.
- The transport of valuable waste to the delivery point.

For the dismantling and restoration of the rest of the facilities not reused in a Rep-WF the following assumptions have been taken into consideration:

- *Foundations:* The dismantling is limited to eliminate protruding parts from the soil surface, usually consisting of the upper anchorage or anchor bolts and the structural concrete. After that, the surface is levelled and refilled with appropriate material; finally, environmental restoration will be performed.
- *Assembly areas*: Due to environmental requirements, the assembly areas were normally restored at the end of the construction of the old WF; thus, no dismantling work is required.
- *Roads:* Restoration is not required because they will be largely reused, and for the parts that are not reused, their removal is not advised because they provide an important service to the community by facilitating access or acting as firebreaks.
- *P&C&G grid*: Reuse is not feasible because the positions of the WTGs will not be the same. In addition, restoration is not necessary because common practice is to bury the cables deeply enough to allow others soil uses (e.g., agriculture) and, according to environmental requirements, trenches must be restored upon completion of the old WF.

The appropriate time for the completion of the aforementioned works of dismantling and restoration is during the construction of new facilities for a Rep-WF. In this case, the cost of all of these construction projects can be considered to be a residual value and should be included in the construction scope for the Rep-WF (Table 5).

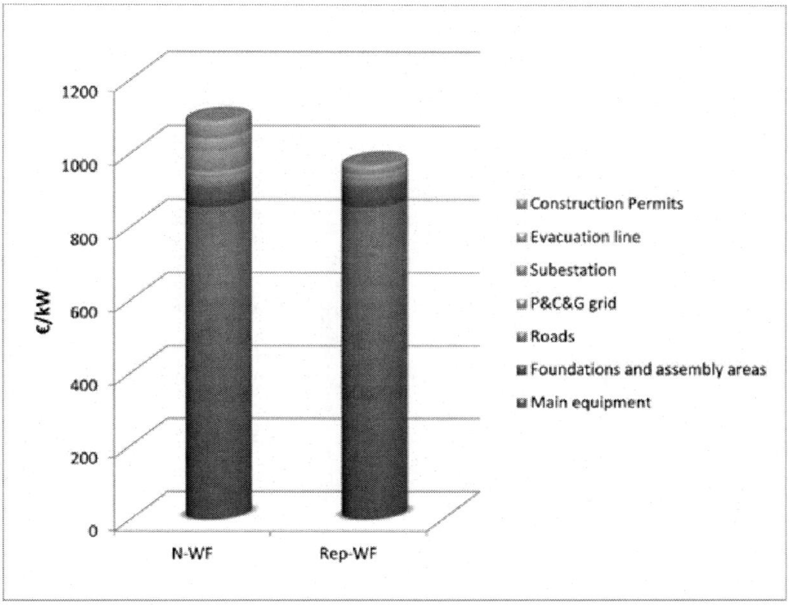


Figure 12. Deployment of the construction costs per kW.

Regarding waste management, there are two different cases, depending on whether the waste is valuable, i.e., whether it is possible to obtain income with their disposal for recycling.

In the case of non-valuable waste (oil, plastic, etc.), the waste items should have their origins classified and be moved directly to qualified organisations authorised for their treatment and recycling. The costs of these tasks are included in the dismantling costs listed in Table 6, which were obtained by means of a request for a proposal issued to ten companies that are specialised in the assembly and transport of WTGs.

In the case of valuable waste, currently, there are stable markets in which these waste items can be traded, thereby obtaining economic incomes [49, 50]. Among all the valuable waste found in a WTG, the most important because of their quantity and price are copper and steel. Table 7 details the amount of copper and steel waste and their origin from a WTG. The amounts indicated are the average value for the three class II WTGs included in Table 2, the individual values were obtained from manufacturer's technical information.

**Table 6. Dismantling and waste disposal costs and valorisation incomes**

| Item | €/kW | € |
|---|---|---|
| *Dismantling of WTG´s and waste disposal* | *67.5* | *1,215,000* |
| *Valorisation of valuable scrap materials* | *-25.9* | *-465,973* |
| Net effect | 41.6 | 749,027 |

Source: Self-elaboration. Prices for dismantling and waste disposal are based on offers received in 2013 from companies that are specialised in the assembly and transportation of WTGs.

**Table 7. Deployment and prices for valuable waste**

| Item | Average valuable waste per WTG | |
|---|---|---|
| | Steel (kg) | Copper (kg) |
| Blades (3#) | 698 | |
| Hub | 2,320 | |
| Anchorage | 3,000 | |
| Tower | 30,000 | |
| Main frame | 6,000 | |
| Shaft (including bearing) | 3,100 | |
| Gearbox | 8,000 | |
| Coupling | 160 | Irrelevant |
| Yaw gear motor | 500 | Irrelevant |
| Yaw ring | 620 | |
| Hydraulic unit | 60 | |
| Generator | | 1,000 |
| Cables | | 866 |
| Quantity per WTG | 54,458 | 1,866 |
| Unitary Price (€/kg) | 0.280 | 1.077 |
| Total Price (€/WTG) | 15,248 | 2,010 |

Source: Self-elaboration. Steel and Copper mass per WTG from manufacturers information. Prices for valuable waste from [51, 52].

The net effect of the dismantling works and the management and valorisation of waste, shown in Table 6, was considered in the costs associated with repowering.

*4.3.2. Operation Works*

The operation works are those necessary to keep the WF that is in commercial operation in optimal technical and safety conditions.

Among the operation works, the most important and expensive corresponds with the WTG O&M. This task is normally provided by the manufacturer of the WTGs and usually includes a mechanical availability guarantee.

A request for an updated WTG O&M price was included in the survey carried out among WTG manufactures for this chapter. The result is included in Table 8, as well as the scope and the unitary costs for the operation works. The operation works, and therefore, the associated costs, are equal in both cases, for an N-WF and for a Rep-WF because the number of WTGs installed and the balance of plant infrastructures will be the same.

**Table 8. Operation works scope and unitary prices**

| Item | Scope, prices and characteristics | | | Comment |
|---|---|---|---|---|
| O&M | *WTG* | *50.000* | *€/(WTG x year)* | |
| | *Balance of plant* | *125.000* | *€/year* | Roads, substation and evacuation line |
| Land easements | *WTG* | *3.000* | *€/(MW xyear)* | Including assembly areas |
| Insurance and administration | *Insurance* | *0,50* | *% Incomes* | All risk and Civil Liability. Typically calculated over incomes |
| | *Administration* | *0,15* | *% Incomes* | Typically calculated over incomes |
| Cost update | *Annual increase of operation costs* | *2* | *%* | Estimate based on the trend in the inflation forecast [53, 54] |

Source: Prices for O&M are based on offers received in 2013 from WTG manufacturers and companies that are specialised in the balance of plant O&M services.

# 5. PROFITABILITY ANALYSIS

The substitution of old WTGs will bring environmental benefits [55]. Repowering would also bring important benefits for the entire electricity system, which would be mainly derived from the improved use of energetic sources by means of more efficient WTGs.

According to [56] a reduction in the spot price is expected with the increase of wind energy integration in the Iberian market (Spain and Portugal), while [57] predicts a very small impact in household electricity prices in EU countries. Integration could be favoured by repowering due to the change of the old WTGs for ones more adapted to the new technical regulations.

Nevertheless, in a liberalised market, as that which exists in Span, the main decision factor for a possible investor to participate in it is the expected profitability, which is assured by regulatory stability.

Currently, the retributive system based on a FIT is suspended both for new plants and for the repowering of WFs in commercial operations [13]. Thus, as the main objective of this study, it is necessary to search for alternatives to ensure sufficient incomes to make repowering projects profitable.

From an economic point of view, the most favourable solution for the entire electric system would be to pay the spot price for the electricity generated by a WF, being able to provide energy in terms of grid parity. This model is currently being successfully applied in countries with emergent economies (e.g., Chile) even to support projects based in photovoltaic and concentrated solar thermal technologies, which are less efficient than wind energy [58, 59].

The retributive system proposed and analysed in this study is based on the following assumptions:

1. Maintain the right to export the entire electricity surplus to the electric grid when it is technically feasible.
2. Participate in the daily electricity market and charge the hourly spot price for the electricity fed into grid.

**Table 9. Cash flow calculation method**

| Shareholder Cash Flow | |
|---|---|
| + | Incomes from electricity remuneration |
| - | Operating expenses |
| = | Gross operating margin (EBITDA) |
| - | Depreciation |
| = | Earnings before interests and taxes (EBIT) |
| - | Financial expenses |
| = | Earnings before taxes |
| - | Taxes |
| = | Earnings after taxes |
| + | Depreciation |
| = | Yearly cash flow $(\boldsymbol{CF_z})$ |
| + | Cumulative previous year cash flow |
| = | Cumulative present year cash flow $(\boldsymbol{CFA_z})$ |

Under these assumptions, the analysis focused in two aspects: first, if a Rep-WF can provide an interesting profitability opportunity for the investors, being a similar or even better choice than constructing a N-WF, and second, if the retributive proposal is sufficient to ensure profitability.

To achieve this objective, two scenarios were analysed and compared:

1. The repowering of an existing WF with the characteristics indicated in Table 1.
2. The construction of a WF in a new site, taking into consideration the development, construction and operation costs calculated in Section 4. To obtain homogeneous results, the installed power considered in this case has been the same as that considered in scenario 1 but the results are applicable to WFs in the same range of installed power[9].

In both scenarios, the profitability for the investor was estimated using the Internal Rate of Return (IRR hereinafter) calculated for a period equal

[9] According with the Spanish regulation there is a limit for the installed power of RES facilities. For WFs this limit is of 50 MW.

to the WTG lifetime – 20 years – considering an external financing source. This return is known as the Shareholder IRR.

Before obtaining the IRR, the cash flows of the investment are calculated following the method described in Table 9.

**Table 10. Common assumptions in IRR calculation for both scenarios**

| Item | Name | Value | Unit | Comment |
|---|---|---|---|---|
| General | WF power | 18 | MW | According to Table 1 |
| | Number of WTG | 6 | | Corresponding with 3 MW nameplate power WTG |
| | Commercial Operation Year | 2015 | | One year of construction period starting in 2014 |
| Tariff | Average spot price 2013 | 50.49 | €/MWh | Average price from 2013. Source: OMIE [60] |
| | Annual increase of spot price | 2 | % | .Typically indexed to inflation. Estimate based on the trend in the inflation forecast [53, 54] |
| Electricity Generation and Losses | Wake losses ($W_L$) | 5 | % | Typical |
| | Electric Losses ($E_L$) | 3 | % | Estimate. It includes from WTG transformers up to grid connection point |
| | WTG Mechanical Availability ($F_{AV}$) | 95 | % | Current market value. Source: WTG Manufacturers |
| Taxes | Profit taxes | 30 | % | |
| | Income taxes | 2 | % | The final value depends on counil and incomes level. Mean value has been taken into consideration |
| | VAT | 21 | % | |
| Financial terms | Financial Leverage (FL) | 70 | % | Over total costs. Estimate from [61] |
| | Interest rate | 5 | % | Annual. Estimate from [61]. |
| | Loan repayment period | 10 | years | Typical |
| | $T_{DEP}$ | 10 | years | Typical |

The Net Present Value (NPV, hereafter) of the investment is calculated from each of the yearly cash flows, discounting back to its present value at the discount rate k, that is

$$NPV = \sum_{z=1}^{n} \frac{CF_z}{(1+k)^z} - I_o \quad (3)$$

where $I_o$ is the portion of the total cost assumed by the investor, with its value depending on the Financial Leverage (FL hereinafter)[10]:

$$I_o = FL\ x\ Total\ Cost = FL\ x\ (Development\ cost + Construction\ cost) \quad (4)$$

IRR is defined as the interest rate k at which the NPV is zero, that is:

$$0 = \sum_{z=1}^{n} \frac{CF_z}{(1+IRR)^z} - I_o \quad (5)$$

The common assumptions in the IRR calculations for both scenarios are detailed in Table 10.

The particular assumptions in the IRR calculations for each scenario are detailed and distinguished in Table 11.

**Table 11. Particular assumptions in IRR calculations**

| Item | Name | New WF | | Rep WF | | Comment |
|---|---|---|---|---|---|---|
| | | Value | Unit | Value | Unit | |
| Costs | Development – Total | 19.28 | €/kW | 14.83 | €/kW | According to calculations in Section 4 |
| | Construction - Total | 1,088.80 | €/kW | 968.57 | €/kW | |
| | Dismantling and valorisations – Net effect | 0.00 | €/kW | 41.61 | €/kW | |
| Electricity generation and losses | Gross CF | 31 | % | 46 | % | According to Section 3 |
| | Net CF | 27 | % | 40 | % | |

[10] The Financial Leverage (FL) is the percentage of the investment financed with an external resource or debt.

Figure 13 shows the variation of IRR as a function of CF in both scenarios analysed. It can be observed that:

1. The IRR for a Rep-WF is independent of the CF and higher than the IRR for an N-WF because the total costs for the development and construction are lower, mainly due to the reuse of the electric infrastructure – substation and evacuation line.
2. The IRR for a Rep-WF, with the increase of the CF after the repowering calculated in Section 3 (approximately 46% of the gross value or 40% of the net value), is greater than the IRR for an N-WF with the current expected value of CF (approximately 31% of the gross value or 27% of the net value).
3. The IRR for a Rep-WF, with the increase of the CF and charging the spot price for the electricity generated, is even better than the IRR for the construction of an N-WF with the current expected value of CF (approximately 31% of the gross value or 27% of the net value).

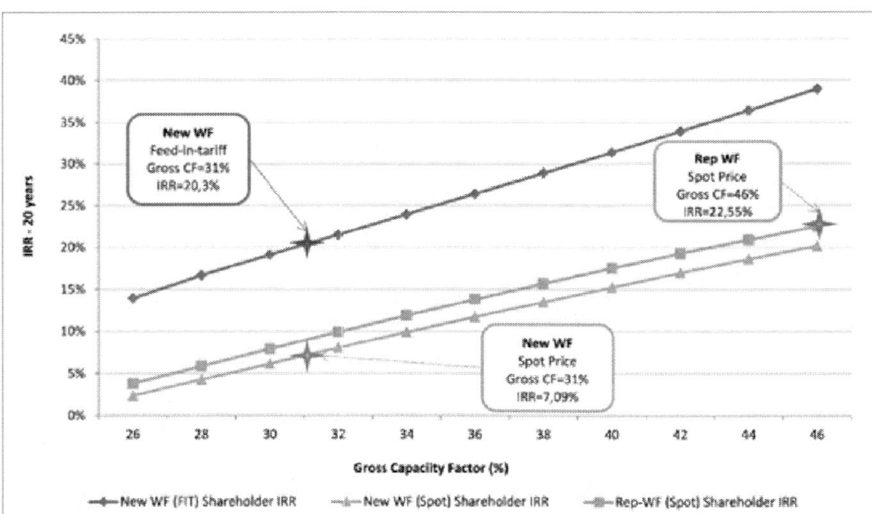


Figure 13. IRR as a function of the gross CF.

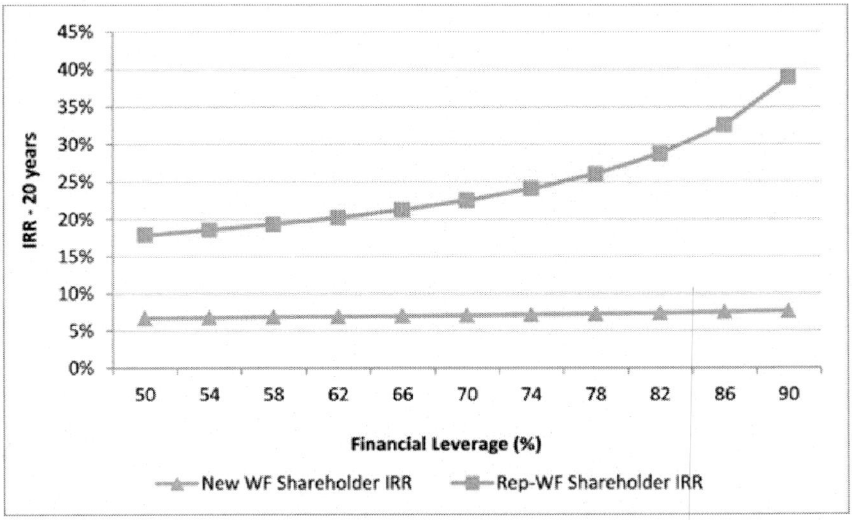


Figure 16. IRR as a function of the FL for both scenarios.

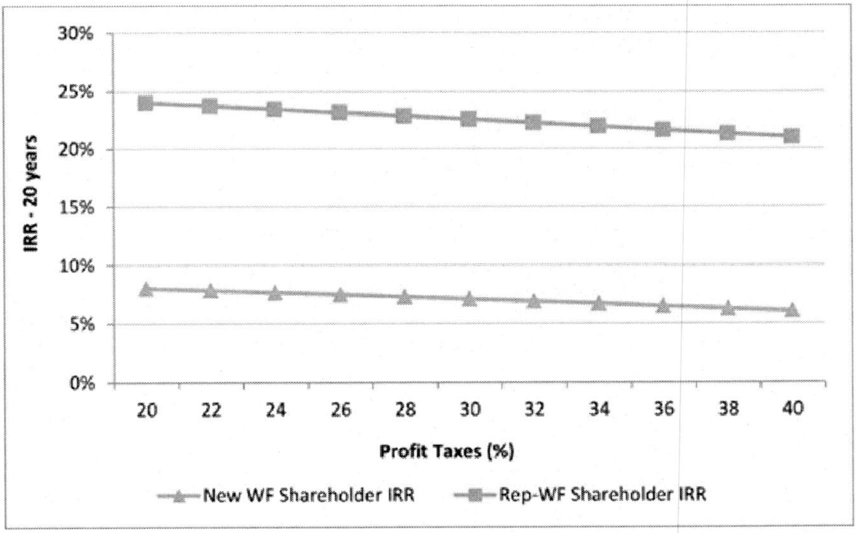


Figure 17. IRR as a function of the profit taxes for both scenarios.

### *5.1.4. Sensitivity in Relation to Taxes*

Among the taxes applied to the activity of the generation of electricity during commercial operation, the most important, due to its value, is the profit tax.

Figure 17 shows the IRR as a function of the profit tax that was obtained by varying its value (Table 10 over the range of 10% and 30% and keeping the rest of assumptions fixed). It can be observed that the sensitivity of the IRR is not significant in this case; therefore, the reduction of the tax burden will not have a relevant effect in the decision to go ahead with repowering.

## CONCLUSION

In light of all that has been accomplished in this work, the investment in the repowering of a WF can be more profitable than investing in a N-WF if the new WTGs are properly selected according to the wind resources at the site and if the facilities for the extraction of the electricity generated are reused.

The total cost is lower for a Rep-WF, even when considering the costs for dismantling the older WTGs because these dismantling costs are considerably reduced by the incomes obtained from the valorisation of the valuable waste.

For an old WF with wind resources greater than those recently made or than the sites currently available, the repowering with new WTGs, which are more efficient and technologically advanced, enables a better use of the wind. This fact allows a Rep-WF to sell the energy produced at the spot price, offering a similar or even better profitability than an N-WF, with a CF that is much lower than that with a retributive system based on a FIT.

Currently, there is a market for repowering of approximately 2.3 GW of installed power in the entire Spanish territory with a period in commercial operation equal to or greater than 13 years. This volume will increase in the next few years – at a rate of approximately 1 GW per year – as electricity generation plants age.

Profitability is strongly subject to the electricity price; thus, the mid-term stability of the price or the possibility to forecast its evolution could make an important contribution in the decision to implement repowering. Nevertheless, there are solutions capable of providing security to investors:

bilateral contracts for selling energy to consumers or traders and new products offered for insurance companies to ensure a minimum value in case the spot price decreases below the minimum value. These solutions are successfully being used in emerging markets to finance RES power plants.

Repowering requires a specific framework that is both technical and retributive, which should be taken into consideration in the next Spanish regulation implemented to allow participation in the daily market and to fit the characteristics of the technology. In addition, legal measurements can be established to improve profitability without an extra cost for the entire electric system by facilitating the access to financial credits for the investors. These measurements should be aimed at providing stability to the incomes and legal security in the long-term for the investment. In this way, investors will be more willing to commit funds to develop Rep-WF projects.

In summary, the repowering approach appears to be a real and feasible option for relaunching the integration of RESs into the Spanish energy mix. In the current situation, due to the lack of specific regulation, an adequate policy framework would enable the development of a large-sized market with sustainable growth in the short term. In addition, the electricity produced by Rep-WFs would be included in the electric market in the condition of grid parity, implying important savings for the entire system with respect to the retributive system based on a FIT applied to the WFs suitable for repowering to date.

## Appendix. Detailed Calculation of the Development and Construction Costs

For the purposes of this chapter, during the year 2013 a survey was carried out among several international companies specializing in the different procedures necessary for the construction and operation of a WF.

## Table A.1. Development scope and unitary prices

| Item | Scope, prices and characteristics | | | | | Comment |
|---|---|---|---|---|---|---|
| Engineering | *Technical project* | | | *30,000* | € | The price includes the definition of the installations (WTG lay-out, Power, Communication and Grounding grid, Substation, Evacuation line) and the technical specifications, drawings and supplements for informing the institutions and the other services affected. |
| | *Environmental report* | | | *10,000* | € | |
| Wind resource assessment | *Meteo tower rent* | | | *40,000* | €/year | Includes the logger and the calibrated sensors. Includes periodic downloads of the raw data. |
| | Height | 90 | m | | | |
| | Measurement heights | 60/75/90 | m | | | |
| | *Final report* | | | *7,000* | € | |
| Personnel | *Hourly average cost* | | | *200* | €/h | |
| | Hours of labour | 600 | h | | | |
| | *Other personnel costs* | | | *100,000* | € | Travel, etc. |

Source: Self-elaboration based on the survey carried out among specialised companies in engineering and wind resource measurement and assessment. Prices valid for H2 2013.

## Table A.2. Construction scope and unitary prices

| Item | Scope, prices and characteristics | | | | | Comment |
|---|---|---|---|---|---|---|
| WTG | *WTG* | | | *850* | *€/kW* | *The price includes:*<br>Assembly and commissioning.<br>SCADA.<br>Systems for controlling reactive power and generation. |
| | WTG Unitary Power | 3 | MW | | | |
| | Number of WTGs | 6 | units | | | |
| | Hub height (HH) | 90 | m | | | |
| | Rotor diameter | 90 | m | | | |
| Meteorological mast | *Meteo mast* | | | *70,000* | *€* | Includes logger and calibrated sensors |
| | Height | 90 | m | | | |
| | Measurement heights | 60/75/90 | m | | | |
| Foundations and assembly areas | *Foundation* | | | *160,000* | *€/WTG* | Structural concrete: 430 $m^3$<br>Steel bars: 25 ton |
| | Foundation dimensions | 15x15x2 | $m^3$ | | | |
| | *Assembly area* | | | *25* | *€/$m^2$* | Includes clearance, excavation, levelling and refilling with the appropriate material. |
| | Dimensions | 40x50 | $m^2$ | | | |
| Roads | *Construction of new roads* | | | *150* | *€/m* | Includes clearance, excavation, levelling and refilling with the appropriate material.<br>Includes storm water trenches |
| | Average distance between WTG (three times rotor diameter) | 270 | m | | | |
| | Access road length | 1,500 | m | | | |
| | Services road length | 1,620 | m | | | |
| | *Adaptation of existing roads* | | | *100* | *€/m* | |
| P&C&G grid | *Civil works* | | | *100* | *€/m* | Includes excavation, signalisation and refilling with the appropriate material. |
| | Trenches length | 2,025 | m | | | |
| | *MV power cables* | | | *15* | *€/m* | Includes terminations, junctions and connections |

**Table A.2. (Continued)**

| Item | Scope, prices and characteristics | | | | | Comment |
|---|---|---|---|---|---|---|
| | Maximum current per sub-circuit | 630 | A | | | |
| | Circuits number | 2 | | | | |
| | Length | 2,430 | | | | |
| | *Optical fibre cables* | | | *6* | *€/m* | Includes terminations, junctions and connections |
| | Type | Single-mode | | | | |
| | Length | 2,025 | m | | | |
| | *Grounding cables* | | | *12* | *€/m* | Includes terminations, junctions and connections |
| | Section | 50 | $mm^2$ | | | |
| | Length | 2,025 | m | | | |
| Substation | *High voltage EPC price* | | | *550,000* | *€* | The price includes an outdoor bay for line and transformer protection fully equipped and a control building |
| | High voltage level | 132 | kV | | | |
| | *Medium voltage EPC price* | | | *150,000* | *€* | Includes the following MV switchgears: Bars measurement, Transformer protection, Line from WTG protection and Auxiliary services transformer protection |
| | Medium voltage level | 33 | kV | | | |
| | *Transformer price* | | | *300,000* | *€* | ONAN/OFAF refrigeration |
| | Main transformer power | 20 | MVA | | | Equipped with on-load tap changer |

| Item | Scope, prices and characteristics | | | | | Comment |
|---|---|---|---|---|---|---|
| Evacuation line | *EPC price* | | | *125,000* | *€/km* | Overhead line single circuit |
| | Voltage level | 132 | kV | | | Includes a fibre optic cable for communications installed in the earth |
| | Length | 5 | km | | | |
| Construction and Land Owners Permits | *County Construction license* | | | *2* | *%* | Percentage to be applied over labor and construction costs |
| | *Urban license* | | | *2,5* | *%* | Percentage to be applied over the total cost |
| | *Land owners permits* | | | | | |
| | Road and trenches | | | 6 | €/$m^2$ | |
| | Typical roads width | 6 | m | | | |
| | Typical trenches width | 1 | m | | | |
| | Evacuation line-Overhead cables | | | 1 | €/$m^2$ | |
| | Typical width | 40 | m | | | |
| | Evacuation line-Pylon | | | 5 | €/$m^2$ | |
| | Used area per pylon | 100 | $m^2$ | | | |
| | Number of pylons | 33 | units | | | One pylon every 150 m |

Source: Self-elaboration based on the survey carried out with WTG manufacturers, electric materials suppliers, civil works and electric contractors and specialised companies for permits management.

Prices valid for H2 2013.

To make the estimations, technical requirements and bills of quantities were delivered to all the requested companies. All of this documentation was performed considering the WF characteristics determined by simulations (Table 1) and the scope of works established in Section 4.

The results of the survey for development and construction costs are included in the following tables for a WF of 20 MW. Nevertheless, the figures are applicable to WFs in the same installed power range.

## REFERENCES

[1] Jefatura del Estado. Gobierno de España. *Ley 54/1997, de 27 de noviembre, del Sector Eléctrico*. 1997 [Head of state. *Government of Spain. Law 54/1997, of November 27, of the Electricity Sector.* 1997

[2] del Río P, Gual MA. An integrated assessment of the feed-in tariff system in Spain. *Energy Policy* 2007;35:994-1012.

[3] Ciarreta A, Gutiérrez-Hita C, Nasirov S. Renewable energy sources in the Spanish electricity market: Instruments and effects. *Renewable and Sustainable Energy Reviews* 2011;15:2510-2519.

[4] Schallenberg-Rodriguez J, Haas R. Fixed feed-in tariff versus premium: A review of the current Spanish system. *Renewable and Sustainable Energy Reviews* 2012;16:293-305.

[5] Iglesias G, del Río P, Dopico JÁ. Policy analysis of authorisation procedures for wind energy deployment in Spain. *Energy Policy* 2011;39:4067-4076.

[6] *Red Eléctrica de España. El mercado eléctrico español en 2012* [*The Spanish electricity market in 2012*] 2013.

[7] Ministerio de Industria, *Energía y Turismo*. Gobierno de España. *La energía en España* 2011 [*Energy in Spain* 2011]; 2012.

[8] IDAE. Ministerio de Industria, Turismo y Comercio. Gobierno de España. [Ministry of Industry, Tourism and Commerce. Government of Spain]. *Plan de Energías Renovables* 2011-2020 [*Renewable Energy Plan* 2011-2020]; 2011.

[9] Rivier Abbad J. Electricity market participation of wind farms: the success story of the Spanish pragmatism. *Energy Policy* 2010;38:3174-3179.

[10] Burgos-Payán M, Roldán-Fernández JM, Trigo-García ÁL, Bermúdez-Ríos JM, Riquelme-Santos JM. Costs and benefits of the renewable production of electricity in Spain. *Energy Policy* 2013;56:259-270.

[11] Gil HA, Gomez-Quiles C, Riquelme J. Large-scale wind power integration and wholesale electricity trading benefits: Estimation via an ex post approach. *Energy Policy* 2012;41:849-859.

[12] CNE. *Comisión Nacional de la Energía* 2013 [*National Energy Commission* 2013];2013.

[13] Jefatura del Estado. Gobierno de España. *Real Decreto-Ley 1/2012 de 27 de enero, por el que se procede a la suspensión de los procedimientos de preasignación de retribución y a la supresión de los incentivos económicos para nuevas instalaciones de producción de energía eléctrica* [*Royal Decree-Law 1/2012 of January 27, which proceeds to the suspension of pre-allocation procedures for compensation and the elimination of economic incentives for new installations for the production of electricity*; 2012.

[14] Ruiz Romero S, Colmenar Santos A, Castro Gil MA. EU plans for renewable energy. An application to the Spanish case. *Renewable Energy* 2012;43:322-330.

[15] Toja-Silva F, Colmenar-Santos A, Castro-Gil M. Urban wind energy exploitation systems: Behaviour under multidirectional flow conditions—Opportunities and challenges. *Renewable and Sustainable Energy Reviews* 2013;24:364-378.

[16] Ruiz-Romero S, Colmenar-Santos A, Gil-Ortego R, Molina-Bonilla A. Distributed generation: The definitive boost for renewable energy in Spain. *Renewable Energy* 2013;53:354-364.

[17] Colmenar-Santos A, Campíñez-Romero S, Pérez-Molina C, Castro-Gil M. Profitability analysis of grid-connected photovoltaic facilities for household electricity self-sufficiency. *Energy Policy* 2012;51:749-764.

[18] Gómez A, Zubizarreta J, Dopazo C, Fueyo N. Spanish energy roadmap to 2020: Socioeconomic implications of renewable targets. *Energy* 2011;36:1973-1985.

[19] Ortegon K, Nies LF, Sutherland JW. Preparing for end of service life of wind turbines. *J Clean Prod* 2013;39:191-199.

[20] del Río P, Calvo Silvosa A, Iglesias Gómez G. Policies and design elements for the repowering of wind farms: A qualitative analysis of different options. *Energy Policy* 2011;39:1897-1908.

[21] Joselin Herbert GM, Iniyan S, Amutha D. A review of technical issues on the development of wind farms. *Renewable and Sustainable Energy Reviews* 2014;32:619-641.

[22] Tabassum-Abbasi, Premalatha M, Abbasi T, Abbasi SA. Wind energy: Increasing deployment, rising environmental concerns. *Renewable and Sustainable Energy Reviews* 2014;31:270-288.

[23] Asociación Empresarial Eólica. [Wind Business Association]. *Eólica* 2013 [*Wind Energy* 2013]; 2013.

[24] International Electrotechnical Commission. IEC 61400-1. *Wind Turbines*. Part 1: Design requirements. 2005.

[25] García López MJ, López Quero M, Avilés Palacios C. *La Articulación De Un Project Finance Como Instrumento De Financiación De Parques Eólicos* [*The Articulation of a Project Finance as an Instrument for the Financing of Wind Farms*; ESIC.; 2008.

[26] Pollio G. Project finance and international energy development. *Energy Policy* 1998;26:687-697.

[27] European Wind Energy Association. Pure Power. *Wind energy* targets for 2020 and 2030 2011.

[28] Staffell I, Green R. How does wind farm performance decline with age? *Renewable Energy 2014*;66:775-786.

[29] Foley AM, Leahy PG, Marvuglia A, McKeogh EJ. Current methods and advances in forecasting of wind power generation. *Renewable Energy* 2012;37:1-8.

[30] Kusiak A, Zhang Z, Verma A. Prediction, operations, and condition monitoring in wind energy. *Energy* 2013;60:1-12.

[31] Gallardo-Calles J, Colmenar-Santos A, Ontañon-Ruiz J, Castro-Gil M. Wind control centres: State of the art. *Renewable Energy* 2013;51:93-100.

[32] Blanco MI. The economics of wind energy. *Renewable and Sustainable Energy Reviews* 2009;13:1372-1382.

[33] Asociación Empresarial Eólica. [Wind Business Association]. *Eólica* 2011 [Wind 2011]; 2011.

[34] Asociación Empresarial Eólica. [Wind Business Association] *Eólica* 2012 [Wind 2012]; 2012.

[35] Asociación Empresarial Eólica. [Wind Business Association] *Eólica* 2004 [Wind 2004]; 2004.

[36] Betz A. *Introduction to the Theory of Flow Machines* G. Braun GmbH: Karlsruhe; 1966.

[37] Carta González JA, Calero Pérez R, Colmenar Santos A, Castro Gil M.; *Centrales De Energías Renovables: Generación Eléctrica Con Energía Renovables* [*Renewable Energy Plants: Electric Generation With Renewable Energy*]. Pearson Educación, S.A.: Madrid; 2009.

[38] IDAE. Ministerio de Industria, Turismo y Comercio. Gobierno de España. [Ministry of Industry, Tourism and Commerce. Government of Spain]. *Spanish Wind Maps*; 2013.

[39] *Gamesa*; 2014.

[40] *Vestas*; 2014.

[41] *Alstom*; 2014.

[42] Windographer. *Windographer*;2013.

[43] González-Longatt F, Wall P, Terzija V. Wake effect in wind farm performance: Steady-state and dynamic behavior. *Renewable Energy* 2012;39:329-338.

[44] Amenedo Rodríguez JL, Burgos Díaz JC, Arnalte Gómez S. *Sistemas Eólicos De Producción De Energía Eléctrica Rueda*; [*Wind Power Production Systems Electric Wheel*]; 2003.

[45] International Renewable Energy Agency (IRENA). Renewable Energy Technologies: Cost Analysis Series. Volume 1: *Power Sector*. Issue 5/5. *Wind Power* 2012;1.

[46] Wiser R, Bolinger M. 2012 *Wind Technologies Market Report* 2013.

[47] Serrano González J, Burgos Payán M, Santos JMR, González-Longatt F. A review and recent developments in the optimal wind-turbine micro-siting problem. *Renewable and Sustainable Energy Reviews* 2014;30:133-144.

[48] Turner SDO, Romero DA, Zhang PY, Amon CH, Chan TCY. A new mathematical programming approach to optimize wind farm layouts. *Renewable Energy* 2014;63:674-680.

[49] European Ferrous Recovery and Recycling Federation. *EU-27 Steel Scrap Statistics 2013.*

[50] Observatorio Industrial del Sector del Metal. *El sector de reciclaje de metales en España* [Industrial Observatory of the Metal Sector. *The metal recycling sector in Spain*]; 2013.

[51] Eurofer. The European Steel Association. *Scrap Price Index*;2013.

[52] Scrapmonster. *European Scrap Prices*;2013.

[53] PWC. *Economic Projections*;2014.

[54] European Commission. Economic and Financial Affairs. *Economic Forecasts*;2014.

[55] Iribarren D, Martín-Gamboa M, Dufour J. Environmental benchmarking of wind farms according to their operational performance. *Energy* 2013;61:589-597.

[56] Pereira AJC, Saraiva JT. Long term impact of wind power generation in the Iberian day-ahead electricity market price. *Energy* 2013;55:1159-1171.

[57] Moreno B, López AJ, García-Álvarez MT. The electricity prices in the European Union. The role of renewable energies and regulatory electric market reforms. *Energy* 2012;48:307-313.

[58] Entrevista a Carlos Barriá del Ministerio de Energía de Chile [Interview with Carlos Barriá of the Ministry of Energy of Chile] *CSP Today* 2013.

[59] La banca local de Chile se abre con cautela a la financiación de proyectos fotovoltaicos [The local bank in Chile opens cautiously to the financing of photovoltaic projects]. *Energías Renovables* 2013.

[60] OMIE. Operador del Mercado Eléctrico. *Polo Español. Resultados del Mercado 2013* [*Electric Market Operator Spanish Polo Market Results 2013*];2013.

[61] Kalamova M, Kaminker C, Johnstone N. Source of Finance, Investment Policies and Plant Entry in the Renewable Energy Sector. *OECD Environment Working Papers* 2011;37.

[62] European Commission. Directorate General for Energy. *EU Energy Trends* to 2030 2009.

# ABOUT THE EDITORS

***Antonio Colmenar Santos, PhD***
Full Professor
Departamento de Ingeniería Eléctrica, Electrónica, Control,
Telemática y Química Aplicada
Universidad Nacional de Educación A Distancia, Madrid, Spain
Email: acolmenar@ieec.uned.es

Dr. Colmenar-Santos has been a senior lecturer in the field of Electrical Engineering at the Department of Electrical, Electronic and Control Engineering at the National Distance Education University (UNED) since June 2014. Dr. Colmenar-Santos was an adjunct lecturer at both the Department of Electronic Technology at the University of Alcalá and at the Department of Electric, Electronic and Control Engineering at UNED. He has also worked as a consultant for the INTECNA project (Nicaragua). He has been part of the Spanish section of the International Solar Energy Society (ISES) and of the Association for the Advancement of Computing in Education (AACE), working in a number of projects related to renewable energies and multimedia systems applied to teaching. He was the coordinator of both the virtualisation and telematic Services at ETSII-UNED, and deputy head teacher and the head of the Department of Electrical, Electronics and Control Engineering at UNED. He is the author of more than 50 papers published in respected journals (http://goo.gl/YqvYLk) and has participated in more than 100 national and international conferences.

***Enrique Rosales Asensio, PhD***
Researcher
Departamento de Física
Universidad de La Laguna, La Laguna, Spain
Email: erosalea@ull.edu.es

Dr. Rosales-Asensio, PhD is an industrial engineer with postgraduate degrees in electrical engineering, business administration, and quality, health, safety and environment management systems. Currently, he is a senior researcher at the University of La Laguna, where he is involved in water desalination project in which the resulting surplus electricity and water would be sold. He has also worked as a plant engineer for a company that focuses on the design, development and manufacture of waste-heat-recovery technology for large reciprocating engines; and as a project manager in a world-leading research centre.

***David Borge Diez, PhD***
Lecturer
Departamento de Ingeniería Eléctrica y de Sistemas y Automática
Universidad de León, León, Spain
Email: dbord@unileon.es

Dr. David Borge-Diez has a PhD in Industrial Engineering and an MSc. in Industrial Engineering, both from the School of Industrial Engineering at the National Distance Education University (UNED). He is currently a lecturer and researcher at the Department of Electrical, Systems and Control Engineering at the University of León, Spain. He has been involved in many national and international research projects investigating energy efficiency and renewable energies. He has also worked in Spanish and international engineering companies in the field of energy efficiency and renewable energy for over eight years. He has authored more than 25 publications in international peer-reviewed research journals and participated in numerous international conferences.

# INDEX

## I



## L



## M



## N



## O



## P



## R